DISPERSION OF POWDERS IN LIQUIDS

With Special Reference to Pigments

THIRD EDITION

DISPERSION OF POWDERS IN LIQUIDS

With Special Reference to Pigments

THIRD EDITION

Edited by

G. D. PARFITT

Department of Chemical Engineering, Carnegie-Mellon University, Pittsburgh, Pennsylvania, USA

APPLIED SCIENCE PUBLISHERS
LONDON and NEW JERSEY

APPLIED SCIENCE PUBLISHERS LTD
Ripple Road, Barking, Essex, England

APPLIED SCIENCE PUBLISHERS, INC.
Englewood, New Jersey 07631, USA

First edition 1969
Second edition 1973
Third edition 1981

British Library Cataloguing in Publication Data

Dispersion of powders in liquids.—3rd ed.
1. Pigments 2. Dispersion 3. Colloids
I. Parfitt, G. D.
667'.29 TP936

ISBN 0-85334-990-8

WITH 35 TABLES AND 109 ILLUSTRATIONS

Printed in Great Britain by Galliard (Printers) Ltd, Great Yarmouth

CONTENTS

LIST OF CONTRIBUTORS

W. Black

Research Department, Imperial Chemical Industries Limited, Organics Division, PO Box No. 42, Hexagon House, Blackley, Manchester M9 3DA, UK

H. Füredi-Milhofer

Laboratory for Precipitation Processes, Ruder Bošković Institute, PO Box 1016, 41001 Zagreb, Croatia, Yugoslavia

M. J. Jaycock

Department of Chemistry, Loughborough University of Technology, Loughborough, Leicestershire LE11 3TU, UK

H. D. Jefferies

Consultant, 4 Tintern Avenue, Billingham, Cleveland TS23 2DE, UK

S. G. Lawrence

Pigments Division, Ciba–Geigy Plastics and Additives Company, Hawkhead Road, Paisley, Renfrewshire PA2 7BG, UK

G. C. Lowrison

Consulting Chemical Engineer, 4 Cranbourne Road, Bradford, West Yorkshire BD9 6BH, UK

R. B. McKay

Pigments Division, Ciba–Geigy Plastics and Additives Company, Hawkhead Road, Paisley, Renfrewshire PA2 7BG, UK

G. D. Parfitt

Department of Chemical Engineering, Carnegie-Mellon University, Schenley Park, Pittsburgh, Pennsylvania 15213, USA

A. L. SMITH
Unilever Research, Port Sunlight Laboratory, Unilever Limited, Port Sunlight, Wirral, Merseyside L62 4XN, UK

F. M. SMITH
Pigments Division, Ciba–Geigy Plastics and Additives Company, Roundthorn Industrial Estate, Wythenshawe, Manchester M23 9ND, UK

A. G. WALTON
Department of Macromolecular Science, Case Western Reserve University, Cleveland, Ohio 44106, USA

D. A. WHEELER
MBW Limited, 11 Sunningdale, Stone, Staffordshire ST15 0LZ, UK

INTRODUCTION

Since the first edition of this book appeared in 1969 there have been some significant developments that are important for a better appreciation of the science and technology of the process of dispersing powders in liquids. Furthermore, there has been an increasing awareness of the value of understanding and applying the fundamental principles of colloid and interface science to practical systems—the gap between theory and practice has narrowed significantly. One area of scientific endeavour has made a major advance, and that concerns the stabilisation of dispersions by macromolecular substances, which is particularly relevant since many practical systems contain molecules of high molecular weight. Our understanding of the chemistry of powder surfaces has also improved, and the importance of such knowledge to the incorporation stage is now being recognised. These and other developments have been incorporated in this third edition.

The fundamentals and terminology of the dispersion process are defined in Chapter 1. It is useful to divide the overall process into three distinct stages. The first involves the wetting of the powder surface, and Chapter 2 considers those aspects of the surfaces of solids and their interactions with liquids that are relevant to this stage. Once the powder is wetted out a milling stage is usually necessary but for pigments this area of mechanical breakdown of aggregated structures is not well defined. The third stage concerns the stability of the dispersion to flocculation, and in Chapter 1 the various theories of stability are described for systems containing either charged or uncharged colloidal particles. The relevant theoretical and practical aspects of electrical phenomena associated with charged solid/solution interfaces, including electrokinetic phenomena and the calculation and significance of zeta potential, are discussed in Chapter 3. These first three chapters contain a number of references to scientific studies of the various aspects of the dispersion process: to the casual reader they may seem too remote from the practical situation to merit further study, but succeeding chapters focus attention on the fundamentals and how they find application in practice, and endeavour to pinpoint the problem areas.

In all three stages it might be said that 'surface activity' predominates, and it is not surprising that in a variety of ways surface active agents play an important role in the process. In the large range of dispersion processes in which pigments are used, the surface active compounds employed are legion. Their types, properties and applications are reviewed in Chapter 4. The alternative technique for the formation of a dispersion of fine particles in a liquid medium, that is by precipitation from supersaturated systems, is discussed in Chapter 5. This is, of course, an important aspect of the manufacture of many inorganic and organic pigments, but the principles involved in the development of precipitate morphology are also relevant to the ageing processes that can occur in certain pigmented systems.

Chapter 6, a new addition to the book, covers the fundamental aspects of the breakdown of solid materials, both amorphous and crystalline, and identifies the parameters that are important in comminution. The techniques used for dispersion in industrial processing are surveyed in Chapter 7, illustrating how the choice of machine is related to the nature of the system involved, and how optimisation can be achieved. The degree of dispersion achieved may be difficult to assess—the ultimate experiment would clearly define the particle size distribution in the dispersed system, but in so many practical cases this experiment is still somewhat remote. Several quite reliable techniques are in regular use and some new ones are introduced; these are based on the principles discussed earlier in the book and something of a link between these principles and the practical problem of assessing the state of dispersion is created in Chapter 8. The final two chapters are concerned with the current position in the dispersion of inorganic (Chapter 9) and organic (Chapter 10) pigments in a variety of media. Where feasible the authors demonstrate the application of fundamentals while focusing attention on the areas where knowledge is limited and empiricism prevails.

The editor is grateful for the numerous constructive suggestions that he has received from readers, and hopes that this new edition goes at least some way to meet their needs.

G. D. PARFITT

CHAPTER 1

FUNDAMENTAL ASPECTS OF DISPERSION

G. D. PARFITT
Carnegie-Mellon University, Pittsburgh, USA

THE DISPERSION PROCESS

The term *dispersion* is used here to refer to the complete process of incorporating a powder into a liquid medium such that the final product consists of fine particles distributed throughout the medium. The dispersion of fine particles is normally termed *colloidal* if at least one dimension of the particles lies between 1 nm and 1 μm and the term *sol* is used for *any* colloidal system in which the dispersion medium is liquid. Solid particles dispersed in a liquid form a *suspension.*

Sols are classified, in terms of the affinity of the colloidal particle for the medium, as *lyophobic* (possessing aversion to liquid) or *lyophilic* (possessing affinity for liquid); *hydrophobic* and *hydrophilic* respectively for aqueous media. Some difficulties may arise with the use of these terms. In the sense of the definition lyophobic implies no affinity between the colloid and the medium, e.g. an insoluble powder, but taken to its logical extreme this would suggest no wetting of the powder by the liquid and hence no dispersion could be formed. But although a metal oxide powder may be insoluble in water it is nevertheless normally wetted by water and hence the powder surface is hydrophilic, but the dispersion is classed as hydrophobic. We might also wish to consider the degree of hydrophobicity of a surface in terms of, say, the heat of wetting in water; there is no hard and fast rule when these terms are used for interfaces, but the application to dispersion is clearly defined. Solutions of macromolecules and association colloids are of the lyophilic type and form spontaneously when the components are brought into contact. They are true solutions and are stable in the thermodynamic sense whereas lyophobic sols do not form spontaneously and hence in principle are thermodynamically unstable. The division between solutions and dispersions is without ambiguity.

Essential to an understanding of the dispersion of powders in liquids is a knowledge of the formation and properties of the solid–liquid interface. In the early stages of the process the solid–air interface is replaced by one between solid and liquid. The powder consists of aggregates/agglomerates of similar particles and to disperse these particles into a liquid mechanical work is performed, and the forces that exist at the interface determine the ease with which the process can be brought about. Once dispersed the particles are free to move in their new environment, and flocculation is prevented by various chemical and physical methods, all of which relate to the character of the solid–liquid interface.

The three stages of the dispersion process

It is useful to consider the overall process of dispersion as consisting of three stages, and take each stage separately in order to assess the importance of the various factors involved. These three stages are quite distinct in their nature, but in practice they overlap. The principles involved in each stage are fairly well established but because of the overlap it is often not easy to recognise any particular aspect in a practical dispersion system. The three stages to be considered are the following:

(*a*) *Wetting of the powder*

In many practical uses of powders the primary particle size is sufficiently small for further subdivision to be unnecessary. But in the dry state the powder usually contains some aggregates of primary particles and these are attached to other aggregates and/or primary particles forming agglomerates. Aggregates are groups of primary particles joined at their faces and having a surface area significantly less than the sum of the areas of their constituent particles. An agglomerate is a collection of primary particles and aggregates which are joined at edges and corners, and the surface area of the whole is not markedly different from the sum of the areas of the individual components.[1] The initial stage of wetting involves both the external surface of particles and the internal surfaces which exist between the particles in the clusters that make up the dry powder. The wetting process is dependent on the nature of the liquid phase, the character of the surface, the dimensions of the interstices in the clusters, and the nature of the mechanical process used to bring together the components of the system. The principles of wetting of external surfaces are well established, but extension to internal surfaces involves geometric factors which for fine particle systems are not readily defined.

(*b*) *Breaking up the clusters to form colloidal particles*

Once the particles are wetted out some mechanical energy is usually required to bring about their complete separation. Aggregates may require considerable energy to break them down completely to the point when the surface of each primary particle is available to the wetting liquid. Presumably agglomerates would normally require less energy than aggregates. Many factors are involved in the powder achieving its particular state so that a knowledge of its manufacture and/or storage conditions is necessary for the problem of breaking the bonds between particles in the clusters to be completely understood.

(*c*) *Flocculation of the dispersion*

Having wetted the surfaces and broken down the clusters into fine particles, these are then dispersed throughout the medium. The problem is then to maintain the dispersed state since the particles have a natural tendency (as a result of Brownian motion) to reduce in number with time due to collisions. An attractive force exists between the particles as they approach, the magnitude of which increases significantly as the distance of approach decreases. The reduction in particle number is here to be termed *flocculation* whatever the mechanism involved. To resist flocculation some sort of repulsive force is necessary and this is usually achieved through the particles being charged or containing adsorbed layers which protect them, or both. The other term, *coagulation*, which is often used to represent this process, has various meanings in technology so is to be avoided in this book; but in the fundamental literature it is usually used either in place of or together with (indiscriminately) the term flocculation. Coagulation implies the formation of compact clusters of particles, leading to the macroscopic separation of a *coagulum*. Flocculation implies the formation of a loose or open network (*floc*) which may or may not separate macroscopically. Hence there might be some justification for making a distinction between the two terms,[2] since the clusters formed would be expected to behave differently, e.g. under shear.

Dispersibility has been defined as the ease with which the particles in the powder are distributed in a continuous liquid medium so that each particle is completely surrounded by the medium, and no longer makes permanent contact with other particles. But Eckhoff[3] has suggested that this definition is too vague and it would be much better if the process could be expressed in terms of the energy expended. It should be possible to define the theoretical minimum energy (E_{min}) which would be necessary to achieve the dispersion of unit mass of powder by a completely efficient process. The theoretical maximum dispersibility would then be defined as $D_{max}=1/E_{min}$. Hence dispersibility has the dimension of mass per unit of energy and is thus a direct measure of the quantity of powder that would be dispersed by the expenditure of one unit of energy. Since no real process can be 100% efficient it is necessary to incorporate an efficiency factor (K) to account for the real situation. Hence $D_{real}=K/E_{min}$ where $0<K<1$. We can then discuss the various factors which determine the dispersibility of a powder in terms of those which affect E_{min} and K separately. In principle E_{min} can be calculated by integrating the work required to break individual interparticulate bonds, although by the very nature of dispersions such a simple integration could be misleading since simultaneous processes take place and depending on the nature of the system they may consume or supply additional energy to that expended on the system from external sources. So apart from the simplest of systems, e.g. the dispersion of powders in gases, the calculation of the minimum energy required to separate the particles is extremely difficult and as yet no satisfactory theoretical treatment has been found. Nevertheless, the term dispersibility can be usefully applied to express the effectiveness of the first two stages of the dispersion process.

The relationships between the factors and principles involved and the practical aspects of dispersibility have received relatively little attention in the past, in contrast to the extensive investigation of the *stability*, i.e. the resistance to flocculation. It is difficult to differentiate between a system that has been dispersed but is unstable and one which cannot readily be dispersed; ill-defined experimental conditions may well lead to confusion in interpretation. Surface active agents often play a leading role in all three aspects of the dispersion process, although they might easily be useful in one but antagonistic in another. There are no simple rules; each case has to be considered in detail.

WETTING

Various terms are used, often loosely, to describe the process that occurs when a solid phase and a liquid phase come into contact so that the solid–air interface is replaced by the solid–liquid interface. In general we refer to this as *wetting* but for practical purposes it is more useful to postulate that more than one kind of wetting may be associated with the formation of a solid–liquid interface, and to describe each type in terms of the changes in free surface energy involved. There are three distinct types of wetting,[4] designated as *adhesional* wetting, *immersional* wetting, and *spreading* wetting, according to the mechanical process taking place.

(i) Adhesional wetting. When 1 cm^2 of a plane solid (S) surface is brought into contact with 1 cm^2 of a plane liquid (L) surface, unit surface area of each phase disappears to form unit area of the new solid–liquid interface. The work W_a involved in this process carried out under isothermal conditions (i.e. the decrease in surface free energy) was given by Dupré[5] as

$$W_a = \gamma_{S/L} - (\gamma_S + \gamma_{L/V}) \tag{1.1}$$

where $\gamma_{S/L}$, γ_S and $\gamma_{L/V}$ are the interfacial free energies at the solid–liquid, solid–vapour and liquid–vapour interfaces respectively, the vapour here being that of the separate phases. ($-W_a$ is the work of adhesion, the reversible work required to restore the initial condition.) The magnitude of this work term was defined by Fowkes[6] in terms of the different kinds of interactions that take place across the interface. He expressed these in the following way:

$$W_a = W_a^d + W_a^h + W_a^\pi + W_a^p + W_a^e \tag{1.2}$$

where the superscripts refer to the type of interaction; d = dispersion, h = hydrogen bond, π = pi-bond, p = other polar interactions, and e = contribution due to charge separation at the interface giving rise to an electric double layer. Dispersion forces are always present and the adhesion between dissimilar materials always depends on and is frequently dominated by these dispersion forces. Even in polar molecules the dipole–dipole interactions are usually insignificant compared with dispersion force interactions. However, with the exception of the aliphatic hydrocarbon solvents, all the components that are normally present in disperse systems possess a certain degree of polarity. This polarity is associated with specific chemical groups that are present both in the liquid phase and on the surface of the solid. These groups may be acidic or basic in nature or even amphoteric (exhibiting both acid and base character), and give rise to those forces other than the dispersive type described above. We may identify specific effects, such as the chemical interaction or hydrogen bonding interaction between an adsorbed molecule and a specific group on the particle surface, as well as non-specific effects, e.g. the interaction between the hydrocarbon chain of a polymer with a carbon black surface.

(ii) Immersional wetting. The total immersion in a liquid of 1 cm^2 of solid surface involves exchange of solid–vapour for solid–liquid interfaces without any change in the extent of the liquid surface, and the immersional work is given by

$$W_i = \gamma_{S/L} - \gamma_{S/V} \tag{1.3}$$

where $\gamma_{S/V}$ is the free energy of the surface at equilibrium with the vapour of the liquid. $\gamma_{S/V}$ is related to γ_S by

$$\gamma_{S/V} = \gamma_S - \pi_e \tag{1.4}$$

where π_e is the equilibrium spreading pressure which takes into account the extra energy associated with the adsorption of liquid vapour on the solid surface.

(iii) Spreading wetting. When a drop of liquid spreads over a plane solid surface, for unit area of solid surface that disappears equivalent areas of liquid surface and solid–liquid interface are found. The work involved is

$$W_s = (\gamma_{S/L} + \gamma_{L/V}) - \gamma_{S/V} \tag{1.5}$$

The three types of wetting differ in that in the first there is unit decrease in liquid–vapour interface, in the second there is no change, and in the third there is a unit increase in this interface.

Values of $\gamma_{S/V}$ and $\gamma_{S/L}$ are not readily accessible by experiment, but they are related by the Dupré equation for contact angle equilibrium (often referred to as Young's equation)

$$\gamma_{S/V} = \gamma_{S/L} + \gamma_{L/V} \cos \theta \tag{1.6}$$

which provides a means of evaluating the wetting functions in terms of the measurable quantities $\gamma_{L/V}$ and θ, where θ is the contact angle between the solid and liquid phases. In this equation the vapour refers to that of the liquid, i.e. the system is at equilibrium with the vapour at its saturated vapour pressure. It is important to remember that eqn. (1.6) only applies to a system at equilibrium and for which $\gamma_{L/V}$ and θ have their equilibrium values.[7] For the wetting of a solid surface which is initially free of liquid molecules but in equilibrium with its own vapour, the π_e term should be included.[8] In principle π_e may be obtained from an adsorption isotherm for the vapour on the solid, but there is a general uncertainty on values of π_e and often the term is neglected leading to the (inaccurate) assumption that $\gamma_S = \gamma_{S/V}$.

The simplest way to describe the wetting phenomena that occur when a solid particle is immersed in a liquid is to consider a cube of side length 1 cm. The three stages, adhesion, immersion and spreading, involved in the complete wetting of the cube are shown in Fig. 1.1 and are represented by the changes (a) to (b), (b) to (c) and (c) to (d) respectively. The energy changes that take place are given by

$$W_a = \gamma_{S/L} - (\gamma_{S/V} + \gamma_{L/V}) = -\gamma_{L/V} (\cos \theta + 1) \tag{1.7}$$

$$W_i = 4\gamma_{S/L} - 4\gamma_{S/V} = -4\gamma_{L/V} \cos \theta \tag{1.8}$$

$$W_s = (\gamma_{S/L} + \gamma_{L/V}) - \gamma_{S/V} = -\gamma_{L/V}(\cos \theta - 1) \tag{1.9}$$

and for simplicity we assume that the solid surface before wetting is in equilibrium with the vapour of the liquid. From these relations we may predict whether or not any particular stage of the process is spontaneous, i.e. when the appropriate W is negative. When W is positive then work must be expended on the system for the process to take place.

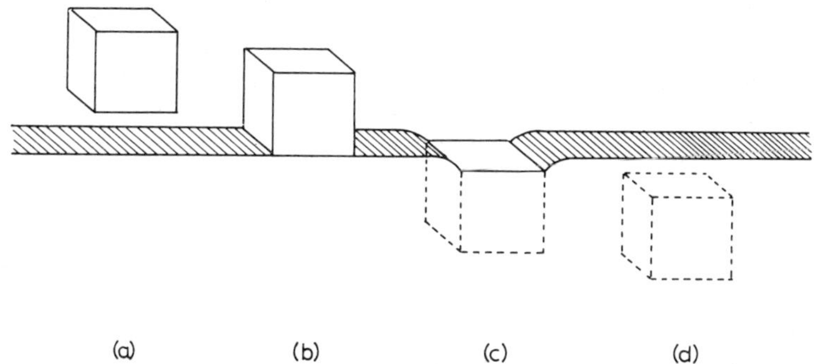

Fig. 1.1. The three stages involved in the complete wetting of a solid cube by a liquid: (a) to (b) adhesional wetting, (b) to (c) immersional wetting, (c) to (d) spreading wetting.

The conclusions are that

(i) adhesional wetting is positive (W_a negative) when the angle of contact is less than 180°; the process is invariably spontaneous;

(ii) immersional wetting is positive and immersion is spontaneous only when θ is less than 90°, and
(iii) spreading wetting is positive only when $\theta=0$, and work must be done to achieve spreading at all larger values of this angle.

The total work W_d for the overall dispersion process is given by the sum of the work terms for the three separate stages

$$W_d = W_a + W_i + W_s = 6\gamma_{S/L} - 6\gamma_{S/V} = -6\gamma_{L/V}\cos\theta \tag{1.10}$$

a result that could have been predicted in terms of the overall energy expenditure. By splitting the operation down into the three stages it is possible to predict whether some intermediate state of the dispersion process may not be spontaneous and therefore retard the overall effect. The requirement that $\theta=0$ for spontaneous spreading explains why certain solids are only partially submerged and still remain at the surface; work must be done to bring about complete submersion. Perhaps the material might be so dense that gravity could furnish sufficient energy for the liquid to spread over the solid surface. Gans has discussed the relevance of wetting phenomena to the paint technologist and his two articles[9,10] are recommended for further reading.

In practice the addition of surface active agents ensures that θ is often close to zero and hence spontaneous dispersion is common. Where θ is zero or so close to zero that spontaneous spreading occurs it is normal to say that the liquid wets the surface; non-wetting means that $\theta > 90°$ so that the liquid tends to form a drop on the surface which runs off easily. From eqn. (1.6)

$$\cos\theta = \frac{\gamma_{S/V} - \gamma_{S/L}}{\gamma_{L/V}} \tag{1.11}$$

from which it is apparent that if $\theta < 90°$ a decrease in $\gamma_{L/V}$ will reduce θ and hence improve wetting. The addition of a surface active agent usually causes a reduction in $\gamma_{L/V}$ and if adsorbed a decrease in $\gamma_{S/L}$. Both effects lead to better wetting. The change in $\gamma_{S/V}$ is probably negligible in most cases so that the dominating factor in wetting is normally $\gamma_{L/V}$, the surface tension of the liquid phase.

Contact angle measurements on powders present many problems; in particular hysteresis is often observed where the angle varies according to whether the interface at which the measurement is made has advanced from or receded to its equilibrium position. This phenomenon has been variously attributed to surface roughness, surface heterogeneity, penetration of liquid into the solid, friction in the vicinity of the three phase line of contact, and the presence of traces of surface active material.[11] Some observations of the measurement of θ for powders have been reviewed,[12] but published data are relatively rare.

The penetration of liquid into the channels between and inside the agglomerates is more difficult to define precisely. The pressure P required to force a liquid into a tube of radius r is

$$P=\frac{-2\gamma_{L/V}\cos\theta}{r} \tag{1.12}$$

and hence penetration is only spontaneous (P negative) when $\theta<90°$. If θ is not zero then, using eqn. (1.6),

$$P=\frac{-2(\gamma_{S/V}-\gamma_{S/L})}{r} \tag{1.13}$$

so that the important requirement is that $\gamma_{S/L}$ should be made as small as possible since $\gamma_{S/V}$ is virtually constant. However, if θ is zero then

$$P=-2\gamma_{L/V}/r \tag{1.14}$$

so that now $\gamma_{L/V}$ should be large. Changes in θ accompany an increase in $\gamma_{L/V}$, and assuming the only variable tension to be $\gamma_{L/V}$, P will ultimately become constant (from eqns. (1.11) and (1.12)), and the limiting value increases proportionally with $(\gamma_{S/V}-\gamma_{S/L})$. Hence it is desirable to maximise $(\gamma_{S/V}-\gamma_{S/L})$ and keep $\gamma_{L/V}$ as small as possible. Since any surface active agent affects both $\gamma_{L/V}$ and $\gamma_{S/L}$, the best agent for promoting such opposing effects may be difficult to find and will vary from one system to another.

When considering changes in $\gamma_{L/V}$ associated with surface active agents it must be remembered that aqueous solutions of these materials contain aggregates (micelles) above a well defined concentration, the critical micelle concentration (c.m.c.). Above the c.m.c. the surface tension remains approximately constant. Similar effects occur with solutions in non-aqueous media, although the c.m.c. values are normally much lower and the range of surface tension lowering is less. Micellisation has been reviewed recently, both for aqueous[13] and non-polar[14] systems.

The relevance of eqns. (1.12)–(1.14) to the practical situation is somewhat tenuous, since the channels between particles in agglomerates are not readily defined in geometric terms. Equation (1.12) indicates that in the absence of air a liquid will enter the channel spontaneously if $\theta<90°$, but agglomerates are normally filled with air and once the liquid has penetrated the channel the air pressure will increase and complete wetting would therefore appear to be impossible. Heertjes and Witvoet[15] have extended the simple treatment to the more complex situation found in practice, and besides demonstrating the relevance to the wetting behaviour of the structure of the agglomerate have shown that when the agglomerate is filled with air complete wetting can only occur when $\theta = 0$.

Another important factor in the wetting process is the rate of penetration of the liquid into the capillaries, and it is desirable that this be as large as possible. For the case of horizontal capillaries or in general where gravity may be neglected, the rate of entry of liquid into the tube is given by the Washburn equation[16]

$$l = \left(\frac{rt\gamma_{L/V} \cos \theta}{2\eta} \right)^{1/2} \tag{1.15}$$

where l is the depth of penetration in time t and η is the viscosity of the liquid. For a packed bed of solid particles the radius may be replaced by a factor K, which contains an effective radius for the bed and a tortuosity factor which takes into account the complex path formed by the channels between the particles and aggregates. Thus

$$l^2 = \frac{Kt\gamma_{L/V} \cos \theta}{2\eta} \tag{1.16}$$

and for any particular (stable) powder bed a plot of l^2 versus t should be linear. For a convenient method of measurement of l for a powder see for example reference 17. To achieve rapid penetration, high $\gamma_{L/V} \cos \theta$, low θ and low η are desirable, with K as large as possible, e.g. a loosely packed powder. High $\gamma_{L/V}$ and low θ are normally incompatible; a low θ is an important requisite, but when $\theta = 0$ further lowering of $\gamma_{L/V}$ will reduce the penetration rate.

The Washburn equation describes a system in which the surface of the capillary is covered with an adsorbed duplex film in equilibrium with the saturated vapour of the wetting liquid (N.B. a duplex film is one that is sufficiently thick for the tensions at the solid–liquid and liquid–vapour interfaces to act independently). Good[18] has extended the argument to the case where the surface is initially free of adsorbed vapour, when the initial driving force for the liquid flow will be larger, and hence the ‘dry’ capillary will wet faster than one already equilibrated with the vapour of the penetrating liquid. He shows that the maximum rate of penetration is given by

$$l^2 = \frac{Kt[\gamma_{L/V} \cos\theta + \pi_e - \pi(t=0)]}{2\eta} \tag{1.17}$$

where $\pi(t=0)$ is the spreading pressure for the film that exists at zero time, but because molecules of the liquid will through diffusion and evaporation processes reach the solid surface before the advancing body of liquid, the actual rate lies between the rates predicted by eqns. (1.16) and (1.17).[19]

Hence the wetting phenomena are directly related to $\gamma_{L/V}$ and θ. In general the overall process is likely to be more spontaneous the lower θ and the higher $\gamma_{L/V}$ although these two factors tend to operate in opposite senses.

An account[20] of Zisman's work on contact angles of liquids on low and high energy solid surfaces in relation to liquid and solid constitution is recommended for further reading. Zisman introduced the concept of critical surface tension for the wetting of a solid, being the intercept at $\cos\theta = 1$ of a plot of surface tension versus $\cos\theta$ for a homologous series of organic liquids in contact with a particular surface, and Gans[21] has used the information to predict the spreading behaviour of liquids on solids as relevant to pigment dispersion.

Reports on studies of dispersion in relation to wetting behaviour are rare. Work from the author's laboratory[22,23] has shown that with a carbon black (Graphon) and rutile in aqueous surfactant solutions there is, as expected from the theory, a marked increase in the degree of dispersion at the surfactant concentration for which θ becomes less than 90°; in these experiments the mixtures were subjected to slow end-over-end action so that under the circumstances spreading wetting was possible. A correlation between contact angle and flotation rate for quartz in aqueous solutions of dodecylammonium acetate, with adsorption from solution and electrokinetic (zeta potential) data, was reported by Fuerstenau.[24] In a later review Fuerstenau discussed the relevance of the adsorption process to flotation technology illustrating the direct relationship between the amount of surfactant adsorbed and the contact angle.[25] With most other published data separation of the effects associated with wetting and surfactant adsorption from flocculation behaviour has not been possible or attempted. Association of dispersion phenomena specifically with surfactant adsorption rather than electrokinetic potential (related to stability) has been assumed for carbon blacks[26–28] and oxides.[29,30]

MECHANICAL BREAKDOWN OF AGGLOMERATES AND AGGREGATES

In the initial stages of incorporation and wetting it is the magnitude of the cohesive forces between the individual particles that determine the extent of disagglomeration before the mechanical action is applied. For example, pigmentary titanium dioxide is intrinsically cohesive, and exists in powder form as loose agglomerates some 30–50 μm in diameter. Surface treatments have a profound (reducing) effect on the cohesive forces, as shown from tensile strength measurements, and therefore assist the disintegration process.[31] Furthermore, surface moisture increases agglomerate strength.[31] Both factors therefore contribute to the efficiency of the wetting stage.

But the most difficult part of the dispersion process to define is that concerned with the breakdown of the aggregates and agglomerates into finer particles after all the available surface has been wetted. Particles held together by weak forces in the dry agglomerate state would presumably require little energy to disperse, once wetted, and charge and surface tension effects would be important. In certain cases penetration of liquid into the channels between particles in the agglomerates might provide sufficient excess air pressure to bring about disintegration.[15] In the milling stage it is more important to disrupt agglomerates and aggregates than to secure a stable dispersion of primary particles. A millbase should therefore be formulated to obtain the fastest breakdown of over-size particles, the final stability of the dispersion being only the secondary consideration at this stage. This means maximising the speed of wetting and the speed of disagglomeration, provided that the rate of flocculation of primary particles is not so rapid that an unfavourable equilibrium results.

Mechanical energy is required to destroy aggregates or break down single crystals into smaller units, and in practice the relation between grinding efficiency and the adjustable parameters in any particular dispersion process has usually been established empirically. A great deal of consideration has been given to this problem by Rehbinder and his colleagues in Moscow.[32–34] They have established that the fine grinding of solids to create new interfaces is facilitated considerably by the adsorption of surface active agents at structural defects in the surface. These defects are normally present in the natural state but they might also appear as microcracks during the milling process. The decrease in surface energy associated with the adsorption may even be so large that the colloidal state becomes more stable than the condensed state, and then the material would break down spontaneously, in the absence of external forces, into smaller units. DiBenedetto[35] has discussed how the presence of submicroscopic defects in the solid can lead to fracture occurring at only a small fraction of the theoretical strength, and the fracture energy is further reduced by a wetting liquid causing embrittlement of the solid. A theory for such spontaneous dispersion has been proposed by Rehbinder,[36] and the dispersions formed in this way should, perhaps, be classed as *lyophilic*.

Of particular interest to the present discussion is a study of the dispersion of various oxides (quartz, rutile, alumina, etc.) in a laboratory eccentric vibratory mill, in dry air and in various liquids (polar and non-polar and solutions of surface active agents) and their vapours.[37] Dispersion efficiency is shown to be related to the degree of lowering of the interfacial tension and is independent of the polarity of the liquid, supporting the theory that the dominating effect is the lowering of surface energy by adsorption. Adsorbed vapours may also facilitate dispersion, the effectiveness depending greatly on surface coverage. For example, in the grinding of quartz sand it was shown that the degree of dispersion is considerably greater with 0·04% moisture than in dry air, and rises sharply up to about 1%. On further increase in water content from 2 to 30% the degree of dispersion falls considerably, with 30–40% it rises sharply again and reaches optimum level (about the same as with 1%) in wet grinding with 50–80% water. The water contents correspond to the following surface coverages: 0·04%, 0·1 monolayer; 1%, 1 monolayer; 30%, 5–10 monolayers. The decrease in efficiency from 2 to 30% is explained by the formation of aggregates of quartz particles held together by thin intermediate layers of water, reaching a maximum in dispersion retardation at 5–10 monolayers. Increasing the water layer thickness weakens the structure until at about 40% further increases have no more effect. Rehbinder remarks that the formation of these structures, the breakdown of which consumes a considerable proportion of the mill energy, is an important factor in dispersion studies with addition of small amounts of surface active materials to dry powders. Similar effects were observed with acetone, ethanol and benzene vapours. The weakening of solids by adsorbed material is generally known as the *Rehbinder effect*. How much it applies to particles of pigmentary size is still an open question.

STABILITY TO FLOCCULATION

A colloidal dispersion may be considered as stable if there is no change in the total number of particles with time, although this definition is to some extent arbitrary since it depends on the time scale over which the observations are made. There are various ways (principally three) by which the number might decrease. One is a consequence of the large interfacial free energy associated with the small particles, which would tend to decrease by crystallisation to a value for which the surface area is minimal, i.e. to one large crystal; the rate at which that occurs in lyophobic systems is insignificant. Sedimentation under gravity would also lead to a reduction in number with time but this is only important for large dense particles since thermal agitation is normally sufficient to keep small colloidal particles dispersed. For dispersions of solid particles in liquid media the primary cause of instability is flocculation in which particles under the influence of random Brownian motion come into close contact and form clusters (flocculates) of aggregates and/or primary particles. In the clusters the particles retain their original identity. The frequency of collisions is determined by the concentration and physical properties of the particles, the viscosity of the medium and the temperature, i.e. on those parameters which determine the rate of diffusion. The factors involved in flocculation will now be considered in some detail.

Maximum rate of flocculation

The continuous, haphazard motion of colloidal particles suspended in a liquid medium was first observed by Robert Brown in 1827. This Brownian motion is the mechanical result of innumerable collisions which take place between the particles and the molecules of the liquid. Einstein showed theoretically that the average square of the displacement $\overline{x^2}$ along any given axis in time t of a particle of radius a in a medium of viscosity η is given by

$$\overline{x^2} = \frac{RTt}{3\pi N_0 \eta a} \text{ or } \overline{x^2} = 2Dt \tag{1.18}$$

where R is the gas constant, T the absolute temperature, N_0 the Avogadro number and D the diffusion coefficient. Using Einstein's law rates of Brownian motion may be evaluated. For spheres in water at 20°C, $D \approx 2{\cdot}15 \times 10^{-13}/a$, from which the following data were calculated.[38]

(i) For $a = 1$ nm, $D = 2{\cdot}15 \times 10^{-6}$ cm^2/s, particle travels a root mean squared distance of 1 nm in $2{\cdot}3 \times 10^{-9}$ s, 100 nm in $2{\cdot}3 \times 10^{-5}$ s and 1 μm in $2{\cdot}3 \times 10^{-3}$ s.

(ii) For $a = 1\ \mu$m, $D = 2{\cdot}15 \times 10^{-9}$ cm^2/s, particle travels a root mean squared distance of 1 nm in $2{\cdot}3 \times 10^{-6}$ s, 100 nm in $2{\cdot}3 \times 10^{-2}$ s and 1 μm in 2·3 s.

The effect of variations in viscosity and temperature are readily estimated; it requires large changes in η and T to have really significant effects on the distances covered by diffusion.

From these ideas we can estimate the minimum time required for two particles to be in collision in the absence of any other force affecting their mutual approach.

Assuming monodisperse spherical particles originally spaced at the corners of a cubic lattice, Mysels[38] shows that the average time of encounter is of the order of

$$t = \frac{5{\cdot}2\eta a^3}{kT\phi^{4/3}} \tag{1.19}$$

where k is the Boltzmann constant and ϕ the volume fraction of the particles in the system. For water at 20°C this becomes

$$t = \frac{1{\cdot}3 \times 10^{12} a^3}{\phi^{4/3}} \tag{1.20}$$

and hence for a dispersion containing a volume fraction of particles of 0·1% with radius 100 nm the time is about 13 s. Hence the time involved in particle collisions may be short but increases rapidly with particle radius and slowly with dilution.

The theory of rapid flocculation was first proposed by Smoluchowski[39] who treated the problem as one of diffusion (Brownian motion) of the spherical particles of an initially monodisperse dispersion, with every collision, in the absence of a repulsive force, leading to a permanent contact. The total number N of particles (singlets, doublets, triplets, etc.) that are present at time t (s) is given by

$$N = \frac{N^0}{(1 + t/t_{1/2})} \tag{1.21}$$

where N^0 is the number initially and $t_{1/2}$ the time required to reduce the total number by one-half. The number concentration of aggregates containing k monomers is

$$N_k = \frac{N^0(t/t_{1/2})^{k-1}}{(1+t/t_{1/2})^{k+1}} \tag{1.22}$$

In general we may express the rate of flocculation, i.e. the rate of decrease of the total number of particles, as

$$\frac{-\mathrm{d}N}{\mathrm{d}t} = k'N^2 \tag{1.23}$$

where k' is the rate constant, and this becomes, when integrated with $N = N^0$ at $t = 0$,

$$1/N = 1/N^0 + k't \tag{1.24}$$

For the most rapid rate of disappearance of particles of all types Smoluchowski's theory gives

$$k'_0 = 8\pi Da \tag{1.25}$$

and a theoretical value for k'_0 may be evaluated using the Einstein expression $D = kT/6\pi\eta a$. Hence the reciprocal of the total number of particles should increase linearly with time during flocculation and the slope of the $1/N$ against t plot for rapid flocculation might be compared with that predicted by the Smoluchowski theory. Such linear plots have been recorded[40-44] and the k'_0 values are of the expected order of magnitude. Experimental k'_0 values are sometimes larger and sometimes smaller than theory predicts. Polydispersity and non-spherical form both lead to greater rates although the effects are not significant unless gross departures from the ideal are involved. However, Smoluchowski's theory does not take into account the long range force of attraction (van der Waals) between the particles (only attraction on contact is considered). Taking account of this force[45] leads to

$$-\frac{\mathrm{d}N}{\mathrm{d}t} = \frac{4\pi DN^2}{\displaystyle\int_{2a}^{\infty} \frac{\exp(V_A/kT)}{R^2}\,\mathrm{d}R} \tag{1.26}$$

where R is the distance between particle centres and V_A the potential energy of attraction; this expression becomes that of Smoluchowski when $V_A = 0$. The factor by which the Smoluchowski rate is increased depends on those parameters which determine the magnitude of the attractive force, i.e. particle radius and nature, and the type of medium. The correction factor would normally be ~ 1–2 but might take values as high as 4 under optimum conditions.[45]

When the experimental rapid rate is appreciably smaller than that predicted, as is frequently found with dispersions in aqueous media, then another mechanism must be invoked. One looks for a possible source of a repulsive force which is not removed under the condition of the experiment; the electrical repulsive force associated with charged particles is annulled by the high concentrations of electrolyte used in rapid-rate experiments with aqueous dispersions. Structuring of the medium around the particles has been proposed,[46] such structured liquid having to be displaced before particles make contact, and this involves energy; the magnitude of this effect has yet to be accurately assessed.

One important observation which may be made from Smoluchowski's theory is that for other than the most dilute dispersions, flocculation would be expected to be fast on any practical time scale, e.g. for aqueous dispersion at 25°C with initially N^0 particles per cm^3, $t_{1/2} \approx 2 \times 10^{11}/N^0$ s, and therefore instant separation of the disperse phase of a practical system (containing ~ 1–10% solids) would occur provided no other factors oppose the collision process.

The foregoing discussion assumes the dispersion to be stationary and all collisions to result from Brownian motion. This process is called *perikinetic* flocculation. For a polydisperse system under the influence of the gravitational field or, of greater importance, when in forced convection (shaking), the disparity in velocity between large and small particles becomes important. The effect is autocatalytic since on flocculation the clusters grow and the disparity increases. Under such conditions we would not expect any theory based strictly on diffusion to apply since shear flow is involved. This effect is called *orthokinetic* flocculation; relatively little effort has been made in its investigation, but some theoretical treatments have been reported.[47–50]

Forces of interaction between particles

At least three major types of interaction are involved in the approach of colloidal particles, namely

(i) the London–van der Waals force of attraction,
(ii) the coulombic force (repulsive or attractive) associated with charged particles, and
(iii) the repulsive force arising from solvation, adsorbed layers, etc.

These forces originate from entirely different sources and therefore may be evaluated separately.

The interplay of (*i*) and (*ii*) form the basis of the classical theory of flocculation of lyophobic dispersions, first proposed by Deryaguin and Landau[51] in Russia and independently by Verwey and Overbeek[52] in the Netherlands, and now known as the DLVO theory. The effects of the adsorbed layer on the attractive force between particles, and also of the repulsive force when two adsorbed layers on approaching particles interact with each other, have been evaluated. More complications arise when the adsorbed species are polymeric ions. The question of whether the medium in the vicinity of the particle plays any significant role in flocculation behaviour, either through electrical effects (e.g. change in dielectric constant) or viscosity effects due to structuring, has yet to be resolved.

In general a colloidal dispersion will be stable for a reasonable period of time if (*ii*) or (*iii*) exceeds (*i*) by such a factor that an energy barrier exists which is of magnitude several orders higher than the thermal energy (kT) of the particles. When the difference is very much larger than kT the flocculation rate will be virtually zero and the system would be considered indefinitely stable, although the time scale is somewhat arbitrary. When (*i*) exceeds (*ii*) or (*iii*) at all distances between particles, the dispersion will flocculate as fast as the particles can diffuse together.

The nature and interplay of these three forces will now be considered in more detail.

London–van der Waals attractive forces

London–van der Waals forces are based on electric interactions; they are due to the interaction of dipoles within the particles and these may be the permanent dipoles of polar particles or the dipoles that may be induced in non-polar particles which are nevertheless polarisable. The interaction between the dipoles is electromagnetic in character, and it can readily be shown that this force is always attractive.

There are two methods of calculating the attractive force between solid particles. The first, the microscopic approach, assumes that the interactions between individual atoms or molecules are additive so that integrating over all pairs of atoms or molecules leads to the total energy for the macroscopic bodies. This method was used by Hamaker[53] and leads to the following expression for the attractive potential energy between two spheres of equal radius in a vacuum

$$V_A = -\frac{A}{6}\left[\frac{2}{s^2-4}+\frac{2}{s^2}+\ln\left(\frac{s^2-4}{s^2}\right)\right] \tag{1.27}$$

where the distance parameter $s = R/a$, and A is termed the Hamaker constant given by $\pi^2 q^2 B$; q is the number of atoms (molecules) per cm^3 and B the London constant in the equation for the attraction between two atoms

$$V = \frac{-B}{r^6} \tag{1.28}$$

where r is the distance between the two atoms. Hence, provided a value for A is available, the relationship of V_A to distance may be calculated. The attraction between two isolated atoms is effective only over a very short range (of the order of atomic dimensions) but on addition, as in this treatment for colloidal particles, the range is much longer and is of the order of the dimensions of the particles.

The evaluation of A is a major problem. The London constant may be estimated (only) theoretically by calculating certain molecular constants from optical data. For this calculation the Slater–Kirkwood expression as modified by Moelwyn-Hughes[54] is probably the most satisfactory

$$B = \frac{3Z^{1/2}}{4h\nu_v \alpha_0^2} \tag{1.29}$$

where Z is the number of electrons in the outer shell of one molecule, h is Planck's constant, ν_v the characteristic frequency of the molecules and α_0 the static polarisability. From optical data ν_v may be calculated using the dispersion formula

$$n^2 - 1 = \frac{C\nu_v^2}{\nu_v^2 - \nu_L^2} \tag{1.30}$$

where n is the refractive index at frequency ν_L and C is a constant. α_0 is calculated from

$$\alpha_0 = \frac{3M}{4\pi N_0 \rho} \cdot \frac{n^2-1}{n^2+2} \cdot \frac{\nu_v^2 - \nu_L^2}{\nu_v^2} \tag{1.31}$$

where ρ is the density of material of molecular weight M. Values of A for most solvents lie in the range 4–8×10^{-20}J, for polymers 6–10×10^{-20}J, and for metals and ionic solids 10–30×10^{-20}J. Indirectly values of A may be obtained from flocculation experiments but agreement between the two methods must not necessarily be taken as indicative of the correctness of the A value. Theoretical and experimental A values have been compared by Lyklema.[55]

Another method of estimating the Hamaker constant A has been proposed by Fowkes[56] based on interfacial tension data. Fowkes demonstrates that the interaction energies due to dispersion forces at an interface may be readily predicted by the geometric mean of the dispersion force contribution to the surface free energy (γ^d) of the two materials. Directly or indirectly (e.g. through contact angle data) values of γ^d for a variety of solids and liquids were obtained using the defining relation

$$\gamma_{12} = \gamma_1 + \gamma_2 - 2\sqrt{\gamma_1^d \gamma_2^d} \quad (1.32)$$

which states that the interfacial tension γ_{12} is equal to the sum of the separate tensions of the components minus a term due to the London interaction between surface molecules of 1 and bulk molecules of 2 and vice versa. γ_1^d is related to the Hamaker constant A_1 for component 1 by

$$A_1 = 6\pi r_{11}^2 \gamma_1^d \quad (1.33)$$

where r_{11} is the intermolecular distance. For a two-phase system

$$A = 6\pi r_{11}^2 \left(\sqrt{\gamma_1^d} - \sqrt{\gamma_2^d}\right)^2 \quad (1.34)$$

and Fowkes has calculated values of A for a variety of solids in contact with water assuming $6\pi r_{11}^2 = 1{\cdot}44 \times 10^{-14}$ for water and systems with volume elements such as oxide ions, metal atoms, $-CH_2$ or $-CH$ groups which have about the same size. Fowkes shows that the agreement between these values and some derived from other experiments is good. Furthermore the basic idea permits accurate calculation of several other important surface chemical parameters which also compare favourably with experiment,[56] lending weight to the validity of the principle.

Methods of evaluating A were reviewed by Gregory,[57] and tables of A values have been reported.[57–59]

The second approach (the macroscopic approach) for the calculation of the attractive force, as proposed by Lifshitz,[60,61] begins with the optical properties of the interacting macroscopic bodies, and derives the attraction from the imaginary parts of their complex dielectric constants particularly in the far ultraviolet. Unfortunately the appropriate data are available for only a very limited number of materials so that at present we are left to make estimates using the London theory. The microscopic and macroscopic approaches have been critically compared by Krupp,[62] and the reader is referred to his article for further details on the Lifshitz theory and examples of the van der Waals constants obtained from it.

When a medium separates the interacting particles the attractive force is weaker than that calculated by eqn. (1.27) because of the interaction between the molecules of the medium and between the particles and the medium. Hamaker showed that for the two phase system the total A is given by

$$A = A_{pp} + A_{mm} - 2A_{pm} \tag{1.35}$$

while the distance function in the brackets of eqn. (1.27) remains the same. A_{pp} and A_{mm} are the constants for the particles and medium (*in vacuo*) respectively and A_{pm} that associated with the particle–medium interaction. In the absence of more precise information it is generally assumed that $B_{pm} = (B_{pp} B_{mm})^{1/2}$, which is predicted by London for two single electronic oscillators having the same characteristic frequency (ν_v) but different polarisabilities; this is normally not strictly true and tends to overestimate the effect. Hence we may write as a good approximation

$$A = (A_{pp}{}^{1/2} - A_{mm}{}^{1/2})^2 \tag{1.36}$$

and the value of A is always positive. Often the values of A_{pp} and A_{mm} are of similar magnitude and since these values are certain to be in error, it is not possible to estimate A for a system to much better than an order of magnitude.

For the interaction between dissimilar particles eqn. (1.35) becomes

$$A = A_{12} + A_{33} - A_{13} - A_{23} \tag{1.37}$$

where the subscripts 1 and 2 refer to the different particle material and 3 to the medium. If $A_{11} < A_{33} < A_{22}$, A becomes negative, i.e. a repulsive force develops.

Two refinements to the Hamaker treatment should be considered. The first involves transmission of the London force through the medium; theoretically we would expect the force to be reduced by a factor equal to the dielectric constant at electronic frequencies, this being to allow for the influence of the dielectric permeability of the medium on the propagation of the force. The factor used is the square of the refractive index of the medium for wavelengths in the visible region of the spectrum at which no light absorption occurs, e.g. $n^2 = 1{\cdot}8$ for water. The second refinement is concerned with the finite time taken for the London force to reach a distant atom, i.e. at distances comparable with, or greater than, the characteristic wavelength of the electrons ($\sim 10^{-5}$ cm). Again a reduction in the force is apparent. For large spheres with distances of separation much smaller than the radius, Schenkel and Kitchener[63] have derived useful approximate formulae for estimating the retarded force. But for other situations the calculation becomes complex[64] and subject to so many uncertainties that it is probably not worthwhile.

The effect of an adsorbed layer having a value of A differing from that of the particle was first considered by Vold,[65] who showed that the presence of an adsorbed layer leads to a decrease in attractive energy. The total effect is the sum of two parts. First, the attractive energy is inversely proportional to the minimum surface separation of the two particles, hence the addition of an adsorbed layer at constant surface separation leads to an increase in separation of the core particles, resulting in a decrease in energy. Second, the change in the A constant for the adsorbed layer leads to a further change in attractive energy, which can be negative or positive, depending on the relative values of the three A constants involved.

Three cases were considered, one being particles covered with a solvate layer of bound medium of thickness δ, one with a homogeneous adsorbed layer of surfactant molecules, and the third with an adsorbed layer of oriented amphipathic molecules. The general expression derived by Vold is

$$V_A = -\frac{1}{12}[(A_{mm}{}^{1/2} - A_{ss}{}^{1/2})^2 H_s + (A_{ss}{}^{1/2} - A_{pp}{}^{1/2})^2 H_p + 2(A_{mm}{}^{1/2} - A_{ss}{}^{1/2})(A_{ss}{}^{1/2} - A_{pp}{}^{1/2}) H_{ps}] \quad (1.38)$$

where the subscript s refers to the solvate layer, and A_{mm}, A_{ss}, and A_{pp} are the Hamaker constants for the medium, solvate layer and particle respectively. H_s, H_p and H_{ps} are the appropriate forms of the distance function in the more general form of the Hamaker equation for two spheres of radii a_1 and a_2:

$$V_A = -\frac{A}{12} H(x, y) \quad (1.39)$$

where

$$H(x, y) = \frac{y}{x^2 + xy + x} + \frac{y}{x^2 + xy + x + y} + 2 \ln\left(\frac{x^2 + xy + x}{x^2 + xy + x + y}\right) \quad (1.40)$$

and $x = \Delta/2a$, and $y = a_2/a_1$. Δ is the shortest distance between the surfaces, which is assumed when the particles are in contact to be the minimum distance between surface atom centres beyond which electronic repulsion becomes important. Thus H_s refers to two spheres of radius $a+\delta$ with the surfaces of the solvate layers separated by Δ, H_p to two unsolvated spheres of radius a separated by $\Delta+2\delta$, and H_{ps} to one sphere of radius a and one of radius $a+\delta$ separated by $\Delta+\delta$. The parameters in eqn. (1.40) are as follows:

$$\text{for } H_s,\ y = 1 \text{ and } x = \Delta/2(a+\delta)$$
$$\text{for } H_p,\ y = 1 \text{ and } x = (\Delta+2\delta)/2a$$
$$\text{for } H_{ps},\ y = (a+\delta)/a \text{ and } x = (\Delta+\delta)/2a$$

and with these functions and eqn. (1.38) values of V_A at contact (Vold assumes $\Delta = 0{\cdot}3$ nm) and as a function of interparticle distance may be calculated.

The reader is referred to the original paper for tabulated examples of the Hamaker functions, and to more recent papers[66,67] which show graphically how the V_A against distance relationship is determined by the value of δ and the relative magnitudes of the Hamaker constants. In general the following conclusions may be drawn, assuming the optimum combination of Hamaker constants.

(i) Only very small particles ($a \sim 10$ nm) can be stabilised by solvate layers of bound medium of thicknesses that are normally associated with such layers.
(ii) A homogeneous adsorbed layer which differs in composition from the medium is, for equal values of δ, more effective in reducing V_A than bound medium alone, but if the adsorbed layer contains oriented amphipathic molecules the effect may be even larger.
(iii) The effect is dependent on particle radius; large values of δ(~ 10 nm) would be required to stabilise particles of radius $\sim 1\ \mu$m.

The topic of adsorbed layers has been thoroughly discussed by Osmond *et al.*[68] and by Vincent,[59] who corrected and extended Vold's treatment by considering retardation and non-homogeneous layers. In particular, they showed that under certain conditions a repulsive energy can develop between particles or between particles and plates coated with differing adsorbed layers.

Despite all the difficulties attractive forces are calculable. For spherical particles of radius $\leqq 100$ nm the Hamaker equation (eqn. 1.27), with A corrected for transmission effects, is probably satisfactory, and Fowkes' values of A seem reliable. For larger particles retardation effects should be considered. Formulae are also available[69] for flat plates which might be more appropriate in certain cases.

The coulombic repulsive force

Since in a medium containing ions a charged particle with its electric double layer is electrically neutral (Chapter 3), no net coulombic force exists between charged particles at large distances from each other. As the particles approach the diffuse parts of the double layers interpenetrate giving rise to a repulsive force which increases in magnitude as the distance between the particles decreases. The distance at which the repulsive force becomes significant increases with the thickness of the double layer ($1/\kappa$) and the force increases with the surface potential (ψ_0).

For particles that are sufficiently far apart for there to be no interaction the characteristics of each individual particle–double layer system are determined by the surface charge (or surface potential) and the ionic strength and properties of the medium. When two particles interact some changes in these parameters must occur, but in the estimation of the interaction energy either constancy in surface potential (implies rapid establishment of adsorption equilibrium) or in surface charge (implies slow desorption) is assumed. Constancy in surface potential is normally considered to be appropriate to systems such as the classical silver halide sols where the potential is determined by the silver and halide ion concentrations in bulk solution, but some doubt has been cast on this approach and it is suggested that the constancy in charge assumption is more appropriate.[70] The latter is also probably more appropriate to systems stabilised by adsorbed surface active agents. Probably neither is strictly true in any one situation but the final results for both are not significantly different. The treatment of interacting double layers is published in detail in Verwey and Overbeek's book,[52] and only a brief account is given here. The theory is not simple; it turns out that the interaction of two flat parallel double layers is the easiest to handle and is the only one to be considered here, but relevant working formulae for spheres will be given.

For a flat double layer we may write

$$\frac{d^2\psi}{dx^2} = \frac{8\pi nze}{\epsilon}\sinh\left(\frac{ze\psi}{kT}\right) \tag{1.41}$$

where ψ is the potential at distance x from the surface, n the concentration of electrolyte containing ions of valency $z_+ = z_- = z$, e is the electronic charge and ϵ the dielectric constant. Substituting

$$y = ze\psi/kT, \quad w = ze\psi_0/kT, \quad \kappa^2 = 8\pi ne^2z^2/\epsilon kT$$

we obtain

$$d^2y/dx^2 = \kappa^2 \sinh y \tag{1.42}$$

the solution of which for $\kappa x > 1$, i.e. at a large distance from the surface, is

$$y = 4\gamma \exp(-\kappa x) \tag{1.43}$$

where $\gamma = \exp[(w/2) - 1]/\exp[(w/2) + 1]$. It may readily be shown that this solution is a very good approximation for distances down to $\kappa x = 1$ for all values of ψ_0.

To derive an exact expression for the potential distribution between two flat plates separated by a distance $2d$, it is necessary to solve eqn. (1.41) to satisfy the boundary conditions $\psi = \psi_0$ when $x = 0$ and $x = 2d$, and $d\psi/dx = 0$ when $x = d$. The solution involves an elliptic integral which can only be solved numerically. However, when $\kappa d > 1$ (i.e. the interaction is not too great) it is sufficiently accurate to assume that the potential may be built up additively from the potentials of the two single double layers. Hence, from eqn. (1.43) the potential ψ_d half-way between the plates is given by

$$u = 8\gamma \exp(-\kappa d) \tag{1.44}$$

where $u = ze\psi_d/kT$ and $\kappa d > 1$.

The simplest way to calculate the force of repulsion between two interacting double layers is to follow Langmuir in considering the force as arising from the osmotic pressure of the excess ions in the space where the double layers overlap. The ionic concentrations (n_+ and n_-) at $x = d$ are given by the Boltzmann equation

$$n_+ = n \exp(-ze\psi_d/kT) \text{ and } n_- = n \exp(ze\psi_d/kT) \tag{1.45}$$

and the osmotic pressure at this plane is

$$kT(n_+ + n_-) = nkT[\exp(u) + \exp(-u)] = 2nkT\cosh u \tag{1.46}$$

Outside the field of the double layers where ψ and $u = 0$, the osmotic pressure is $2nkT$. Hence the difference between these two pressures is a measure of the force p acting on the unit area of each plate:

$$p = 2nkT(\cosh u - 1) \tag{1.47}$$

The repulsive potential energy is given by the work done against this force when the plates are brought together from a large distance. This work per unit area of plate is given by

$$V_R = -\int_{\infty}^{d} p\mathrm{d}x = -2\int_{\infty}^{d} 2nkT(\cosh u - 1)\mathrm{d}x \tag{1.48}$$

but in the general case this integral cannot be readily evaluated because the relation between u and x is complex. However, if we again assume small interaction, i.e. u is small and $(\cosh u-1)=u^2/2$, then

$$V_R = -2\int_{\infty}^{d} nkT64\gamma^2 \exp(-2\kappa x)\mathrm{d}x \tag{1.49}$$

or

$$V_R = 64nkT\gamma^2[\exp(-2\kappa d)]/\kappa \tag{1.50}$$

The exact numerical solutions for the interaction of parallel plates have been computed and published by Devereux and de Bruyn.[71] The corresponding tables for spherical particles are not generally available and it is normal practice to use approximate formulae which are valid for limited conditions. The two equations most commonly used were derived for small ψ_0 and prove most useful in many practical situations.

For systems in which $\kappa a \gg 1$ (large particles in aqueous systems with moderate electrolyte concentrations) and there is weak interaction (shortest distance Δ between surfaces of spheres is large compared with $1/\kappa$),

$$V_R = \tfrac{1}{2}\epsilon a \psi_0^2 \ln[1+\exp(-\kappa\Delta)] \tag{1.51}$$

The equation gives values which are close to those from the exact treatment (using graphical integration) provided $w(=ze\psi_0/kT) \ngtr 2$ and $\kappa a \nless 10$.

For systems in which $\kappa a \ll 1$ (small particles in non-aqueous media or in aqueous systems with very low electrolyte concentration)

$$V_R = \frac{\epsilon a^2 \psi_0^2}{\Delta + 2a}\beta \exp(-\kappa\Delta) \tag{1.52}$$

and is quite accurate up to $\kappa a = 1$ and $\psi_0 \sim 50$–60 mV. β is a factor to allow for loss of spherical symmetry in the double layers as they overlap and is defined in Verwey and Overbeek's book.[52] For the treatment of the intermediate region of κa and for higher potentials the reader is referred to the original literature.

The total interaction energy

Since the double layer repulsion and the van der Waals attraction operate independently the total potential energy V_{tot} for the system is given by the sum of the two terms; the form of the resulting potential energy against distance relationship is dependent upon the relative magnitudes of the two forces. V_R decreases exponentially with distance while V_A shows an approximate inverse relationship with the square of the distance. Attraction predominates at short distances (other than immediately adjacent to the surface at atomic distances when repulsion dominates—little is known about this effect and in colloid problems it is usually ignored). Otherwise the form of the V_{tot} curve depends to a large extent on the V_R term. Figure 1.2 illustrates the type of plot we might

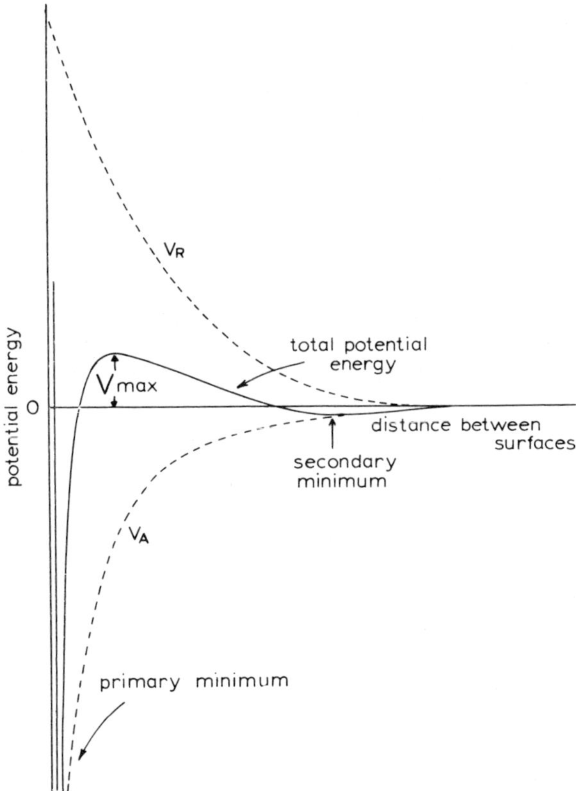

Fig. 1.2. Potential energy curves for the interaction of two charged surfaces.

expect for particles of radius 0·1–1 μm in an aqueous system containing about 0·01 mol/dm^3 of 1 : 1 electrolyte for which the range of attractive and repulsive forces are similar.

Three important characteristics are shown in Fig. 1.2 and these are directly related to flocculation behaviour. The potential energy barrier must be surmounted before the particles make lasting contact in the primary minimum. Provided the barrier is considerably larger than the thermal energy of the particles relatively few will make contact and the system should be stable. But if the secondary minimum is of depth $\gg kT$ then the particles would flocculate with a liquid film between them in the cluster. Since both the attractive and repulsive forces are approximately proportional to the particle radius, the secondary minimum should become increasingly significant with increasing particle size, and particularly so with parallel plates. The effect will also increase with increasing electrolyte concentration which reduces the distance over which the repulsive forces operate, but this also reduces the energy barrier which again would promote flocculation. Systems which have flocculated into the secondary minimum tend to be *reversible*, i.e. they can be readily redispersed (peptised) with shaking; those in the primary minimum need considerably more energy to redisperse. In some cases particles might on standing pass from the secondary to the primary minimum.

The effect of reducing the electrolyte concentration (increasing $1/\kappa$ at constant ψ_0) on the total potential energy is shown in Fig. 1.3 and illustrates the difference between dispersions in aqueous and non-aqueous solutions of a surface active agent of the same stoichiometric concentration, e.g. sodium di-2-ethylhexyl sulphosuccinate, which dissolves in both water and hydrocarbons, but ionises to markedly different extents in the two solvents. When $1/\kappa$ is large the secondary minimum disappears since the range of V_R is then considerably greater than that of V_A, and under these conditions the accuracy of the V_A values becomes much less important.

Rate of slow flocculation

The energy relationships tell us whether or not we might expect a system to be stable, but give no information on the rate of flocculation. We now consider the kinetics of the flocculation of a system for which there is an energy barrier. The rate is a function of the probability of particles having sufficient energy to overcome this barrier. The problem

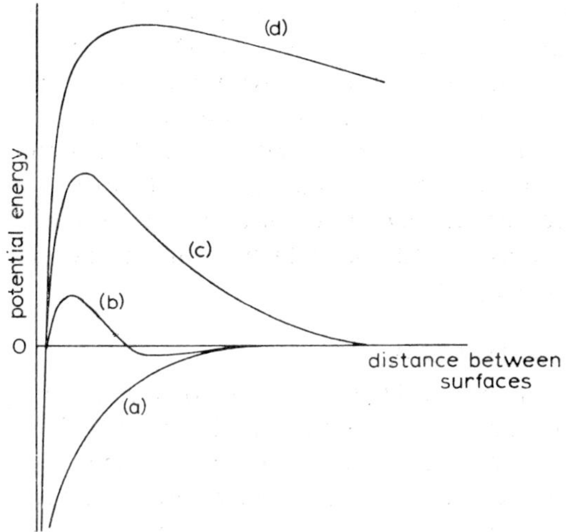

Fig. 1.3. Influence of electrolyte concentration on the total potential energy of interaction of two spherical particles of radius 100 *nm in aqueous media.* (a) $1/\kappa = 10^{-7}$ *cm*, (b) $1/\kappa = 10^{-6}$ *cm*, (c) $1/\kappa = 10^{-5}$ *cm*, (d) $1/\kappa = 10^{-4}$ *cm.*

is again one of diffusion and was first solved by Fuchs[72] in 1934 for aerosols and later for aqueous dispersions by Deryaguin.[73]

Fuchs' theory leads to a factor W by which the rapid rate (Smoluchowski) is reduced by the presence of a repulsive force; W is called the *stability ratio* and is related to the height of the potential energy barrier by

$$W = 2\int_2^\infty \frac{\exp(V_{\text{tot}}/kT)}{s^2}\,\mathrm{d}s \tag{1.53}$$

However, Smoluchowski's value for the rapid rate does not assume the existence of long range attractive forces. Introduction of the appropriate term leads to the following equation.[45]

$$W = \frac{\displaystyle\int_2^\infty \frac{\exp(V_{\text{tot}}/kT)}{s^2}\,\mathrm{d}s}{\displaystyle\int_2^\infty \frac{\exp(V_{\text{A}}/kT)}{s^2}\,\mathrm{d}s} \tag{1.54}$$

so that $W=1$ when $V_{tot}=V_A$ at all distances. Given the appropriate potential energy diagrams from, say, the DLVO theory, the stability ratio may be calculated by graphical or numerical integration and then compared with experimental values of $W=k'_0/k'$, the ratio of the experimental rate constants for rapid and slow flocculation. Such a comparison is a severe test of the applicability of theory to practical cases, and the observed deviations, although often not appreciable, reflect the assumptions and approximations which are necessary in the calculation of the potential energy terms.

Application of the DLVO theory

The application of the DLVO theory to flocculation problems has become common practice despite all the limitations. There is no doubt that in principle the theory is correct but refinements are obviously necessary in many cases. We shall consider here some examples of theory related to practice for the purpose of illustrating the applicability and shortcomings of the theory.

In making theoretical predictions of stability it is necessary to assign values to three important parameters:

(i) the Hamaker constant,
(ii) the Stern potential (see Chapter 3) which determines the characteristics of the diffuse part of the double layer, the region particularly relevant to colloid stability, and
(iii) the particle dimensions.

Evaluation of the Hamaker constant has already been discussed. Since the Stern potential cannot be determined by direct measurement it is usual to use the zeta potential obtained from an electrokinetic experiment (Chapter 3), and this has been successful in a number of cases.[63,74–76] Measurement of particle dimensions is itself a major problem, hence the desirability of using monodisperse systems when the validity of the theory is being assessed. Nevertheless, for a dispersion containing only one type of particle, the deviations from monodispersity and spherical geometry may effectively be small enough so that an appreciable error is not involved when the theory for monodisperse spheres is applied. For powders, electron microscopy is the most attractive technique for assessing particle dimensions. The problems of aqueous and non-aqueous systems will now be treated separately. For aqueous systems the potential energy relationships are sensitive to all three parameters, but in hydrocarbons an absolute value of A is not required.

(*a*) *Aqueous systems*

It is well known that the addition of sufficient indifferent electrolyte to a stable aqueous hydrophobic sol usually leads to flocculation. The familiar Schulze–Hardy rule states that the effect is determined by the counterion and increases markedly with the valency. A great deal of early work supported this hypothesis and it is also predicted by the DLVO theory.

Although in principle the transition from slow to rapid flocculation is continuous, i.e. there is no sharp flocculation point, the relation between W and electrolyte concentration usually shows a fairly sharp transition at a *critical* concentration value and in practice such a sharp transition is observed; it is a question of rate plus energy barrier and not of the barrier alone. The critical electrolyte concentration n' at which rapid flocculation takes place is taken as that for which the maximum in the total potential energy curve is zero, i.e. when $V_{tot}=0$ and $dV_{tot}/dd=0$ are satisfied by the same value of d. Hence, for two flat plates

$$\frac{64n'kT\gamma^2[\exp(-2\kappa d)]}{\kappa}-\frac{A}{48\pi d^2}=0 \qquad (1.55)$$

and

$$\frac{dV_{tot}}{dd}=-2\kappa V_R+2V_A/d=0 \qquad (1.56)$$

The term $A/48\pi d^2$ is the appropriate form of the attractive potential energy.[69]

So $\kappa d=1$ and

$$64n'kT\gamma^2\exp(-2)/\kappa=A\kappa^2/48\pi \qquad (1.57)$$

or

$$n'=107\epsilon^3k^5T^5\gamma^4/A^2e^6z^6 \qquad (1.58)$$

which approaches $n'=\text{constant}/z^6$ when ψ_0 is high ($\gamma\rightarrow 1$), i.e. the critical electrolyte concentration is inversely proportional to the sixth power of the valency of the counterion. Mean flocculation concentration of mono-, di-, and tri-valent cations for a negative silver iodide sol are 142, 2·43 and 0·068 mmol/dm^3 respectively, which are in the ratio 1:0·017:0·0005. The theoretical z^{-6} ratio is 1:0·016:0·0013. Deviations for ions of high valency may be attributed to neglect of adsorption and ion-size effects, besides chemical changes (hydrolysis) that may have occurred with the electrolyte. A secondary, but not negligible, effect is that associated with varying the counterion at a given valency[77] (the lyotropic series), the concentration decreasing as the unhydrated ionic radius increases. Such specific effects are not explicable using the Gouy theory, which is non-specific, so Stern theories or modifications to the Poisson–Boltzmann relation must be introduced.

Some interesting observations have been made comparing experimental W values with those predicted by the theory. Rate constants are measured for slow and rapid flocculation of aqueous dispersions of polymer latices, which may be prepared in a virtually monodisperse spherical form and make ideal systems for such comparison. Obviously it is necessary to study the initial stages of the process where only doublets are formed, since the appearance of larger aggregates would alter the potential energy relationships. The experiment involves adding indifferent electrolyte (containing no potential determining ions) to the stable dispersion and following changes in turbidity, particle number, etc. W is then the relative rate of flocculation compared with that observed with high concentrations of electrolyte corresponding to rapid flocculation (when $W=1$). Plots of log W against log n for the experimental data are compared with the linear relationship predicted by the DLVO theory. Published data[75,78–80] on a variety of systems meet the requirement and usually extrapolate at $W=1$ to values of the critical electrolyte concentration n' that are in agreement with the theory. However, on more critical examination, involving variations in particle size or surface potential, it has become apparent that the theory does not fully account for the facts. For example, theory predicts an increase in slope with particle size but Ottewill and Shaw[79] found that for polystyrene latices covering a range of particle diameters from 60 to 423 nm the slopes are almost independent of particle size. Experiments carried out in the author's laboratory[81] with silver iodide sols also fail to confirm the expected increase of slope with surface potential, but the observed effect can be explained by introducing the discreteness of charge effect.[82]

In practical systems particles consisting of aggregates of smaller primary particles are often involved, and it is of particular interest to know which value to assign to the effective radius, that of the primary particle or that of the total aggregate. This would be a crucial factor in the case where the double layer thickness is small, i.e. at reasonably high electrolyte concentrations. A study[76] of the flocculation behaviour of two carbon blacks in aqueous solutions of sodium dodecyl sulphate containing 0·1 mol/dm^3 NaCl (for which $1/\kappa = 1$ nm) admirably demonstrates this point. Graphon and Sterling MTG are two graphitised blacks with surface areas of 78·9 and 7·7 m^2/g and mean primary particle radius of approximately 12·5 and 125 nm, respectively. In dispersion Sterling MTG exists primarily as individual particles but Graphon is mainly aggregated with a peak in the size distribution at about 100 nm. The adsorption of the surface active agent is identical on both carbons when put on a unit area basis. Theory predicts that under the conditions prevailing the Sterling MTG particles should flocculate into a secondary minimum even at high zeta potentials, and indeed the systems were all unstable. Similar behaviour would be expected for Graphon if the aggregate size determines the stability, but in fact the experimental log W against zeta potential relationship closely followed that predicted by theory for $a = 12{\cdot}5$ nm (the primary particle radius).

(*b*) *Non-aqueous systems*

Extension of the DLVO theory to non-aqueous systems has attracted relatively little attention. For media of intermediate dielectric constant, e.g. alcohols, ketones, etc., we might expect the behaviour to be similar to that already established with aqueous systems, but with the actual ionic concentrations one or two orders of magnitude lower. Studies of a fundamental nature on such systems are scarce,[83] but some attention has been paid to dispersion in hydrocarbon solutions of surface active agents for which the ionic concentrations are extremely small, e.g. $\sim 10^{-10}$ mol/dm^3 corresponding to $1/\kappa \sim 10\,\mu$m.[44] Hence κa falls almost to zero, the exponential term in eqn. (1.52) tends to unity and V_R is given, to a very good approximation, by

$$V_R = \frac{\epsilon a^2 \psi_0^2 \beta}{\Delta + 2a} \tag{1.59}$$

Since $1/\kappa$ is large the capacity of the double layer is small and only a small surface charge density is necessary to obtain an appreciable surface potential. Furthermore, the slow decay in potential from the surface means that the zeta potential, readily obtained from electrophoresis experiments,[84,85] may be equated with considerable accuracy to the surface potential. Another simplification compared with aqueous systems is that an absolute value of A is not required since the V_R term dominates the V_{tot} against distance relationship. Only the particle radius presents problems (as usual). A secondary minimum is theoretically not feasible for colloidal particles in media of low dielectric constant, unless factors other than charge, e.g. force (iii), are relevant.

It is therefore possible to predict, with some confidence, the stability of a dispersion from readily obtainable parameters, and this has been done for hydrocarbon media for a range of particle size and surface potential.[86] Computations based on eqns. (1.27), (1.54) and (1.59) using β values calculated from Verwey and Overbeek's equation[87] assuming $\kappa a = 0$ and $\epsilon = 2{\cdot}24$ are illustrated in Fig. 1.4 for a range of a and ψ_0, and Fig. 1.5 shows how the half-life of $t_{1/2}$ of a dispersion is related to the stability ratio and potential energy maximum V_{max}. The half-life was calculated from the expression

$$t_{1/2} = \frac{6\eta \displaystyle\int_2^\infty \frac{\exp V_{tot}/kT}{s^2} \mathrm{d}s}{4kTN^0} \qquad (1.60)$$

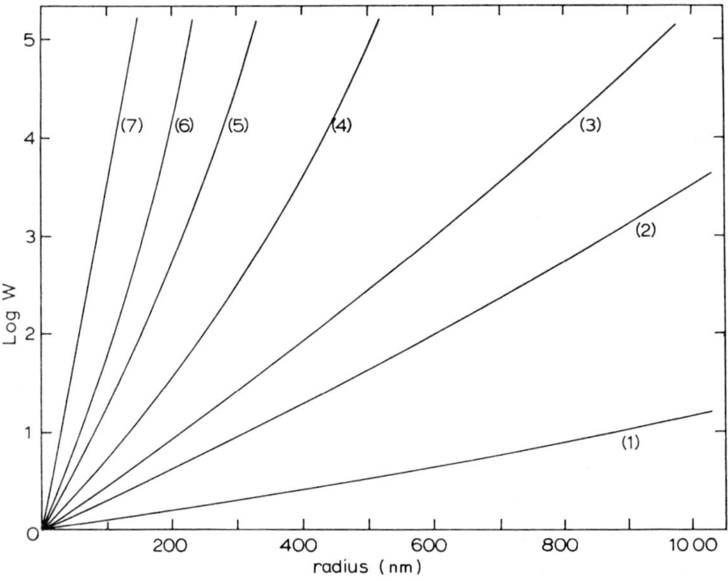

Fig. 1.4. Theoretical curves of stability ratio against particle radius for systems in which $\kappa a \ll 1$. $A = 2{\cdot}5 \times 10^{-19}$ J, $\psi_0 = 15$ mV (1), 25 mV (2), 30 mV (3), 40 mV (4), 50 mV (5), 60 mV (6), 80 mV (7).

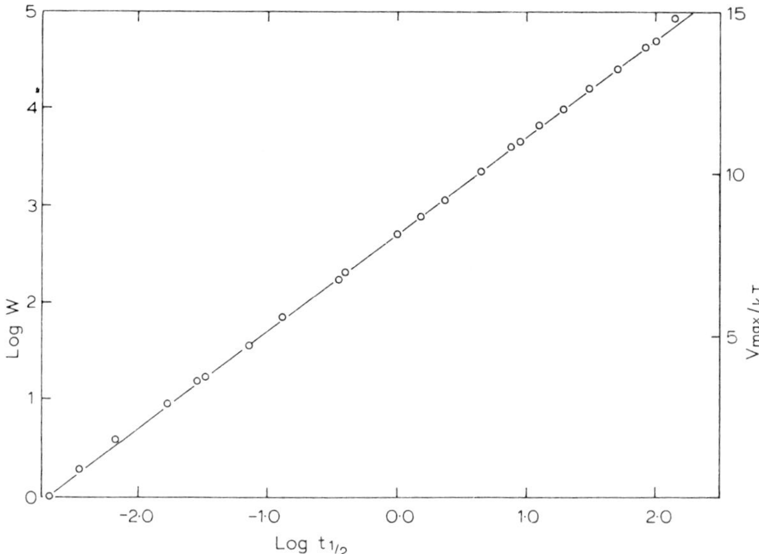

Fig. 1.5. Theoretical values of the stability ratio in relation to the half-life (in hours) of a dispersion and the potential energy maximum, for various radii in the range 20–1000 nm and $\kappa a \ll 1$. $A = 2{\cdot}5 \times 10^{-19}$ *J and particle concentration* 10^{10} *per ml.*

using $N^0 = 10^{13}$ particles/dm^3 for Fig. 1.5. Figure 1.4 demonstrates the importance of particle size, and Fig. 1.5 shows how the stability depends on V_{max}. From Fig. 1.5 the stability of any particular dispersion may be estimated by making the appropriate allowance for particle concentration (eqn. 1.60). Table 1.1 shows how for 10^{13} particles/dm^3 the various factors, commonly used to express colloid stability, are related.

A correlation between zeta potential and stability for dispersions in non-polar media and the criteria for accurate measurement of electrophoretic mobility was first established by van der Minne and Hermanie.[84,88] The validity of the DLVO theory has been demonstrated both by qualitative[89,90] and by rigorous experimental tests[44,91] on systems in which there is no significant influence of the interaction of adsorbed layers of surface active material, i.e. only the charge mechanism is operative. The origin of the charge on the particles is still subject to

TABLE 1.1

V_{max}/kT	$log\ W$	$t_{1/2}(h)$
0	0	0·002
5	1·6	0·1
10	3·3	7
15	5·0	400
25	8·3[a]	10^{6}[a]

[a]Extrapolated.

debate. Fowkes[92,93] and Tamarabuchi and Smith[94] have suggested an acid–base mechanism, involving transfer of charge between the adsorbed molecules and the surface, followed by desorption of the charged species. The presence of trace amounts of water has a profound effect on the stability of hydrophilic powders in non-polar media,[83,95,96] presumably through changes in surface energy and/or charge, but the details have not yet been worked out. The subject of charge stabilisation in non-polar media has been reviewed.[97]

Concentrated dispersions

For practical systems, when the particle concentration is high, the validity of the DLVO theory is open to question. Particle–particle interaction might not begin at 'infinite' distance of separation, i.e. one particle is, on average, sufficiently close to another to reduce the effective energy barrier which must be overcome for the particles to come into contact (not considering at this time the secondary minimum effect). Albers and Overbeek[98] made a preliminary attempt to solve this problem, their purpose being to explain their flocculation data for the charged droplets of water-in-oil emulsions. The two-particle interaction calculations were abandoned and replaced by a model based on twelve nearest neighbours, situated on a spherical shell around a central particle, with all other particles randomly distributed outside this shell. It was shown that the calculation of the interaction energy of the one particle with all other particles leads to an energy barrier which is significantly lower than that for the simple interaction of two particles, and the effect increases with particle concentration. The magnitude of the effect will be directly related to the thickness of the double layer, since double-layer overlap is the cause of the repulsion. Hence it will be much more significant in non-aqueous media. The calculations show, for example, that when $1/\kappa = 10\,\mu m$ a particle concentration of 10% would correspond to an energy barrier about 50% of that for an infinitely dilute dispersion and at 50% concentration the barrier would be reduced by an order of magnitude.

Using an alternative approach Levine and his coworkers[99–101] calculated the force between two charged spherical particles of a concentrated dispersion in a hydrocarbon medium, and demonstrated that the electric double layer force is not sufficient to promote stability to flocculation, in contrast to its effect in dilute dispersions; this is borne out in practice. Furthermore, Rehbinder[102] suggested that neither in aqueous nor in non-aqueous media will the electrostatic charge mechanism assure stabilisation of concentrated dispersions, and to prevent flocculation some sort of protective action by polymeric species is required. This protective action could arise from the effects associated with adsorption or through the polymer in the bulk forming a gel structure through which diffusion of the particles is restricted. In principle there is a range of possible mechanisms from that directly associated with the properties of the adsorbed layer to that which is entirely dependent on bulk solution properties, and hence the whole problem becomes complex and lacks quantitative precision.

Rehbinder[103] devoted considerable effort to the study of concentrated disperse systems, and believes that effective stabilising action by the adsorbed layer is only possible if the layer is structured, i.e. it has a much higher viscosity than the medium and thus serves as a mechanical barrier. It is, of course, necessary for the cohesive energy between the external surfaces of the layers to be low.

Unfortunately adequate quantitative experimental data are not available for a worthwhile comparison to be made with these ideas but practical experience tends to support the general conclusions.

Heteroflocculation

So far we have considered only the interaction between particles which are identical in nature, size and surface potential, i.e. only one value for each of a, A and ψ_0 is applicable. When only one type of particle is present it is often adequate to assume that the deviations from monodispersity and spherical geometry may be effectively small enough that little error is involved in applying the DLVO theory. When more than one type of particle is present or if there is a large disparity in size, the situation becomes more complex. Interaction between dissimilar particles is termed *heteroflocculation*.

A theory of heteroflocculation was proposed originally by Deryaguin,[104] and Devereaux and de Bruyn[71] included the interaction of dissimilar double layers in their calculations on flat plates. A more recent theoretical treatment[105] is mathematically simpler because it is based on the approximate DLVO theory for spheres with $\kappa a > 10$ and values of ψ_0 less than 50 mV. The potential energy of repulsion for two particles of radius a_1 and a_2 and potentials ψ_{01} and ψ_{02} is given by

$$V_R = \frac{\epsilon a_1 a_2 (\psi_{01}^2 + \psi_{02}^2)}{4(a_1 + a_2)} \left[\frac{2\psi_{01}\psi_{02}}{(\psi_{01}^2 + \psi_{02}^2)} \ln \left(\frac{1 + \exp(-\kappa\Delta)}{1 - \exp(-\kappa\Delta)} \right) + \ln(1 - \exp(-2\kappa\Delta)) \right] \quad (1.61)$$

which becomes eqn. (1.51) when $a_1 = a_2$ and $\psi_{01} = \psi_{02}$. Combining this with the appropriate expression of Hamaker[53] for the attraction between two spheres gives the total potential for the interaction from which energy barriers, stability ratios, etc., may be evaluated. The effects of changes in radius and potential on the potential energy diagrams are illustrated in Fig. 1.6 and Fig. 1.7. Two facts emerge from the theoretical calculations.

Firstly, it is the particle with the smaller radius and/or potential that determines the height of the potential energy barrier. Secondly, although the different particles may have the same sign of surface charge and potential (but different magnitude) they may nevertheless attract each other when the double layers overlap; this is a consequence of the relative changes which occur in the charge and potential as the particles approach.[106]

The interaction of oppositely charged particles has been recently analysed by Prieve and Ruckenstein.[107] Constant charge is assumed, i.e. one type has strongly acidic (negative) surface sites and the other strongly basic (positive) sites, and it is shown that when the ratio of the absolute charge densities is significantly different from unity ($\sim$10:1), stability ratios $> 10^4$ are obtained over a significant range of ionic strength. At large ionic strength ($\sim 10^{-1}$ mol/dm^3) rapid heteroflocculation occurs; similarly at very low ionic strength ($\sim 10^{-4}$ mol/dm^3), whilst in the intermediate range the system is stable despite the fact that the particles are oppositely charged. So the conditions for initia-

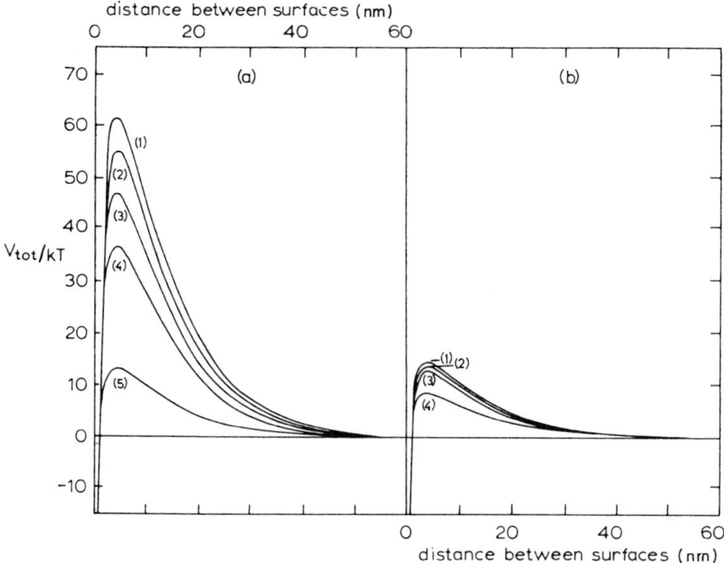

Fig. 1.6. Theoretical curves of total potential energy against distance of separation of two spherical particles of radius a_1 and a_2 and equal surface potential (35·86 *mV*)
$A = 5 \times 10^{-20} J$, $1/\kappa = 10^{-6}\ cm^{-1}$, $\epsilon = 78{\cdot}5$.
(*a*) $a_1 = 125\ nm$, $a_2 = 125\ nm$ (1), 100 *nm* (2), 75 *nm* (3), 50 *nm* (4), 12·5 *nm* (5).
(*b*) $a_1 = 12{\cdot}5\ nm$, $a_2 = 125\ nm$ (1), 100 *nm* (2), 75 *nm* (3), 12·5 *nm* (4).

ting flocculation by adding a sol whose particles carry an opposite charge need to be carefully analysed—the simple rule of thumb may not apply.

The potential energy of interaction between two dissimilar particles in non-polar media has been evaluated by applying Maxwell's equations for charged conducting spheres.[108] The presence of an electric double layer was ignored, which would seem justified in view of the very small concentrations of ions involved in media of low dielectric constant; and for this case eqn. (1.52) becomes equivalent to Coulomb's law for point charges when $\beta = 0$. It was shown that the use of the simple Coulombic law leads to significant error in the V_R term. Furthermore, it was demonstrated that it is the particle of lower radius and potential that controls the energetics of the flocculation process, a similar conclusion to that reached for aqueous media, as described above.

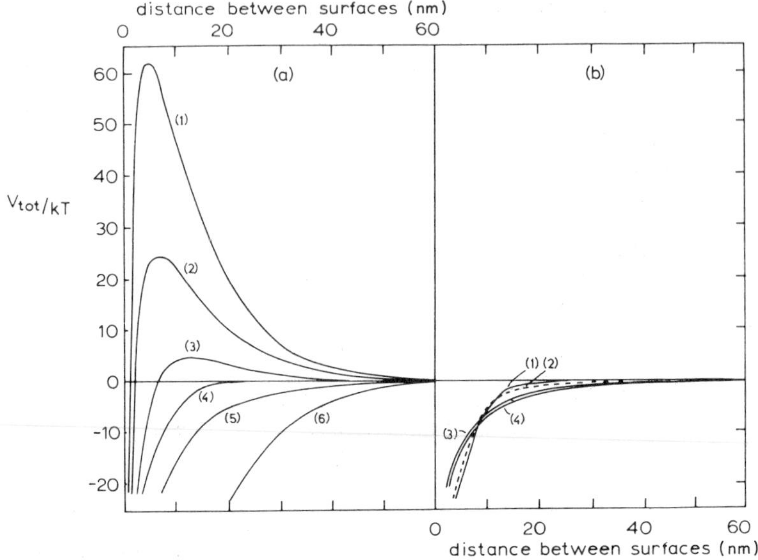

Fig. 1.7. *Theoretical curves of total potential energy against distance of separation of two spherical particles of radius* $a_1 = a_2 = 125$ *nm and varying surface potential.*

$A = 5 \times 10^{-20}$ *J*, $1/\kappa = 10^{-6}$ *cm*$^{-1}$, $\epsilon = 78{\cdot}5$.

(*a*) $\psi_{01} = +35{\cdot}86$ *mV*, $\psi_{02} = +35{\cdot}86$ *mV*(1), $+20{\cdot}48$ *mV*(2), $+10{\cdot}24$ *mV*(3), $+5{\cdot}12$ *mV*(4), 0 *mV*(5), $-35{\cdot}86$ *mV*(6).

(*b*) $\psi_{01} = +5{\cdot}12$ *mV*, $\psi_{02} = +35{\cdot}86$ *mV*(1), $+20{\cdot}48$ *mV*(2), $+10{\cdot}24$ *mV*(3), $+5{\cdot}12$ *mV*(4).

All these theoretical treatments relate to dilute dispersions because only the interaction between two particles is considered.

Stability arising from adsorbed layers

The presence of an adsorbed layer of surface active agent on the surface of solid particles dispersed in a liquid medium may affect the stability to flocculation in a number of different ways, each differing in principle but often overlapping in practice, hence making difficult the interpretation of observed phenomena.

This discussion will be arbitrarily divided into two sections, one dealing with completely non-ionic (uncharged) systems and the other with ionic systems. Some of the steric factors involved in the former will also be important in the latter, e.g. with an ionised polymer adsorbed on a charged surface in an aqueous medium.

(a) Uncharged systems

Essentially there are two factors which must be considered to explain the stability of uncharged systems. The first concerns the effect of the adsorbed layer on the attractive force between particles, which has already been considered. With the appropriate choice of surface active agent a significant reduction in the magnitude of the attractive force is possible. When particles collide the distance between the surfaces is increased by approximately twice the thickness of the adsorbed layer so that provided the adsorbed layer is compact and no desorption occurs, and that there is negligible attraction between the material of the adsorbed layers, the stability will depend on the magnitude of V_A corresponding to this distance. The second factor arises when the attractive force at this distance is still sufficiently large that interaction of the adsorbed layers may occur. Two extreme cases can be envisaged, one relating to the mixing of the two polymer layers and the other in which no mixing occurs but the layers become compressed on close approach of the surfaces. Both involve a reduction in the configurational freedom of the adsorbed polymer molecules which results in a repulsive force. In any particular system one or both of these effects could be present, depending on the nature of the polymer and solvent, and the structure of the adsorbed layer. During the last two decades much has been written on the theoretical and experimental aspects of this *steric* hindrance mechanism for stability to flocculation. The following discussion will review briefly the basic ideas that have been put forward.

(1) *Repulsion due to mixing.* Since an adsorbed polymer immobilises solvent within its coils, as with free polymer in solution, it is not surprising that the proposed mixing process when two polymer-coated surfaces approach each other can be discussed in terms of polymer solution theory. Polymer–solvent interactions determine the behaviour of solutions and dispersions. Phase separation that occurs in polymer solutions at the θ temperature is paralleled by dispersion flocculation at the *critical flocculation temperature*, at which the free energy increase on mixing (the repulsive free energy) disappears. A similar effect occurs when a non-solvent is added to the system, for which there is a *critical flocculation volume*. Several papers by Napper *et al.*[109,110] illustrate these phenomena.

The (Gibbs) free energy of mixing (ΔG_M) is related to the enthalpy (ΔH_M) and entropy (ΔS_M) charges by the relation

$$\Delta G_M = \Delta H_M - T\Delta S_M \tag{1.62}$$

The sign of ΔG_M is determined by the relative magnitudes of the enthalpy and entropy terms—if negative, flocculation is (energetically) promoted; if zero, flocculation proceeds as if no adsorbed layers were involved; if positive, the particles are protected from flocculation. For ΔG_M to be positive we require that

(i) both the entropy ($T\Delta S_M$) and enthalpy terms are negative, with the former being the larger—termed *entropic* stabilisation indicating a dominating effect due to reduction in the number of configurations of the molecules in the adsorbed layer, or
(ii) both terms are positive with the enthalpy change predominating—termed *enthalpic* stabilisation, or
(iii) ΔH_M is positive and ΔS_M negative, and therefore both terms contribute to stability—termed *enthalpic–entropic* stabilisation, or
(iv) ΔH_M is negative and ΔS_M is positive, both terms contribute to a negative ΔG_M and no stability is possible.

Entropic and enthalpic effects may be distinguished in practice. In (i) the magnitude of the entropy term decreases as the temperature is reduced, hence entropically stabilised dispersions would flocculate on cooling. Enthalpically stabilised systems normally flocculate on heating, since in this case the $T\Delta S_M$ term is less than ΔH_M but the difference between them decreases as the temperature is raised. But both entropy and enthalpy terms can be temperature dependent, so the temperature change criterion is not definitive, and there is no unambiguous way of experimentally defining the controlling factor.

Examples of the three types of stabilisation were given by Napper.[110] His work demonstrates in a convincing manner that adsorbed layers of the type that confer stability to flocculation obey the thermodynamic principles of polymer solutions, and that the process may be studied using polymer solution theory. The application of the (outdated) Flory–Krigbaum theory of polymer solutions[111] has been reviewed,[97] and it is claimed that it provides an excellent description of all the observed qualitative features of steric stabilisation.[112]

(2) *Repulsion due to compression.* The existence of a repulsive force arising from loss of configurational entropy when two adsorbed layers interact was first recognised by Mackor,[113] who treated the simple case of two flat plates to which are attached rigid rods (representing the adsorbed molecules) anchored at one end by a freely hinged joint. All orientations of the rod are assumed to have equal probability and

therefore the model is only appropriate to low surface coverages. When the plates are brought close together the gyration of the rod on the surface is hindered and the total number of configurations reduced, and this, Mackor demonstrated, would lead to a repulsive force. The idea was extended by Mackor and van der Waals[114] to higher surface coverages by use of a quasi-lattice model, and by assuming desorption occurs on interaction leading to an increase in free energy which is equivalent to a repulsion. They showed that the magnitude of the repulsive term was sufficient, when added to the attractive energy, to give rise to a potential energy of such a magnitude that stabilisation would occur. However, the applicability of this approach is limited to short chains and even then the parameters are difficult to evaluate from experimental data. Furthermore, for adsorbed layers of thickness δ no entropic repulsion would exist until $\Delta < 2\delta$, and for short chains ($\delta \sim 3$ nm) the value of V_A at this distance would in most cases be sufficiently large for there to be a secondary minimum in the potential energy diagram deep enough for flocculation to be guaranteed (theoretically).

A more tractable method for longer chains, therefore applicable to polymeric adsorbates, has been worked out by Clayfield and Lumb.[115] The calculation applies to an irreversibly adsorbed molecule with an adsorbable group at one end only, contained in a solvent having no net interaction with the unadsorbed part of the molecule; the authors suggest that a hydrocarbon solution of a block copolymer comprising a hydrocarbon chain and a polar chain, is an appropriate basis for the model. The reduction in entropy of the adsorbed polymer molecule on compression is calculated, not the interpenetration of molecules on opposing surfaces. Sphere–sphere and sphere–flat plate interactions are considered and the relative orders of magnitude of the effect compared. The theory predicts that the greater the freedom of movement of the polymer chain, the more effective it is as a stabilising agent. Furthermore, there is an optimum size of molecule which will prevent adhesion of particles of a given size, and increasing the length of the polymer chain with the object of increasing δ may actually increase the possibility of flocculation. Experimental support for this conclusion is at present limited.[116] The computer studies were extended to cover adsorbed random copolymers,[117,118] for which the protective action is seen to increase (within limits) with decreasing strength of adsorption which is paralleled by increasing effective thickness of the adsorbed layer. At the same layer thickness, the adsorbed random copolymer can produce a larger repulsive energy than that of a terminally adsorbed chain.

There has been considerable controversy over the relative merits of the two approaches to steric stabilisation. Perhaps one model is more relevant to one type of system, e.g. the mixing model for 'good' solvents and the volume restriction theory for 'poor' solvents. Napper has suggested[119] from studies of incipient flocculation, measurements of the repulsive interactions between stable particles, and of stability in polymer melts that two separate regions of close approach must be distinguished. The first corresponds to a separation between the approaching surfaces of one to two adsorbed layer thicknesses, within which interpenetration of the polymer chain occurs. At closer approach there will be a region where both interpenetration and compression occur. His all-embracing theory, although in its early stages, appears to provide at least a qualitative explanation for many of the experimental data he considered which cover a variety of stability situations. Napper's critical review is commended to the reader as an introduction to the subject, which over the last two decades had advanced considerably although its interpretation for practical systems has hardly been attempted.

To assess the magnitude of the repulsive potential energy arising from these effects it is necessary to have information on the adsorbed layer, such as surface coverage, fraction of polymer segments attached directly to the surface, effective thickness, etc. Such data have been made available by the application of infrared absorption and ellipsometry. The use of infrared in the study of polymers adsorbed on solid surfaces was first reported by Fontana and Thomas,[120] and makes use of the fact that on adsorption there is a shift in the frequency of the characteristic band which permits the fraction of segments adsorbed to be estimated. With a knowledge of the fraction of groups bound directly to the surface and adsorption data it is possible to obtain information on the structure of the adsorbed layer.[121] Many structures are possible for a looped or coiled macromolecule attached in part to a surface, ranging from those which yield a relatively flat and compressed layer to those which give adsorbed layers highly extended away from the interface. The number and arrangement of attached portions and the size and distribution of the unattached loops define the conformation of the polymer molecule. Ellipsometry provides a method for estimating the extension of the loops away from the surface. The change in the state of polarisation of light upon reflection from a film-covered surface is measured and used to calculate the thickness and refractive index of the film, the latter giving information on the concentration of polymer in the adsorbed layer.[121,122]

(*b*) *Charged systems*

The action of adsorbed high-molecular-weight materials is somewhat more complex in this case since both ionic and non-ionic mechanisms may be operative. The situation that arises in any particular system will depend on the magnitude of the surface potential and ionic strength, the ionic character of the adsorbed material, the amount adsorbed and the thickness of the adsorbed layer. The variety of phenomena observed may be roughly divided into two categories, one associated with stabilisation (protective action) and the other with flocculation (flocculating action) brought about by the material itself.

(1) *Protective action.* Gums, starch, and cellulose derivatives, proteins and polyacrylates are effective agents for protecting charged colloidal particles in aqueous media against flocculation by electrolytes. Several factors are involved, including those already discussed (DLVO theory, entropic repulsion, reduced attraction by adsorbed layer).[123] Mathai and Ottewill[67] studied the addition of electrolytes to negative aqueous silver iodide sols stabilised by various homogeneous non-ionic surfactants of the polyoxyethylene type. From plots of log W against log n critical electrolyte flocculation concentrations n' were obtained (see above) which for any particular polymer increased with concentration, the increase being considerable close to the critical micelle concentration, in which range there was a rapid increase in adsorption. Above the critical micelle concentration the critical electrolyte concentration remains almost constant and the sols are very stable. Increasing the length of the ethylene oxide chain for the same alkyl chain length results in a decrease in the extent of adsorption; hence for the same molar concentration $C_{16}E_9$ is less effective in stabilising the sol against flocculation by cations than $C_{16}E_6$ (C_{16} indicates a hydrocarbon chain containing 16 carbon atoms and nine ethylene oxide units are indicated by E_9). Hence, the protection is related directly to the adsorption and the hydrocarbon chain length.

The protective action of polyoxyethylene glycols on gold sols is also related to the molecular weight. Heller and Pugh[124] followed the colour change (red to blue) on the addition of potassium chloride to sols protected by the polymers and suggested that the flexible polymer chains sterically prevented the particles approaching close enough for the attractive forces to lead to flocculation.

The fundamental parameters of the electric double layer will also be changed by adsorption of high-molecular-weight materials. Experiments on the modification of the Stern potential at the mercury–electrolyte solution interface by the adsorption of polyoxyethylene compounds indicates[125] that there is a change in the potentials at short distances from the surface which results in a decrease in the magnitude of the electrostatic repulsion term. Furthermore, for the adsorption of amphipathic surfactants with the charged end-group on the solution side of the adsorbed layer, e.g. the sulphate ion of sodium dodecyl sulphate adsorbed on a carbon black from aqueous solution,[76] it is necessary to modify the distance parameters used in the V_R calculation since the polar group giving rise to the surface potential does not physically coincide with the surface. This is particularly important at moderately high ionic strengths, when the double layer thickness is of the same order of magnitude as the adsorbed layer thickness.

(2) *Flocculating action.* The same additives that at high concentration (and strongly adsorbed) give protective action, might give rise to flocculation at lower surface coverages. For example, suspensions of clay in solutions of gelatin of concentration around 1–10 ppm flocculate, but 0·1% solutions stabilise such materials. In certain cases when the colloidal particles and agent are oppositely charged the dispersion may become unstable as a result of charge neutralisation associated with the absorption of the additive whereas at higher concentration (and adsorption) protection results, e.g. silica with polyethyleneimine of low molecular weight. Another mechanism of action of polymeric materials is that in which parts of the adsorbed molecules are attached to two or more particles bringing about flocculation by a bridging mechanism; this may only occur when the surface coverage is low leaving parts of the surface available for further adsorption, hence bridging, when the particles come into close proximity. At higher additive concentrations protective action is obtained.

Some considerable effort is being put into the understanding of the flocculating action of high-molecular-weight polymers, a process which is of great importance in water clarification. The important question is often the nature of the interaction of the polymer with the surface of the particles. The appropriate points of attachment both on the polymer molecule and the surface must be determined, and to this end such studies as the effect of pH and ionic strength on adsorption, flocculation, electrokinetic effects and viscosity of polymer in solution, and studies on the character of the surface by infrared spectroscopy, etc., have been directed. Each situation has its own peculiarities and controlling factors; in most cases the details of the behaviour are not well understood. The reader is referred to papers by Healy and La Mer,[126] and by Kitchener *et al.*[127] for further discussion on the polymeric flocculants.

REFERENCES

1. E. Honigmann and J. Stabenow, *VI FATIPEC Congress*, (1962) 89.
2. V. K. La Mer, *J. Colloid Sci.*, **19** (1964) 291.
3. R. K. Eckhoff, *Chr. Michelsens Institutt*, (3) **39** (1976) 1.
4. H. J. Osterhof and F. E. Bartell, *J. Phys. Chem.*, **34** (1930) 1399.
5. A. Dupré, *Ann. Chimie*, (4) (**6**) (1865) 274.
6. F. M. Fowkes, *Wetting*, SCI Monograph, **25** (1967) 3.
7. J. A. Kitchener, *Proc. 3rd Int. Congress on Surface Activity*, **2** (1960) 426.
8. W. J. Dunning, *J. Oil Colour Chem. Assoc.*, **48** (1965) 509.
9. D. M. Gans, *J. Paint Technol.*, **38** (1966) 322.
10. D. M. Gans, *J. Paint Technol.*, **39** (1967) 501.
11. A. W. Adamson, *Physical Chemistry of Surfaces*, 3rd edition, Wiley-Interscience, New York, 1976, Chapter 7.
12. P. M. Heertjes and N. W. F. Kossen, *Powder Technol.*, **1** (1967) 33.
13. B. Lindman and H. Wennerström, *Top. Curr. Chem.*, **87** (1980) 1.
14. H. -F. Eicke, *Top. Curr. Chem.*, **87** (1980) 85.
15. P. M. Heertjes and W. C. Witvoet, *Powder Technol.*, **3** (1970) 339.
16. E. D. Washburn, *Phys. Rev.*, **17** (1921) 374.
17. V. T. Crowl and W. D. S. Wooldridge, *Wetting*, SCI Monograph, **25** (1967) 200.
18. R. J. Good, *Chem. Ind.* (*London*), (1971) 600.
19. R. J. Good, *J. Colloid Interface Sci.*, **42** (1973) 473.
20. W. A. Zisman, *Contact Angle, Wettability and Adhesion*, Adv. Chem. Series No. 43, ACS, Washington, 1964, p. 1.
21. D. M. Gans, *J. Paint Technol.*, **41** (1969) 515.
22. F. G. Greenwood, G. D. Parfitt, N. H. Picton and D. G. Wharton, *Adsorption from Aqueous Solution*, Adv. Chem. Series No. 79, ACS, Washington, 1968, p. 135.
23. G. D. Parfitt and D. G. Wharton, *J. Colloid Interface Sci.*, **38** (1971) 431.
24. D. W. Fuerstenau, *Trans. AIME, Mining Eng.*, (1957) 1367.
25. D. W. Fuerstenau, *Pure Appl. Chem.*, **24** (1970) 135.
26. T. M. Doscher, *J. Colloid Sci.*, **5** (1950) 100.
27. K. Meguro, *J. Chem. Soc. Jpn* (*Ind. Chem. Sect.*), **58** (1955) 905.
28. K. Tamaki, *J. Jpn Oil Chem. Soc.*, **9** (1960) 426.
29. F. F. Aplan and D. W. Fuerstenau, *Froth Flotation*, Ed. D. W. Fuerstenau, American Institute of Mining and Metallurgical Engineering, New York (1962).
30. G. A. H. Elton, *Proc. 2nd Int. Congress on Surface Activity*, **3** (1957) 161.
31. G. D. Parfitt, *XIV FATIPEC Congress Book*, 1978, p. 107.
32. P. A. Rehbinder, *Colloid J. USSR*, **20** (1958) 493.
33. P. A. Rehbinder and V. I. Likhtman, *Proc. 2nd Int. Congress on Surface Activity*, **3** (1957) 563.
34. E. D. Shchukin and P. A. Rehbinder, *Colloid J. USSR*, **20** (1958) 601.
35. A. T. DiBenedetto, *The Structure and Properties of Materials*, McGraw-Hill, New York, 1967.
36. G. M. Bartenev, I. V. Iudena and P. A. Rehbinder, *Colloid J. USSR*, **20** (1958) 611.
37. G. S. Khodakov and P. A. Rehbinder, *Colloid J. USSR*, **22** (1960) 375.

38. K. J. Mysels, *Introduction to Colloid Chemistry*, Interscience, New York, 1959, Chapter 5.
39. M. Von Smoluchowski, *Z. Physik. Chem.*, **92** (1917) 129.
40. A. E. van Arkel and H. R. Kruyt, *Rec. Trav. Chim.*, **39** (1920) 656.
41. P. Tuorilla, *Kolloidchen. Beih.*, **22** (1926) 191.
42. W. I. Higuchi, R. Okada, G. A. Stelter and A. P. Lemberger, *J. Am. Pharm. Ass.*, **52** (1963) 49.
43. R. H. Ottewill and D. J. Wilkins, *Trans. Faraday Soc.*, **58** (1962) 608.
44. K. E. Lewis and G. D. Parfitt, *Trans. Faraday Soc.*, **62** (1966) 1652.
45. D. N. L. McGown and G. D. Parfitt, *J. Phys. Chem.*, **71** (1967) 449.
46. G. A. Johnson, S. M. A. Lecchini, E. G. Smith, J. Clifford and B. A. Pethica, *Disc. Faraday Soc.*, **42** (1966) 120.
47. D. L. Swift and S. K. Friedlander, *J. Colloid Sci.*, **19** (1964) 621.
48. G. M. Hidy, *J. Colloid Sci.*, **20** (1965) 123.
49. G. M. Hidy and D. K. Lilly, *J. Colloid Sci.*, **20** (1965) 863.
50. D. S. Jovanovic, *Kolloid Z.*, **203** (1965) 42.
51. B. V. Deryaguin and L. D. Landau, *Acta Physicochim. URSS*, **14** (1941) 633.
52. E. J. W. Verwey and J. Th. G. Overbeek, *Theory of the Stability of Lyophobic Colloids*, Elsevier, Amsterdam, 1948.
53. H. C. Hamaker, *Physica*, **4** (1937) 1058.
54. E. A. Moelwyn-Hughes, *Physical Chemistry*, 2nd edition, Pergamon, London, 1961.
55. J. Lyklema, *Pont. Acad. Sci. Scr. Varia*, **31** (1967) 181.
56. F. M. Fowkes, *Ind. Eng. Chem.*, **56** (1964) 40.
57. J. Gregory, *Adv. Colloid Interface Sci.*, **2** (1970) 396.
58. J. Visser, *Adv. Colloid Interface Sci.*, **3** (1972) 331.
59. B. Vincent, *J. Colloid Interface Sci.*, **42** (1973) 270.
60. E. M. Lifshitz, *Soviet Phys. JETP*, **2** (1956) 73.
61. E. E. Dzyaloshinkskii, E. M. Lifshitz and L. P. Pitaevskii, *Soviet Phys. JETP*, **37** (1960) 161.
62. H. Krupp, *Adv. Colloid Interface Sci.*, **1** (1967) 111.
63. J. H. Schenkel and J. A. Kitchener, *Trans. Faraday Soc.*, **56** (1960) 161.
64. H. B. G. Casimir and D. Polder, *Phys. Rev.*, **73** (1948) 360.
65. M. J. Vold, *J. Colloid Sci.*, **16** (1961) 1.
66. V. T. Crowl, *J. Oil Colour Chem. Assoc.*, **50** (1967) 1023.
67. K. G. Mathai and R. H. Ottewill, *Trans. Faraday Soc.*, **62** (1966) 759.
68. D. W. J. Osmond, B. Vincent and F. A. Waite, *J. Colloid Interface Sci.*, **42** (1973) 262.
69. E. J. Verwey and J. Th. G. Overbeek, *Theory of the Stability of Lyophobic Colloids*, Elsevier, Amsterdam, 1948, p. 101.
70. G. Frens, D. J. Engels and J. Th. G. Overbeek, *Trans. Faraday Soc.*, **63** (1967) 418.
71. O. F. Devereux and P. L. de Bruyn, *Interaction of Plane Parallel Double Layers*, M.I.T. Press, Cambridge, USA, 1963.
72. N. Fuchs, *Z. Physik.*, **89** (1934) 736.
73. B. V. Deryaguin, *Trans, Faraday Soc.*, **36** (1940) 203.
74. S. N. Srivastava and D. A. Haydon, *Trans. Faraday Soc.*, **60** (1964) 971.
75. A. Watillon and A. M. Joseph-Petit, *Disc. Faraday Soc.*, **42** (1966) 143.

76. G. D. Parfitt and N. H. Picton, *Trans. Faraday Soc.*, **64** (1968) 1955.
77. J. Th. G. Overbeek, *Colloid Science*, Ed. H. R. Kruyt, Elsevier, Amsterdam, 1948, Vol. 1, Chapter 8.
78. H. Reerink and J. Th. G. Overbeek, *Disc. Faraday Soc.*, **18** (1954) 74.
79. R. H. Ottewill and J. N. Shaw, *Disc. Faraday Soc.*, **42** (1966) 154.
80. R. H. Ottewill and D. J. Wilkins, *Trans. Faraday Soc.*, **58** (1962) 608.
81. G. D. Parfitt and A. L. Smith, unpublished data.
82. S. Levine, J. Mingins and G. M. Bell, *J. Electroanalyt. Chem.*, **13** (1967) 280.
83. L. A. Romo, *Disc. Faraday Soc.*, **42** (1966) 232.
84. J. L. van der Minne and P. H. J. Hermanie, *J. Colloid Sci.*, **7** (1952) 600.
85. G. D. Parfitt, *J. Oil Colour Chem. Assoc.*, **51** (1968) 137.
86. D. N. L. McGown and G. D. Parfitt, *Kolloid Z.*, **219** (1967) 48.
87. E. J. W. Verwey and J. Th. G. Overbeek, *Theory of the Stability of Lyophobic Colloids*, Elsevier, Amsterdam, 1948, p. 155.
88. J. L. van der Minne and P. H. J. Hermanie, *J. Colloid Sci.*, **8** (1953) 38.
89. H. Koelmans and J. Th. G. Overbeek, *Disc. Faraday Soc.*, **18** (1954) 52.
90. D. N. L. McGown, G. D. Parfitt and E. Willis, *J. Colloid Sci.*, **20** (1965) 650.
91. D. N. L. McGown and G. D. Parfitt, *Disc. Faraday Soc.*, **42** (1966) 225.
92. F. M. Fowkes, F. W. Anderson and R. J. Moore, *Abstracts*, 150th ACS Meeting, September 1965.
93. F. M. Fowkes, *Disc. Faraday Soc.*, **42** (1966) 246.
94. K. Tamarabuchi and M. L. Smith, *J. Colloid Sci.*, **22** (1966) 204.
95. F. J. Micale, Y. K. Lui and A. C. Zettlemoyer, *Disc. Faraday Soc.*, **42** (1966) 238.
96. D. N. L. McGown and G. D. Parfitt, *Kolloid Z.*, **220** (1967) 56.
97. G. D. Parfitt and J. Peacock, *Surface and Colloid Science*. Ed. E. Matijevic, Plenum Press, New York, 1978, Vol. 10, p. 163.
98. W. Albers and J. Th. G. Overbeek, *J. Colloid Sci.*, **14** (1959) 510.
99. C. S. Chen and S. Levine, *J. Colloid Interface Sci.*, **43** (1973) 599.
100. G. R. Feat and S. Levine, *J. Chem. Soc. Faraday Trans. II*, **71** (1975) 102.
101. G. R. Feat and S. Levine, *J. Colloid Interface Sci.*, **54** (1976) 34.
102. P. A. Rehbinder and A. B. Taubman, *Colloid J. USSR*, **23** (1961) 301.
103. P. A. Rehbinder, *Colloid J. USSR*, **20** (1958) 493.
104. B. V. Deryaguin, *Disc. Faraday Soc.*, **18** (1954) 85.
105. R. Hogg, T. W. Healy and D. W. Fuerstenau, *Trans. Faraday Soc.*, **62** (1966) 1638.
106. A. Bierman, *J. Colloid Sci.*, **10** (1955) 231.
107. D. C. Prieve and E. Ruckenstein, *J. Colloid Interface Sci.*, **73** (1980) 539.
108. G. D. Parfitt, J. A. Wood and R. T. Ball, *J. Chem. Soc. Faraday Trans. I*, **69** (1973) 1908.
109. D. H. Napper, *Trans. Faraday Soc.*, **64** (1968) 1701.
110. R. Evans and D. H. Napper, *J. Colloid Interface Sci.*, **52** (1975) 260.
111. P. J. Flory and W. R. Krigbaum, *J. Chem. Phys.*, **18** (1950) 1086.
112. J. B. Smitham, R. Evans and D. H. Napper, *J. Chem. Soc. Faraday Trans. I*, **71** (1975) 285.
113. E. L. Mackor, *J. Colloid Sci.*, **6** (1951) 492.
114. E. L. Mackor and J. H. van der Waals, *J. Colloid Sci.*, **7** (1952) 535.
115. E. J. Clayfield and E. C. Lumb, *J. Colloid Interface Sci.*, **22** (1966) 269.

116. E. J. Clayfield and E. C. Lumb, *Disc. Faraday Soc.*, **42** (1966) 285.
117. E. J. Clayfield and E. C. Lumb, *Macromolecules*, **1** (1968) 233.
118. E. J. Clayfield and E. C. Lumb, *J. Colloid Interface Sci.*, **47** (1974) 6, 16.
119. D. H. Napper, *J. Colloid Interface Sci.*, **58** (1977) 390.
120. B. J. Fontana and J. R. Thomas, *J. Phys. Chem.*, **65** (1961) 480.
121. R. R. Stromberg, E. Passaglia and D. J. Tutas, *J. Res. Natl. Bur. Std.*, **67A** (1963) 43.
122. R. R. Stromberg, E. Passaglia and D. J. Tutas, *Ellipsometry for the measurement of surfaces and thin films*, NBS Misc. Publication, **256** (1964) 281.
123. R. H. Ottewill, *Non-ionic Surfactants*, Ed. M. Schick, M. Dekker, New York, 1967, Chapter 19.
124. W. Heller and T. L. Pugh, *J. Polymer Sci.*, **47** (1960) 203.
125. A. Watanabe, F. Tsuji and S. Veda, *Kolloid Z.*, **193** (1963) 39.
126. T. W. Healy and V. K. La Mer, *Rev. Pure Appl. Chem.* (*Australia*), **13** (1963) 112.
127. R. W. Slater, J. P. Clark and J. A. Kitchener, *Proc. Brit. Ceramic Soc.*, 1969, p. 1.

CHAPTER 2

PROPERTIES OF THE SOLID–LIQUID INTERFACE OF RELEVANCE TO DISPERSION

M. J. JAYCOCK
Loughborough University of Technology, Leicestershire, UK

FORCES BETWEEN ATOMS, IONS AND MOLECULES[1–3]

The physical and chemical characteristics of any interface, or for that matter of any other state of matter, are primarily determined by the chemical and physical forces present. These not only determine such properties as surface tension, but also such parameters as the extent of adsorption of any species at the liquid–solid, solid–gas or liquid–liquid interfaces, as well as adhesion and friction between surfaces. It will be readily appreciated that knowledge of such forces is of considerable importance in understanding the nature of any interface, and they therefore make a convenient starting point for this chapter. The theoretical description of these chemical and physical forces represents one of the few ideas common throughout the various branches of physical science.

Attractive and repulsive forces between molecules

Let us consider, for the sake of simplicity, a pair of molecules of a monatomic gas. The forces which exist between the pair are usually termed van der Waals' or dispersion forces. If there is a preferred separation of the molecules it will be at the distance corresponding to a minimum in the potential energy φ of the pair. At short range repulsive forces pre-dominate, increasing with decreasing distance of separation a causing an increase in the potential energy. These are associated with electron shell interaction and are sometimes called Born repulsive forces. At longer range attractive forces predominate, also increasing with decreasing distance of separation, causing a decrease in the potential energy of the pair. These result from the effects of instantaneous dipoles upon polarisable molecules. The position of the potential energy minimum will be determined by the relative magnitudes of the attractive and repulsive energies, and will correspond to the position of zero force. A diagrammatic representation of the total potential energy as a function of the distance of separation is given in Fig. 2.1.

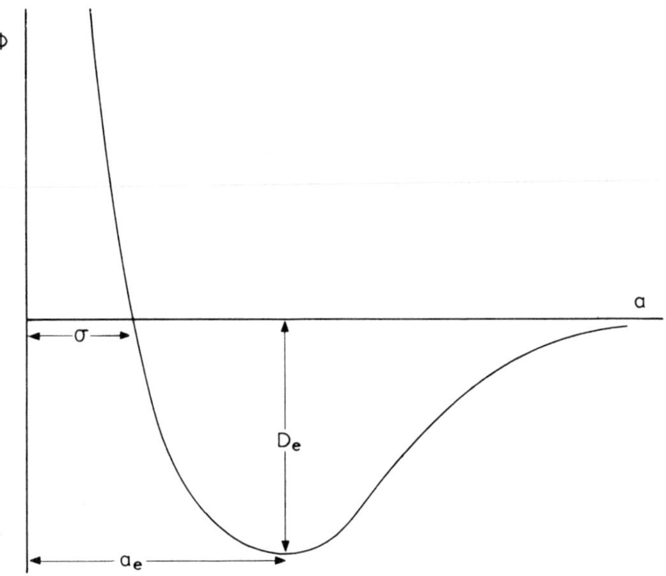

Fig. 2.1. The total potential energy curve of a pair of molecules.

The total potential energy may be represented as[4,5]

$$\varphi_{\text{total}} = \varphi_{\text{repulsive}} + \varphi_{\text{attractive}} \tag{2.1}$$

or

$$\varphi = \frac{A}{a^n} - \frac{B}{a^m} \tag{2.2}$$

where A and B are positive constants, and n and m are integers where n is greater than m.

For a pair of molecules, eqn. (2.2) is often rewritten as either

$$\varphi = 4D_{\text{e}}\left[\left(\frac{\sigma}{a}\right)^{12} - \left(\frac{\sigma}{a}\right)^{6}\right], \tag{2.3}$$

or
$$\varphi = \frac{27}{4} D_e \left[\left(\frac{\sigma}{a}\right)^9 - \left(\frac{\sigma}{a}\right)^6 \right] \tag{2.4}$$

where D_e is the minimum value of φ, and σ is the value of a when $\varphi = 0$. These expressions are often called the 12:6 and 9:6 potentials, respectively, but eqn. (2.3) is even more commonly called the Lennard-Jones potential. The general equations (2.1) and (2.2) may be applied to systems other than a pair of molecules, providing the correct expression for the type of force involved is used. The expressions presented here are not necessarily the most accurate equations to describe these forces, and for a detailed consideration of the problem the reader is referred to ref. 1.

Coulombic forces between ions

Coulomb experimentally established that the force f between two charges, e_1 and e_2, a distance a apart in a medium of dielectric constant ϵ, is given

by*
$$f = \frac{e_1 e_2}{\epsilon a^2} \tag{2.5}$$

if we adopt the cgs system of units, and the potential energy by

$$\varphi(a) = -\int \frac{e_1 e_2}{\epsilon a^2}\, da = \frac{e_1 e_2}{\epsilon a}. \tag{2.6}$$

The comparative strength of the electrostatic forces between ions in aqueous solution can be estimated from the fact that at 25°C the potential energy of attraction between two univalent ions of opposite sign is approximately equal to kT when their distance is 0·7 nm and approximately ten times this value if the separation is reduced to 0·225 nm. It is this type of force which is the major factor at comparatively long range.

* If SI units (rationalised MKS) are used, equation (2.5) becomes

$$f = \frac{e_1 e_2}{4\pi\epsilon\epsilon_0 a^2} \tag{2.5a}$$

where ϵ_0 is dielectric permittivity of free space. The factor 4π is included so that its occurrence or otherwise in derived expressions would accord with geometric expectations and not vice versa, e.g. the equation for the capacity of a parallel plate condenser becomes $C = \epsilon\epsilon_0/d$ and that for an isolated spherical condenser $C = 4\pi\epsilon\epsilon_0 a$.

Interaction between a permanent dipole and an ion

The interaction between a permanent dipole and an ion may be treated quite simply by adding the coulombic terms due to the interaction of the ion with each pole-charge in turn, making allowance for the orientation of the dipole. When the length of the dipole is small compared with the distance of separation, the force between the ion and the dipole, of moment μ, is given by

$$f = \frac{2e\mu \cos\theta}{a^3} \tag{2.7}$$

and the potential energy by

$$\varphi(a) = \frac{e\mu \cos\theta}{a^2} \tag{2.8}$$

where θ is the angle between the axis of the dipole and the line joining the ion to the centre of the dipole. Thus the interaction may be either repulsive or attractive according to the sign of the ionic charge and the orientation of the dipole.

Interaction between two permanent dipoles

The energy of interaction between two permanent dipoles may be obtained by applying Coulomb's Law to the four fractional charges concerned. The resulting expression is more complicated because of the necessity of describing the orientation of one dipole with respect to the other. If the distance apart is large compared with the length of either dipole, then

$$\varphi = \frac{\mu_1 \mu_2}{a^3}[2 \cos\omega_1 \cos\omega_2 - \sin\omega_1 \sin\omega_2 \cos(\chi_1 - \chi_2)] \tag{2.9}$$

where ω_1 and ω_2 are the angles of inclination of the dipolar axes to the lines of centres, reflected in some plane, and χ_1 and χ_2 are the angles subtended between the dipolar axes and the perpendiculars to the plane at the dipolar centres. Equation (2.9) predicts that the system will have maximum potential energy when the axes of the dipoles are in the same plane with like poles facing each other, of $2\mu_1\mu_2/a^3$. The position of minimum energy will also be when the dipoles are in the plane but with unlike poles facing each other, when the potential energy will be $-2\mu_1\mu_2/a^3$.

Polarisability

A symmetrical molecule such as methane or argon possesses no permanent dipole moment. However, in the presence of an external electric field electrons may be displaced from their usual positions and the molecule acquires an induced dipole. It is usually postulated that at moderate field strengths the induced moment μ_i is directly proportional to the field strength F thus

$$\mu_i = \alpha F \tag{2.10}$$

This equation defines the polarisability α. A whole series of relationships may be derived to describe the interaction of an induced dipole with an ion, or a dipole, or an induced dipole using Coulomb's Law, and for further information one of the advanced texts should be consulted.

DESCRIPTION OF AN INTERFACE

With the problem of dispersing a bag of hopefully dry pigment in either water or an organic solvent, a number of interfacial contacts and areas are going to undergo change during the dispersion process. The solid–gas interface is eventually going to disappear if good dispersion is achieved. The solid–liquid interface will have been created. The liquid–vapour interface will have fluctuated in area and probably also in composition. If in this book we are to deal adequately with these interfaces, then certain fundamental definitions must be clearly stated.

Some are easiest to understand in relation to liquid interfaces so these will also be considered, but first we should start with the definition of the term *interface*. The boundary between two homogeneous bulk phases should not be thought of as a plane of zero thickness, but more realistically as a separate surface phase of definite thickness. The definition of a surface phase as described by Gibbs[6,7] can be readily understood with the aid of Fig. 2.2 which represents a section perpendicular to the surface. At all points in and above plane AA' the physical and chemical properties are those associated with the bulk phase α, whereas at all points in and below plane BB' the physical and chemical properties are those associated with the bulk phase β. In the surface phase σ the properties gradually change from those of phase α at plane AA', to those of β at plane BB'; in any plane in the surface phase parallel to AA' or BB' the properties will everywhere be the same.

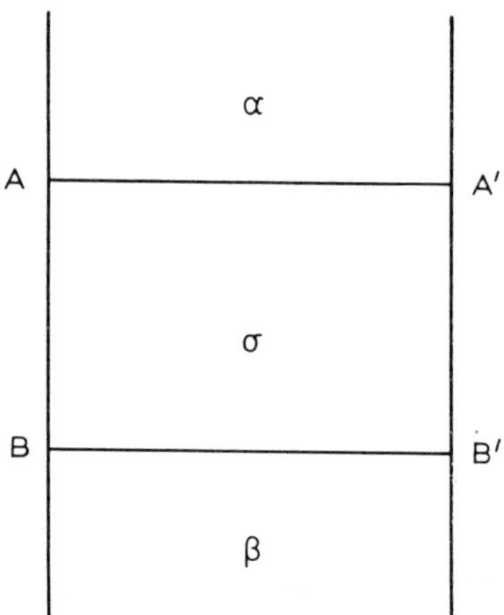

Fig. 2.2. Definition of a surface phase.

Surface tension

Drops of liquid behave as if they are surrounded by an elastic skin which prevents them from spreading. Young[8] attempted an explanation of this *surface tension* in terms of the forces existing between the molecules of the liquid. The cohesive forces must be greater than those due to thermal motion otherwise the drop would vaporise. The molecules at the surface will be less attracted to their neighbours than molecules in the bulk, since a molecule at the surface has fewer neighbours. Therefore since the energy of a surface molecule will be greater than that of a similar molecule in the bulk phase, the surface of such a pure phase will tend to contact spontaneously, reducing the free energy of the system to a minimum.

The surface tension γ_0 is the force per unit length contracting such a surface.[9] If P, V, A, S, and T represent the pressure, volume, surface area, entropy and absolute temperature of the system of n moles of chemical potential μ, then

$$dA = -SdT - PdV + \gamma_0 d\mathrm{A} + \mu dn \tag{2.11}$$

where A is the Helmholtz free energy or work function of the system. At constant values of T, V and n this reduces to

$$\gamma_0 = \left(\frac{\partial A}{\partial \mathrm{A}}\right)_{T,V,n} \tag{2.12}$$

If we define ${}^sA$ as the Helmholtz free energy per unit area, then $\mathrm{d}A = \mathrm{d}(\mathrm{A}{}^sA) = {}^sA\mathrm{dA} + \mathrm{A}\mathrm{d}{}^sA$, and therefore on substitution.

$$\gamma_0 = {}^sA + \mathrm{A}\left(\frac{\partial^s A}{\partial \mathrm{A}}\right)_{T,V,n} \tag{2.13}$$

However, in a one-component system ${}^sA$ is only dependent upon the configuration of the surface molecules, therefore it must be constant, hence

$$\gamma_0 = {}^sA \tag{2.14}$$

when T and V are constant. It can also be shown that

$$\gamma_0 = {}^sG \tag{2.15}$$

the Gibbs free energy per unit area, when T and P are constant. Since surface changes are accompanied by only small changes in pressure and volume, ${}^sA$ and ${}^sG$ are virtually identical.

Total surface energy

The total surface energy per unit area ${}^sU$ may be calculated from

$${}^sA = {}^sU - T{}^sS \tag{2.16}$$

where the superscript s has the same significance as before. The entropy term ${}^sS$, may be found by using the relation

$${}^sS = -\left(\frac{\partial^s A}{\partial T}\right)_{n,V} = -\left(\frac{\partial \gamma_0}{\partial T}\right)_{n,V} \tag{2.17}$$

so that

$${}^sU = \gamma_0 - T\left(\frac{\partial \gamma_0}{\partial T}\right)_{n,V} \tag{2.18}$$

The quantity ${}^sU$ is practically temperature independent.

Surface entropy

As has been mentioned previously, a molecule in the surface has fewer neighbours than a molecule in a bulk phase. So that compared with a bulk liquid there is a new possibility of randomness. Davies[9] has shown that the two possibilities of a molecule being near the surface or in the bulk phase, give rise to an entropy increase of approximately $R \ln 2 = 5{\cdot}8$ JK^{-1}/mol, this being a standard entropy change associated with forming a surface. The temperature coefficient of surface tension must consequently be negative.

Interfacial tension

Although we have talked about surface tension we mean by this the interfacial tension between a gaseous phase and the liquid phase. Whether the gaseous phase is the pure vapour of the liquid, or air saturated with vapour, makes no measurable difference to the value of the interfacial tension, and consequently it may be called surface tension since, for practical purposes, we can consider the measured value of γ_0 as being a property solely of the liquid. This surface tension or interfacial tension is fairly readily measured by a number of standard methods.[9,10]

The situation in respect of liquid–liquid interfaces is somewhat similar since, for example, a drop of water in oil behaves similarly to a drop of water in air; the surface area being reduced to a minimum under the influence of the interfacial tension γ_i. Furthermore, interfacial tensions may be fairly readily measured by methods[9,10] essentially similar to those for the measurement of surface tension. However, in this case both liquids often influence the interfacial tension. For example, for butan-1-ol against water at 25°C, γ_i is only 1·6 mN/m, a low value being characteristic of oils with polar or hydrophilic groups. Since the surface tension of butan-l-ol at 25°C is 24 mN/m the butan-1-ol molecules must have their $-$OH groups oriented towards the interface, where the repulsion between these molecules prevents γ_i attaining a high value.

Further examples illustrating this point can be found in Table 2.1, namely octan-1-ol and diethyl ether, although the values of γ_i are higher. As Bikerman[12] has noted the mutual insolubility of the oil and water runs parallel with the interfacial tensions.

TABLE 2.1

SURFACE TENSIONS OF PURE LIQUIDS AGAINST AIR,[11] AND INTERFACIAL TENSIONS BETWEEN WATER AND PURE LIQUIDS[11]

Liquid	*T(°C)*	$\gamma_0(mN/m)$	$\gamma_i(mN/m)$
Water	20	72·8	
	25	72·0	
Bromobenzene	25	35·75	38·1
Benzene	20	28·88	35·0
	25	28·22	34·71
Octan-l-ol	20	27·53	8·5
Carbon tetrachloride	20	26·9	45·1
Diethyl ether	20	17·01	10·1

In this section we have confined attention to liquid surfaces and have not considered the solid surface or interfacial tensions involving a solid, which is the subject of the next section.

THE SOLID SURFACE[13]

The nature of a solid prevents the application of the methods suitable for a gas–liquid or liquid–liquid interface to the problem of measuring the interfacial tension* when one of the phases is a solid. However, under special conditions the methods may occasionally be applied. For example, a copper wire at temperatures near its melting point tends to shorten, being sufficiently plastic to flow under the influence of surface tension. By applying stress to reduce the strain to zero, a surface tension of 1370 mN/m was calculated,[14] which is much higher than the values quoted for γ_0 and γ_i in Table 2.1.

Surface mobility

Sintering is possible with certain solids because under the correct conditions they show a certain amount of bulk and surface mobility. In the formation, say, of a sintered glass disc, powdered glass is heated under some pressure to a temperature just below its melting point, when the particles of glass tend to fuse where they touch each other and form a porous disc. Surface tension forces certainly play a part in what is a rather complicated process. Because of irregularities in the surface of the particles on a submicroscopic or microscopic scale, the area of contact is likely to be very small, and quite low external pressures applied to the mass as a whole will develop local pressures great enough to cause plastic deformation.

Furthermore, bulk and surface diffusions are likely to be significant factors at temperatures high enough to cause sintering.[15] Scratches on a silver surface disappear at temperatures near the melting point,[16] or if a silver ball is placed on the surface of the metal at this temperature then the area of contact enlarges and a weld results.[17] Sintering becomes

* The use of the term *surface tension* and also the term *surface energy* for an interface involving a solid has given rise to considerable confusion. The equation relating γ to ${}^sA$ is $\gamma = {}^sA - \Sigma\mu_i\Gamma_i$. If a dividing surface is chosen such that $\Sigma\mu_i\Gamma_i = 0$, then and only then would it be legitimate to term γ the *specific surface free energy* but this situation is not usual (see ref. 15, p. 8, and ref. 91).

important at temperatures above about three-quarters of the melting point expressed in degrees Kelvin; it is characterised by an increase in apparent density,[18,19] and a decrease in absorptive power and chemical reactivity.[20]

It is possible to get some idea of surface mobility by applying gas kinetic theory to the equilibrium between evaporation and condensation. Adamson[21] has calculated that for the equilibrium between water and water vapour at room temperature the average life-time of a molecule in the surface is 10^{-6} s. Similar calculations for a solid surface, say of a metal, show that if the saturation vapour pressure is 10^{-40} atmos the average lifetime of a surface atom would be 10^{37} s. On the other hand[22] for copper at three-quarters of its melting temperature (i.e. at 725°C) the estimated vapour pressure is 10^{-8} mm Hg which leads to an average surface lifetime of about 1 s. Similar circumstances apply to bulk diffusion, and for copper at 725°C[23] the bulk self-diffusion coefficient is about 10^{-11} cm^2/s, which would correspond to a mean displacement of 45 nm in one second, whereas at room temperature the mean displacement per second is negligible, since there is an apparent activation energy of 225 kJ/mol.

When considering surface diffusion the picture is rather different. This has been studied directly by means of the field-emission microscope.[24] The tip of a thin metal wire cathode is located at the centre of a hollow glass sphere, the inner surface of which is coated with a suitable fluorescent material. When a high potential is applied, electrons are emitted from the tip and cause a fluorescent pattern of light and dark areas to appear representing a magnified image of the emission pattern of the tip. Magnifications of up to 10^7 may be obtained. It has been shown by this method that surface migration becomes important at about 40% of the melting point in degrees Kelvin. Noticeable surface diffusion has been found on copper[25] at 700°C, when the activation energy is low.

For most high melting point solids at room temperature the surface must be considered essentially static with the atoms or molecules merely vibrating around equilibrium or quasi-equilibrium positions. However, as the melting point is approached, firstly there is an increase in lateral mobility, and then an increase in bulk diffusion, and finally vaporisation.

Conditioning of solid surfaces

The comparative immobility of the surface of a solid under normal conditions implies that the physical state of a surface will be largely dependent on the immediate history of that surface. Thus one would expect to find differences between a freshly cleaved surface, a polished surface, a ground surface, and a heat-treated surface.

Polishing surfaces drastically affects their nature. The layer on the surface formed during this process is usually known as the Beilby layer, and appears to be amorphous under the microscope,[26] having the appearance of a liquid viscous film which has flowed into all irregularities in the surface.[27] Electron diffraction studies of this surface layer show diffuse rings typical of the amorphous state or of the unordered liquid state. Raether[28] has concluded that the Beilby layer is in fact microcrystalline, but that the crystallites are so small that they give rise to diffuse rings with electron diffraction. However, there is also evidence of oxide being dragged into the surface region[29] together with impurities from the polishing media.[30] The fact that the Beilby layer is not stable has been illustrated by Cochrane[31] who deposited a thin layer of gold on nickel. The gold layer was polished and then removed and studied by electron diffraction, which showed a gradual transition from microcrystalline to crystalline. Electropolished surfaces are more nearly normal than polished surfaces, usually showing crystalline electron diffraction patterns.[32]

The above discussion illustrates some of the reasons for surface variation, not forgetting that contamination might be more important, in an individual case, than any of them.

Surface tension and surface free energy

The surface free energy of a solid is not necessarily equatable to the surface tension. Whereas a newly formed liquid surface will rapidly take up a uniform equilibrium state, the same is not true of a solid. The newly formed solid is likely to possess a considerable range of values of surface free energy, varying from region to region of the surface. Moreover, for a liquid the surface tension is the same in all directions, but this may not be true for a solid. If we resolve the surface tension into two directions at right angles, we can represent these partial surface tensions by γ_1 and γ_2. For an anisotropic solid, if the area is increased in two directions by dA_1 and dA_2, as shown in Fig. 2.3 then the total increase in free energy is given by the reversible work[33] against the stresses γ_1 and γ_2

$$\mathrm{d}(\mathrm{A}^{s}A)=\gamma_1\mathrm{dA}_1+\gamma_2\mathrm{dA}_2 \tag{2.19}$$

where ${}^{s}A$ denotes the free energy per unit area. If $\gamma_1=\gamma_2$ then

$$\gamma=\frac{\mathrm{d}(\mathrm{A}^{s}A)}{\mathrm{dA}}={}^{s}A+\mathrm{A}\left(\frac{\mathrm{d}^{s}A}{\mathrm{dA}}\right) \tag{2.20}$$

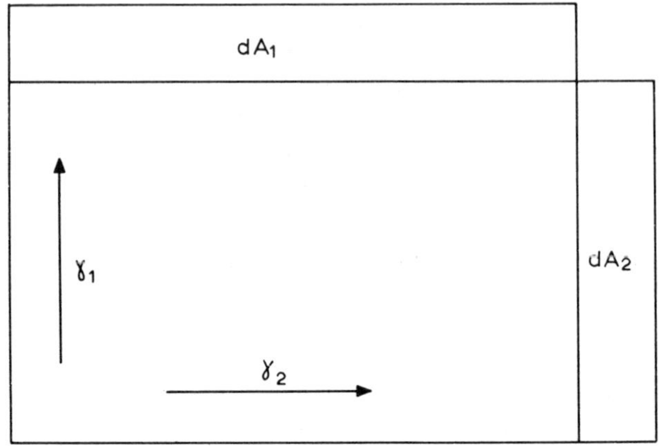

Fig. 2.3.

which is the same as eqn. (2.13) and implies the same condition. In this case, however, $(d^sA/dA) \neq 0$ as a general rule: it could only be so for a homotattic (perfectly uniform) surface.[34,35] Thus if the surface achieved some uniform equilibrium state ${}^sA = \gamma$; in all other cases ${}^sA$ and γ will be different from their equilibrium values and different from each other.[36]

The Laplace equation describes the pressure gradient ΔP across a curved ellipsoidal surface, in terms of the two radii of curvature, r_1 and r_2, thus

$$\Delta P = \gamma\left[\frac{1}{r_1} + \frac{1}{r_2}\right] \tag{2.21}$$

For a nearly spherical isotropic crystal, the pressure difference causes compression. If β is the compressibility then the volume change ΔV is approximately

$$\frac{\Delta V}{V} = 3\frac{\Delta r}{r} = -\Delta P.\beta$$

or

$$\Delta r = -2\beta\gamma/3 \tag{2.22}$$

This effect has been investigated[37] by means of X-ray powder photographs. For magnesium oxide values of Δr near the expected value of 0·06 nm, i.e. a 0·1% change in lattice distance for a 60 nm crystal, were found using a calculated value of 6573 mN/m for the surface tension. The results, however, were thought to be affected by adsorbed gases and surface contamination. The problem has been reinvestigated by Guilliatt and Brett[38] who have confirmed that adsorbed films produce marked effects. Traces of moisture produced lattice dilation, whereas in the absence of water vapour or adsorbed gases a lattice contraction was observed, the magnitude of which increased with decreasing crystallite size, and was substantially in agreement with the theoretical results of Anderson and Scholtz.[39]

Calculated values of surface tension and surface energies

The surface energy can be considered to have important implications in relation to wettability,[40] since this is determined by the extension across the interface of the forces defining the surface energy. However the surface tension and surface energy of a solid are not directly measurable, hence calculated values have added significance.

The method of calculating surface energies varies with the nature of the forces that exist between the atoms or ions or molecules which constitute the surface. These forces were considered separately earlier in this chapter. The simplest case is that of a covalently bonded crystal, such as the diamond, since in this case it is possible to ignore long range forces and merely consider valency forces. Harkins[41] calculated the surface energy at 0K as one-half of the energy of the number of bonds broken in forming the surface.

$$^{s}U = \tfrac{1}{2}U_{\text{cohesion}} \tag{2.23}$$

If the cleaved surface of diamond is the (111) plane, and if the interplanar atomic separation is 0·232 nm, then $1{\cdot}83 \times 10^{15}$ bonds are broken per cm^2 of surface. The covalent bond energy is 5·4 J/m^2. The corresponding value for the (100) plane is 9·14 J/m^2. It should be noted that since we are calculating for 0K these values will be equal to the surface free energy at that temperature. Harkins calculated by means of eqn. (2.17) and the Eötvös equation[42]

$$\gamma(M/\rho)^{2/3} = K(T_c - T) \tag{2.24}$$

where M is the molecular weight, ρ the density, T_c the critical temperature, and K is a constant, that the entropy contribution at 25°C to the surface energy was negligible; so that the above values are also approximately those of the surface free energy at 25°C. This method of calculation can only be approximate since no allowance is made for surface distortion, and the surface considered is an ideal one.

The simplest crystals involving long range forces for which calculations are possible are those of the rare gases, which are normally face-centred cubic lattices. The principle of the calculation is essentially the same as for the diamond crystal, namely, the calculation of the net energy of interaction across the cleavage plane, that is the surface energy at 0K is taken as being the excess potential energy of the molecules near the surface. Shuttleworth's[43] calculations (Table 2.2) were based on a form of eqn. (2.3) but included allowance for surface distortion by allowing the atoms of the surface plane to move to thc position of minimum energy. Although this lattice distortion term was less than 1% of the total it represented a considerable change in the first plane position.

TABLE 2.2
SURFACE ENERGIES AT 0K FOR RARE GAS CRYSTALS[43]

Rare gas	${}^{s}U$ *(mJ/m²)* (111) *plane*	(100) *plane*
Ne	17·2	17·9
A	41·1	42·7
Kr	53·3	55·3
Xe	60·7	63·0

The calculation of surface energies for ionic crystals is more complex since there are both long range forces and coulombic forces to consider, and a more complicated potential energy function must be employed. Early calculations employing once more the calculation of the potential energy function for a cleavage plane of alkali halide crystals were made by Born *et al.*[44] This type of calculation was developed by Lennard-Jones, Taylor and Miss Dent[45-47] who allowed for surface distortion by allowing movement of atoms in the surface plane, which they found represented a contraction of the distance of the outer plane from that beneath it of about 5%, to the position of minimum energy. They found ${}^{s}U=77$ mJ/m² for sodium chloride at 0K.

Subsequently developments have been made in several directions. The potential energy function has been refined,[48]

$$\varphi(a)=z_1 z_2 \frac{e^2}{a}-\frac{A'}{a^6}-\frac{B'}{a^8}+C' \exp\,(-a/\rho) \tag{2.25}$$

where z_1 and z_2 are the charges on the ions and A', B' and C' are constants; also a more complete treatment of surface distortion has been proposed, associated with the unsymmetrical electric field at the surface. Verwey[49] stated that besides surface polarisation, the outer plane distortion could not be treated by movement of the plane as a whole, but that there were differences between positive and negative ions. The positive ions move in and the negative ions out from the bulk lattice plane separation. This leads to a distortion energy of about 80 mJ/m^2 tending to decrease the surface energy, and giving a value[50] of ^{s}U of about 50 mJ/m^2. The situation has been well reviewed[50] and it was concluded that the calculated values of the surface energy of the (100) plane of sodium chloride vary from 33 to 129 mJ/m^2 depending on which assumptions are made. This would seem to underline the current unsatisfactory position, which is due in no small measure to uncertainty in the exact form of the potential functions. However, in order to give some idea of the order of magnitude and of the difference between compounds, some values for the alkali halides[43] and the alkaline earth oxides and sulphides[45] are listed in Table 2.3. The experimental values are for the liquids at their melting points. The results for the divalent ions should be treated with just as much caution as the alkali halide results. Benson and McIntosh[51] have shown that by using different potential energy functions to calculate the surface energy of magnesium oxide, values vary from −298 to 1362 mJ/m^2.

TABLE 2.3

SURFACE ENERGIES FOR IONIC CRYSTALS (mJ/m^2)

Cation		*Anion*							
		F^-	Cl^-	Br^-	I^-	O^{2-}		S^{2-}	
		(110)	(110)	(110)	(110)	(110)	(100)	(110)	(100)
Na^+	calc.	171	155	145	132				
	expt.	335	190	178	138				
K^+	calc.	160	134	124	111				
	expt.	242	173	159	139				
Ca^{2+}	calc.					1030	2850	359	1440
Mg^{2+}	calc.					1360	3940	358	1730

Calculations of the distortion in the surface region of a 100 face of a sodium chloride crystal have been extended by Benson *et al.*[52] The model employed allowed ions in the first five layers at a free 100 face of a hemi-crystal to relax in a direction normal to the face and to be polarised by the electric field in the surface region. The equilibrium positions were determined by minimising the energy of the system, and corresponded to a value of $-107{\cdot}4\,mJ/m^2$ for the total correction to the surface energy due to surface distortion. The calculated equilibrium positions show several interesting characteristics. The negative ions in each of the first five layers are displaced outwards towards the surface from their expected bulk crystalline positions, the displacement increasing as the surface is approached. The displacement of the positive ions on the other hand alternates. Those in the first, third and fifth layers are displaced inwards, whereas those in the second and fourth layers move outwards. Thus these calculations suggest that there is considerable difference between the surface atomic conformation of a real crystal and that existing in the bulk of the material. Clearly this type of approach should be extended to adsorption potential calculation, but because of the considerable extra complexity involved in also introducing lateral relaxation, this has not so far been attempted.

The calculations so far considered have assumed that the crystals of ionic solids are completely ionic, that is that the bonds possess no covalent character. In an attempt to allow for this with lithium fluoride[53] a value of ${}^{s}U = 557$ mJ/m^2 for the (110) plane was obtained, a value markedly higher than approximately 200 mJ/m^2 obtained by a purely ionic model calculation. The calculated values of surface energy usually are very low compared with experimental estimates.[13] The experimental values quoted in Table 2.3 are those for the liquid halides at their freezing points, and one would expect the values for the solids to be higher than for the liquids, yet the tabulated calculated values are lower.

In view of the uncertainty of the values of surface energy of solids it is scarcely surprising that calculated values of the surface tension differ widely, since in general the method of calculation depends on eqn. (2.20). Values for the surface tension of alkali halide crystals vary from positive values[45,46] of several thousand mN/m to negative values of several hundred.[43] As mentioned earlier the change in X-ray spacings for fine magnesium oxide and sodium chloride at least indicates that the surface tension values should be positive.

Energies associated with edges and corners

Those atoms or molecules or ions which are nearest an edge obviously have fewer nearest neighbours than those on a surface or those in the bulk of the solid. By similar methods to those used to calculate surface energies, it is possible to calculate edge energies. Edge energies for two cubes having a common edge (Fig. 2.4) have been calculated[45] and a

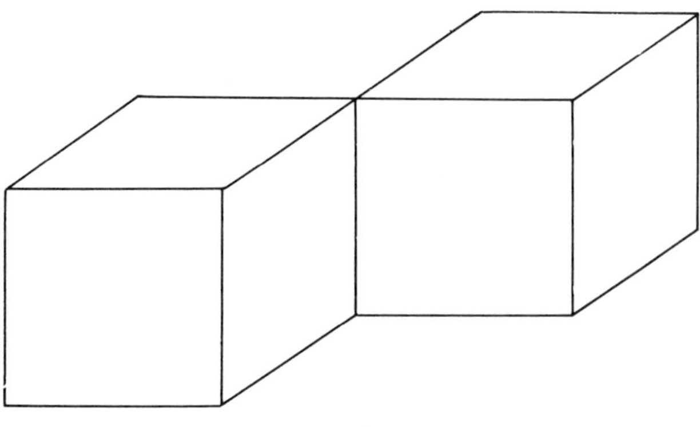

Fig. 2.4.

value of 100 pJ/m obtained. More recently[54] a value of 29 pJ/m for the edge energy between (110) planes of sodium chloride has been obtained. Little confidence can be placed in the absolute magnitude of these quantities since no attempt was made to allow for the distortion which one would expect to be present at an edge.

Energies associated with single edges and corners can be calculated, but the problem of allowing for distortion is even more acute, and at present one would have no confidence in the answers.

Specific surface area

The specific surface area is defined as the total surface area of 1 g of material, and is a frequently determined parameter of powdered substances. In effect it is a way of expressing the state of subdivision, and in many respects it is a method of expressing particle size without the need to determine the size distribution.

The BET method

The most popular experimental method of measuring specific surface area is that of gas adsorption.[55] In essence, if the amount of gas required to cover the surface with a complete monolayer can be evaluated, the specific surface area can be calculated, if the area occupied by a single molecule is known. Brunauer[56] classified experimental isotherms into five types according to their shape, as shown in Fig. 2.5, where the volume adsorbed V (i.e. the equivalent volume at STP of the number of moles of gas adsorbed) is plotted as a function of the relative pressure, P/P_0, where P is the equilibrium pressure and P_0 the saturation vapour pressure of the adsorbate at the

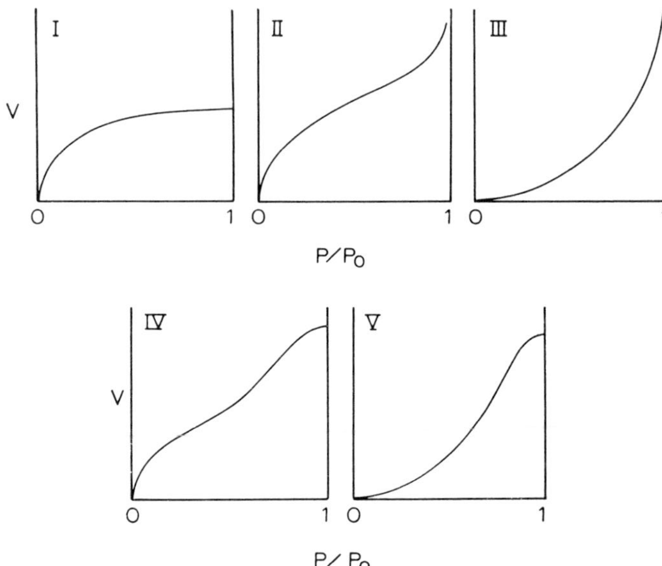

Fig. 2.5. Brunauer's five types of gaseous adsorption isotherm.

isotherm temperature. The majority of adsorption isotherms for gases such as nitrogen on non-porous solids have the form of Type II, where the equivalent volume at STP of the gas adsorbed is plotted as a function of the relative pressure. In 1935 Brunauer and Emmett[57] considered that the central nearly linear portion of a Type II isotherm corresponded to the formation of a second layer, and that the monolayer equivalent volume could be evaluated by extrapolating this linear portion back to the V axis, this value being subsequently termed point A. In a later paper Emmett and Brunauer[58] considered a number of other points on the isotherm which might be used to evaluate the monolayer volume and concluded that the point B, the beginning of the linear section of the isotherm was more consistent, if a range of gases were considered, in the value of the specific surface area. In Table 2.4 some of their values for the surface area of catalysts are given assuming each point in turn is equivalent to the monolayer volume, and calculating the area occupied by a single molecule from both the solid (s) and liquid (L) densities. The point B method offered the most concordant series of values, and on this evidence together with further evidence from the heats of adsorption, the point B method became a frequently used method. Practical identification of this point is not always easy, and the subsequent development of the

TABLE 2.4

Gas adsorbed	Temp. of isotherm (°C)	B.p. of gas (°C)	Point A V_m (cm^3 at STP)	Point A Surface area(m^2) (S)	Point A Surface area(m^2) (L)	Point B V_m (cm^3 at STP)	Point B Surface area(m^2) (S)	Point B Surface area(m^2) (L)
N_2	−183	−195·8	4·9	18·3	22·5	5·5	20·5	25·3
N_2	−195·9	−195·8	4·8	17·9	21·0	5·5	20·5	24·1
CO	−183	−192·0	5·0	18·5	21·5	5·7	21·1	25·9
Ar	−183	−185·7	5·2	18·0	20·3	5·8	20·1	22·6
O_2	−183	−183·0	4·7	15·4	17·9	5·5	18·0	21·0
CO_2	− 78·5	− 78·5	4·7	17·9	23·4	5·95	22·7	27·4
C_4H_{10}	0·0	0·3	2·0	17·3	17·3	2·2	19·0	19·0

Brunauer–Emmett–Teller[59] (BET) equation represents an analytical way of locating point B,[60] despite the considerable theoretical shortcomings of the equation. The equation expresses the equivalent volume adsorbed at relative pressure P/P_0, using a constant C, and the monolayer capacity V_m

$$V=\frac{V_mCP}{(P_0-P)\,(1+(C-1)\,P/P_0)} \tag{2.26}$$

which may be rearranged to give the practically useful form

$$\frac{P}{V(P_0-P)}=\frac{1}{V_mC}+\frac{C-1}{V_mC}\,\frac{P}{P_0} \tag{2.27}$$

Hence, if $P/V(P_0-P)$ is plotted against P/P_0, the value of V_m can be calculated from

$$V_m=1/(slope+intercept) \tag{2.28}$$

The BET equation is capable of describing more or less adequately most Type II and Type III isotherms, the type being a function of the value of the constant C, but in general only Type II isotherms corresponding to high C values yield reliable values of V_m.

The mathematical character of the BET equation can perhaps be more readily seen if it is rearranged in the form

$$\frac{V}{V_m}=\frac{1}{1-x}-\frac{1}{1+(C-1)x} \tag{2.29}$$

where $x=P/P_0$. Thus the adsorption isotherm of V/V_m against x consists of the difference between the upper branches of two rectangular hyperbolae representing the first and second terms on the right hand side of eqn. (2.29). The first has asymptotes at $V/V_m=0$ and $x=1$, while the second has asymptotes at $V/V_m=0$ and $x=-1/(C-1)$, and also cuts the V/V_m axis at $V/V_m=1$. The difference between these two curves gives the BET isotherm, the shape of which is strongly dependent on C. For C values greater than 2 an isotherm of Type II results, whereas if C is less than 2 the resulting isotherm is of Type III. The point of inflection on the Type II curve is not necessarily at a value of $V/V_m=1$. It can be shown that this is the case when $C\cong 9$ and when $C\rightarrow\infty$. However, the discrepancy is only marked when $C<9$.

The calculation of the specific surface area from V_m requires a value for the area occupied per molecule in the complete monolayer σ_m. The assumption is made that the value of σ_m for a particular adsorbate at a specific adsorption temperature is independent of the adsorbent. Obviously this represents an oversimplification of any practical situation; however the value of 0·162 nm^2/molec for nitrogen at $-195{\cdot}8$°C, based on the liquid density, has become a 'standard' value for this type of measurement, and values of σ_m for other adsorbates are usually obtained by reference to this 'standard'. The variability of the values of σ_m obtained in this way is due to the disturbance of the ideal liquid structure of the adsorbed film, which has been assumed, by the surface of the adsorbent. Table 2.5 illustrates this point for the case of krypton adsorption at -195°C, and it should be noted that variations of up to $\pm 0{\cdot}02$ nm^2/molec were found in several groups of solids[61,66] This degree of uncertainty in σ_m due to various degrees of pseudolocalisation, should be borne in mind in assessing the absolute validity of any calculated

TABLE 2.5

VALUES OF σ_m OBTAINED BY COMPARISON WITH NITROGEN ADSORPTION AT -195°C BY THE BET METHOD ASSUMING $\sigma_m=0{\cdot}162$ nm^2/MOLECULE

$\sigma_m(Kr)$ ($nm^2/molec$)	*No. of solids*	*Reference*
0·195	5	61
0·195	3	62
0·195	2	63
0·208	6	64
0·218	1	65
0·226	2	66
0·152 from liquid density		61
0·140 from solid density		61

value of specific surface area,[67] despite the reproducibility which can be achieved in the BET value of V_m with one solid sample and one adsorbate at one adsorption temperature.

The BET equation (2.26) applies to those cases where the adsorbed film can build up to an infinite thickness, but in the case of porous solids, the pore size may restrict the possible thickness of the adsorbed film. In this case the usual form of the BET equation (2.27) does not fit the experimental data at high pressures, and a satisfactory value for V_m cannot always be obtained by this method. If adsorption at saturation is restricted to n layers, the BET methods lead to the isotherm[59,68]

$$V = \frac{V_m C x}{(1-x)} \frac{(1-(n+1)x^n + nx^{n+1})}{(1+(C-1)x - Cx^{n+1})} \tag{2.30}$$

where $x = P/P_0$ and the remaining terms have their previous significance. This equation is not easily applied since there are three unknown constants. A graphical method of solution has been described by Joyner, Weinberger and Montgomery.[69] But it is more common to deal with the problems of porous solids by the methods which will be discussed later.

Area from electron microscopy

The electron microscope offers an obvious way of determining the size and shape of particles and hence their surface area. If powdered materials existed normally in precise geometric shapes, and all the particles in a sample were of identical size, then the problem of calculating the surface area would be simple. In practice the irregularity of the particles and the wide distribution of particle size frequently encountered, make the problem of calculating surface area a very difficult one. Usually a large number of particles have to be investigated in order to gain some idea of the size distribution, and considerable approximations made in the shape in order to calculate an area. The method is comparatively time consuming, but being a direct method is still frequently used. Because of the difficulties of allowing for surface roughness, areas calculated by this method are frequently lower than those obtained by other methods.

Surface area by measurement of adsorption from solution[70]

Although adsorption from solution is only of secondary importance in determining surface areas, it has the attractive features of comparative simplicity of technique and quickness. However, if the method is to be reliable it must be possible to determine the monolayer coverage unambiguously, and to assign an accurate value to the area occupied by the adsorbed species, which normally means that the orientation of the adsorbed species must be known and this is usually in doubt: thus in many respects the problems are similar to those encountered with the BET method. Furthermore it is likely that the phenomenon of localisation is common, and therefore σ_m will have an individual value for each adsorbing surface. Generally sparingly soluble solutes are used, but adsorption from solution may give a complete monolayer on one solid substrate, but not on another at any practicable concentration.

The most formidable objection to solution methods lies in the problem of assigning a satisfactory value to the area occupied by a single adsorbed molecule, which is commonly done by using a 'standard' solid whose surface area has been determined by the BET method. Thus the determined value has all the uncertainties of the BET method together with those of adsorption from solution.

The adsorption of long-chain fatty acids from organic solution has been frequently used since it has been thought that the molecular areas are accurately known from work on insoluble monolayers on aqueous substrates, in other words, it is assumed that a close-packed vertically oriented monolayer is formed. The method has been used with a modicum of success to measure the surface area of titania,[71] alumina[72] and some metals.[73,74] Dyestuffs are attractive because of the ease of their colorimetric determination. Methylene blue has been quite widely used, but many different values have been ascribed to the area occupied by an adsorbed molecule, 0·54 nm^2/molec,[75] 0·78 or 1·38 nm^2/molec,[76] 1·02 and 1·08 nm^2/molec,[77] and 1·97 nm^2/molec.[78] The range of these values obviously makes this unsuitable as a general method.

Giles[79] has considered the requirements for a satisfactory general solute, and has claimed considerable success with *p*-nitrophenol[80] from water, or from hydrocarbons. This method should be approached with caution. It should be remembered that the size of an adsorbed molecule determines the magnitude of the surface imperfections which will be followed; for example a pore that will be penetrated by a nitrogen molecule in a BET determination, may be too small to allow a methylene blue molecule to enter.

The surface of real solids

Impurities and heterogeneity

Ideally the surface of a solid is considered to be smooth, but in actual fact[81] it is probable that although the total surface energy would then be a minimum, this situation would also correspond to minimum entropy. Thus the normal equilibrium surface with the minimum attainable energy is rough with irregularities perhaps several tens of nm in height representing a compromise between entropy and energy considerations. Even if we can exclude the possibility of contamination, which can obviously give rise to surface variation, the surface of a real solid is irregular because of surface defects, growth spirals, edges, and corners.

Earlier when considering the effects of edges and corners we considered the effect on surface energy of a decrease in the number of nearest neighbours of a particular atom or molecule. The surface equivalents of Shottky and Frenkel defects will produce similar effects, and a growth spiral effectively represents a spiral edge on a face. All these factors and a number of others[82] contribute to surface heterogeneity. If the surface of an uncontaminated real crystal is likely to be heterogeneous, then it is scarcely surprising that impurities may also be a cause of surface heterogeneity. In fact surface heterogeneity means that calculated values of surface energy can only approximate to experimental values. Thus different preparations of the same substance are likely to give rise to different experimental values of surface energy, as are different methods of evaluation.

Before leaving the subject of surface energy, the effect of variations in surface tension and particle size on solubility might be mentioned. The Kelvin equation may be written as

$$RT\ln\left(\frac{a}{a_0}\right)=\frac{2\gamma\bar{V}}{r} \tag{2.31}$$

for a spherical, isotropic particle where $\bar{V}$ is the molar volume, and a the activity of the solid. The value of a determines the solubility of the solid in relation to a_0, the value for a plane surface. The increased solubility of small particles is considerable. The results obtained by Dundon and Mack[83] are shown in Table 2.6, and it is worth noting the parallel between the surface tension, calculated from the increase in solubility, and hardness.

Porosity

An extreme case of surface irregularity is the pore. Porous solids show certain differences from non-porous ones in terms of their surface properties. A porous solid has a high specific surface area without necessarily having a small particle size. Thus practical adsorbents are

TABLE 2.6
EFFECT OF SURFACE TENSION ON THE SOLUBILITY OF SALTS

Salt	*Increase in solubility (%)*	*Particle size (μm)*	*Surface tension (mN/m)*	*Relative hardness on Moh's scale*
PbI_2	2	0·4	130	Very soft
$CaSO_4 2H_2O$	4–12	0·2–0·5	370	ca. 2
Ag_2CrO_4	10	0·3	575	ca. 2
PbF_2	9	0·3	900	ca. 2
$SrSO_4$	26	0·25	1 400	3·0–3·5
$BaSO_4$	80	0·1	1 250	2·5–3·5
CaF_2	18	0·3	2 500	4

often porous materials such as charcoals, and offer the advantage of large specific surface area without the difficulties caused by small basic particle size.

It is convenient to divide the pores in an adsorbent into three types in terms of size: macropores, transitional pores and micropores.[84] In principle the upper limit of the radius of curvature of a macropore is large, but the lower limit may be taken conventionally as 100 to 200 nm. This lower limit is chosen so that the volume filling of macropores by the capillary condensation of vapour takes place only at relative pressure values very close to unity. Thus conventional gas adsorption methods cannot detect the difference between the surface of a macropore and a plane surface. Experiments on the forcing of mercury into macropores permit an estimate of the volume distribution of pores and their specific surface area to be obtained for the range of effective radii from 100 to 0·1 μm.

The dimensions of transitional pores are still considerably larger than molecular dimensions, but they are small enough for capillary condensation to occur at relative pressure values easily distinguishable from unity in a gas adsorption experiment. Thus the adsorption isotherm usually shows hysteresis, that is the desorption branch is above the adsorption branch at higher values of relative pressure (Fig. 2.6). The reason for this can be seen by considering the liquid meniscus which will be formed when the pore is full. The Laplace equation (2.21) suggests that a pressure drop will exist across the meniscus at the neck of a capillary, and that if the pressure inside the capillary is to drop below P_0

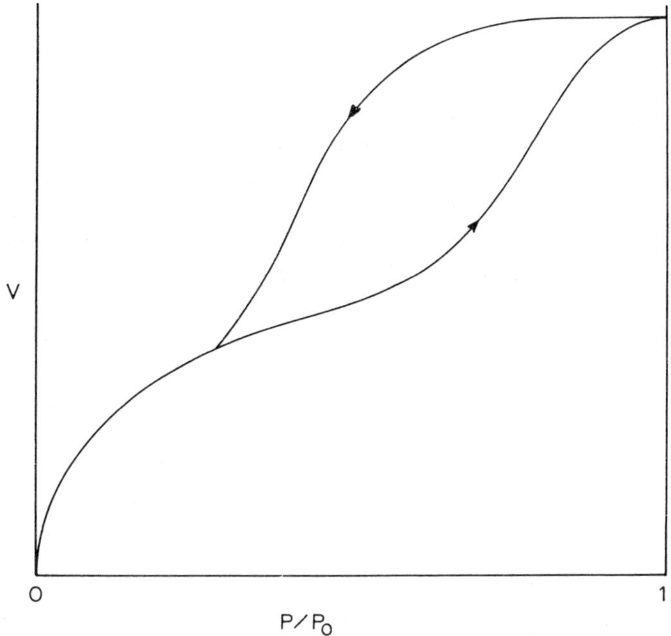

Fig. 2.6. Brunauer Type IV isotherm showing hysteresis.

then the external pressure will have to be less than $(P_0 - \Delta P)$. Thus the height of the hysteresis loop will give an indication of the pore volume, and the width of the pore.[85] The effective radii of transitional pores at the maxima of the distribution curves usually fall within the range 4 to 20 nm.

The mercury porosimeter will also give information about transitional pores as well as the macropores previously mentioned. If a liquid is added to a porous solid, menisci will be formed in the pores. For contact angles less than 90° the liquid enters the pores spontaneously, but for contact angles greater than 90° a positive pressure will be necessary to force the liquid into the pores. If the pores are assumed to have circular cross-sections then the Laplace equation may be written as

$$\Delta P = \frac{2\gamma}{r}\cos\theta \qquad (2.32)$$

where r is the radius of the pore and θ the contact angle. Thus by knowing the volume of mercury forced into the pores and the pressure it is possible to calculate a distribution of pore volume as a function of pore radius.[86]

The smallest pores are usually known as micropores and have dimensions commensurate with molecular dimensions. It is known that the adsorption energy in micropores is considerably higher than the corresponding values in transitional pores or on a plane surface. The absence of layer-by-layer filling and capillary condensation means that the concept of surface area has little meaning and pore volume cannot be obtained by hysteresis measurements. Dubinin[84] has developed a method for obtaining the pore volume from adsorption data, and de Boer's '*t*-plot' has permitted somewhat similar information to be obtained.[87]

Recently considerable attention has been devoted to determining experimental standard isotherms for non-porous adsorbents. At present these have been established for a number of oxides using nitrogen as adsorbate at 77·5 K, and methods have been developed to use these as a basis for evaluating the surface area and porous characteristics of other samples of the same oxides. The standard curve is established by determining the adsorption isotherm on a number of non-porous samples and expressing them all in terms of the plot V/V_m against P/P_0, where V_m is usually determined by the BET method.

This isotherm may also be expressed in terms of the statistical thickness, t, of the adsorbed layer. De Boer[87] calculated that the adsorbed film would be 0·354 nm thick for a complete monolayer of nitrogen at 77·5 K, so the V/V_m scale may be directly converted into one in terms of t (Fig. 2.7a). Use is made of the standard in the following manner. For each experimental point on the experimental isotherm, V_a vs. P/P_0, a value of t is read off from the standard isotherm for each experimental value of P/P_0, and the experimental data replotted in terms of V_a versus t. If the sample is non-porous then a straight line results (Fig. 2.7b), and the slope of the line is directly proportional to the surface area. Microporous samples are characterised by plots of the type shown in Fig. 2.7c, and those with transitional pores give rise to curves of the type shown in Fig. 2.7d.

Clearly the method cannot be used unless it is known that the standard isotherm is appropriate for the sample being investigated. The fact that the standard isotherm is inappropriate can often be recognised by a failure of the extrapolation of the experimental region of the t-plot to pass through the origin.

Certain objections to the t-plot procedure have been put forward by

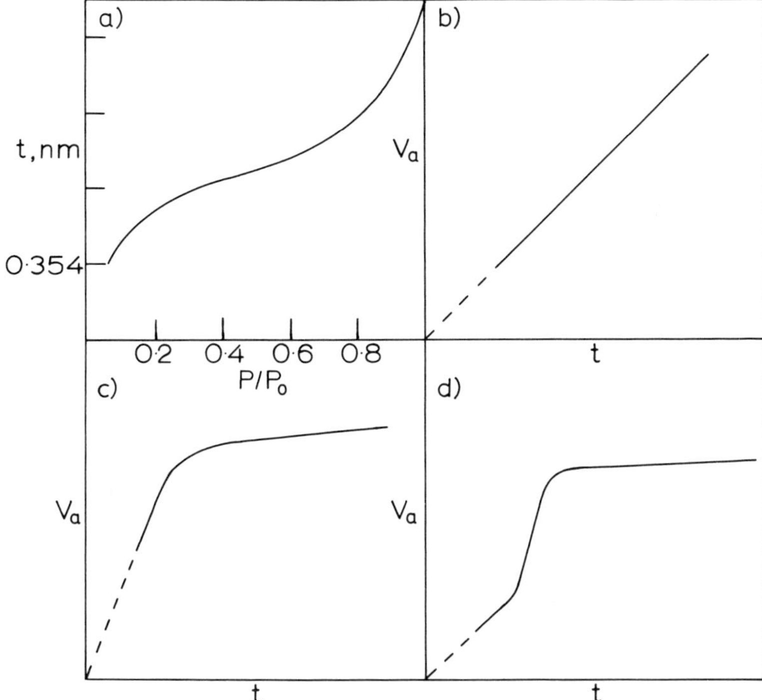

Fig. 2.7. The de Boer 't'-plot method. (a) The standard isotherm, (b) t-plot for a non-porous sample, (c) for a microporous sample and (d) for a sample with transitional pores.

Sing.[88] Most significant among these are that the method is not independent of the BET method, and cannot be readily applied to cases where the non-porous standard sample gives a Type III isotherm or a Type II isotherm with a very small knee at low pressure. He proposes that t be replaced by $(V/V_x)_s$, which is termed α_s, where V_x is the amount adsorbed at a selected value $(P/P_0)_x$. The location of $(P/P_0)_x$ is not completely arbitrary, usually being chosen at a value of 0·4 since micropore filling and monolayer coverage usually occur at lower pressures, and capillary condensation, at least when in association with hysteresis, takes place at higher pressures.

Chemical characterisation[89]

More often than not it is a mistake to assume that the chemical nature of the surface of pigment will resemble the chemical characteristics of the bulk material. For example oxide surfaces frequently hydroxylate in the presence of water or water vapour. If we consider the surface of titanium(IV) oxide, then two types of hydroxyl groups are formed:

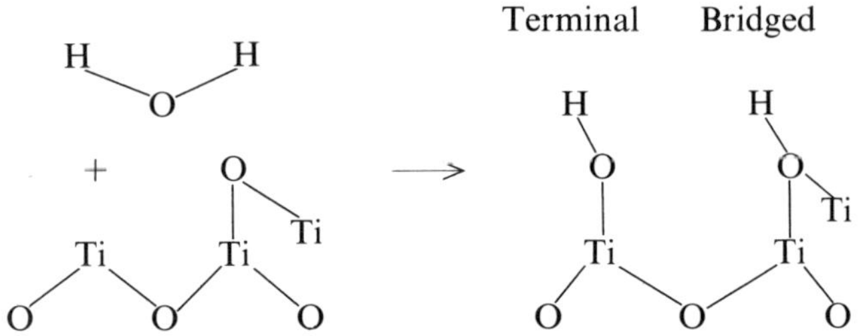

Terminal Bridged

It is the chemical nature of surface functional groups which primarily determines the dispersibility of a powdered pigment. That is to say it influences the adsorption, wetting behaviour and stability of pigment to be prepared as a dispersion. In consequence the chemical characterisation of a surface is of considerable significance.

A wide range of techniques have been employed to identify the chemical composition of a surface, but the various spectroscopic phenomena are probably the most widely employed, covering the full range from ultraviolet, visible, infrared and Raman spectroscopy, to n.m.r., e.s.r., Mössbauer, Auger, ESCA, LEED and neutron scattering. Straightforward reaction processes with specific functional surface groups have been used quite frequently to study oxides and carbons. Temperature programmed desorption offers an interesting way of studying certain specific types of surface group (e.g. surface hydroxyl) both quantitatively and qualitatively. Finally we might mention the variety of electrochemical titration techniques such as pH titration and Karl Fischer titration as well as the information that can be gained from the isoelectric curves derived from electrokinetic measurements.

WETTING AND CONTACT ANGLES

If a drop of liquid is placed on a perfectly flat and uniform solid surface, it need not necessarily spread completely over the surface, but its edge may make a contact angle θ with the solid. Figure 2.8 illustrates three possible cases in order of increasing attraction between molecules of the liquid and solid. If the equilibrium tensions are resolved horizontally,

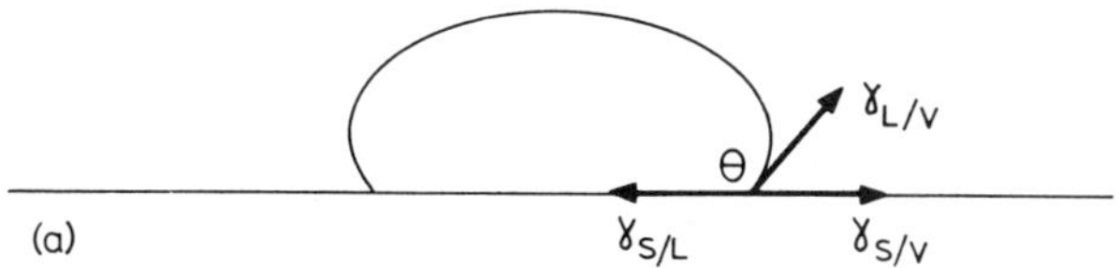

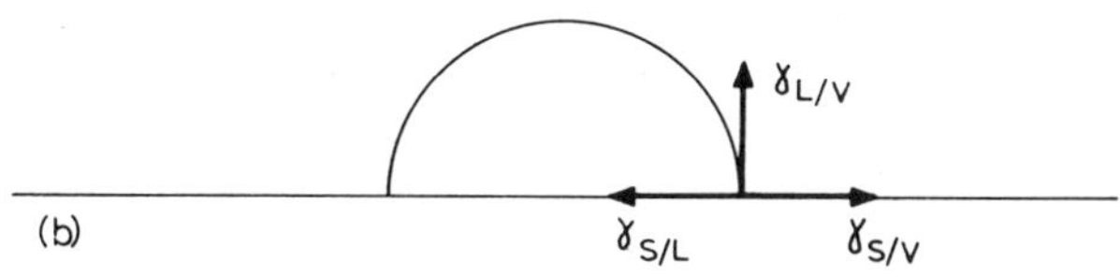

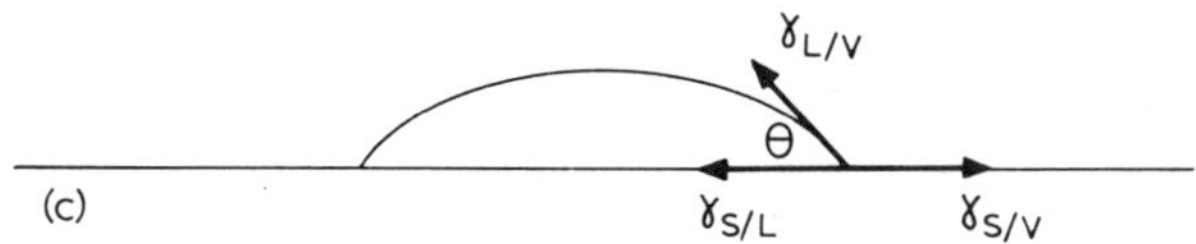

Fig. 2.8. Drop of liquid on a solid.

then

$$\gamma_{S/V} = \gamma_{S/L} + \gamma_{L/V} \cos \theta \tag{2.33}$$

where the subscripts S, L, V refer to the solid, liquid and vapour phases respectively. If we apply Dupré's methods[90] to define the work of adhesion between solid and liquid W_{SLV} we have

$$W_{SLV} = \gamma_{S/V} + \gamma_{L/V} - \gamma_{S/L} \tag{2.34}$$

and hence by combining the two equations we arrive at Young's equation:[91]

$$W_{SLV} = \gamma_{L/V}(1 + \cos \theta) \tag{2.35}$$

This derivation has assumed that the vertical force due to $\gamma_{L/V}$, that is $\gamma_{L/V} \sin \theta$, produces no deformation of the solid surface. In certain cases the presence of this force has been demonstrated, using a drop of mercury on a sheet of mica 1 μm thick, the sheet of mica deforming all round the edge of the drop.[92]

Magnitude of contact angles

Young's equation indicates that the magnitude of the contact angle will depend on the relative values of the adhesion between solid and liquid, and the mutual cohesion of the liquid which is related to $\gamma_{L/V}$. An angle of 180° would indicate zero adhesion, but this angle has never been observed in practice even with hydrophobic materials and water. The contact angle[9] between water and polythene is 94°, and between water and paraffin wax up to 110°. The values just quoted refer to smooth surfaces soon after formation. The contact angle is also dependent on the roughness of the surface and on any contamination of the solid surface or of the liquid surface, or possible rearrangement of the solid surface caused by the presence of the liquid phase.

Surface roughness has the effect of making the contact angle further from 90°. If the smooth solid gives a contact angle greater than 90°, the angle is increased by roughening the surface; if the contact angle is less than 90°, then roughness will decrease it. For example, rough paraffin wax[93] has a contact angle of 132°. Wenzel[94] has proposed that surface roughness may be measured by

$$R = \cos \theta' / \cos \theta \tag{2.36}$$

where R is the roughness ratio, and θ' is the average contact angle on a rough surface as opposed to θ on a smooth surface. This relation is only applicable to submicroscopic roughness, since for coarse roughness the edge becomes ragged and R is no longer a constant.

Contact angles are best measured where possible by the inclined plate method in which a plate of solid is immersed in the liquid and tilted so that the meniscus on one side of the plate is horizontal, as illustrated in Fig. 2.9. The angle of inclination of the plate is taken as being equal to the contact angle.

It is often difficult to obtain reproducible results, and in particular the contact angle when the solid surface is advancing into the liquid is frequently greater than that when the solid is being withdrawn, so that there is hysteresis of the contact angle. It has been found that very clean and smooth surfaces show no hysteresis of the contact angle[93,95,96] and therefore it seems probable that this phenomenon is associated with surface contamination and roughness.

It is also possible to get some idea of the contact angle for a powdered sample by measuring the pressure required to force a liquid which will wet the powder through a sample ($\theta = 0°$), and then the pressure required to force a liquid with unknown contact angle through the powder. Thus

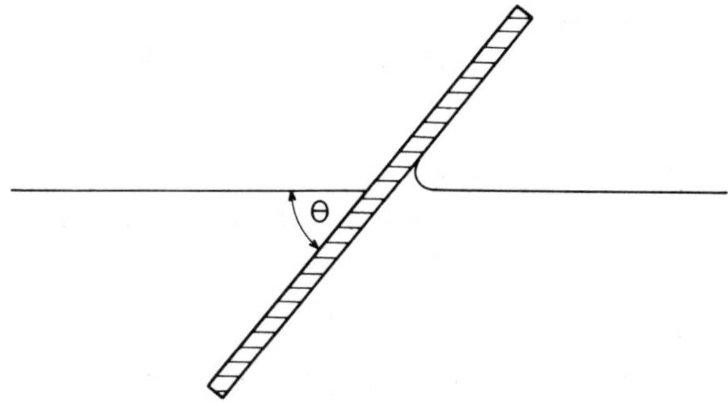

Fig. 2.9. The measurement of contact angle.

by applying eqn. (2.32) to both cases and eliminating r, the contact angle may be calculated. The approximate nature of the method is obvious. This and other methods have been described,[97] each subject to uncertainties.

Spreading coefficients

The balance of surface tensions at equilibrium has been described by eqn. (2.33). When a liquid is spreading and wetting a solid then the advancing contact angle of the very thin film must tend to zero and there is a resulting force causing spreading. This is usually described in terms of a spreading coefficient σ_{SLV}

$$\sigma_{SLV} = \gamma_{S/V} - \gamma_{S/L} - \gamma_{L/V} \tag{2.37}$$

If an equilibrium contact angle can be maintained then

$$\sigma_{SLV} = \gamma_{L/V}(\cos\theta - 1) \tag{2.38}$$

which is valid when $\theta > 0°$, and is often used to calculate spreading coefficients from values of θ and $\gamma_{L/V}$.

Oils spread readily on chromium and σ_{SLV} is positive. For 7-butyltridecane σ_{SLV} is 25 mN/m although by increasing the chain branching σ_{SLV} may be reduced to about 17 mN/m.[9] The importance of the surface history can be shown since if the chromium surface is cleaned by ion bombardment and at once wetted with water, oil will not spread and therefore σ_{SLV} must be negative.[98] Lubricating oils spread readily on steels particularly if they are somewhat oxidised to polar compounds.[99]

Many metal surfaces adsorb a monolayer of the spreading liquid and hence the value of θ measured, and the value of σ_{SLV} so derived, will be for the liquid on a monolayer coated solid.[98,100,101] To obtain a value for the clean surface a correction must be applied for the work of adhesion.

The adhesion of liquids to solids

The work of adhesion of a liquid is given by equation (2.34). If this is combined with equation (2.37), then

$$W_{SLV}=\sigma_{SLV}+2\gamma_{L/V} \tag{2.39}$$

The value for the adhesion of water to paraffin wax is about 46 mJ/m^2. If measured values are used then if the solid is likely to be covered with an adsorbed monolayer, and if values for a clean surface are required, correction must be made.[100,101]

Flotation[102,108]

If a bag of pigment is added to an aqueous solution, often there is a tendency for some at least of the pigment to float on the surface of the solution. Wetting agents are frequently added to prevent flotation; but in the mineral processing industries the problem of flotation is exploited rather than suppressed, so we will consider these processes in order to illustrate the phenomenon in general.

The essential step in froth flotation is the attachment of a particle, already completely wetted by the solution around it, to an air bubble of sufficient size to carry it from the bulk to the surface of the solution. In practice the particles are small in comparison to the air bubbles, and many of them may be attached to each bubble.

Let us consider the gravity-free model illustrated in Fig. 2.10. The particles of solid are assumed to be spherical of radius r and all three phases of the same density. If the particle is entirely in phase A, then

$$A_A=4\pi r^2\gamma_{S/A} \tag{2.40}$$

where A_A is the Helmholtz free energy of the surface, and $\gamma_{S/A}$ the specific surface free energy. Similarly if the particle is completely in phase B, then

$$A_B=4\pi r^2\gamma_{S/B} \tag{2.41}$$

Finally the particle may be at the interface, in which case both $\gamma_{S/A}$ and $\gamma_{S/B}$ contribute to the surface free energy, and part of the interface between phases A and B has been eliminated. If there is to be an equilibrium position for the particle at the interface then the total surface free energy must be at a minimum, therefore

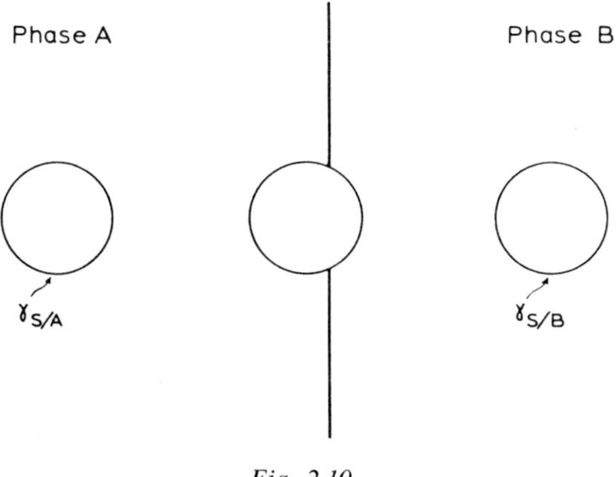

Fig. 2.10.

$$dA = \gamma_{S/A} dA_{S/A} + \gamma_{S/B} dA_{S/B} + \gamma_{A/B} dA_{A/B} = 0 \tag{2.42}$$

Considering Fig. 2.11 shows that

$$\gamma_{S/A}(2\pi r dh) + \gamma_{S/B}(-2\pi r dh) - \gamma_{A/B}\pi(2r - 2h)dh = 0 \tag{2.43}$$

which simplifies to

$$\gamma_{S/A} - \gamma_{S/B} = (1 - h/r)\gamma_{A/B} \tag{2.44}$$

but

$$\cos\theta = 1 - h/r$$

therefore

$$\gamma_{S/A} - \gamma_{S/B} = \gamma_{A/B} \cos\theta \tag{2.45}$$

which is the Dupré equation (2.33) for contact angle equilibrium. Consequently a particle will seek a stable position in the interface such that the angle θ becomes the equilibrium contact angle. In the ordinary circumstances when gravity is present, there will be a slight displacement from this position until the surface tension restoring forces are equal to the gravitational pull. Thus particles forming a finite contact angle at the air–liquid contact angle can be floated, whereas those wetted by the solution cannot.

Separation of ores by flotation involves adding reagents to control the contact angle so that it has a large value on one ore and the other is wetted.

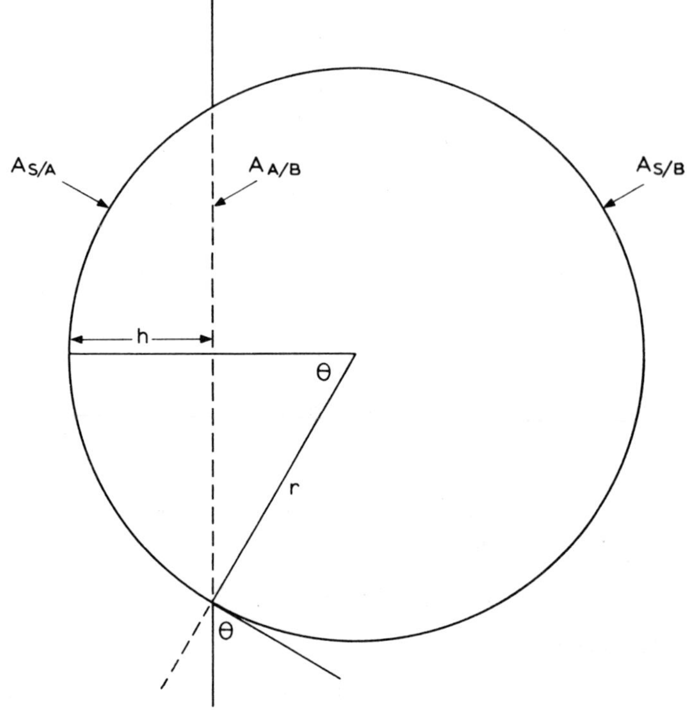

Fig. 2.11.

Wetting and non-wetting

A solid is completely wet with a liquid if the contact angle is zero. If an equilibrium is established it will be described by eqn. (2.33), which may be rearranged as

$$\cos\theta = \frac{\gamma_{S/V} - \gamma_{S/L}}{\gamma_{L/V}} \tag{2.46}$$

If the right hand side is equal to unity, or exceeds unity under non-equilibrium conditions, the liquid will wet the solid. Thus $\gamma_{S/V}$ should be large and $\gamma_{S/L}$ and $\gamma_{L/V}$ small. Often $\gamma_{S/L}$ is large when surfaces are rough or particulate. Wetting agents not only lower $\gamma_{L/V}$ but also allow the water to penetrate the surface lowering $\gamma_{S/L}$ also. Many wetting agents, such as sodium di-*n*-octylsulphosuccinate, have irregularly shaped molecules.[104]

Non-wetting is a term usually applied to the case where the contact angle is greater than, or equal to, 90°. On rough or hairy surfaces many liquids have high contact angles and are non-wetting. If we have a composite surface where f_1 and f_2 are the fractions which are liquid–solid and liquid–air respectively then[105]

$$\cos\theta = f_1 \cos\theta_1 - f_2 \tag{2.47}$$

Ducks have a structure of fine hairs on their feathers which makes them difficult to wet because f_2 is large as well as θ_1. In a surface active agent solution θ_1 has a low value and hence the term $f_1 \cos\theta_1$ changes from negative to positive, and if f_2 is not too large, the duck is wetted.

ADHESION[104,106,107]

Adhesive forces

It is possible to show[104] that for perfect surfaces adhesive strengths should be very high. If one substitutes in Young's equation (2.35) the value $\theta = 0°$, that is the adhesive just wets the solid, and $\gamma = 30$ mN/m, the force required to separate the surfaces by a molecular dimension is equal to approximately 600 kg/cm^2 or 4 ton/in^2. However, the values obtained in practice are usually of the order of 30 kg/cm^2, which is attributed to plastic and viscous flow which may take place before fracture, and the presence of flaws and dislocations in the surface of the solid.[106,107]

If two pieces of hard metal are carefully cleaned and then pressed together, the adhesive force is very small if the air is clean and dry. This is because under these conditions the surfaces undergo elastic deformation at the points where they come into contact, resulting in a force which tends to separate the surfaces again. However, in a humid atmosphere, the adhesion is affected by the extent of the adsorption of water vapour at the surface, and increases towards a limit at the saturation vapour pressure, which is the adhesion in the presence of liquid water.

This type of system is usually studied experimentally by using a sphere in contact with a plane surface. If it is assumed that the sphere and the plane surface are smooth, and that the liquid collects around the point of contact, then we have the situation as shown in Fig. 2.12. The pressure inside the liquid will be less than atmospheric, and by applying the Laplace equation (2.21), the pressure difference will be approximately γ/r_m. The total force over the whole liquid pool is thus $\pi r^2 \gamma/r_m$, and since

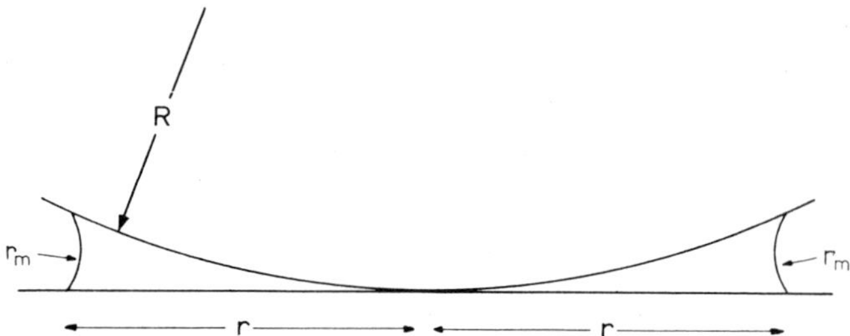

Fig. 2.12. Adhesion of a hard sphere, radius R, to a hard solid plane surface, with water present at the point of contact as a pool of radius r. The radius of curvature of the meniscus is r_m, the separation of the solid surfaces at the meniscus being approximately $2r_m$.

as a first approximation $r^2 = 2R.2r_m$, we have

$$\text{adhesive force} = 4\pi R\gamma \tag{2.48}$$

This result shows that the adhesion should be directly proportional to R, and independent of the thickness of the liquid film.

Flat, parallel surfaces may also adhere strongly by means of a liquid film. If we consider two flat circular discs, each 1 cm^2 in area, then if a film of oil which wets the discs separates them by 100 nm, the radius of curvature of the meniscus at the edge of the discs will be 50 nm. The Laplace equation indicates a pressure difference of 0·5 MPa if the surface tension of the oil is 25 mN/m. The adhesive force would be equal to 5·1 kg which may cause distortion of quite thick discs, for example microscope cover glasses.

The adhesive forces between pieces of indium are very considerable, even in the absence of a liquid film. This is attributed to the lack of elastic stresses in the solid. The formation of junctions between such surfaces may be reduced by liquid lubricants, and reduced often to negligible proportions by covering the surfaces with a monolayer of a substance such as lauric acid.[106]

The stickiness of particles

Powders frequently adhere strongly when they are damp as they often are on delivery. With comparatively small amounts of water present interparticle adhesion may become comparable with the weight of the particles, since the weight varies as the cube of the radius, and the adhesive force as the radius. For two spherical particles, each of radius R, eqn. (2.48) must be modified[104] so that

$$\text{adhesive force} = 2\pi R\gamma \tag{2.49}$$

The 'stickiness' of particles completely immersed in a medium is a different phenomenon since no 'lenses' of a third phase around the points of contact are involved. There are two main reasons for this phenomenon. Firstly, high molecular weight, polymeric materials may be adsorbed on to the surface of the particles, thereby increasing the collision radius and decreasing the electrical repulsion between the particles. Such materials are added to fine suspensions to increase their filterability, and to electrically stabilised emulsions to cause creaming without coalescence. The macromolecules may form bridges between the surface, preventing direct contact and also independent movement. The whole subject of stability is more fully discussed in Chapter 1. The second mechanism involves bridging by polyvalent ions between charged groups on the surface of two particles. Not only do these ions form bridges but they also reduce the Stern potentials of the individual particles, and hence the mutual repulsion between them. The particle surfaces may approach very closely, but the bridging is sensitive to the presence of sequestering and complexing agents which react with the polyvalent ions concerned. If the surfaces are strongly hydrophobic, the reduction of the electrical charges by specific absorption of polyvalent ions will result in very strong adhesion. The theory behind the electrical double layer phenomena just mentioned will be considered in Chapter 3.

Sliding friction and lubrication

Sliding friction is the force opposing the movement of one surface over another. This is of importance in dispersion since it affects the amount of work required to disperse a solid in a liquid, and is in turn affected by the lubricating properties of the medium. This process involves the shearing of any welded junctions formed at points of contact, and the energy of elastic deformation is not recovered. The coefficient of friction μ is defined as the ratio of the frictional force to the normal force holding the surfaces together. In many circumstances μ is a constant; this is known as Amonton's Law. The tangential force required to induce sliding is known as the static friction, and the lesser force required to maintain the sliding motion is known as the kinetic friction.

The presence of an oxide film on the surface of metals such as nickel, tungsten or copper reduces the coefficient of friction from the high value of 5 to a value of the order of 0·5. A thick layer of oil between the surfaces, fluid lubrication, will reduce the value still further to 0·002; this very low value being due to the fact that the only resistance to motion is due to the viscosity of the oil itself. Usually in practice it is not possible to maintain a thick lubricating layer between the surfaces, and the film will be reduced until the surfaces are separated by adsorbed films of molecular thickness, which is known as boundary lubrication. In this case the coefficient of friction is about 0·1, the exact value depending on the nature of the surface and the chemical composition of the lubricant, the bulk viscosity of the lubricant being of lesser significance. The addition of a little fatty acid to a non-polar oil can reduce the coefficient of sliding friction of two cadmium surfaces by about two orders of magnitude, effective lubrication occurring with boundary films of only one or two molecular thicknesses of fatty acid. However, single layers are rapidly worn away, and must be rapidly replaced; therefore clearly the rate of adsorption is important.

Rolling friction

The coefficient of rolling friction is generally less than the coefficient of sliding friction. Interfacial slip scarcely contributes to the resistance. Consequently lubricants have little effect on the coefficient of rolling friction, but may greatly reduce the wear occurring during rolling.

ADSORPTION FROM SOLUTION

Many early experiments on adsorption from solution were concerned with adsorbents of technological importance. Isotherms were measured for dilute solutions and only recently have more concentrated solutions been investigated.[108] For solutions of solids such as octadecanoic acid in benzene a certain modicum of success has been obtained in interpreting experimental isotherms in terms of the Freundlich equation.

$$w/m = Ac^{1/n} \tag{2.50}$$

where w is the weight of adsorbate adsorbed, m the weight of adsorbent, c the concentration of adsorbate in solution at equilibrium, and A and n are constants. The Langmuir equation has been similarly employed

$$w/m = \frac{Bc}{1+Dc} \tag{2.51}$$

where B and D are constants. The equations are of use in suggesting suitable ways of representing experimental data as linear plots ($\ln(w/m)$ vs. $\ln c$ for the Freundlich equation and $c/(w/m)$ vs. c for the Langmuir equation) for interpolation and evaluating the monolayer capacity. They are often applied to isotherms of the shape illustrated in Fig. 2.13(a), but they cannot describe the shapes shown in Fig. 2.13(b) and (c). These two shapes are characteristic of adsorption from solutions of completely miscible liquids. Adsorption on charcoal of ethanoic acid in water gives the former shape, while ethanol in benzene gives the latter.

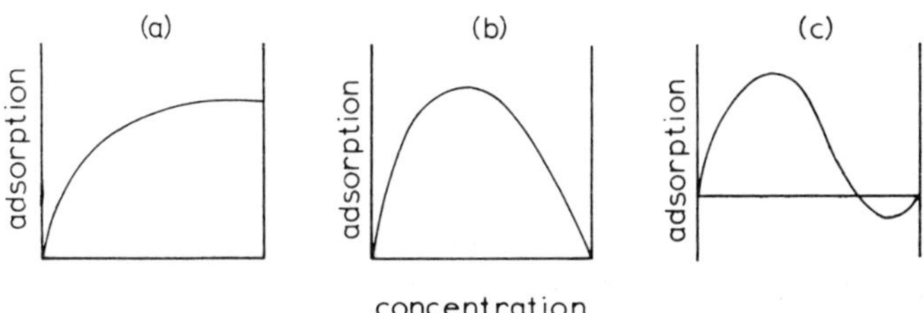

Fig. 2.13. Shapes of adsorption isotherms from solution.

Experimentally isotherms are determined by measuring the change in concentration of a solution when an adsorbent is added, due to adsorption. It is possible to express the adsorption in two different ways. The first is given directly by the experimental method just described, and is usually termed preferential or selective adsorption, which corresponds to the term surface excess, frequently employed for the liquid–vapour interface. The second is absolute adsorption, and is defined as the surface concentration of the adsorbed component.

The composite isotherm

Since adsorption from solution of necessity involves more than one component, the composite isotherm represents the result of combining the individual isotherms of the components.[108]

When a weight m of adsorbent is brought into contact with n_0 moles of a two-component solution, the mole fraction of component 1 decreases by Δx. This change in concentration is the result of the adsorption on the surface of ${}^{S}n_1$ moles of component 1 and ${}^{S}n_2$ moles of component 2 per unit weight of adsorbent. At equilibrium there are n_1 and n_2 moles of the two components in the liquid phase giving fraction x of component 1 whose initial mole fraction was x_0. Then

$$n_0 = n_1 + n_2 + {}^{S}n_1 m + {}^{S}n_2 m \qquad (2.52)$$

and

$$x_0 = \frac{n_1 + {}^{\mathrm{S}}n_1 m}{n_0},\quad x = \frac{n_1}{n_1 + n_2} \text{ and } (1-x) = \frac{n_2}{n_1 + n_2}$$

therefore

$$\Delta x = x_0 - x$$
$$= \frac{n_2\, {}^{\mathrm{S}}n_1 m - n_1\, {}^{\mathrm{S}}n_2 m}{(n_1 + n_2) n_0}$$

Thus

$$\frac{n_0 \Delta x}{m} = {}^{\mathrm{S}}n_1 (1-x) - {}^{\mathrm{S}}n_2 x \tag{2.53}$$

The quantity $n_0 \Delta x/m$ is the quantity usually plotted against x to give the composite isotherm. Many examples of composite isotherms are given by Kipling.[108]

Adsorption from completely miscible liquids

The study of adsorption from this type of solution is characterised by the multiplicity of models which have been used to interpret isotherms. In fact a considerable amount of data have been published without detailed analysis which may reflect the current uncertainty which is felt about the theoretical basis of such analysis.[108] A more rigorous treatment for ideal[109] and regular[110] solutions has been published but there is still much to be done.

For adsorption by porous solids an approach has been made based on the postulate that adsorption is a process of pore filling, in which the available volume of the pores is the controlling factor.[111,112]

Then

$${}^{\mathrm{S}}n_1 V_1 + {}^{\mathrm{S}}n_2 V_2 = V \tag{2.54}$$

where V_1 and V_2 are the respective partial molar volumes of the adsorbates and V the pore volume per unit weight of adsorbent. For monomolecular adsorption on a plane surface eqn. (2.54) would be written in terms of areas where

$${}^{\mathrm{S}}n_1 A_1 + {}^{\mathrm{S}}n_2 A_2 = A \tag{2.55}$$

where A_1 and A_2 are the partial molar areas occupied at the surface by the two components and A is the specific surface area. However, since it is thought that adsorption is often, if not usually, as multilayers, no analysis can be attempted until an appropriate relation between ${}^{S}n_1$ and ${}^{S}n_2$ has been developed.

Adsorption from partially miscible liquids

The isotherm normally found for adsorption from a partially miscible system of non-electrolytes is usually S-shaped, resembling BET Type II classification of gaseous isotherms. The adsorption increases rapidly as the miscibility limit is approached, and it seems likely that this usually corresponds to multilayer formation, that is incipient phase separation under the influence of a solid surface.[113] Satisfactory theoretical descriptions of such isotherms are still lacking. The BET equation has been applied because of the analogous shape, with P/P_0 replaced by c/c_0 where c_0 is the concentration at the miscibility limit: there has been no attempt at an adequate theoretical justification of this procedure.

Adsorption of dissolved solids from solution

The majority of isotherms for the adsorption of uncharged dissolved solids from solution are of the Langmuir form. A complete shape classification has been attempted by Giles *et al.*[79] The Langmuir equation (2.51) with pressure replaced by concentration has been remarkably successful in fitting isotherms of adsorption from dilute solution. This is particularly the case where the competition between solute and solvent strongly favours the solute. Many teaching laboratories use the adsorption of ethanoic acid on charcoal from dilute aqueous solution to illustrate the non-rigorous application of the equation, usually successfully. However, the monolayer capacity evaluated by this means is often open to considerable doubt, and on solids of known specific surface has often been found to correspond to a completely unreasonable value of the area occupied by an adsorbed molecule. Therefore the treatment of experimental results has varied widely from system to system and it is difficult to generalise.

Certain isotherms show pronounced steps in their shape, which do not seem to be interpretable in terms of the formation of first and second layers in the majority of cases. These discontinuities have been attributed to phase changes in the adsorbed film, or to changes in orientation of the adsorbed molecules, and it is often impossible on the basis of the isotherm alone to make any statement as to the cause.

The adsorption of polymers

For most systems involving the adsorption of a polymer the adsorption isotherm either reaches some plateau or tends towards a limiting value, and usually multilayer adsorption is not thought to occur. Nevertheless, the amount of polymer adsorbed often exceeds that corresponding to a complete monolayer of monomer units; thus models postulating that polymer molecules always lie flat on the adsorbent surface would seem to be inappropriate, particularly since there is evidence that the adsorbed film is often several monomer units thick. This could be explained if the adsorbed molecules are attached at only a few points to the surface (anchor segments) and that the remainder of the molecule is effectively free from the surface.[114]

Molecules of high molecular weight are thought to exist in the form of random coils which are approximately spherical when the molecular weight is sufficiently high. These coils occupy a larger volume when the polymer is in a good solvent than when it is in a poor solvent. Simha, Frisch and Eirich[115-118] assume that when a flexible polymer is adsorbed from dilute solution, a localised monolayer is formed, but that it is extremely unlikely that each segment of the polymer chain will be attached to the surface. Normally there will only be a few anchor segments per molecule, the remainder of the polymer chain forming loops or bridges extending out into solution.

The adsorption of electrolytes

The adsorption of electrolytes from aqueous solution is the result of electrostatic forces existing between ions in solution and the surface charge on the solid.[119] Adsorption of this type may be considered in terms of electric double layer theory, which is the subject of Chapter 3. If the simple Stern[120] model of the double layer is accepted then the charge density, σ_1, in the Stern plane is given by

$$\sigma_1 = \frac{N_1 z_+ e}{1+\dfrac{N_0}{n_+ M}\, exp\dfrac{z_+ e\psi_\delta + \varphi_+}{kT}} - \frac{N_1 z_- e}{1+\dfrac{N_0}{n_- M}\, exp\dfrac{-z_- e\psi_\delta + \varphi_-}{kT}} \tag{2.56}$$

where N_1 is the available number of adsorption sites per cm^2, z_+ and z_- are the valencies of the respective ions having concentrations n_+ and n_-,

ψ_δ is the potential in the Stern plane, φ_+ and φ_- the specific adsorption potentials of the ions, M the molecular weight of the solvent, and k the Boltzmann constant. N_0 is the Avogadro number. A diagrammatic representation of the double layer is shown in Fig. 2.14.

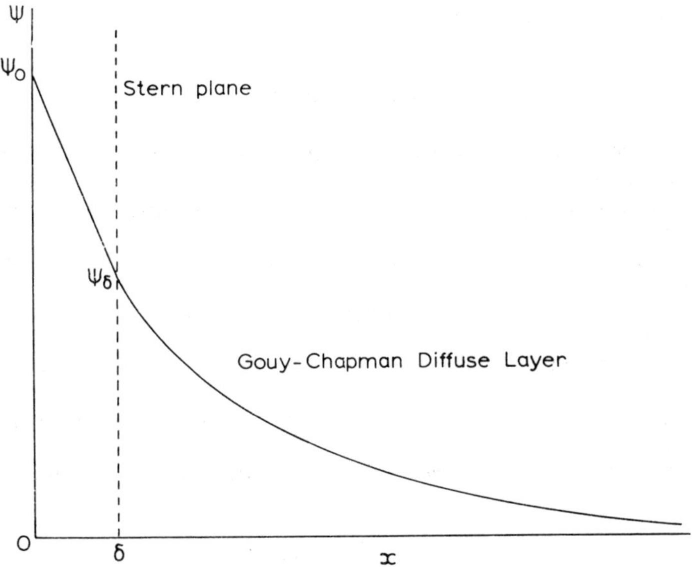

Fig. 2.14. Diagrammatic representation of the double layer (*Stern*).

Stern's model is essentially analogous to the Langmuir adsorption isotherm applied to both types of ions. Experimental methods have not reached a state where Stern's theory can be rigorously tested for the solid–liquid interface, although it has been applied a number of times.[121] It should be noted that even if the charge density in the Stern plane has a definite sign, there will normally be a very considerable number of ions of opposite sign also in the plane. This can perhaps be more readily appreciated if we look at the magnitude of $ze\psi_\delta$ for a potential of 100 mV, when for a univalent ion it corresponds only to 9·6 kJ/mol, which is less than the values usually ascribed to a hydrogen bond.

Work carried out on the mercury–solution interface has indicated that the Stern model is too simple, and it has been modified by Grahame.[122] Unfortunately the increased number of parameters involved make it difficult to apply to adsorption at the solid–liquid interface.

In conclusion one might say that the study of the solid–liquid interface represents a study of the forces existing between atoms, molecules and ions. Most of our experimental studies do not penetrate this far, being concerned mainly with the observable results of such forces. With simple academic systems the existing theories can be comparatively readily applied, but with far more complex practical systems the application is often more difficult and less obvious.

BIBLIOGRAPHY

M. J. JAYCOCK and G. D. PARFITT, *Chemistry of Interfaces*, Ellis Horwood, Chichester, 1981.

A. W. ADAMSON, *Physical Chemistry of Surfaces*, 3rd edition, Interscience, New York, 1976.

S. J. GREGG and K. S. W. SING, *Adsorption, Surface Area and Porosity*, Academic Press, London, 1967.

J. T. DAVIES and E. K. RIDEAL, *Interfacial Phenomena*, 2nd edition, Academic Press, London, 1963.

D. M. YOUNG and A. D. CROWELL, *Physical Adsorption of Gases*, Butterworths, London, 1962.

J. J. KIPLING, *Adsorption from Solutions of Non-Electrolytes*, Academic Press, London, 1965.

H. R. KRUYT, Ed., *Colloid Science*, Vol. 1, Elsevier, Amsterdam, 1952.

REFERENCES

1. J. O. Hirschfelder, C. F. Curtiss and R. B. Bird, *Molecular Theory of Gases and Liquids*, Wiley, New York, 1954.
2. E. A. Moelwyn-Hughes, *Physical Chemistry*, 2nd edition, Pergamon, Oxford, 1961.
3. D. M. Young and A. D. Crowell, *Physical Adsorption of Gases*, Butterworths, London, 1962, Chapter 2.
4. G. Mie, *Ann. Physik*, **11** (1903) 657.
5. J. E. Lennard-Jones in *Statistical Mechanics*, Ed. R. H. Fowler, 2nd edition, Cambridge U.P., Cambridge, 1936, Chapter 10.
6. J. Willard Gibbs, *The Collected Works of J. Willard Gibbs*, Longmans, Green, London, 1928.
7. E. A. Guggenheim, *Thermodynamics, an Advanced Treatment for Chemists and Physicists*, North-Holland, Amsterdam, 1949, p. 46.
8. T. Young, *Phil. Trans. Roy. Soc. (London)*, **95** (1805) 65, 82.
9. J. T. Davies and E. K. Rideal, *Interfacial Phenomena*, 2nd edition, Academic Press, New York, 1963, Chapter 1.
10. A. W. Adamson, *Physical Chemistry of Surfaces*, 3rd edition, Interscience, New York, 1976, Chapter 1.

11. W. D. Harkins, *The Physical Chemistry of Surface Films*, Reinhold, New York, 1952.
12. J. J. Bikerman, *Surface Chemistry*, 2nd edition, Academic Press, New York, 1958.
13. Ref. 10, Chapter 5.
14. H. Udin, A. J. Shaler and J. Wulff, *J. Metals*, **1** (1949) 186.
15. C. Herring, in *Structure and Properties of Solid Surfaces*, Eds. R. Gomer and C. S. Smith, University of Chicago Press, Chicago, 1953, p. 5.
16. B. Chalmers, R. King and R. Shuttleworth, *Proc. Roy. Soc.* (*London*), **A193** (1949) 465.
17. G. C. Kuczynski, *J. Metals*, **1** (1949) 96.
18. J. K. MacKenzie and R. Shuttleworth, *Proc. Roy. Soc.* (*London*), **B62** (1949) 833.
19. G. C. Kuczynski, *Acta Met.*, **4** (1956) 58.
20. G. F. Huttig, *Kolloid Z.*, **98** (1942) 263.
21. Ref. 10, p. 60.
22. H. N. Hersch, *J. Am. Chem. Soc.*, **75** (1953) 1529.
23. G. Cohen and G. C. Kuczynski, *J. Appl. Phys.*, **21** (1950) 1339.
24. R. Gomer, *Advan. Catalysis*, **7** (1955) 93.
25. E. Menzel, *Z. Physik*, **132** (1952) 508.
26. G. Beilby, *Aggregation and Flow in Solids*, Macmillan, New York, 1921.
27. R. C. French, *Proc. Roy. Soc.* (*London*), **A140** (1933) 637.
28. H. Raether, *Z. Physik*, **124** (1948) 286.
29. A. R. Bailey, *Rep. Progr. Appl. Chem.*, **34** (1954) 85.
30. T. S. Renzema, *J. Appl. Phys.*, **23** (1952) 1142.
31. W. Cochrane, *Proc. Roy. Soc.* (*London*), **A166** (1928) 228.
32. J. J. Trillat, *Compt. Rend.*, **224** (1947) 1102.
33. R. Shuttleworth, *Proc. Phys. Soc.* (*London*), **A63** (1950) 444.
34. S. Ross and J. P. Olivier, *On Physical Adsorption*, Interscience, New York, 1964, p. 8.
35. C. Sandford and S. Ross, *J. Phys. Chem.*, **58** (1954) 228.
36. C. Gurney, *Proc. Phys. Soc.* (*London*), **A62** (1949) 639.
37. M. M. Nicholson, *Proc. Roy. Soc.* (*London*), **A228** (1955) 490.
38. I. F. Guilliatt and N. H. Brett, *Trans. Faraday Soc.*, **65** (1969) 3328.
39. P. J. Anderson and A. Scholtz, *Trans. Faraday Soc.*, **64** (1968) 2972.
40. F. M. Fowkes in *Chemistry and Physics of Interfaces*, ACS, Washington, D.C., 1965, p. 1.
41. W. D. Harkins, *J. Chem. Phys.*, **10** (1942) 268.
42. R. Eötvös, *Wied Ann.*, **27** (1886) 456.
43. R. Shuttleworth, *Proc. Phys. Soc.* (*London*), **A62** (1949) 167; **A63** (1950) 447.
44. M. Born and W. Heisenberg, *Z. Physik*, **23** (1924) 338; M. Born and J. E. Mayer, *Z. Physik*, **75** (1932) 1.
45. J. E. Lennard-Jones and P. A. Taylor, *Proc. Roy. Soc.* (*London*), **A109** (1925) 476.
46. J. E. Lennard-Jones and B. M. Dent, *Proc. Roy. Soc.* (*London*), **A121** (1928) 247.

47. B. M. Dent, *Phil. Mag.*, **8** (1929) 530.
48. M. L. Huggins and J. E. Mayer, *J. Chem. Phys.*, **1** (1933) 643.
49. E. J. W. Verwey, *Rec. Trav. Chim.*, **65** (1946) 521.
50. G. C. Benson, H. P. Schreiber and D. Patterson, *Can. J. Phys.*, **34** (1956) 265.
51. G. C. Benson and R. McIntosh, *Can. J. Chem.*, **33** (1955) 1677.
52. G. C. Benson, P. I. Freeman and E. Dempsey, *Adv. Chem. Series No. 33*, ACS, Washington, 1961, p. 26.
53. B. M. E. van der Hoff and G. C. Benson, *Can. J. Phys.*, **32** (1954) 475.
54. H. P. Schreiber and G. C. Benson, *Can. J. Phys.*, **33** (1955) 534.
55. Ref. 3, Chapter 6.
56. S. Brunauer, *The Adsorption of Gases and Vapours*, Princeton University Press, Princeton, N.J., 1945.
57. S. Brunauer and P. H. Emmett, *J. Am. Chem. Soc.*, **57** (1935) 1754.
58. P. H. Emmett and S. Brunauer, *J. Am. Chem. Soc.*, **59** (1937) 1553.
59. S. Brunauer, P. H. Emmett and E. Teller, *J. Am. Chem. Soc.*, **60** (1938) 309; for errata see P. H. Emmett and T. W. DeWitt, *Ind. Eng. Chem.* (*Anal.*), **13** (1941) 28.
60. G. D. Halsey, *Disc. Faraday Soc.*, **8** (1950) 54.
61. R. A. Beebe, J. B. Beckwith and J. M. Honig, *J. Am. Chem. Soc.*, **67** (1945) 1554.
62. A. C. Zettlemoyer, A. Chand and E. Gamble, *J. Am. Chem. Soc.*, **72** (1950) 2752.
63. R. J. Johansen, P. B. Lorenz, C. G. Dodd, F. D. Pidgeon and J. W. Davis, *J. Phys. Chem.*, **57** (1953) 40.
64. R. T. Davis, T. W. DeWitt and P. H. Emmett, *J. Phys. Chem.*, **51** (1947) 1232.
65. H. L. Pickering and H. C. Eckstrom, *J. Am. Chem. Soc.*, **74** (1952) 4775.
66. R. A. W. Haul, *Angew. Chem.*, **68** (1956) 238.
67. S. Brunauer, L. S. Deming, W. E. Deming and E. Teller, *J. Am. Chem. Soc.*, **62** (1940) 1723.
68. M. J. Jaycock in *Particle Size Analysis*, Ed. M. J. Groves, Heyden, London, 1978, p. 308.
69. L. G. Joyner, E. B. Weinberger and C. W. Montgomery, *J. Am. Chem. Soc.*, **67** (1945) 2182.
70. K. S. W. Sing in *Characterisation of Powders*, Eds. G. D. Parfitt and K. S. W. Sing, Academic Press, London, 1976, p. 47.
71. W. D. Harkins and D. M. Gans, *J. Am. Chem. Soc.*, **53** (1931) 2804.
72. A. S. Russell and C. N. Cochran, *Ind. Eng. Chem.*, **42** (1950) 1332.
73. C. Orr, H. G. Blacker and S. L. Craig, *J. Metals*, **4** (1952) 657.
74. H. A. Smith and J. F. Fuzek, *J. Am. Chem. Soc.*, **68** (1946) 229.
75. F. Paneth and A. Radu, *Ber.*, **B57** (1924) 1221.
76. A. Clauss, H. P. Boehm and U. Hofman, *Z. Anorg. Allgem. Chem.*, **290** (1957) 35.
77. J. J. Kipling and R. B. Wilson, *J. Appl. Chem.* (*London*), **10** (1960) 109.
78. D. Graham, *J. Phys. Chem.*, **59** (1955) 896.
79. C. H. Giles, T. H. MacEwan, S. N. Nakhwa and D. Smith, *J. Chem. Soc.* (1960) 3973.

80. C. H. Giles and S. N. Nakhwa, *J. Appl. Chem. (London)*, **12** (1962) 266; C. H. Giles, A. P. D'Silva and A. S. Trivedi in *Proc. Int. Symposium on Surface Area Determination*, Eds. D. H. Everett and R. H. Ottewill, Butterworths, London, 1970, p. 317.
81. H. N. V. Temperley, *Proc. Cambridge Phil. Soc.*, **48** (1952) 683.
82. I. Stranski, *Z. Physik. Chem.*, **136** (1928) 259.
83. M. L. Dundon and E. Mack, Jr., *J. Am. Chem. Soc.*, **45** (1923) 2479; M. L. Dundon, *J. Am. Chem. Soc.*, **45** (1922) 2658.
84. M. M. Dubinin, *Chemistry and Physics of Carbon*, Ed. P. L. Walker, **2** (1966) 51.
85. P. H. Emmett, *Catalysis*, **1** (1954) 31; H. Mark and E. J. W. Verwey, *Advan. Colloid Sci.*, **3** (1950) 1.
86. H. L. Ritter and L. C. Drake, *Ind. Eng. Chem. (Anal.)*, **17** (1945) 782, 787; L. C. Drake, *Ind. Eng. Chem.*, **41** (1949) 780.
87. J. H. de Boer, B. C. Lippens, B. G. Linsen, J. C. P. Broekhoff, A. van der Heuvel and Th. J. Osinga, *J. Colloid Interface Sci.*, **21** (1966) 405.
88. K. S. W. Sing, *Proc. Int. Symposium Surface Area Determination*, Butterworths, London, 1970, p. 25.
89. G. D. Parfitt and C. H. Rochester in *Characterisation of Powder Surfaces*, Eds. G. D. Parfitt and K. S. W. Sing, Academic Press, London, 1976, Chapter 2.
90. A. Dupré, *Theorie Mechanique de la Chaleur*, Paris, 1869, p. 369.
91. F. M. Fowkes, *Ind. Eng. Chem.*, **56** (1964) 40.
92. A. L. Bailey, *Proc. 2nd Int. Congress on Surface Activity (London)*, **3** (1957) 189.
93. B. R. Ray and F. E. Bartell, *J. Colloid Sci.*, **8** (1953) 214.
94. R. N. Wenzel, *Ind. Eng. Chem. (Anal.)*, **28** (1936) 988.
95. G. Macdougall and C. Ockrent, *Proc. Roy. Soc. (London)*, **A180** (1942) 151.
96. B. V. Derjaguin, *Proc. 2nd Int. Congress on Surface Activity (London)*, **3** (1957) 446.
97. Ref. 10, p. 270.
98. W. D. Harkins and E. H. Loeser, *J. Chem. Phys.*, **18** (1950) 556.
99. E. W. Washburn and E. A. Anderson, *J. Phys. Chem.*, **50** (1946) 401.
100. W. D. Harkins and H. K. Livingston, *J. Chem. Phys.*, **10** (1942) 342.
101. N. K. Adam and H. K. Livingston, *Nature*, **182** (1958) 128.
102. A. M. Gaudin, *Flotation*, McGraw-Hill, New York, 1957.
103. K. L. Sutherland and I. W. Wark, *Principles of Flotation*, Aus. Inst. M. M., Melbourne, 1955.
104. Ref. 9, Chapter 8.
105. A. B. D. Cassie and S. Baxter, *Trans. Faraday Soc.*, **40** (1944) 546.
106. F. P. Bowden and D. Tabor, *The Friction and Lubrication of Solids*, Oxford University Press, Oxford, 1954.
107. D. D. Eley, Editor, *Adhesives*, Oxford University Press, Oxford, 1961.
108. J. J. Kipling, *Adsorption from Solutions of Non-Electrolytes*, Academic Press, London, 1965, Chapters 1–8.
109. D. H. Everett, *Trans. Faraday Soc.*, **60** (1964) 1803.
110. D. H. Everett, *Trans. Faraday Soc.*, **61** (1965) 2478.
111. R. S. Hansen and R. D. Hansen, *J. Colloid Sci.*, **9** (1954) 1.

112. D. C. Jones and G. S. Mill, *J. Chem. Soc.* (1957) 213.
113. R. S. Hansen, Y. Fu and F. E. Bartell, *J. Phys. Chem.*, **53** (1949) 769.
114. E. Jenckel and B. Rumbach, *Z. Elektrochem.*, **55** (1951) 612.
115. R. Simha, H. L. Frisch and F. R. Eirich, *J. Phys. Chem.*, **57** (1953) 584.
116. H. L. Frisch and R. Simha, *J. Phys. Chem.*, **58** (1954) 507.
117. H. L. Frisch, *J. Phys. Chem.*, **59** (1955) 633.
118. H. L. Frisch and R. Simha, *J. Chem. Phys.*, **27** (1957) 502.
119. H. R. Kruyt, Ed., *Colloid Science*, Elsevier, Amsterdam, 1952, Vol. 1, Chapter 4.
120. O. Stern, *Z. Elektrochem.*, **30** (1924) 508.
121. M. B. Abramson, M. J. Jaycock and R. H. Ottewill, *J. Chem. Soc.* (1964) 5034, 5041.
122. D. C. Grahame, *Chem. Rev.*, **41** (1947) 441.

CHAPTER 3

ELECTRICAL PHENOMENA ASSOCIATED WITH THE SOLID–LIQUID INTERFACE

A. L. SMITH
Unilever Limited, Wirral, Merseyside, UK

INTRODUCTION

The state of electric charge of the particles of a colloidal dispersion is always an important factor governing the stability to flocculation. In the case of particles which do not greatly influence the dispersion medium in their vicinity and in the absence of adsorbed material of high molecular weight, it is the dominant factor. The correlation between the magnitude of the electrokinetic (zeta) potential and thc stability of a lyophobic dispersion has long been recognised and resulted in the approximate working rule[1] that a zeta potential of at least 30 mV is necessary for long-term sol stability. While any such universal rule cannot stand up to close scrutiny it illustrates the importance of an understanding of the relationship between interparticle repulsion and the various potentials which it is possible to define, though not necessarily directly measure, at or near the particle surface.

The whole of the region in the vicinity of the solid–solution interface in which any discontinuity or change in mean potentials or electrical properties occurs is usually termed the electric double layer. It will be the purpose of this chapter to discuss the origin and structure of this double layer, how far it is possible to obtain quantitative information about it, and particularly the technique of microelectrophoresis.

THE ELECTRIC DOUBLE LAYER

Origin of the electric double layer

The redistribution of charge which is implied by the formation of an electric double layer when electrically neutral particles are placed in a solution which is itself (overall) electrically neutral will be governed by the following factors.

(i) The dissociation of any ionogenic groups present in the particle surface, e.g. —COOH groups which would give the particle a negative charge at high pH.

(ii) The unequal dissolution of oppositely charged ions of which the particle may be composed. An example of this is the negative charge remaining on silver halides when suspended in water.

(iii) The adsorption by the particle of one of the ion-types in the solution in which it is suspended, or alternatively the unequal adsorption of oppositely charged ions. Adsorption must be understood generally here to include negative adsorption.

(iv) The adsorption and/or orientation of dipolar molecules at the particle surface. It is evident that such dipoles will not directly contribute to the net charge of the particle but they may have an important effect on the electric double layer as demonstrated below. The dipoles may be the result of the deformation of polarisable molecules in the electric field at the interface.

The disperse particles as normally encountered are more often negatively charged than positively. This is partly because, in aqueous solution, anions are usually less strongly hydrated, and therefore more readily specifically adsorbed, than cations. That this is not the whole explanation, however, is shown by the observation that air bubbles in most aqueous solutions migrate towards the anode when an electric field is applied, despite the fact that surface tension data, interpreted by the Gibbs equation, show that in these solutions there is an overall negative adsorption of electrolyte. Evidently the negative adsorption of cations is more pronounced than that of anions in these cases.

The case of suspended air bubbles illustrates a possible ambiguity in statements about the charge of a particle. The air bubble itself, if by this it is meant to exclude any part of the solution or any ion partly in the solution, is electrically neutral; any observation of a negative electrophoretic mobility indicates that what may be called the electrokinetic unit is negatively charged. The electrokinetic unit will include some solvent and ions close to the surface and may well have a net electric charge. In an extreme case these two interpretations of the charge in the particle may even give rise to opposite signs, e.g. AgI particles in a solution containing a slight excess of I^- ions and sufficient cationic surfactant. This should be made clearer by a more detailed discussion of the inner part of the double layer later in this chapter.

The nature and definition of electric potentials in the double layer

In any attempt to discuss the electric potential difference between a point in the solution and a point within a suspended particle the same difficulty arises as is familiar in treatments of electrode potentials. It was pointed out, not first but perhaps most clearly, by Guggenheim[2] that such a potential difference is not accessible to direct measurement. Since disembodied electric charge does not exist, the work done in transporting unit electrical charge from a point in the solution to a point within the particle will necessarily include a contribution from the difference between what may be called the chemical potentials of the charge-carrying particle in the two situations. The latter contribution may be looked upon as arising from the change in the short range interactions of the charged particle with its surroundings.

It is not possible to separate unambiguously the electrostatic and chemical contributions to the total work since all the interactions involved are fundamentally electrical in nature. The total work done in the transport of unit electric charge associated with some body, e.g. an electron, from a point in one phase to a point in another is an unambiguous quantity, termed by Guggenheim the *electrochemical potential difference* $\Delta\eta$.

Despite the considerations outlined above it is nevertheless still useful, especially for theoretical purposes, to construct a model in which the components of $\Delta\eta$ are supposed to be separated. The model outlined here is that due to Lange (1933).

In Fig. 3.1 a charged spherical particle, considered initially to be situated in a vacuum, is imagined to be separated into (a) a sphere of dimensions equal to that of the real particle, homogeneous and of permittivity ϵ but with no net charge and no oriented dipoles in the interface and (b) the excess charge and oriented dipoles concentrated into a thin shell of radius equal to that of the real particle.

In case (a) the work done in taking a species i of charge q_i from a point A distant from the particle to a point B within it is termed μ_i the chemical potential. This is not, as might first appear, independent of q_i

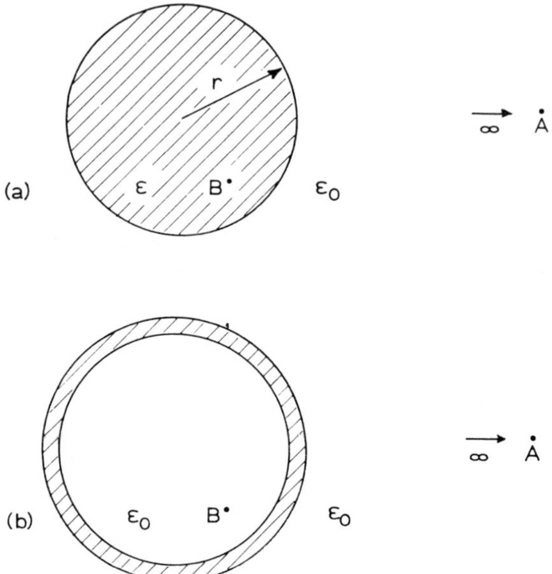

Fig. 3.1 Model for a charged spherical particle in a vacuum. The particle is separated into (a) a homogeneous sphere of permittivity ϵ with no net charge and no oriented dipoles, and (b) a hollow shell carrying the net charge and oriented dipoles.

since if the species i is taken as a sphere of radius a, it will include a term

$$\frac{q_i^2}{2\epsilon a} - \frac{q_i^2}{2\epsilon_0 a} \quad \text{or} \quad \frac{q_i^2}{2a}\left[\frac{1}{\epsilon} - \frac{1}{\epsilon_0}\right] \tag{3.1}$$

which vanishes if $q_i = 0$. Thus the term chemical potential for μ_i is not ideal but is, nevertheless, usually employed.

In case (b) the work done in taking the charge q_i from A to B is given by $q_i\varphi$ where φ is the *inner* potential (Lange) of the particle phase. Thus

$$\eta_i = \mu_i + q_i\varphi \tag{3.2}$$

The inner potential φ may conveniently be subdivided into a part ψ due to the net charge on the shell and a part χ due to oriented dipoles in the shell. Thus

$$\varphi = \psi + \chi \tag{3.3}$$

Although the φ potential cannot, as explained above, be directly measured, the ψ potential is accessible to experimental measurement (at least in principle) and for an isolated spherical particle of radius r with excess charge q will be given by $q/\epsilon_0 r$. ψ is termed the *outer* potential (Lange) and may be looked upon as the electric potential just outside the particle surface. In this respect ψ_0 is to be seen as a mean potential which can, for example, be put into electrical capacity relations but is not the potential experienced by an ion just outside the surface and therefore should not appear in adsorption isotherm relations. The *surface* or *wall* potential which occurs in discussion of colloid stability, etc., is to be identified with this potential.

Although the χ potential was termed by Lange the surface potential, this term is frequently used for ψ and it is perhaps better to refer to the *ki potential.* For a shell consisting of n dipoles per unit area, each of effective moment μ, with positive poles directed inwards, i.e. towards the particle, then if potentials are reckoned relative to point A (Fig. 3.1), the dipole contribution to φ will be given by

$$\chi = 4\pi n\mu/\epsilon_0 \tag{3.4}$$

where ϵ_0 is the (unrationalised) permittivity of free space.

It must be noticed that for typical dipole layers the effective moment μ may be very much smaller than the gas phase dipole moment of the molecule concerned. This arises because

(a) the dipoles may not be perpendicular to the surface,
(b) the dipoles will tend to point in different directions, and
(c) imaging in the solid surface may produce a depolarising field.

Of the quantities occurring in eqns. (3.2) and (3.3) it is seen, therefore, that η_i and ψ are accessible to direct measurement whereas μ_i, φ and χ are not. Experiments can, however, give some information on the latter quantities, particularly changes in them, and they can be the subject of a theoretical calculation.

If it is now supposed that the particle is situated not in a vacuum but in some other medium, e.g. a solution, eqns. (3.2) and (3.3) become respectively

$$\Delta\eta_i = \Delta\mu_i + q_i\Delta\varphi \tag{3.5}$$

$$\Delta\varphi = \Delta\psi + \Delta\chi \tag{3.6}$$

Here $\Delta\varphi$, the difference in inner potentials of the two phases, is termed the Galvani potential difference and $\Delta\psi$, the difference of the outer potentials, is termed the Volta or contact potential difference. It must not be assumed, however, that $\Delta\chi$ is the difference between the two separate χ potentials of the free surfaces (to a vacuum) of particle and solution, and this factor, together with the obvious and universal practice of reckoning potentials of particles relative to bulk solution far from any particle, makes it desirable to write eqns. (3.5) and (3.6) as

$$\eta_i = \mu_i + q_i\varphi$$

and

$$\varphi = \psi + \chi$$

i.e. identical in form to eqns. (3.2) and (3.3) though with slightly adjusted significance. Symbolism such as $\Delta\chi$ can now be used to indicate a change in the χ potential at the particle–solution interface. Some authors prefer to retain χ for the air–solution case and use the symbol g_{dip} for the interphase situation. Such nomenclature emphasises the non-additivity of χ values as between different phases.

A more thorough treatment of the matters raised in this section has been given by Parsons.[3]

Potential determining ions and the Nernst equation

Consider a silver iodide particle suspended in a solution containing (inevitably) silver and iodide ions. An exchange will take place between the silver and halide ions in the particle and those in solution until at (a postulated) equilibrium the electrochemical potential η of each ion is the same both in the solid and in solution.

For the Ag^+ ion with a charge equal to e the electronic charge:

$$\eta_{Ag^+}(\text{solid}) = \eta_{Ag^+}(\text{solution})$$

$$\mu_{Ag^+}(\text{solid}) + e\varphi = \mu^0_{Ag^+}(\text{solution}) + kT \ln a_{Ag^+}(\text{solution})$$

$$[\varphi(\text{solution}) = 0 \text{ by convention}]$$

so that

$$\varphi = \text{constant} + \frac{kT}{e} \ln a_{Ag^+}$$

which is a form of the Nernst equation.

At 25°C

$$\varphi = \text{constant} - 59{\cdot}2(p\text{Ag})\,\text{mV}$$

where pAg is used to indicate $-\log_{10} a_{Ag^+}$

or

$$\Delta\varphi = -59{\cdot}2\Delta(p\text{Ag})\,\text{mV} \tag{3.7}$$

The corresponding equation for the iodide ion will be

$$\Delta\varphi = +59{\cdot}2\Delta(p\text{I})\,\text{mV}$$

Thus for a 10-fold change in the silver (or iodide) ion activity (which in dilute solution may be taken as equal to the concentration) the φ potential changes in the appropriate sense by 59·2 mV. If the χ potential is taken as constant, and it must be recognised that this is an assumption, then the ψ potential changes by 59·2 mV also. The charge transfer necessary to achieve this will depend on the capacity of the electric double layer which is discussed later in this chapter.

In this and similar cases the ions concerned are termed the *potential determining ions*. It is not appropriate to regard such ions as inhabiting the *Inner Helmholtz Plane* (see below) since, after adsorption, they are indistinguishable from the ions which constitute the bulk of the particle.

In the case of silver halide sols the potential determining ions are particularly well defined and this is one reason for their historical popularity in studies of the electric double layer and its influence on colloid stability.

In the case of metal oxide and hydroxide sols and in sols of biological interest H^+ and OH^- can be regarded to some extent as potential determining ions. For gold sols even the Cl^- ion, which forms stable complexes at the particle surface, has been regarded as a potential determining ion, though this is not a usage to be recommended.

Applicability of the Nernst equation to surfaces

In the above derivation of the Nernst equation, applied to the AgI-solution case in order to illustrate the concept of potential determining ions, it was assumed that a state of equilibrium exists between the bulk of the solid and the surface. It is by no means self-evident that this is justified even for the apparently straightforward silver halide situation. For the oxide–solution interface and other cases where the H^+ and OH^- ions are seen as potential determining, the difficulty is even more apparent since neither H^+ nor OH^- may exist at all in the bulk solid.

Clearly a more generalised treatment of the Nernst equation is necessary in which the idealised ionic solid, with or without internal equilibrium, is a limiting case. In the following an H^+/OH^- potential determining mechanism in an aqueous system will be used as illustration.

Suppose that the mechanism by which the surface gains a negative charge is

$$AH \rightarrow A^- + H^+$$

and that by which it becomes positively charged is

$$AH + H^+ \rightarrow AH_2^+$$

where A represents some surface site and H^+ ions are present (and at equilibrium) in solution. The negative-going mechanism, represented as the loss of a proton, might equally be the addition of an OH^- ion so far as the following is concerned. In terms of electrochemical potentials the two mechanisms give respectively

$$\mu_{AH}(s) = \eta_{A^-}(s) + \mu_{H^+} \tag{3.8}$$

and

$$\mu_{AH}(s) = \eta_{AH_2^+}(s) - \mu_{H^+} \tag{3.9}$$

where (s) indicates the surface and η has been used in place of μ only where these differ. $\eta_{H^+}(s)$ is clearly equal to μ_{H^+} in bulk solution.

The total, random, number of configurations g available to the three types of site, viz. positively charged (+), negatively charged (−) and uncharged (0) will be given by

$$g = \frac{N_s!}{n_+!n_-!n_0!}$$

where N_s is the total number of sites per unit area. Thus the configurational part of the (surface) chemical potential of, for example, the positive sites will be given by:

$$-kT\frac{\mathrm{d}\ln g}{\mathrm{d}n_+} = +kT\ln\frac{\theta_+}{1-\theta_+-\theta_-}$$

where $\theta = n/N_s$.

The remaining part of μ will be taken, in this approximate treatment, as independent of θ and can therefore be written μ with an appropriate implied standard state.

The electrical part of η should involve the micro-potential at the site concerned but here the mean surface (wall) potential ψ_0 will be used giving $e_0\psi_0$ for the positive site and $-e_0\psi_0$ for the negative site.

Elimination of $\mu_{AH}(s)$ between (3.8) and (3.9) then gives

$$\frac{(\mu^0_{\mathrm{AH}_2^+}(\mathrm{s}) - \mu^0_{\mathrm{A}^-}(\mathrm{s}) - 2\mu^{0+}_{\mathrm{H}})}{kT} = 2 \ln a_{\mathrm{H}^+} - \ln \frac{\theta_+}{\theta_-} - \frac{2e_0\psi_0}{kT} \tag{3.10}$$

where the LHS of eqn. (3.10) has the form of the ratio of two 'equilibrium constants' representing the tendencies of the surface sites to become negatively charged via eqn. (3.8) or positively charged via eqn. (3.9), termed K_- and K_+ respectively. Equation (3.10) now becomes

$$\ln \frac{K_+}{K_-} = 4{\cdot}606\,\mathrm{pH} + \ln \frac{\theta_+}{\theta_-} + \frac{2e_0\psi_0}{kT} \tag{3.11}$$

Elimination of μ_{H^+} between eqns. (3.8) and (3.9) yields

$$K_+K_- = \frac{\theta_+\theta_-}{\theta_0^2} \tag{3.12}$$

At the zero point of charge θ_+ and θ_- will take some common value θ_c and, in the absence of specific adsorption of other than H^+ and OH^- ions, ψ_0 will be zero. Equations (3.11) and (3.12) then become

$$\log \frac{K_+}{K_-} = 2\,\mathrm{pH}_0 \tag{3.13}$$

and

$$K_+K_- = \frac{\theta_c^2}{(1-2\theta_c)^2} \tag{3.14}$$

respectively. At $\mathrm{pH} \gg \mathrm{pH}_0$ where θ_+ will be negligible eqn. (3.11) becomes

$$-\ln K_- = 2{\cdot}303\,\mathrm{pH} - \ln \frac{\theta_-}{1-\theta_-} + \frac{e_0\psi_0}{kT} \tag{3.15}$$

If K_- is known or estimated, from the analysis of potentiometric titrations or otherwise, then K_+ can be obtained from an experimental pH_0 and eqn. (3.13). Equations (3.11) and (3.12) then allow deviations from an ideal Nernst equation (3.7) to be estimated.

As an example consider a surface only slightly removed from the zero point with $\theta_+ = \theta_- + \delta$; then

$$\ln \frac{\theta_+}{\theta_-} \simeq \frac{(\theta_+ - \theta_-)}{\theta_-} - \ldots \simeq \frac{(\theta_+ - \theta_-)}{\theta_c}$$

If this term is negligible, i.e. $(\theta_+ - \theta_-) \ll \theta_c$, then eqn. (3.11) reduces to an ideal Nernst equation relating ψ_0 to ΔpH. Deviations from the Nernst equation are seen to be most serious when θ_c is very small as would happen when K_+ or K_- or both are small. The zero point is then characterised by the virtual absence of charge rather than by the presence of equal numbers of positively and negatively charged sites.

At the other extreme θ_c takes its maximum value (0·5) for a surface such as the silver halides discussed above, and the Nernst equation is obeyed even at the zero point.

It will be noticed that the potential involved in this (very approximate) treatment is ψ_0 rather than the inner potential φ. The implication of this is that any potential drop within the solid would not be 'seen' from the solution side in a case such as the silver halides whether there was equilibrium between the bulk solid and the surface or not.[4]

In the case of oxide surfaces where the electrical properties are commonly observed to vary slowly with time it seems reasonable to attribute such drifts to solid state effects. Provided such drifts are slow the above treatment can be applied 'instantaneously' and the slow changes can be seen as drifts of the zero point.

Adsorption of non-potential determining ions and the inner part of the double layer

It is clear that the finite volume of ions places a limit on the closeness with which the centre of an ion in solution can approach a solid surface. To take account of this Stern[5] (1924) proposed a model for the inner part of the double layer in which the ions were taken as point charges in all respects except their inability to approach closer to the solid surface than some distance *d*. Assuming for a moment a flat interface, this plane, later termed the *Stern Plane*, was at first taken to indicate not only the closest distance of approach but also the centres of any ions specifically adsorbed on to the solid surface.

Specific adsorption is often defined as adsorption which takes place (even) at the zero point of charge of the adsorbent surface, but a more general concept is adsorption which depends on the nature, rather than merely the charge, of the ion.

The Stern model for the inner part of the double layer was refined by Grahame[6] who distinguished between the Stern Plane, also called the *Outer Helmholtz Plane* (OHP) to indicate the closest distance of approach of hydrated ions in solution, and an *Inner Helmholtz Plane* (IHP) to indicate the centres of ions (if any) which are specifically adsorbed on to the solid surface.

A distinction between the two planes is in general necessary since even though the specifically adsorbed ions may be the same size as (or the same ions as) the ions governing the position of the OHP they will probably be dehydrated, at least in the direction of the surface, allowing closer approach to the surface than the OHP. Another factor mentioned by Grahame is that the cations and anions in solution will not be the same size. In aqueous solution simple inorganic cations are more strongly hydrated than corresponding anions leading to a large hydrated radius for the cations and more ready specific adsorption of anions.

It must be remembered that Fig. 3.2 represents only one model, probably as simple as possible, for the inner part of the double layer which will need to be modified in special cases such as the adsorption of large surfactant ions where small counterions may approach closer to the surface than the specifically adsorbed ions. In the following treatment the model represented by Fig. 3.2 will be used to illustrate typical procedures.

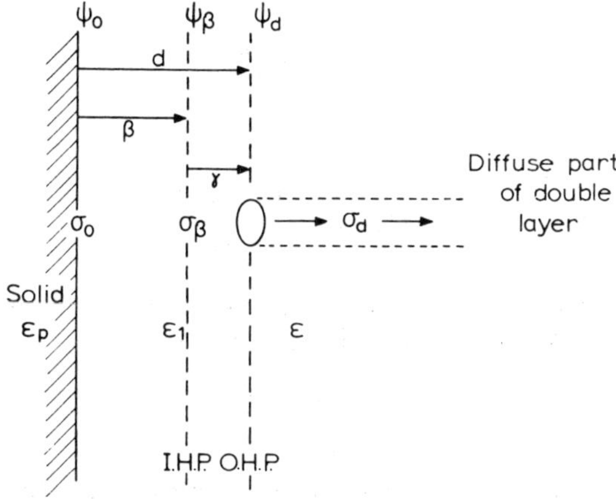

Fig. 3.2. Model for the electric double layer at the solid–solution interface showing potentials ψ and charge densities σ.

The potential referred to as ψ above becomes ψ_0 in Fig. 3.2 so that ψ_0 represents the mean potential just outside the solid surface and will be called the *wall potential.* ψ_β and ψ_d, the mean potentials at the IHP and OHP respectively are, of course, also Volta potentials accessible (in principle) to experimental measurement. σ_0 and σ_β are the charges per unit area of the particle surface and IHP respectively; σ_β will be zero in the absence of specific adsorption. σ_d is the net charge contained in a cylinder of unit cross-sectional area extending outwards from the OHP (theoretically to infinity).

The effective permittivity ϵ_1 of the inner region of the double layer has not been assumed equal to that of bulk solution ϵ. Available evidence shows that in aqueous solution the dielectric constant (ϵ/ϵ_0) near both the mercury and silver iodide interface at 25°C is only of the order of 10 compared with 78 in bulk solution. This is to be expected in view of the special state of solvent dipoles near the interface.

The condition of overall electrical neutrality gives

$$\sigma_0+\sigma_\beta+\sigma_d=0 \tag{3.16}$$

The capacity per unit area of the whole of the inner region K is given by

$$K=\frac{\epsilon_1}{4\pi d} \tag{3.17}$$

so that

$$\psi_0-\psi_\beta=\frac{\sigma_0\beta}{Kd} \tag{3.18}$$

and

$$\psi_\beta-\psi_d=\frac{(\sigma_\beta+\sigma_0)\gamma}{Kd} \tag{3.19}$$

If it is not wished to assume that ϵ_1 is uniform over the inner region then Kd/β can be written K_β and Kd/γ as K_γ so that

$$\frac{1}{K}=\frac{1}{K_\beta}+\frac{1}{K_\gamma} \tag{3.20}$$

and the quantities d, β and γ lose their significance as simple distances.

If the electrolyte in solution around the particles is assumed to be of symmetrical $z{:}z$ valency type with n ions of each type per unit volume, the Gouy–Chapman theory (below) gives

$$\sigma_d=-\frac{\epsilon kT\kappa}{2\pi ze}\sinh\frac{ze\psi_d}{2kT} \tag{3.21}$$

where κ is the Debye–Hückel parameter defined by

$$\kappa^2=\frac{4\pi e^2\sum nz^2}{\epsilon kT}$$

which, with the assumption above, becomes

$$\kappa^2 = \frac{8\pi e^2 n z^2}{\epsilon k T} \tag{3.22}$$

Combining eqns. (3.18) and (3.19)

$$\psi_0 - \psi_d = \frac{\sigma_0}{K} + \frac{\sigma_\beta \gamma}{Kd} \tag{3.23}$$

If the *zero point of charge* (*pzc*) is defined as the condition when $\sigma_0 = 0$, as at the mercury–solution interface, then at the *pzc*

from eqn. (3.18)

$$\psi_0 = \psi_\beta$$

from eqn. (3.23)

$$\psi_0 - \psi_d = \frac{\sigma_\beta \gamma}{Kd}$$

and from eqn. (3.16)

$$\sigma_\beta + \sigma_d = 0 \tag{3.24}$$

Figure 3.3 shows the possible situation at the *pzc* with σ_β negative. Only in the absence of specific adsorption, i.e. with $\sigma_\beta = 0$, does it follow that

$$\psi_0 = \psi_\beta = \psi_d = 0$$

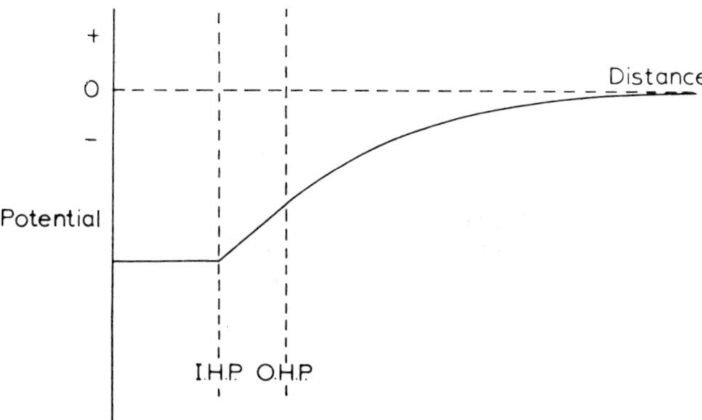

Fig. 3.3. Idealised variation of electric potential with distance into a solution for a surface at its pzc but with specific adsorption of anions into the Inner Helmholtz Plane.

The electrokinetic or zeta potential ζ is presumably to be identified either with ψ_d or the potential some distance further out into the diffuse part of the double layer. At low potentials $\zeta \approx \psi_d$ and certainly when $\zeta = 0$ then $\psi_d = 0$ so that the zero point of electrophoretic mobility corresponds to $\psi_d = 0$, not to the *pzc* as defined above, though the two conditions are sometimes confused. When $\psi_d = 0$ then from eqn. (3.21) $\sigma_d = 0$ so that $\sigma_0 = -\sigma_\beta$; from eqn. (3.19) $\psi_\beta = \psi_d$ $(=0)$ and from eqn. (3.18)

$$\psi_0 = \frac{\sigma_0 \beta}{Kd}$$

Figure 3.4 shows a possible situation at the zero point of electrophoretic mobility with σ_β negative.

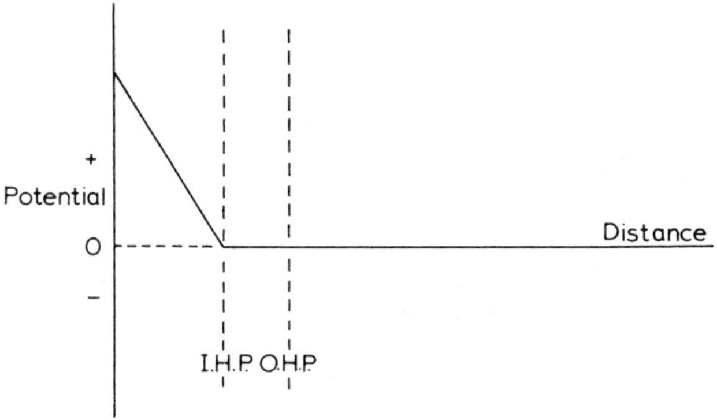

Fig. 3.4. Variation of electric potential with distance into solution for a surface at the zero point of electrokinetic potential but with specific adsorption of anions.

The zero point of electrophoretic mobility termed the isoelectric point, does, of course, correspond to the zero point of total charge within the OHP and, being a directly measurable quantity, is perhaps a more convenient parameter than the *pzc* defined as $\sigma_0 = 0$, especially for systems without well defined potential determining ions. In the absence of specific adsorption the two zero points coincide.

Adsorption of dipolar molecules

It might be thought that since an adsorbed neutral dipolar molecule cannot contribute to the net charge of a particle it will have no effect on the electric double layer, except perhaps to displace it to slightly greater distances from the surface, and no effect on observed quantities such as the electrophoretic mobility. This, however, is not necessarily the case.

Consider a particle suspended in a certain (constant) concentration of potential determining ions. Thus, according to the Nernst equation, the φ potential is fixed and

$$\varphi = \psi_0 + \chi$$

where χ is due to the orientation of dipoles in the particle surface. Suppose now that neutral dipolar molecules are introduced into the solution and adsorbed at the particle solution interface with their positive poles towards the solid. This introduces a positive contribution to the χ potential which will be written $\Delta\chi$. If it is assumed that the activity of the potential determining ions is unchanged at the (perhaps small) concentration of neutral molecules used, then φ is unaltered and

$$\Delta\psi_0 = -\Delta\chi$$

Thus ψ_0 will move to more negative values and with it, provided that $|\psi_0|$ was not already large, ψ_d and the electrokinetic potential. If the particles were at the *pzc* (specific adsorption of ions may be temporarily ignored) before the dipolar adsorption, then to restore the *pzc* after this adsorption a higher concentration of positively charged potential determining ions will be needed. In a similar way, at the mercury–solution interface the same adsorption would require more positive polarisation of the mercury to restore the electrocapillary maximum.

These considerations indicate that non-ionic stabilising or sensitising agents, for example, do not necessarily act solely by non-ionic mechanisms. In any attempt to relate $\Delta\chi$ to the orientation and adsorbed density of dipoles it must be remembered that in addition to the factors influencing the effective dipole moment in any adsorbed layer, the dipole moment and orientation of any displaced solvent layer must be considered. The latter could even predominate.

As has been noted above, it is often assumed that the χ potential remains constant during variations of surface charge and potential. That this is not universally justifiable is seen by considering the case where dipolar molecules are adsorbed on a surface at the *pzc* with the dipoles oriented parallel to the surface. As the surface charge is increased (say positively) it is likely that the adsorbed dipole direction will rotate somewhat to give a negative $\Delta\chi$.

The Stern adsorption isotherm and the discreteness of charge effect

The charge in the IHP is not smeared out but consists of discrete ionic charges. When an ion from solution is adsorbed into the IHP the discrete charges will be rearranged and this must be taken into account when writing the electrostatic work of adsorption. The rearrangement will impose a *self-atmosphere* potential on the adsorbed ion which is the two-dimensional analogue of the three-dimensional self-atmosphere potential occurring in the Debye–Hückel theory of electrolytes. This effect, which has been termed the *discreteness of charge* effect or *discrete ion* effect, was introduced by Esin and Shikov[7] after an original suggestion by Frumkin,[8] to explain the Esin–Markov[9] effect at the mercury–solution interface.

The importance of the effect in colloid stability theory has been stressed by Levine, Bell and collaborators and a review[10] has appeared. The main consequence of the theory to colloid stability is that it predicts that under suitable conditions the OHP potential $|\psi_d|$ can reach a maximum as $|\psi_0|$ is increased so that further increase in $|\psi_0|$ leads to a decrease in $|\psi_d|$ and therefore to a decrease in interparticle repulsion. Note, however, that the frequently observed decrease of electrophoretic mobility as (the implied) $|\psi_0|$ is increased may have a quite different origin (see p. 145).

For the adsorption into the IHP of a cation of charge ze, the electrostatic work of adsorption will be not $ze\psi_\beta$ but

$$ze\psi_A = ze(\psi_\beta + \varphi_\beta) \tag{3.25}$$

where φ_β is the self-atmosphere potential referred to above and ψ_A is the *micro-potential* in the *hole* which rearrangement of other adsorbed ions has provided. In the treatment of Ershler[11] the adsorbed ions are assumed to form a hexagonal array, but the more convenient method of Grahame[12] will be followed here in which an adsorbed ion is regarded as at the centre of a circular area in the IHP of radius r_0 such that $\pi r^2 \sigma_\beta = ze$ from which the uniform charge density has been removed. φ_β is now the potential at the centre of this disc due to a uniform charge density of $-\sigma_\beta$ over its whole area. The calculation of φ_β on this model is in general a matter of some complexity[13] involving multiple electrical images of the disc in the particle wall and OHP, but can be simplified if certain assumptions are acceptable.

If $\epsilon \gg \epsilon_1$, and $\epsilon_p \gg \epsilon_1$, which are reasonable approximations at the mercury–solution interface, and if $r_0 \gg d$, i.e. low charge density in the IHP, then φ_β becomes the potential of an infinite plane of charge density

$-\sigma_\beta$ between two earthed conducting plates, and simple electrostatics gives

$$\varphi_\beta = -\frac{\beta\gamma\sigma_\beta}{Kd^2} \tag{3.26}$$

If $\epsilon \gg \epsilon_1$ and $\epsilon_p = \epsilon_1$, a reasonable approximation at the interface between a dielectric and aqueous solution, and as before $r_0 \gg d$, then

$$\varphi_\beta = -\frac{\gamma\sigma_\beta}{Kd} \tag{3.27}$$

Following Levine and Bell[14] eqns. (3.26) and (3.27) may be combined in

$$\varphi_\beta = -\frac{\beta\gamma\sigma_\beta}{Kd^2}g \tag{3.28}$$

where the factor g is d/β in the conditions leading to eqn. (3.27), though this treatment is greatly oversimplified. If $n_\beta(=\sigma_\beta/ze)$ is the number of adsorbed cations per unit area of the IHP, each occupying p of the N_s adsorption sites available per unit area, then equating the electrochemical potentials for an adsorbed cation and a cation in bulk electrolyte

$$kT \ln \frac{n_\beta}{(N_s - pn_\beta)^p} + ze\psi_A = \text{constant} + kT \ln nf \tag{3.29}$$

where nf is the activity of cations in bulk solution and the constant term includes the chemical energy of adsorption.

If $p=1$ and the activity coefficient f is taken as unity then eqn. (3.29) becomes the Stern adsorption isotherm which can be written in the form

$$\frac{N_s}{n_\beta} = 1 + \frac{n_0}{n} \exp\left[\frac{ze\psi_\beta + \Phi}{kT}\right] \tag{3.30}$$

where n_0 is the number of solvent molecules per unit volume, and Φ is the specific adsorption potential.

If eqns. (3.17), (3.19), (3.20), (3.25), (3.28) and (3.29) are combined the following equations result[14]

$$r - \ln r + p \ln(1 - pbr) = \frac{ze\psi_d}{kT} + \frac{2v\gamma}{d} \sinh\left(\frac{ze\psi_d}{2kT}\right) - \ln x + \text{constant} \tag{3.31}$$

and
$$\frac{e\psi_0}{kT}=\frac{e\psi_d}{kT}+\frac{2v}{z}\sinh\left(\frac{ze\psi_d}{2kT}\right)-\frac{rd}{\gamma gz} \tag{3.32}$$

where

$$r=\frac{\sigma_\beta}{\sigma_\beta^0},\qquad \sigma_\beta^0=\frac{kTKd^2}{ze\beta\gamma g}$$

$$b=\frac{kTd^2K}{(ze)^2\beta\gamma gN_s},\qquad v=\frac{\kappa\epsilon}{4\pi K}.$$

It follows that if ψ_0 is varied at constant electrolyte concentration, $|\psi_d|$ will have a maximum at a value of r given, if κ is assumed constant, by

$$1-\frac{1}{r}-\frac{p^2b}{1-prb}=0 \tag{3.33}$$

This maximum in $|\psi_d|$ does not occur if the self-atmosphere potential φ_β is omitted. Figure 3.5 shows[14] the variation of ψ_d with ψ_0 for a negatively charged surface at concentrations of 1:1, 2:2 and 3:3 electrolytes corresponding to their flocculation concentrations, using the above equa-

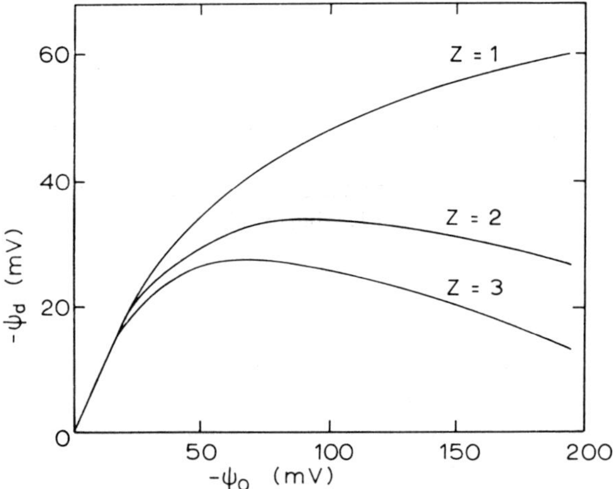

Fig. 3.5. Influence of discrete-ion effect on the variation of Stern potential ψ_d with wall potential ψ_0 at a negatively charged surface, in the presence of 1:1 ($p=1$), 2:2 ($p=3$) and 3:3 ($p=5$) valent electrolytes at their respective flocculation concentrations. $K=40\ \mu F/cm^2$, $N_s=5\times10^{14}$, $\gamma/d=1/4$ and $g=4/3$.

tions with reasonable values for the parameters concerned. The discreteness of charge effect has been able to give a satisfactory explanation for results at the mercury–solution interface,[10,15] and at the silver iodide–solution interface has been used to explain maxima in the co-ion contribution to the capacitance[16] and flocculation kinetic results.

Correction of the Stern adsorption isotherm to include the self-atmosphere potential has been shown[17] to reduce the specific adsorption of potassium ions on silver iodide to only $-0{\cdot}5kT$ which agrees with the observation that potassium ions have only an insignificant effect on the *pzc* of this surface.

The diffuse region of the double layer in solution

In terms of the model for the double layer adopted in Fig. 3.2 the part extending outwards from the OHP or Stern plane is termed the *diffuse* region. Here the thermal motion of the ions in solution results in an approximately exponential decay in potential and in the classical treatment of the diffuse layer the concentration of ions is taken to be governed by a Boltzmann distribution in the form

$$n'_{\pm} = n_{\pm} \exp\left[\mp \frac{ze\psi}{kT}\right] \tag{3.34}$$

where $n'_{\pm}$ is the number of ions of either charge per unit volume at a point in the diffuse layer where the mean potential is ψ, $n_{\pm}$ is the corresponding concentration in bulk solution and z is the valency of the ion concerned, not taking a sign.

If this distribution law is combined with the Poisson equation

$$\nabla^2\psi = -\frac{4\pi}{\epsilon}\rho \tag{3.35}$$

where ρ is the charge density $(\sum n'_{+} z_{+} e - \sum n'_{-} z_{-} e)$, ϵ the permittivity of the solution (taken as independent of ψ) and the differential operator ∇^2 is given in Cartesian co-ordinates by

$$\left[\frac{\partial^2}{\partial x^2} + \frac{\partial^2}{\partial y^2} + \frac{\partial^2}{\partial z^2}\right]$$

then the Poisson–Boltzmann equation results, namely

$$\nabla^2\psi = -\frac{4\pi}{\epsilon}\left[\sum z_{+} e n_{+} \exp\left(-\frac{z_{+}e\psi}{kT}\right) - \sum z_{-} e n_{-} \exp\left(+\frac{z_{-}e\psi}{kT}\right)\right] \tag{3.36}$$

For a plane interface ∇^2 becomes $\partial^2/\partial x^2$ and for a single symmetrical $z\!:\!z$ electrolyte in solution, eqn. (3.36) simplifies to

$$\frac{d^2\psi}{dx^2} = +\frac{8\pi}{\epsilon}\, zen \sinh\left(\frac{ze\psi}{kT}\right) \tag{3.37}$$

which is readily solved (Gouy, 1910), with the boundary conditions $\psi=\psi_d$ at $x=0$ and $\psi=d\psi/dx=0$ at $x=\infty$ to give

$$\gamma=\gamma_d \exp(-\kappa x) \tag{3.38}$$

and

$$\sigma_d = -\left[\frac{2n\epsilon kT}{\pi}\right]^{1/2} \sinh\left(\frac{ze\psi_d}{2kT}\right) \tag{3.39}$$

where

$$\gamma=\tanh\left(\frac{ze\psi}{4kT}\right) \tag{3.40}$$

The assumption of a symmetrical electrolyte in the above solution is less restricting than might at first be supposed since it is well known in colloid phenomena that the co-ion (i.e. ion of the same charge as the surface) is of little importance. This is indeed shown by splitting σ_d into σ_{d+}, the contribution due to an excess of cations, and σ_{d-}, the contribution due to an excess of anions, excess being understood algebraically.

Then

$$\sigma_{d+} = ze\int_0^\infty (n'_+ - n)\,dx$$

$$= \left[\frac{\epsilon nkT}{2\pi}\right]^{1/2}\left[\exp\left(-\frac{ze\psi_d}{2kT}\right)-1\right] \tag{3.41}$$

and

$$\sigma_{d-} = -ze\int_0^\infty (n'_- - n)\,dx$$

$$= \left[\frac{\epsilon nkT}{2\pi}\right]^{1/2}\left[1-\exp\left(\frac{ze\psi_d}{2kT}\right)\right] \tag{3.42}$$

It can be seen that if ψ_d is for example large and positive, then the major part of σ_d is σ_{d-} and the nature of the positively charged co-ion is relatively unimportant.

At a spherical interface eqn. (3.36) cannot be integrated analytically unless some approximation to the exponential term is made. If the linear approximation is justified, i.e. $ze\psi/kT<1$, which implies $z\psi<25$ mV, the treatment becomes that of Debye and Hückel (1923). The equation corresponding to eqn. (3.37) is then

$$\frac{1}{r^2}\frac{\mathrm{d}}{\mathrm{d}r}\left(r^2\frac{\mathrm{d}\psi}{\mathrm{d}r}\right)=\kappa^2\psi$$

which, with the boundary conditions $\psi=\psi_d$ at $r=a$ and $\psi=\mathrm{d}\psi/\mathrm{d}r=0$ as $r\to\infty$, leads to

$$\psi=\psi_d\,\frac{a}{r}\exp\left[-\kappa(r-a)\right] \tag{3.43}$$

This equation gives the potential ψ at a distance r from the centre of the sphere, where a is the radius at the inner boundary of the diffuse layer.

If the Debye–Hückel linear approximation is applied to eqn. (3.37) the solution, valid only at low potentials, is

$$\psi=\psi_d\exp\left(-\kappa x\right) \tag{3.44}$$

a result which can be obtained from eqn. (3.38) with ψ_d small. The same approximation applied to eqns. (3.41) and (3.42) leads to the result that the net charge in the diffuse layer is due equally to an excess of counterions and deficit of co-ions. It is perhaps here that the deficiencies of the Debye–Hückel approximation in the treatment of colloid problems are most apparent.

More exact solutions of the Poisson–Boltzmann equation for the spherical case have been obtained by Müller (1928) graphically and by Loeb, Overbeek and Wiersema[18] who have published a very full compilation using numerical methods.

The validity of all these treatments, however, including that of the plane diffuse layer, depends upon the applicability of the Poisson–Boltzmann equation in the form of eqn. (3.36). It is beyond the scope of this discussion to review the very considerable literature in which corrections to this equation have been treated but the use of the uncorrected Poisson–Boltzmann equation probably gives rise to little error, in the case of 1:1 electrolytes, even at concentrations of 0·1 M for potentials up to 75 mV.

Capacity of the diffuse region of the double layer

From the way in which σ_d has been defined it is evident that as ψ_d becomes more positive σ_d becomes more negative, so that the capacity per unit area of the diffuse region of the double layer is given by $-\sigma_d/\psi_d$ and the differential capacity by $-\mathrm{d}\sigma_d/\mathrm{d}\psi_d$.

Differentiating eqn. (3.39) for a plane double layer

$$C_d = -\frac{\mathrm{d}\sigma_d}{\mathrm{d}\psi_d} = \frac{\kappa\epsilon}{4\pi}\cosh\left(\frac{ze\psi_d}{2kT}\right) \tag{3.45}$$

At low potentials, where cosh $(ze\psi_d/2kT)$ tends to unity

$$C_d = \frac{\kappa\epsilon}{4\pi} \tag{3.46}$$

and since this expression is independent of ψ it is also, under these conditions, the integral capacity $-\sigma_d/\psi_d$. For a 1 mM aqueous solution of a 1:1 electrolyte at 25°C this becomes 7·22 μF/cm^2.

The capacity of the diffuse part of the double layer is not accessible to direct measurement and it is indeed only in certain cases that the capacity of the whole double layer σ_0/ψ_0, which consists of an inner capacity in series with the diffuse capacity, can be measured. At the mercury–solution interface, as a result of the mercury being conducting and polarisable, the capacity may be measured by a direct a.c. bridge method and such measurements, after allowing for the capacity of the diffuse region calculated as above, have given valuable information[15] on the inner region which may be applied, with suitable reservations, to the inner regions at solid particle–solution interfaces.

Capacity measurements at the silver halide–solution and oxide–solution interfaces have also been measured by a.c. methods but most estimations have been made by potentiometric titration[19,20] in which, using silver halides as examples, $\Delta\sigma_0$ and $\Delta\varphi$ can be obtained from the change in emf of a silver halide–calomel electrode pair when a known quantity of potassium halide or silver nitrate is introduced into the silver halide dispersion of known surface area. If the χ potential is assumed constant $\Delta\varphi$ becomes $\Delta\psi_0$ and, knowing the ionic concentrations corresponding to the *pzc*, σ_0 can be plotted against ψ_0 to give integral (σ_0/ψ_0) and differential $(\mathrm{d}\sigma_0/\mathrm{d}\psi_0)$ capacities of the double layer.

Some results of Lyklema[21] are shown in Fig. 3.6, in which the differential capacity $\mathrm{d}\sigma_0/\mathrm{d}\psi_0$ is shown as a function of ψ_0 for aqueous silver iodide sols in the presence of various nitrates at an ionic strength of 1 mM. The form of these curves is similar to results obtained at the mercury–solution interface though there is greater specificity of cation effect on silver iodide. This specificity provides an explanation for the widely observed sequence of flocculation concentrations for the alkali

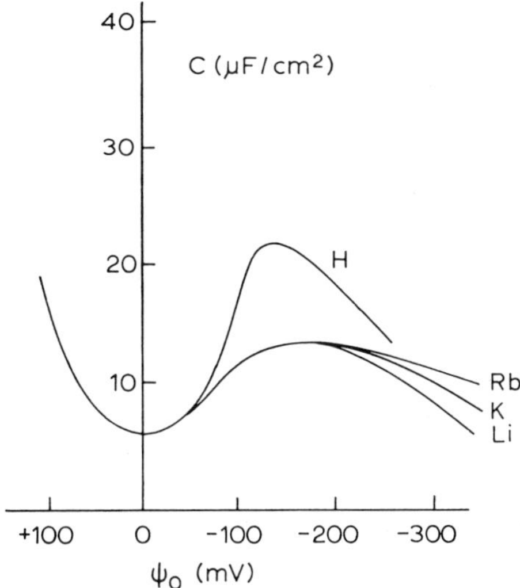

Fig. 3.6. Total differential capacities at the AgI–solution interface in aqueous solutions of various nitrates at 10^{-3} M. *(after Lyklema[21]).*

metal cations on negatively charged sols, viz. $Li^+ > Na^+ > K^+$. The Li^+ ion, being more strongly hydrated, is less specifically adsorbed than the succeeding members of the series so that $|\psi_d|$ remains higher and with it the flocculation concentration. Some later experiments of Lyklema[20] indicate that at temperatures in excess of 50°C the specificity largely disappears both from the capacity curves and flocculation concentrations.

The limited value of the capacity at elevated negative potentials as shown in Fig. 3.6 is largely due to the effect of the capacity of the inner part of the double layer which is effectively in series with that of the diffuse part. If the Gouy–Chapman expressions are used to give the capacity of the diffuse region, then the measured total capacities indicate a value for the integral capacity of the inner part of the double layer varying from ~30 μF/cm² near the zero point to ~10 μF/cm² at elevated negative charges. These figures are similar to those found for the mercury–aqueous solution interface and for any reasonable thickness of the inner layer indicate a substantial lowering of the effective dielectric constant of water near the interface. At both interfaces the capacities rise at positive polarisations, an effect which is probably partly due to the weaker hydration and therefore greater adsorbability of anions compared to cations, and partly due to water dipole re-orientation. There has been considerable, and largely unresolved, discussion of the origin of a maximum in the differential capacity of mercury at positive polarisations which persists through many electrolytes and many non-aqueous solvents (of sufficient dielectric constant to dissociate electrolytes).

At the oxide–aqueous solution interface a rather different capacity pattern emerges. There is virtual symmetry about the zero point and the capacities seem markedly higher than for Hg or AgI. It is difficult, however, to derive capacities at the zero point from experimental data because of the extra factor θ_c (eqn. (3.14)) which occurs. It seems reasonable to attribute the higher capacities to a higher effective dielectric constant of water near oxide interfaces, which will be in a different state from that at the hydrophobic Hg or AgI. The symmetry about the zero point could well arise from considerably less dehydration of adsorbed ions which would certainly diminish differences between anions and cations.

Double layer on the solid side of the interface

In the case of a conductor such as mercury or a solid metal there can be no potential gradient (i.e. electric double layer) within the conductor except possibly very near the interface. In such a case the whole of any electric double layer is on the solution side of the interface. In the case of an ionic solid, however, in which the presence of lattice defects allows the migration of charge it will be quite possible to have a substantial fraction of the total potential drop from bulk solution to bulk solid within the solid. A general treatment of the distribution of an electric double layer between two phases has been given by Verwey and Niessen.[22]

The situation in Fig. 3.7 has been simplified by ignoring the inner parts of the electric double layers near the interfaces, i.e. the Gouy diffuse layers have been assumed to continue right up to the interface. The χ potential has been omitted and $(\psi_0)_1 + (\psi_0)_2$ written as D. The condition of overall electrical neutrality gives

$$\int_{-\infty}^{0} \rho_1 \mathrm{d}x + \int_{0}^{\infty} \rho_2 \mathrm{d}x = 0 \qquad (3.47)$$

The Poisson equation may be applied on each side of the interface to give

$$\rho = -\frac{\epsilon}{4\pi}\frac{\mathrm{d}^2\psi}{\mathrm{d}x^2}$$

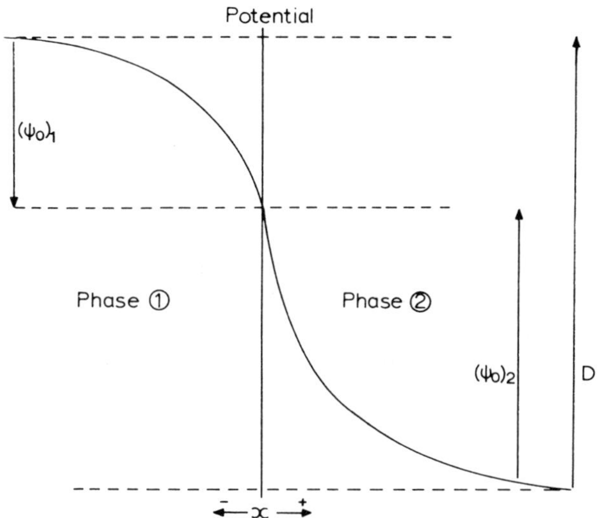

Fig. 3.7. Electric double layer on the two sides of a phase boundary, potential and inner parts of the double layers omitted.

and substitution in eqn. (3.47) yields

$$\epsilon_1 \int_{-\infty}^{0} \left(\frac{d^2\psi}{dx^2}\right)_1 dx + \epsilon_2 \int_{0}^{\infty} \left(\frac{d^2\psi}{dx^2}\right)_2 dx = 0$$

With the boundary conditions $(d\psi/dx)_1 = 0$ as $x \to -\infty$ and $(d\psi/dx)_2 = 0$ as $x \to \infty$ we therefore have, at $x = 0$

$$\epsilon_1 \left(\frac{d\psi}{dx}\right)_1 = \epsilon_2 \left(\frac{d\psi}{dx}\right)_2 \tag{3.48}$$

showing that there is a discontinuity in potential gradient at the interface.

Equation (3.37) gives for $d\psi/dx$ at any point in the diffuse layer

$$\frac{d\psi}{dx} = -\left[\frac{8\pi nkT}{\epsilon}\right]^{1/2} 2 \sinh\left(\frac{ze\psi}{zkT}\right)$$

and substitution in eqn. (3.48) gives

$$(n_1\epsilon_1)^{1/2} \sinh\left[\frac{ze(\psi_0)_1}{2kT}\right] = (n_2\epsilon_2)^{1/2} \sinh\left[\frac{ze(\psi_0)_2}{2kT}\right] \tag{3.49}$$

Thus the division of the total potential drop D into its components $(\psi_0)_1$ and $(\psi_0)_2$ is governed by the ratio

$$\alpha = \left(\frac{n_1\epsilon_1}{n_2\epsilon_2}\right)^{1/2} = \frac{\kappa_1\epsilon_1}{\kappa_2\epsilon_2} \tag{3.50}$$

In most cases the medium with the lower dielectric constant will also have the lower concentration of ions so that the greater part of the potential drop D is in the medium of lower dielectric constant.

Experimental values for $\mathrm{d}\zeta/\mathrm{d}p\mathrm{Ag}$ measured at $\zeta \rightarrow 0$ for AgI particles have been held[23,24] to indicate the presence of a double layer within the solid in so far as the numerical value for this quantity falls short of the Nernst value $2{\cdot}303kT/e_0$, i.e. 59 mV at 25°C. The interpretation then rests on

$$\left[\frac{\mathrm{d}\zeta}{\mathrm{d}(p\mathrm{Ag})}\right]_{\zeta\rightarrow 0} = \left[\frac{\mathrm{d}(\psi_0)_2}{\mathrm{d}(p\mathrm{Ag})}\right]_{(\psi_0)_2\rightarrow 0} = -59\left[1+\frac{1}{\alpha}\right]^{-1} \mathrm{mV} \tag{3.51}$$

However, the ζ potential is more correctly related to ψ_d rather than ψ_0 and in this case, assuming no specific adsorption, ζ will fall short of ψ_0 even with no double layer in the solid, to the extent

$$\left[\frac{\mathrm{d}\zeta}{\mathrm{d}(p\mathrm{Ag})}\right]_{\zeta\rightarrow 0} = -59\left[\frac{K^0}{K^0 + C_d^0}\right] \mathrm{mV} \tag{3.52}$$

where K^0 is the inner capacity at the zero point and $C_d^0 = (\kappa\epsilon/4\pi)$ is the diffuse layer capacity at the same point. In 0·1 M 1:1 electrolyte $C_d^0 \cong 72$ μF/cm^2 and if K^0 is taken as 30 μF/cm^2 then $[\mathrm{d}\zeta/\mathrm{d}p\mathrm{Ag}]_{\rightarrow 0}$ becomes $\sim$18 mV, which is precisely the sort of magnitude observed. Of course the value for K^0 may include a contribution from a capacity within the solid but a value for $[\mathrm{d}\zeta/\mathrm{d}p\mathrm{Ag}]_{\zeta\rightarrow 0}$ lower than 59 mV is not in itself evidence for double layer within the solid. Comparison with mercury, where there can be no capacity in the metal, suggests that any capacity in solid AgI would have to be at least $\sim$70 μF/cm^2 which, as a diffuse capacity, would involve a fraction of defects in the solid as high as 0·4%, which seems much too large.

To summarise, there seems to be no experimental evidence for dispersions of solids in liquids in which the surface electrical properties are influenced by any double layer within the solid. Since some such solids definitely contain finite concentrations of mobile charge carriers (e.g.

silver halides) it becomes attractive to suppose that a surface screening mechanism applies.

Free energy of the diffuse double layer

The electric double layer forms spontaneously around a charged particle or interface in solution so that the free energy of formation of the double layer must be negative. When two such particles or interfaces approach each other so that the respective double layers overlap it is evident that a part of the previously existing double layers has been destroyed. This will give a positive contribution to the free energy which is manifested as a repulsion between the particles.

It is perhaps not immediately apparent how the free energy of the double layer comes to be negative. In this connection it is useful to consider an analogous case—the charging of a simple parallel-plate condenser. In Fig. 3.8 when the switch S is closed the condenser C (capacitance C) will charge from the cell of emf E through the resistance R. If an increment of charge dq flows into the condenser when the potential difference between its plates is V, then the work done is $V\mathrm{d}q$ and for the whole of the charging process the work will be

$$\int_{q=0}^{q=Q} V\mathrm{d}q \tag{3.53}$$

Writing $V = q/C$ and taking C to be constant this evaluates to $\frac{1}{2}Q^2/C$, i.e. $\frac{1}{2}QE$. This is the familiar expression for the potential energy of a charged condenser and it will be noted that it is positive. To obtain the total free energy change in the charging process, however, it is necessary to

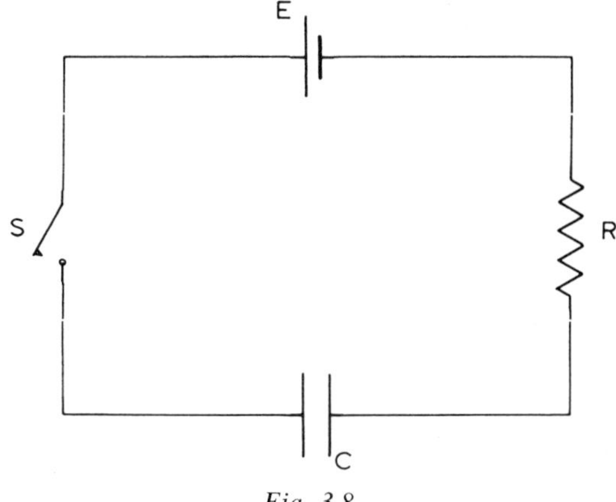

Fig. 3.8.

consider the state of the cell which supplied the charge to the condenser. Since the emf remains constant at E while a total charge Q is extracted, the decrease in free energy of the cell (as a result of the chemical reaction proceeding within it) is QE. Thus the total increase of free energy on closing S is

$$\tfrac{1}{2}QE - QE = -\tfrac{1}{2}QE \tag{3.54}$$

The contribution $+\frac{1}{2}QE$ may be looked upon as the electrical part of the free energy change and $-QE$ as the chemical part.

In the case of the electric double layer, where it must be noted that the capacitance is not constant but depends upon the wall potential and electrolyte concentration, the free energy of formation per unit area of (plane) double layer will be, by analogy with eqns. (3.53) and (3.54),

$$\int_0^\sigma \psi' \mathrm{d}\sigma' - \psi\sigma = -\int_0^\Psi \sigma' \mathrm{d}\psi' \tag{3.55}$$

where the primes indicate values of σ and ψ during the charging process and eqn. (3.55) gives the free energy (of formation) of a double layer, or part of a double layer, extending from a potential ψ out into bulk solution ($\psi = 0$). It can be seen from eqn. (3.55) that the free energy will always be negative as the argument at the beginning of this section indicated.

For a diffuse double layer of inner Gouy potential ψ_d, $\sigma = -\sigma_d$ and substitution of eqn. (3.39) into eqn. (3.55) gives for the free energy

$$-\left[\frac{2n\epsilon kT}{\pi}\right]^{1/2} \int_0^{\psi_d} \sinh\left(\frac{ze\psi'}{2kT}\right) \mathrm{d}\psi'$$

$$= -\frac{8nkT}{\kappa}\left[\cosh\left(\frac{ze\psi_d}{2kT}\right) - 1\right] \tag{3.56}$$

which can also be expressed as

$$-\frac{16nkT}{\kappa} \sinh^2 \frac{ze\psi_d}{4kT}$$

which is clearly negative.

When particles approach to such distances that the double layers overlap, some electric double layer is destroyed, so that there will be a positive contribution to the total free energy which is manifested as a repulsion between the particles. This question is considered more fully in Chapter 1.

ELECTROKINETIC PROPERTIES OF DISPERSIONS

Classification

Electrokinetic properties include all phenomena in which there is relative tangential motion between charged phases. This may be the result of an applied electric field or, conversely, may be mechanically achieved and produce an electric field. A convenient and usual subdivision of these phenomena is into

(i) electrophoresis,
(ii) electro-osmosis,
(iii) streaming potential and
(iv) sedimentation potential (Dorn effect).

In electrophoresis experiments an electric field is applied to a dispersion and the velocity of the particles measured, either directly as in microelectrophoresis or as a moving boundary between the sol and pure dispersion medium. Electro-osmosis differs only in that the solid phase is kept stationary as a porous plug or single capillary while the velocity of the liquid is measured. As an alternative to a direct measurement of the liquid velocity, the volume of liquid transported may be measured or the reverse pressure necessary to prevent electro-osmotic flow.

Phenomena (iii) and (iv) represent the reverse situation where the motion is achieved mechanically and the resulting potential difference is measured between suitable electrodes. In the case of streaming potentials the solution is streamed through a porous plug of the solid phase. If the solid phase is particulate it is convenient to contain the plug by means of the electrodes themselves which may be of blacked platinum or of the reversible type. An electrometer should be used to record the resulting potentials to minimise the current passed between the electrodes. In the case of sedimentation potentials particles are allowed to move through the dispersion medium in a gravitational or centrifugal field with suitably placed electrodes and an electrometer to record the resulting potential difference. The sedimentation potential technique presents the greatest experimental difficulties of the four alternatives listed above and would not normally be used to obtain information on dispersions. It will not therefore be discussed further.

The zeta potential

The potential yielded by any of the electrokinetic effects is referred to non-committally as the zeta (ζ) potential and will be expected to be the potential at the surface of shear between the phases in relative motion. The concept of such a surface of shear is possibly complicated by any variation of viscosity with distance from the surface. If electrokinetic experiments are interpreted in terms of charge then this will be the total charge within the effective surface of shear. In terms of Fig. 3.2 the ζ potential will not be expected to refer to any surface nearer to the particle than ψ_d and could well in some cases be the potential further out into the diffuse layer if the particle immobilises a considerable amount of solvent.

When it is further borne in mind that electrophoretic mobilities are not easy to measure with accuracy and that the conversion of the mobilities into zeta potentials is still open to some doubt, especially at high mobilities, it is perhaps understandable that zeta potentials have been described as 'difficult to measure and impossible to interpret'. This is, however, an over-pessimistic assessment of the situation and considerable information can be obtained from electrokinetic measurements which is relevant to the problem of colloid stability.

If silver halides are typical of lyophobic surfaces then certainly for such surfaces ζ is equal to ψ_d within normal experimental accuracy when both are small, and there is evidence that the two potentials do not differ by more than about 2 or 3 mV up to 85 mV in 0·01 M 1:1 electrolyte and up to 30 mV in 0·1 M 1:1 electrolyte. This range covers that most often of experimental interest.

In order to investigate the relationship between ζ and ψ_d suppose that ζ refers to a plane distant Δ from the Stern plane out into the diffuse layer (Fig. 3.2). Then in the limiting case of low potentials

$$\zeta = \psi_d \exp(-\kappa\Delta)$$

Taking silver halides as examples since here there are well established potential determining ions, then at a given κ, i.e. at a given electrolyte concentration,

$$\left[\frac{d\zeta}{d(pAg)}\right]_{\zeta\to 0} = S = \left[\frac{d\zeta}{d\psi_d}\,\frac{d\psi_d}{d\psi_0}\,\frac{d\psi_0}{d(pAg)}\right]_{\zeta\to 0} \tag{3.57}$$

On the right-hand side of eqn. (3.57) the factor

$$\left[\frac{d\psi_0}{d(pAg)}\right]$$

is the Nernst factor which can be treated as experimentally determinable and termed N. The factor $(d\psi_d/d\psi_0)$ can be simplified, in the absence of specific adsorption to $K^0/(K^0+C_d^0)$ where K^0 and C_d^0 are the inner layer and diffuse layer capacities respectively at the zero point (eqn. 3.52), so that

$$S=N\left[\frac{K^0}{K^0+C_d^0}\right]\exp(-\kappa\Delta) \tag{3.58}$$

or

$$S^{-1}=N^{-1}\left[1+\frac{C_d^0}{K^0}\right]\exp(+\kappa\Delta) \tag{3.59}$$

Figure 3.9 shows an experimental plot of S^{-1} versus C_d^0 (i.e. $\kappa\epsilon/4\pi$) for AgI at 25°C in KNO_3 (aq).

The intercept at $C_d^0=0$ (i.e. $K=0$) is seen to be $1/59\,\mathrm{mV}^{-1}$ as expected and the plot is linear within experimental error. This indicates that Δ is small and the slope gives $K^0=30\ \mu\mathrm{F/cm}^2$. Since this latter figure is that which results from the now extensive potentiometric titration and directly measured capacity data, it can be taken as reinforcing the conclusion from the linearity of Fig. 3.9 that Δ is small. Any specific

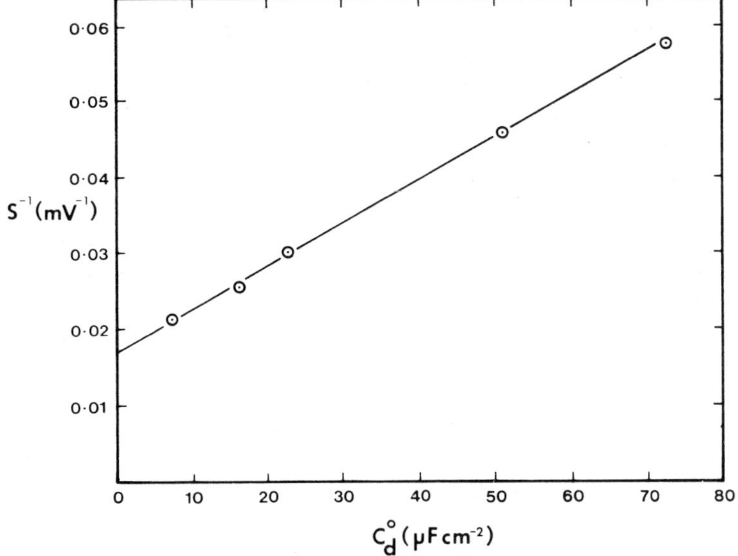

Fig. 3.9. Experimental plot of S^{-1} against C_d^0 $(=\kappa\epsilon/4\pi)$ for AgI in KNO_3 (aq) at 25°C. At $C_d^0=0$, $S=59\,mV$; slope yields $K^0=30\ \mu F/cm^2$.

adsorption, which could be allowed for formally in eqn. (3.59), will clearly have the effect of reducing S, and therefore increasing S^{-1}, at large C_d^0. Since this is in the same sense as $\Delta > 0$ the conclusion that Δ is small is again reinforced. At 100 mM KNO_3, with $C_d^0 = 72{\cdot}2\ \mu F/cm^2$, a value of Δ as small as 1 Å would increase S^{-1} by 10%, which seems the limit of experimental error.

This experimental indication that $\zeta = \psi_d$ even at high electrolyte concentration (0·1 M), is restricted to the limiting case of low potentials. However, if Δ is assumed to remain zero as the potentials are increased, a good fit is obtained between experimental ζ values and those calculated by assuming that specific adsorption follows a Stern isotherm (eqn. (3.30)), up to potentials around 85 mV. To achieve such potentials it is, however, necessary to reduce the 1 : 1 electrolyte concentration to 0·01 M.

The interpretation of S^{-1} plots for oxide[25] and similar surfaces is complicated by the factor θ_c (eqn. 3.14).

Experimental determination of electrophoretic mobilities

Since electrophoresis is the electrokinetic technique of widest application to dispersion stability problems, experimental details will be discussed at some length. The first choice to be made is between the moving-boundary method and microelectrophoresis in which the motion of individual particles is followed.

The microelectrophoretic technique may be said to have the following advantages.

(i) The particles are observed while in their normal sol environment.
(ii) Very dilute sols can be studied so that flocculation rates, even at high electrolyte concentration or near the *zpc* are negligible.
(iii) The high magnification leads to very short observation times and high sensitivity.
(iv) In a polydisperse sol the particles in a chosen (though necessarily wide) size range can be observed while others are ignored.

These advantages are so extensive that the moving-boundary method will probably only be used in cases where it is desired to study colloidal dispersions in a concentrated state or the particles are too small or too similar in refractive index to the dispersion medium to be visible or detectable by photomultiplier. In this connection it may be noted that polystyrene latex particles of radius 400 Å dispersed in water are readily visible in suitable illumination and particles down to a few tens of

Ångströms can be timed in laser–Doppler systems. Some workers[26] have preferred the moving-boundary method where the effect of surfactants on a dispersion is to be studied, on the grounds that non-uniform adsorption of the surfactant on the walls of the cell makes the interpretation of microelectrophoretic experiments difficult or impossible. While this is true it must be noted that any such non-uniform adsorption is readily detected and does not seem to give trouble in practice if the cells are kept scrupulously clean. Acidified sodium fluoride solution (i.e. dilute HF aq.) has been recommended[27] for cleaning electrophoresis cells where there is a danger of interference from silicate ions from the glass.

Stationary levels in microelectrophoresis cells

Before discussing the factors involved in the choice of type of microelectrophoresis cell it is necessary to realise that when an electric field is applied between the ends of a microelectrophoresis cell, of whatever cross-section, the solution contained in the cell will, in general, experience electro-osmotic flow on account of surface charge on the cell walls. Only where this electro-osmotic flow velocity is zero, the so-called *stationary levels*, can the electrophoretic mobility of a particle be observed directly.

Figure 3.10 shows the situation in a microelectrophoresis cell which is assumed, in the first instance, to be of circular cross-section. The electro-osmotic effect alone will give rise to a solution velocity v_{eo} uniform across the cell cross-section, towards the electrode of the same polarity as the charge on the cell wall. If the cell is closed it is evident that a reverse flow

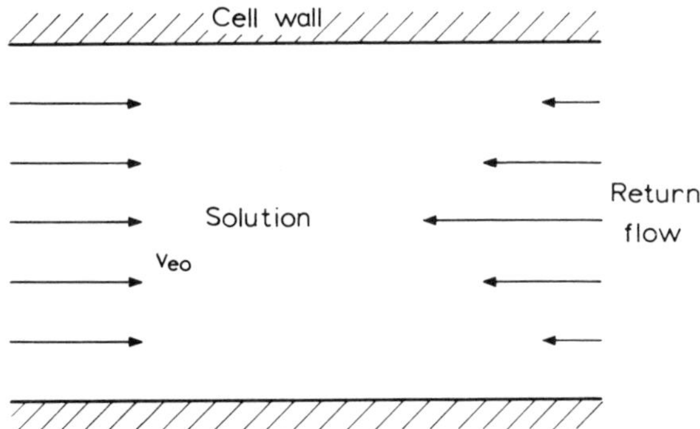

Fig. 3.10. Flow conditions within a closed cylindrical electrophoresis cell with electric field applied.

will be set up so that there is no net transport of liquid. This will be true even if the cell is not closed once the necessary hydrostatic reverse pressure is built up. Since the reverse flow will follow Poiseuille's law, the condition of no overall liquid transport gives

$$\int_{r=0}^{r=a} 2\pi v r \mathrm{d}r = 0 \tag{3.60}$$

with

$$v = v_{eo} - C(a^2 - r^2) \tag{3.61}$$

where v is the liquid velocity at a distance r from the centre of the tube of radius a, and C is a constant.

Equations (3.60) and (3.61) solve to give $C = 2v_{eo}/a^2$ so that

$$v = v_{eo}\left[2\frac{r^2}{a^2} - 1\right] \tag{3.62}$$

At the cell wall $r = a$ and $v = v_{eo}$ as expected; at the centre $r = 0$ and $v = -v_{eo}$ so that the velocity of liquid flow is equal in magnitude but opposite in direction to that at the wall. The condition for zero liquid velocity is given by eqn. (3.62) as $r = a(1/2)^{1/2}$. Thus the stationary level, which in this case will be cylindrical, is 0·707 of the radius from the centre, or as more usually expressed, 0·146 of the (internal) diameter from

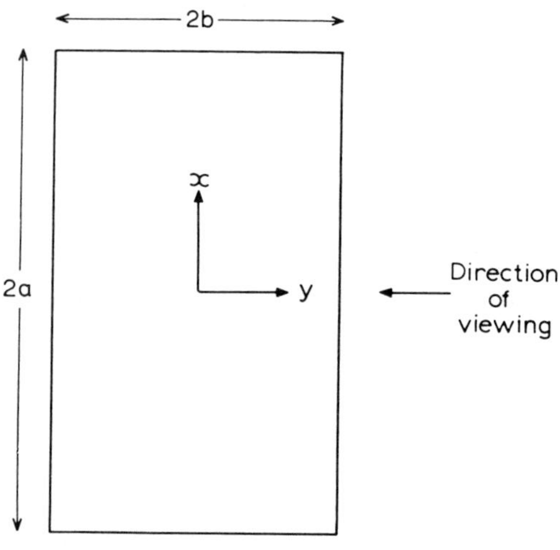

Fig. 3.11.

the wall. Looking into a cylindrical cell along a diameter there will, therefore, be two such levels at which particles show their electrophoretic velocity.

For cells with rectangular cross-section the situation is more complex.[28,29] Using the nomenclature illustrated by Fig. 3.11

$$\frac{v(x=0)}{v_{eo}}=1-\frac{3}{2}\left[1-\frac{y^2}{b^2}\right]\left[1-\frac{192b}{\pi^5 a}\right]^{-1} \tag{3.63}$$

If the ratio a/b is taken as infinite then $v(x=0)$ is zero when $y/b=0{\cdot}5774$ so that the stationary levels are 0·211 of the cell thickness ($2b$) from both the front and back walls. The position of the stationary levels for other values of a/b are shown in Table 3.1.

TABLE 3.1
POSITION OF STATIONARY LEVELS IN ELECTROPHORESIS CELL OF RECTANGULAR CROSS-SECTION
Nomenclature shown in Fig. 3.11, $x=0$ in all cases

a/b	$(b-y)/2b$
∞	0·211
50	0·208
20	0·202
10	0·194

It is not necessary to restrict observations to the stationary levels in a cell. Observations can be made across the cell (across a diameter for the cylindrical cell) from wall to wall and use made of eqns. (3.60) and (3.61) in deducing the particle electrophoretic velocity from the condition of zero overall liquid transport.

Choice of microelectrophoresis cell

Single channel microelectrophoresis cells are usually of rectangular or circular cross-section and the latter may be thick- or thin-walled. Each type has its advantages. Those of the flat cell may be listed as follows.

(i) There is no optical correction to apply (as would be the case with curved walls) in the sense that fractional depths in the cell are not affected by the refractive index of the solution within the cell or of the cell wall or of the surrounding liquid (if any).

(ii) When the observing microscope is focused on a particular level within the cell the whole of the field of view is at this level and velocity measurements are valid even at the extremities of the field of view. This can be quite important when observing very dilute sols.

(iii) Provided that the cell is used with the dimension *a* (Fig. 3.11) vertical, any particles falling under the influence of gravity will remain in the stationary level while in view, probably for long enough to measure their velocity. The particles will, furthermore, fall to a part of the cell where their presence is relatively unimportant both with regard to interference with illumination and to distortion of the electro-osmotic flow.

(iv) The cell is readily constructed and silica can be used as easily as glass.

However, the large cross-sectional area of the flat cell compared with cylindrical cells, leads to larger currents for the same electric field within the cell, leading to gassing and polarisation if reversible electrodes are not used; these are much less convenient in use than blacked platinum electrodes. Furthermore, when used with the dimension *a* vertical in order to take advantage of (iii) above, convection currents can be troublesome, especially with intense illumination. The advantages of the cylindrical cell are as follows.

(i) The small cross-sectional area leads to small electric currents for a given field strength so that large-area blacked platinum electrodes can be used with negligible polarisation errors and the more troublesome reversible electrodes are avoided.

(ii) The volume of sol required to fill the cell can be made very small.

(iii) The cell can be illuminated (e.g. by laser) from a direction perpendicular to that of observation in the traditional ultramicroscope configuration which is much more satisfactory when very small particles have to be observed than the dark ground condensers used with flat cells.

(iv) Thermostatting can be made very effective, especially with thin-walled cells, leading to the virtual elimination of convection currents and a great reduction in the self-heating effect resulting from the current passed.

The situation may be summarised by saying that for electrophoresis in aqueous or other highly conducting solutions the cylindrical cell is to be preferred, but for most non-aqueous (low conductance) electrophoresis the flat cell, constructed of silica, is usually better.

Cylindrical cells may be thick-walled as with the Mattson[30] cell or thin-walled as the van Gils[29] cell. Thick-walled capillary tubing of uniform bore is readily purchased and forms a mechanically strong cell, but it is necessary to grind a flat on the cell for observation (and usually another for illumination) and even with this flat and the tube surrounded by thermostatting fluid, the optical correction which must be considered for apparent positions within the cell is large, and distortion makes it difficult to observe particles beyond the centre of the cell. The latter point is of some importance since reliable mobility measurements should include observations at both stationary levels.

Considerable skill is required both to draw thin-walled capillary tube of accurately circular cross-section and to join such tubing to more substantial end tubing. The resulting cell is also rather fragile but apart from this it has every advantage over the thick-walled cell particularly with regard to efficient thermostatting and negligible optical corrections at the very small thickness which become possible (e.g. 20 μm).

An interesting cell is that due to Smith and Lissie[31] in which the electrode compartments are connected by two separate capillaries. The advantage of this cell is that, provided the capillaries are of identical material and that their lengths (l) and radii (a) satisfy the relationship

$$\frac{l_2}{l_1}=\left(\frac{a_2}{a_1}\right)^2\left[\left(\frac{a_2}{a_1}\right)^2-2\right] \tag{3.64}$$

the stationary level is at the centre of the tube of smaller cross-section where the electro-osmotic velocity gradient is zero.

That this cell has been so little used is probably due to the difficulty of satisfying the necessary condition given above. The only satisfactory test would presumably be to check it against a more conventional cell.

Experimental detail

A convenient manual form of microelectrophoresis apparatus is obtainable from Rank Bros, Bottisham, Cambridge, UK. However, for those constructing their own equipment a few points will be made here.

It is essential that electrophoretic cells are thermostatted since the temperature coefficient of particle mobility in aqueous solution, as for ion conductance, is of the order of 2% per °C, this figure arising mainly from the variation of solution viscosity with temperature. In addition, the thermostatting arrangements will facilitate the removal of heat produced within the cell as a result of the applied field. The heat produced per unit volume within the cell is $E^2\kappa$ where E is the applied field (volt cm^{-1}) and κ is the solution conductivity (ohm^{-1} cm^{-1}), yielding the heat in J cm^{-3}.

In this expression κ should not be confused with the Debye–Hückel reciprocal distance parameter for which the same symbol is used.

A total magnification of 200–400 is sufficient even for small particles since these will be detected by the light scattered at 90° in the dark-ground ultramicroscope mode. In the ultramicroscope mode a laser constitutes a very convenient illumination source; for example polystyrene latex particles of 400Å radius are readily 'seen' in the focused beam of a 5mW helium–neon laser. There must, however, be some compromise between focusing the beam for maximum intensity of illumination and having a sufficient area illuminated to time particles over a reasonable distance.

The electric field within the cell, which must be known in order to obtain the particle mobility from observed velocities, is of form V/l where V is a potential difference and l a distance. However, l is not the physical distance between electrodes but the 'effective' distance. This distance is given by $R\kappa A$ where R is the resistance, conveniently measured by a conductance bridge, between the electrodes of the cell when it contains a solution of known conductivity, κ, and A is the cross-sectional area of the cell at the point of observation. It is not necessary that the cell cross-section be uniform for more than a short distance around the point of observation and it is indeed desirable that the electrode compartments should be enlarged both to accommodate substantial electrodes and to ensure that slight irreproducibility in the position of the electrodes has a negligible effect on l.

It is usual to measure about 20 particle transit times across the observing microscope graticule successively in opposite directions, so that the reciprocal times, and hence velocities, can be averaged. This process compensates for any no-field drift of the particles and also helps to minimise polarisation in two-electrode cells.

For a new cell it is good practice to verify that eqns. (3.62) and (3.63) are obeyed. In the case of the cylindrical cell it is most convenient to plot observed particle velocity against $(r/a)^2$, which should result in symmetrical straight lines on each side of the centre $(r/a=0)$. The symmetry of the electro-osmotic flow should be checked each time a sol is investigated either graphically in the above manner, or by making measurements at both stationary levels or, perhaps most sensitively and conveniently, by checking that the velocity of particles just within the near and far walls is the same.

If the electro-osmotic velocity at the cell wall were zero (i.e no double layer at the wall–solution interface), then valid particle electrophoretic velocities could be observed anywhere in the cell. If the electro-osmotic velocity at the wall were small but finite then at least errors due to mislocation of the stationary levels would be minimised. For this reason some workers, especially those always using the same conditions of pH, etc., prefer to coat the cell walls with a polysaccharide gel which has a zero or small electro-osmotic velocity under the conditions used. If this procedure is adopted the integrity of the coating and its zero electro-osmotic velocity should be regularly checked and not assumed to persist over long times or under different pH and electrolyte conditions. There is also the possibility that coating material will adsorb on to particles in the cell.

Electrodes

The best arrangement is to have four electrodes; two larger outer electrodes to conduct the current to and from the cell, accompanied by two smaller inner electrodes placed near the ends of the uniform cross-section portion of the cell. The inner electrodes of blacked platinum wire are 'potential sensing' electrodes whose potential difference is measured by an electrometer or high resistance digital voltmeter so that there is no polarisation effect. In the calibration discussion above V and l now refer to these inner electrodes.

Gassing and polarisation at the outer electrodes is now of little importance. Their effect can be minimised by using large-area blacked platinum electrodes, or palladium electrodes, or suitable reversible electrodes which will not gas. Alternatively, the outer electrodes can be separated by a membrane filter from the cell proper and allowed to gas freely without mechanical effect.

In order to take full advantage of the four-electrode system a constant current (rather than constant voltage) supply should be connected to the outer electrodes, though it is not necessary to measure this current since calibration is at the inner electrodes. This technique has the advantage that any change in the voltage between the inner electrodes indicates changes within the cell, e.g. a decrease in that voltage will usually be the result of a temperature rise within the cell. If it is wished to monitor the constant current this is best done by incorporating a series of standard resistance and measuring the potential difference across it with the electrometer or digital voltmeter otherwise used at the inner electrodes.

When using a two-electrode system constant current has the property of hiding changes of temperature in that a rise of temperature lowers the solution resistance and therefore the field strength but also lowers the viscosity by about the same fraction so that the observed particle velocity remains virtually constant.

Rotating prism

A distinct aid to electrophoretic mobility observations is to insert into the optical path of the microscope a glass prism which can be rotated by a suitable small electric motor at controlled speed. A typical size for the flat-sided prism is 1–2 in across, and typical required rotation speeds are 0–0·25 rpm. With the microscope focused on a stationary object, rotation of the prism makes the object appear to move steadily. In use the prism is rotated in such a direction and at such a speed as to make actually migrating particles appear stationary. Suitable calibrations then yield the actual velocity.

This technique has the following advantages:

(i) The eye can simultaneously judge many particles (say 5–10) when required only to detect that they are stationary, whereas when measuring velocities only one particle at a time can be watched.
(ii) The occasional particle, perhaps of dust, moving at a very different speed from the rest is easily detected and, if appropriate, ignored.
(iii) A mixed population of two particle types with different mobilities is readily detected as such and both mobilities obtained separately, rather than averaged.

Laser–Doppler methods

Manual microelectrophoresis techniques are tedious, time-consuming and subject to operator bias in selecting particles to observe. Fortunately the observer can be removed and the process automated by using a laser as the illuminator and detecting the small change in frequency of laser light scattered by a moving particle, i.e. utilising the Doppler effect. The frequency shift is very small, e.g. 100 Hz, but this is readily detected by 'beating' it on a photomultiplier cathode against a reference beam split off from the illuminating beam. Alternatively, a set of interference fringes can be produced across which the particles must move. In both cases the amplified photomultiplier signal is passed to a spectrum analyser (or correlator) from which it can be displayed as a mobility distribution. The line width with zero field can give the particle size; the line width with field applied will normally be greater, reflecting the spread of mobilities.

The instrument applied by Malvern–Rank in the UK is shown in block diagram form in Fig. 3.12.

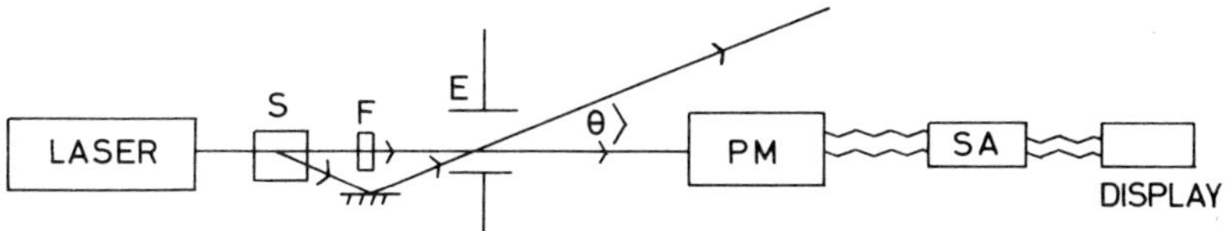

Fig. 3.12. Block diagram of a laser–Doppler electrophoresis. S is a beam splitter producing a straight-through reference beam with intensity controlled by the filler F, together with the main scattering beam scattering light from particles between the electrodes E, at angle θ, into the photomultiplier PM. SA is a spectrum analyser.

The laser–Doppler technique can, of course, be used with the cylindrical and flat cells discussed for the manual method but can also, with advantage, be used with a dip type cell as implied in Fig. 3.12 having electrodes only 1 mm apart.

Optical corrections

In the case of flat-walled cells there is no optical correction to apply in the sense that apparent fractional depths within the cell as measured by the observing microscope correspond to the true fractional depths. For the Mattson cell with, in effect, a plano-concave cylindrical lens on top of the observation tube it is necessary to apply an appreciable correction to the apparent depths for at least one of the points of focus of an inevitably distorted image, as seems first to have been pointed out by Henry.[32] The correction for thin-walled cylindrical cells is much less serious than for Mattson cells and can be made insignificant.

Figure 3.13 shows a cross-section of a cylindrical cell with wall of thickness δ and refractive index n_2 containing liquid of refractive index n_1 and immersed in a thermostatting liquid of refractive index n_3. Assuming a vanishingly small aperture for the observing objective then for a point such as P at a distance u from the top inside face, successive refractions at the inner and outer wall respectively yield the relations

$$\frac{n_1}{u} - \frac{n_2}{v'} = -\frac{(n_2 - n_1)}{a} \tag{3.65}$$

$$\frac{n_2}{v' + \delta} - \frac{n_3}{v + \delta} = \frac{(n_2 - n_3)}{(a + \delta)} \tag{3.66}$$

where all quantities have their numerical values (i.e. are positive). After

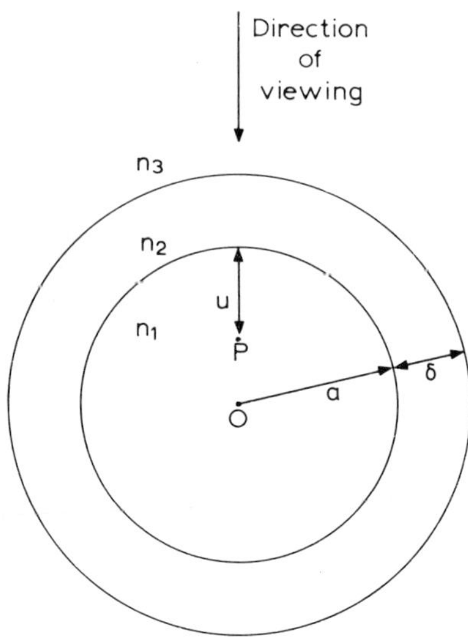

Fig. 3.13. Cross-section of a cylindrical microelectrophoresis cell.

elimination of v' these relations give v, the apparent distance of P from the top inside face as seen by the observing microscope.

A point at the centre of the capillary O gives (obviously) $v=u$ and is unshifted. If for simplicity it is assumed that $n_1=n_3$ $(=n)$ then points in the top inside edge $(u=0)$ are shifted upwards by a distance t given by

$$t=\delta a(n_2-n)/(n_2a+n\delta) \tag{3.67}$$

while points in the lower inside edge are shifted upwards by $at/(a+2t)$ which is very nearly equal to t for a typical thin-walled cell. Since the inside face of the capillary is used as a reference point for the observing microscope this means that, in effect, the upper and lower inside faces are unshifted (i.e. the apparent inside diameter is correct) and the centre shifted (relatively) downwards by t. The stationary levels will be shifted by intermediate amounts. Since the cell wall is cylindrical rather than spherical the correction to the stationary level position will be zero in the axial direction; thus if the eye is assumed to choose the 'circle of least confusion' the corrections to stationary level positions as calculated above should be halved. For a well designed thin-walled cell with $\delta<40$ μm the correction is in any case negligible.

Accuracy of electrophoretic mobilities

The chief sources of error, not already discussed, in a well-designed and calibrated cell are probably (a) the depth of focus of the observing microscope and (b) Brownian motion of the particles.

The gradient of electro-osmotic liquid velocity with distance along a diameter of a cylindrical cell is given from eqn. (3.62) by

$$\mathrm{d}v/\mathrm{d}r = 4v_{\mathrm{eo}}(r/a^2)$$

so that at the stationary level

$$\frac{\mathrm{d}v}{\mathrm{d}r} = \frac{2\sqrt{2}}{a}\, v_{\mathrm{eo}}$$

For a typical arrangement with $a \approx 2$ mm and particles with mobility $\approx v_{\mathrm{eo}}$ then an observing microscope with a depth of focus of 30 μm implies a possible error in a single particle mobility measurement of $\approx 2\%$.

The error due to Brownian motion of the particles will depend upon the size of particle and also upon the distance over which the particles are timed. The latter is commonly around 250 μm with a convenient transit time of ~ 10 s achieved typically for particles with mobility 2 to 4×10^{-4} cm/s per V/cm by applying a potential gradient of 5 to 10 V/cm. Particles of radius 0·1 μm will have a diffusion coefficient of $2{\cdot}5 \times 10^{-8}$ cm^2/s in water at 25°C giving Brownian displacements of the order of 5 μm in 10 s, again leading to errors in single mobility measurements of the order of 2%.

While the error in mobility measurements can be minimised by careful attention to the factors discussed above and by averaging over large numbers of particles, it seems reasonable to estimate that the probable error in electrophoretic mobility measurement is seldom less than 2%.

Conversion of mobilities to zeta potentials

Historically the first equation relating the observed mobility of a colloid particle to the ζ potential was due to Helmholtz and Smoluchowski and is usually referred to as the Smoluchowski equation

$$v_e = \frac{E\epsilon\zeta}{4\pi\eta} \tag{3.68}$$

where v_e is the velocity of the particle under the influence of an electric field of intensity E in a solution of viscosity η and permittivity ϵ. The ratio (v_e/E) is usually termed the electrophoretic mobility and has magnitudes typically in the range 0 to 6×10^{-4} cm^2 V^{-1} s^{-1} (cf. $7{\cdot}6\times10^{-4}$ cm^2/s per V/cm for the ion K^+). It is therefore convenient to record mobilities in $\mu m/s$ per V/cm.

If the permittivity of the dispersion medium ϵ is put equal to the dielectric constant (e.g. 78·35 for water at 25°C), this implies that $\epsilon_0=1$ and the use of corresponding esu/cgs units for other quantities, so that if E is to be understood as in V/cm and ζ is in volts a numerical factor must be introduced and

$$v_e=\frac{E\epsilon\zeta}{4\pi\eta}\times\frac{1}{299{\cdot}8^2}$$

where η is in Poise. For aqueous sols at 25°C, therefore,

$$\zeta\ (\mathrm{mV})=12{\cdot}83\times(\text{mobility in }\mu\text{m/s per V/cm})$$

The use of MKS units makes it possible to use volts directly for both E and ζ but then of course η and ϵ will take different numerical values.

The Smoluchowski equation is often derived by a method of doubtful validity which assumes a parallel-plate condenser model for the electric double layer at the particle–solution interface. A more satisfactory derivation equates the force $E\rho\mathrm{d}x$ exerted by the external field on an element in the double layer of unit cross-sectional area and length $\mathrm{d}x$ perpendicular to the (plane) particle surface, with the frictional force on the element due to the velocity gradient across it

$$E\rho\mathrm{d}x=\frac{\mathrm{d}}{\mathrm{d}x}\left(\eta\frac{\mathrm{d}v}{\mathrm{d}x}\right)\mathrm{d}x$$

Substituting for ρ from the Poisson equation in one dimension

$$\rho=-\frac{1}{4\pi}\frac{\mathrm{d}}{\mathrm{d}x}\left(\epsilon\frac{\mathrm{d}\psi}{\mathrm{d}x}\right)$$

gives

$$\frac{\mathrm{d}}{\mathrm{d}x}\left(\epsilon\frac{\mathrm{d}\psi}{\mathrm{d}x}\right)=-\frac{4\pi}{E}\frac{\mathrm{d}}{\mathrm{d}x}\left(\eta\frac{\mathrm{d}v}{\mathrm{d}x}\right)$$

Integrating once with $\mathrm{d}\psi/\mathrm{d}x=\mathrm{d}v/\mathrm{d}x=0$ at $x=\infty$

$$\frac{\mathrm{d}\psi}{\mathrm{d}x}=-\frac{4\pi}{E}\left(\frac{\eta}{\epsilon}\right)\frac{\mathrm{d}v}{\mathrm{d}x} \tag{3.69}$$

If η and ϵ are taken to be constant (i.e., independent of position outside the shear plane), eqn. (3.69) may be integrated again with $\psi=\zeta$ at the shear plane where $v=0$, and $\psi=0$ at $x=\infty$ where the solution velocity is equal to the electrophoretic velocity v_e (strictly the electro-osmotic velocity on this model of stationary solid and moving solution) to give

$$v_e=\frac{E\epsilon\zeta}{4\pi\eta}$$

This derivation makes it clear that the Smoluchowski equation only applies to plane double layers or to those in which the extent of the diffuse double layer (measured qualitatively by κ^{-1}) is small compared with the radius of curvature a of the particle, i.e. κa is large (say >500). Subject to this limitation, which should apply to all parts of the particle, the Smoluchowski equation is valid for all particle shapes.

For spherical particles (only) with small values of κa (say $<0{\cdot}05$) the Hückel equation may be used

$$v_e=\frac{E\epsilon\zeta}{6\pi\eta} \tag{3.70}$$

This equation is readily derived from the Debye–Hückel expression for the potential ψ at a distance r from an ion of radius a and charge q

$$\psi=\frac{q}{\epsilon r}\,\frac{1}{1+\kappa a}\exp\left[-\kappa(r-a)\right] \tag{3.71}$$

In an applied field E the force Eq will be balanced in the steady state by the frictional force $6\pi\eta v$ if the particle is spherical and Stokes law applies. Substituting for q from eqn. (3.71) with $\psi=\zeta$ at $r=a$ and $\kappa a\ll 1$, the Hückel equation results.

In aqueous or other solutions in which electrolytes are largely dissociated the use of the Hückel equation is seldom justified since, even for particles as small as 200 Å radius, the electrolyte concentration will have to be only $\sim 10^{-6}$ M to give $\kappa a<0{\cdot}1$, and such low ionic concentrations are difficult to control. For non-ionising solvents the values of κ will be extremely small (despite the direct effect on κ of the low ϵ) and the use of the Hückel equation may be justified.

Henry[33] showed that the Hückel and Smoluchowski equations were limiting forms, at low and high values of κa respectively, of the equation

$$v_e=\frac{E\epsilon\zeta}{6\pi\eta}\left[1+\lambda f(\kappa a)\right] \tag{3.72}$$

in which

$$\lambda = \frac{\sigma_0 - \sigma}{2\sigma_0 - \sigma}$$

where σ_0 is the electrical conductivity of the electrolyte and σ that of the particle.

In principle the effect of particle conductivity is to allow movement of the charge with respect to the particle under the influence of the applied field, an effect which should be negligible for small particles with extended double layers (κa small) as is borne out in the Henry treatment. Even where a correction is indicated by eqn. (3.72) for particle conductivity (i.e. $f(\kappa a) \neq 0$) it is probably best to omit it since the particle will quickly become polarised in the applied field, preventing any further charge transport so that the particle behaves as if it were non-conducting.

TABLE 3.2
LIMITING VALUES FOR $[1+\frac{1}{2}f(\kappa a)]$ IN THE HENRY EQUATION (eqn. (3.72))

	Low κa	*High* κa
Sphere	1·0	1·5
Cylinder parallel to field	1·5	1·5
Cylinder perpendicular to field	0·75	1·5

Henry[33] gave values for $f(\kappa a)$ for particles of spherical and other shapes, the limiting values of which are summarised in Table 3.2. Table 3.3 shows values of the correction factor for spherical particles over a range of κa values. For the reasons given above λ has been taken as $\frac{1}{2}$ in these tables.

TABLE 3.3
VALUES FOR $[1+\frac{1}{2}f(\kappa a)]$ IN THE HENRY EQUATION FOR A RANGE OF κa VALUES

κa	0·01	0·1	0·3	1	3	5	10	20	50	100	1000
$[1+\frac{1}{2}f(\kappa a)]$	1·000	1·001	1·004	1·027	1·100	1·160	1·239	1·340	1·424	1·458	1·495

The Henry equation, while taking into account the *electrophoretic effect*, i.e. the retardation caused by the movement of counterions in the opposite direction to that of the particle under the influence of the applied field, does not include the *relaxation effect* familiar in the theory of electrolyte conductance. This is again a retardation and results from the deformation of the otherwise symmetrical double layer around a charged particle when it is moving in the applied field. It is evident that the relaxation retardation will be most serious when the counterions in solution have low mobilities and when ζ is high.

The relaxation correction is most serious in the range $0{\cdot}2 < \kappa a > 50$ which is unfortunately that commonly encountered in colloid problems. The analytical treatments of Overbeek[34] and Booth,[35] especially the former, have been widely used to interpret mobilities in this range, both yielding power series in ζ but restricted in the number of terms by the mathematical difficulties. In both treatments exponential terms have been linearised with the implication $e\zeta/kT < 1$, i.e. $\zeta < 25$ mV though, as in other applications of the Poisson–Boltzmann equation, the actual range of validity seems to be considerably greater than this.

The equations of Overbeek and Booth were superseded both in range of validity and convenience of use first by the work of Wiersema, Loeb and Overbeek[36] and, more recently, by that of O'Brien and White,[37] who have published numerical computations for various electrolyte types. Before discussing the use of these calculations it is opportune to list the main assumptions on which they rest.

(i) The particles are not sufficiently close to influence each other.
(ii) The particles are rigid non-conducting spheres with the charge uniformly distributed over the surface.
(iii) The dielectric constant of the particle and the dielectric constant and viscosity in the solution are not functions of position.
(iv) Only one type of positive ion and one type of negative ion are present in the ionic atmosphere of the particle.

The Henry equation can be looked upon as the limiting form of this treatment at low ζ, the Smoluchowski equation at high κa and the Hückel equation at low κa. In these extreme cases the conditions of applicability are as already discussed. It may be noted that the effect of surface conductance, at least outside the surface of shear, is automatically included in the results of this work.

Table 3.4 shows the results of computations from the programmes of O'Brien and White[37] for the case of particles in an aqueous solution of KCl at 25°C. These could be used for any 1:1 electrolyte with similar ionic mobilities ($\lambda \cong 75$ ohm^{-1} cm^2/eq). Table 3.5 shows a selection of results from the same programmes for particles in aqueous solutions of

TABLE 3.4

PARTICLE MOBILITIES IN μm/s PER V/cm AS A FUNCTION OF THE ZETA POTENTIAL AT VARIOUS VALUES OF κa FOR PARTICLES SUSPENDED IN AQUEOUS SOLUTION AT 25°C CONTAINING ONLY 1:1 ELECTROLYTE WITH IONIC MOBILITIES CORRESPONDING TO THOSE FOR KCl

$\zeta(mV)$ \ κa	0	0·05	0·1	0·5	1·0	2·0	5·0	10	20	50	100	500	∞
25·7	1·33	1·33	1·32	1·31	1·38	1·39	1·52	1·63	1·75	1·86	1·92	1·97	2·0
51·4	2·67	2·64	2·63	2·48	2·58	2·57	2·77	3·04	3·35	3·66	3·81	3·94	4·00
77·1	4·00	3·93	3·88	3·50	3·47	3·40	3·56	3·98	4·52	5·24	5·59	5·91	6·00
102·8	5·34	5·17	5·04	4·33	4·10	3·94	4·00	4·45	5·17	6·32	7·00	7·76	8·00
128·5	6·67	6·35	6·08	4·96	4·53	4·24	4·19	4·58	5·36	6·85	7·93	9·43	10·00
154·1	8·00	7·46	7·00	5·41	4·79	4·38	4·23	4·47	5·24	6·89	8·39	10·89	12·00

TABLE 3.5

PARTICLE MOBILITIES (μm/s PER V/cm) AS A FUNCTION OF THE ZETA POTENTIAL AS FOR TABLE 3.4 BUT WITH PARTICLES IN $Ba(NO_3)_2$ AND IN $MgSO_4$ (THE LATTER IN PARENTHESES)

The particles are taken to have negative potentials and the mobilities given are therefore negative.

$\zeta(mV)$ \ κa	0	0·05	0·10	0·5	1·0	2·0	5·0	10	20	50	100	500	∞
25·7	1·33	1·31	1·30	1·25	1·24	1·25	1·35	1·50	1·66	1·82	1·90		2·00
		(1·33)	(1·32)	(1·29)	(2·29)	(1·31)	(1·41)	(1·54)	(1·69)	(1·83)	(1·90)	(1·97)	
51·4	2·67	2·58	2·51	2·18	2·02	1·93	2·00	2·27	2·67	3·24	3·56		4·00
		(2·62)	(2·57)	(2·28)	(2·15)	(2·06)	(2·12)	(2·34)	(2·70)	(3·24)	(3·55)	(3·90)	
77·1	4·00	3·75	3·49	2·69	2·35	2·11	2·02	2·22	2·70	3·66	4·43		6·00
		(3·81)	(3·61)	(2·93)	(2·55)	(2·31)	(2·21)	(2·37)	(2·77)	(3·64)	(4·38)	(5·56)	
102·8	5·34	4·74	4·24	2·91	2·46		1·88			3·11	4·11		8·00
		(4·84)	(4·40)	(3·26)	(2·69)		(2·12)		(2·37)	(3·23)	(4·16)	(6·19)	

$Ba(NO_3)_2$ and $MgSO_4$ at 25°C. The surface is taken to be negatively charged, i.e. in the case of $Ba(NO_3)_2$ it is the counterion which is doubly charged.

These tables show that for intermediate values of κa the mobility values go through a maximum as ζ is increased, an effect which was foreshadowed in the earlier work[34-36] but not then regarded as substantiated. There is considerable experimental evidence for the occurrence of such maxima and, if they are accepted, their observation might be regarded as an indication (through κa) of the particle size being observed. The programmes of O'Brien and White, illustrated in Tables 3.4 and 3.5, can be regarded as the best available mobility to zeta potential conversions.

REFERENCES

1. F. Powis, *Z. Physik. Chem.*, **89** (1915) 186.
2. E. A. Guggenheim, *J. Phys. Chem.*, **33** (1929) 842.
3. R. Parsons, *Modern Aspects of Electrochemistry*, Ed. J. M. Bockris, Butterworths, London, 1954, p. 103.
4. P. L. Levine, S. Levine and A. L. Smith, *J. Colloid Interface Sci.*, **34** (1970) 549.
5. O. Stern, *Z. Elektrochem.*, **30** (1924) 508.
6. D. C. Grahame, *Chem. Revs.*, **41** (1947) 441.
7. O. A. Esin and V. M. Shikov, *Zh. Fiz. Khim.*, **17** (1943) 236.
8. A. N. Frumkin, *Phys. Z. Sovjetunion*, **4** (1933) 256.
9. O. A. Esin and B. F. Markov, *Zh. Fiz. Khim.*, **13** (1939) 318.
10. S. Levine, J. Mingins and G. M. Bell, *J. Electroanalyt. Chem.*, **13** (1967) 280.
11. B. V. Ershler, *Zh. Fiz. Khim.*, **20** (1946) 679.
12. D. C. Grahame, *Z. Electrochem.*, **62** (1958) 264.
13. S. Levine, G. M. Bell and D. Calvert, *Canad. J. Chem.*, **40** (1962) 518.
14. S. Levine and G. M. Bell, *J. Colloid Sci.*, **17** (1962) 838.
15. D. C. Grahame and B. A. Soderberg, *J. Chem. Phys.*, **22** (1954) 449.
16. S. Levine, A. L. Smith, J. Mingins and G. M. Bell, *Proc. 4th Int. Cong. Surface Active Substances, Brussels* (1964).
17. S. Levine and A. L. Smith, *Disc. Faraday Soc.*, **42** (1966) 97.
18. A. L. Loeb, J. Th. G. Overbeek and P. H. Wiersema, *The Electric Double Layer around a Spherical Colloid Particle*, M.I.T. Press, Cambridge, Mass., USA, 1960.
19. E. L. Mackor, *Rec. Trav. Chim.*, **70** (1951) 763.
20. J. Lyklema, *Disc. Faraday Soc.*, **42** (1966) 81.
21. J. Lyklema, *Kolloid Z.*, **175** (1961) 129.
22. E. J. W. Verwey and K. F. Niessen, *Phil. Mag.*, **28** (1939) 435.
23. R. H. Ottewill and R. F. Woodbridge, *J. Colloid Sci.*, **19** (1964) 606.

24. E. P. Honig, *Trans. Faraday Soc.*, **65** (1969) 2248.
25. A. L. Smith, *J. Colloid Interface Sci.*, **55** (1976) 525.
26. C. J. van Oss and J. M. Singer, *J. Colloid Interface Sci.*, **21** (1966) 117.
27. K. N. Davies and A. K. Holliday, *Trans. Faraday Soc.*, **48** (1952) 1061.
28. S. Komagata, *J. Electrochem. Soc. Japan*, **1** (1933) 97.
    J. Allen, *Phil. Mag.*, **18** (1934) 488.
29. G. E. van Gils and H. R. Kruyt, *Kolloid-Beih.*, **45** (1936) 60.
30. S. Mattson, *J. Phys. Chem.*, **37** (1933) 223.
31. M. E. Smith and M. W. Lissie, *J. Phys. Chem.*, **40** (1936) 339.
32. D. C. Henry, *J. Chem. Soc.* (1938) 997.
33. D. C. Henry, *Proc. Roy. Soc.*, **A133** (1931) 106.
34. J. Th. G. Overbeek, *Kolloid-Beih*, **54** (1943) 287 and *Adv. Colloid Sci.*, **3** (1950) 97.
35. F. Booth, *Proc. Roy. Soc.*, **A203** (1950) 514.
36. P. H. Wiersema, A. L. Loeb and J. Th. G. Overbeek, *J. Colloid Sci.*, **22** (1966) 78.
37. R. W. O'Brien and L. R. White, *J. Chem. Soc. Faraday II*, **2** (1978) 1607.

CHAPTER 4

SURFACE ACTIVE COMPOUNDS AND THEIR ROLE IN PIGMENT DISPERSION

W. Black
Imperial Chemical Industries Limited, Manchester, UK

INTRODUCTION

Surface active compounds (substances which alter the conditions prevailing at interfaces) have been obtained from natural products by extraction or modification since prehistoric times. The cave dwellers in their drawings and paintings used egg albumen and other natural gums to disperse their pigments in liquid media and to give some degree of preservation to their works of art.

The early pigments were natural earths, charcoal or, occasionally, naturally occurring pigments and in the main egg albumen was used as a binder. It was not until the Renaissance that naturally occurring varnish extracts and rosins came into general use. For several hundred years only minor modifications were made in the use of pigments and binders in paint making but with the advent of the industrial revolution greater demands were made on the properties of paint and other film-forming binders.

It is clear that many of the pigments used up to 70 years ago were not well dispersed as we would interpret that word today, and that the dispersion properties depended to a large extent on the naturally occurring surface active components in the film-forming binders. The compounds were among the earlier macromolecular dispersing agents although this was not realised at that time. The presence of these surface active compounds was the probable explanation of the rather limited and very specific success of certain natural agents such as tallow, stearine, pitches and waxes which were added in small amounts in attempts to improve the working properties of the paints or inks.

The introduction of new pigments, such as the synthetic organic pigments which seem to be inherently more difficult to disperse in many media, led to a greater interest in dispersion techniques. The new film-forming binders and the problems associated with improved application and drying times, in addition to their use in motor-car and other production lines, placed even greater demands on dispersion and led to an increasing interest in surface behaviour and surface active compounds.

Synthetic surface active agents, and by this term we understand substances which have been especially synthesised to obtain surface active effects, represent a fairly modern development, which stemmed from demands on the technology required in modern industrial nations. The application of these surface active materials is very widespread in everyday life; examples which at once come to mind are washing and other applications in textile materials, the preparation of dispersions and emulsions, the application of agricultural sprays, and other specialised uses, the number of which increases steadily year by year. Many books and articles have been written on the use of such substances in recent years but the main theme of this chapter will be on their employment in the dispersion of pigments and their application in the pigment industry.

The author considers surface active compounds in this chapter under three main headings: (i) surface activity, the physical requirements, general properties and selection of surface active agents; (ii) the use of such agents in aqueous media and (iii) their use in non-aqueous media. In this review considerable reference is made to patents, which have been treated purely as literature — no regard is paid to their validity as patents nor to their value as monopoly instruments. The term 'surface active compound' is interpreted perhaps more loosely than many purists would desire. This is done deliberately, since it is the author's belief that restriction of this term to conventional surface active agents could not adequately cover the surface-chemical effects, such as adsorption, especially with regard to dispersion of pigments in non-aqueous media. Accordingly there are several references in the text to altering the adsorption behaviour of the surface of pigments by various coating techniques which deposit a layer or layers of a substance which may be completely insoluble in the medium into which the pigment is to be dispersed. If this coating alters the adsorption of any species in the media to enhance dispersion stability then the coating agent can also be considered to be a surface active or perhaps better a surface activating compound: as such, reference to it in this chapter is considered justified.

SURFACE ACTIVITY

The general requirement for surface activity is, of course, the adsorption of a solute, usually from a liquid phase, at one or more interfaces of the system under consideration. There are several ways in which surface active agents can be adsorbed but the most general manner and the way which can be most generally predicted lies in the amphipathic behaviour of certain organic molecules.

This amphipathic behaviour occurs most frequently in aqueous solutions. It is exhibited by a wide variety of organic types and can be most easily understood by reference to some examples of amphipathic surface active agents such as:

$C_{17}H_{35}COO^-Na^+$	sodium stearate
$C_{16}H_{33}OSO_3{}^-Na^+$	sodium cetyl sulphate
$C_{16}H_{33}N^+(CH_3)_3Br^-$	cetyl trimethylammonium bromide
$C_{16}H_{33}(OCH_2CH_2)_{20}OH$	cetyl alcohol adducts with 20 molecular proportions of ethylene oxide
$(C_3H_7)_2C_{10}H_5SO_3{}^-Na^+$	sodium diisopropylnaphthalene sulphonate
$C_8H_{17}OOC{\cdot}CH{\cdot}SO_3{}^-Na^+$ $C_8H_{17}OOC{\cdot}CH_2$	sodium dioctyl sulphosuccinate

All these compounds have amphipathic character, i.e. they all contain a hydrocarbon group which is expelled by water, and a polar group which is water-liking and which tends to seek to remain in the water. This gives rise to amphipathic adsorption in which the hydrophobic groups are oriented away from the water and the polar groups towards it.

The nature of the solubilising group on the surface active species gives rise to a convenient classification of the various agents into anionic, cationic and non-ionic types. Surface charge is a very important property of colloid systems as shown in earlier chapters, and can lead to very important technical effects. An example of this is specific adsorption, which occurs when strong attractions are set up between the polar groups of a surface active agent and specific groups in the surface of a solid phase; then adsorption with reversed orientation can occur, in which the hydrophobic groups are forced to orientate towards the aqueous phase. This is a highly specific phenomenon which is more difficult to predict; it occurs, for instance, when cetyl trimethylammonium bromide is adsorbed from an aqueous medium on to Prussian Blue and is really a case of chemisorption. This type of technique has, of course, been used in the pigment industry to render pigments more oleophilic, in flushing processes for example. In non-aqueous media specific adsorption tends to play a very important role in surface treatments and this makes the selection of surface active agents for particular technological improvements more difficult.

The distinction between amphipathic and specific adsorption is that the former is determined mainly by the general tendency of the medium to control the adsorption process by an expulsion of one part of the molecule or ion, whereas the adsorption process in specific adsorption occurs by interactions between the surface and the agent molecule.

The adsorption behaviour of polymeric surface active agents does not fall conveniently into either of the two types of adsorption, i.e. amphipathic and specific, which we have already examined. Certain water-soluble surface active polymers such as sulphonated polystyrene, hydrolysed styrene/maleic anhydride copolymers, and condensation products of naphthalene-2-sulphonic acid and formaldehyde, might be considered as having a hydrophobic backbone and hydrophilic side groups, but it seems unlikely that such a simple concept can adequately interpret the surface activity of such complicated macromolecules. Specific interactions between certain sites on the solid phase and groups in the polymeric surface active agent certainly occur but, because of the cumulative effect of several attachment points between the surface and one macromolecule, the specific bonds do not require to be as strong as with the simpler surface active species which are not polymeric.

The adsorption of polymeric materials is discussed in Chapter 2. It seems likely that because of the solution properties of such materials, a contributory factor in their surface active behaviour may be related to a sort of configurational entropy, the most probable configuration depending on both solution properties and behaviour at the surface.

Because of the difference in behaviour it would seem desirable, in addition to anionic, non-ionic and cationic types, to consider macromolecular surface active agents in another classification.

It is also possible to obtain surface active materials which belong to more than one class, e.g. they may contain both cationic and anionic groups, as in the betaine structures such as

$$C_{12}H_{25}N^{+}\begin{matrix}(CH_3)_2\\ CH_2COO^-\end{matrix} \qquad \text{dodecyl betaine,}$$

or both ionising and non-ionic hydrophilic groups as in structures of the type

$$R\text{—}(OCH_2CH_2)_nOSO_3^-Na^+$$

or

$$R\text{—}N\begin{matrix}(CH_2CH_2O)_nH\\ (CH_2CH_2O)_mH\end{matrix} \qquad \text{(in strongly acid media).}$$

Ampholytic surface active agents which can be anionic or cationic depending on the pH of the medium are also known.[1] An example of such a surface active agent in its various forms is

$C_{12}H_{25}NHCH_2CH_2COO^-Na^+$
sodium dodecylaminopropionate

$C_{12}H_{25}N^+H_2CH_2CH_2COOHCl^-$
dodecylaminopropionic acid hydrochloride

and $C_{12}H_{25}N^+H_2CH_2CH_2COO^-$
dodecylaminopropionic acid.

PROPERTIES OF SURFACE ACTIVE AGENTS

It is only within the scope of this chapter to discuss briefly the properties of thc various types of surface active agent. More extensive treatments can be found elsewhere as indicated in the bibliography at the end of the chapter. The theoretical development in earlier chapters has indicated the importance of surface tension, the thermodynamics of adsorption processes and wetting, and the problems associated with the forces between colloid particles. All these can be affected by surface active agents as indeed can precipitate formation and possibly one or two examples of how such effects arise might be advantageous. Before examining these aspects, however, it seems desirable to discuss the broad chemical characteristics of surface active compounds.

Surface active agents are organic compounds and are subject to the same factors which affect the stability of other non-surface-active organic molecules. This is an important factor to be taken into consideration when selecting surface acting compounds for use in various technological applications. For example, although sulphated alcohols are stable in alkaline or neutral aqueous solution, they are hydrolysed under acid conditions to the free alcohols and sulphuric acid, and because of this, can only be used in acid conditions if they have fulfilled their purpose before or very shortly after being exposed to the acid. Sulphonic acids do not suffer this hydrolytic attack and are better in this respect and in their thermal stability. Cationic agents can be used in acid solution but quaternary ammonium compounds, especially the pyridinium types, are liable to decompose in alkaline solution. In general it is preferable to use anionics under neutral or alkaline conditions and cationics neutral or slightly acid. The non-ionic agents are relatively insensitive to acids or alkalis, except the esters which are subject to hydrolysis under alkaline conditions.

Chemical reactions are possible between surface active agents and other species in the formulations. An example of this is the specific adsorption of cetyl trimethylammonium bromide on Prussian Blue which has already been mentioned. Anionic agents are generally incompatible with cationic agents in aqueous media but non-ionic agents are compatible with both these types.

Polymeric surface active compounds usually exist in various molecular weight ranges, and have different adsorption and solubility characteristics. This makes it difficult to predict the effects of other chemicals, both organic and inorganic, incorporated into a common solvent. Many papers have been published[2] concerning interactions between ionic surface active agents and polymers in aqueous media, and possibilities of interactions of this kind must be considered in emulsion paint formulations.

All surface active agents tend to form oriented molecular aggregates of micelles at certain concentrations. The formation of micelles is a result of the tendency of the solvent to expel certain parts of the surface active molecule, and occurs in both aqueous and non-aqueous media. In fairly dilute solutions the micelles are probably spherical although at higher concentrations they may exist in rod-shaped aggregates. In aqueous media the expulsion of the non-polar part of the molecule gives rise to the interior of the micelle having an oily nature with the polar groups orientated towards the water. The reverse is the case in non-aqueous media. The formation of micelles reduces the total free energy of the system and is therefore based on firm thermodynamic grounds. Indeed, the thermodynamics of micelle formation has been the subject of many papers in the literature.

One of the important properties of micellisation is that of solubilisation. Solutions of surface active agents, at or above the critical micelle concentration (c.m.c.), have the property of dissolving considerable amounts of added chemicals which have only very limited solubility in the solvent. For example, certain simple azo compounds are considerably more soluble in aqueous solutions of surface active agents above the c.m.c. than in water, and the solubility of styrene is considerably increased in the presence of micelles, this being the basis of emulsion polymerisation, which is examined in more detail later. In non-aqueous media the presence of micelles can lead to the solubilisation of water and this is of considerable importance in the application and performance of paints.

The concentration at which micelles form depends on the polarity and size of both radicals which can be considered as forming the surface active agent, and these are the factors which determine the efficiency of the surface active material. The concept of the balance of polar groups and non-polar groups being related to performance has always played a part in the selection of particular agents, and the introduction of a measurement of this quantity, the hydrophile–lipophile balance (HLB), which is considered in more detail later, has considerably improved the selection of surface active agents for particular technological requirements.

Adsorption and wetting are dealt with in Chapters 1 and 2 but it is perhaps useful here to summarise the conclusions especially as they relate to the application of surface active agents.

The cohesive nature of liquids and solids arises from the existence of attractive forces such as van der Waals forces. The forces are non-symmetrical on the molecules in the surface layer and this non-uniformity of forces at the surface gives rise to a boundary tension acting parallel to the surface. The tension is numerically equal to the work done to increase the interface reversibly by unit area, at constant temperature, pressure, composition, and areas of all other interfaces, and can be defined as

$$\gamma_i = \left(\frac{\partial G}{\partial \mathrm{A}_i}\right)_{p,T,n,A_j,\ldots\ldots} \tag{4.1}$$

where γ_i is the interfacial tension, A_i the area of the i-th interface and G the total Gibbs free energy of the system. When a surface active compound is reversibly adsorbed γ_i is reduced according to the Gibbs equation

$$\Gamma_r = -\frac{\mathrm{d}\gamma_i}{\mathrm{d}\mu_r} = -\frac{a_r}{RT}\cdot\frac{\mathrm{d}\gamma_i}{\mathrm{d}a_i} \tag{4.2}$$

where μ_r is the chemical potential and a_r the activity of a surface active species r, and Γ_r the surface excess relative to the solvent, which in the majority of cases can be taken as equal to the adsorption at the ith interface. From eqn. (4.2) it follows that positive adsorption leads to a reduction in the interfacial tension. If r is an ionised material or there are mixtures of micelle-forming surface active materials, special considerations have to be applied but this statement is still qualitatively valid.

Both eqns. (4.1) and (4.2) assume that equilibrium is established but in many processes, especially in the adsorption of polymeric material in viscous media, this can take a considerable time. In certain wetting processes also, the interfacial tension is lowered, but not to the equilibrium value in the time available, so that dynamic values are the important ones in many practical systems.

In the pigment industry wetting is a pre-requisite for all dispersion processes in both aqueous and non-aqueous media, and is also important in paint and other applications with reference to the final gloss, texture and adhesion of the product.

Young's equation

$$\gamma_{13} = \gamma_{23} + \gamma_{12}\cos\,\theta \tag{4.3}$$

is usually applied to describe wetting processes in a system of three phases (1, 2 and 3), where each γ is the relevant interfacial tension and θ is the contact angle provided $180° < \theta > 0°$. If $\gamma_{13} > \gamma_{23} + \gamma_{12}$ then phase 2 spreads completely over the surface and wetting is complete. In this case there is no equilibrium contact angle and the driving force behind the process is expressed by the spreading coefficient σ, given by

$$\sigma = \gamma_{13} - \gamma_{23} - \gamma_{12}\,. \tag{4.4}$$

Surface active compounds will therefore be of assistance in processes described by eqns. (4.3) and (4.4) if they can lower θ or increase σ.

Special considerations apply when we consider the wetting of porous masses such as a large quantity of pigment powder. Here two different situations can arise. The first is when the mass of pigment is thrown on the surface of the liquid. In this case the liquid has to penetrate into the fine capillaries contained between the pigment particles and displace the air. The equations for capillary rise predict that the force driving the liquid into the system of capillaries is proportional to γ_{12} cos θ, where

γ_{12} is the boundary tension between the liquid entering the capillaries and the air which is hoped will be expelled from the capillaries. This means that γ_{12} should be as *high* as possible, consistent with a small value of θ. These two variables γ_{12} and θ are not independent and it is $\gamma_{12}\cos\theta$ which has to be maximised for the best result. If, on the other hand, the liquid is poured over the mass of powder, air has to escape from the capillaries through the liquid. This case therefore demands a *low* value of γ_{12}.

These requirements can be used to select agents to improve wetting but rather than measure contact angles (a very difficult and laborious process) and interfacial tensions, it is invariably easier, especially recalling that the kinetics of such processes may be involved, to devise a simple test procedure based on the particular type of wetting which is required.

The use of surface active agents for the dispersion of pigments in a variety of media is becoming of increasing importance. In addition to deflocculation which is effected by reversibly adsorbed surface active agents, many irreversible surface-coating treatments may be said to be the application of a surface active compound in that a considerable modification of the pigment surface occurs and this can lead to improved dispersion in certain organic binders. In all these cases, the prime requirement for dispersion of solids in liquids is adsorption, reversible or irreversible, and unless this occurs dispersion cannot be improved. Although adsorption is a necessary condition, it is not a sufficient condition as certain other factors involving the forces between the particles have also to be satisfied. These forces and the effect on dispersion stability are discussed in Chapter 1. Surface active agents modify the electric charge on surfaces, they can introduce steric barriers which increase the stability and they can alter the adsorption characteristics of the surface to increase the adsorption of polymeric materials in paint and other pigment systems, thus increasing the barriers to flocculation.

In aqueous systems the stability of a suspension is due to an energy barrier which is set up by one or more of three complementary mechanisms, viz. an electric double layer (see Chapter 3), a steric protective layer and a solvation energy effect. In many cases it is impossible to separate these factors, since the electric double layer contributes to the solvation effect in water, and to the possible hydration sheath.[3] Steric protection might be considered to be exactly the same as a hydrated sheath, and here again it is difficult to separate the various factors involved, but the importance of the steric effect can be discerned in the preparation of redispersible powders, which will re-disperse into water to give highly disperse, stable colloidal dispersions.

Since the above mechanisms are independently operative, to a certain extent, it follows that there are several possible types of disperse systems which are stabilised by (i) an electric double layer only, (ii) a steric mechanism, (iii) a solvated sheath, and (iv) a combination of two or more of these effects. Systems which are stabilised by combinations of some of these mechanisms tend to be the most stable. These concepts of dispersion stability are most useful when we consider very fine dispersions where Brownian motion is sufficiently powerful to overcome any tendency towards sedimentation. The great majority of technically useful dispersions contain some material which is coarser than this size range, and this fact has two main consequences; (i) the adhesion between flocculated particles is more easily broken down mechanically, and (ii) the dispersed particles will settle even if they are completely deflocculated. The latter consequence leads to an effect which is rather paradoxical at first sight. If such particles are deflocculated or only very slightly flocculated, either in aqueous or non-aqueous media, their mutual approaches during sedimentation will only very occasionally result in adhesion, consequently the particles will glide past each other during sedimentation and the final volume of a given mass of sediment will be small. If the disperse phase is a solid, an extremely stiff clay-like mass may be formed which is very difficult to stir into the supernatant liquid. If, however, the suspended particles are brought into a more flocculated condition, so that an appreciable number of particles are adhering together as a result of collisions during settling, the final sediment tends to be more voluminous and is more easily broken up and stirred into the supernatant liquid. Such flocculated suspensions tend to exhibit rather anomalous viscosity behaviour and may even be thixotropic, i.e. they may form gel structures which break down on stirring and re-form when they are left undisturbed.

In aqueous media the adsorption of ionogenic surface active agents leads to increases in the capacity of the electric double layer provided that the adsorption process does not destroy the ionisation. Anionic agents will tend to make the stabilising potential more negative, and cationic agents will make the potential more positive. In view of the tendency for ions to flocculate dispersions of opposite ionic charge it is not surprising to find that anionic agents are more effective deflocculators in alkaline solutions, or in solutions containing multivalent anions, while the opposite is true of cationic surface active agents.

The electric double layer as an explanation of stability is satisfactory for fairly dilute aqueous systems, but in most technological dispersions the effect of concentration and presence of inorganic electrolytes is such that it seems unlikely to play a predominant role in stabilising such dispersions. It cannot account for the enhancement of stability shown by many non-ionic surface active compounds in aqueous media, nor can it account for the very much greater than expected stability of certain ionic surface active agents to added electrolytes. It seems likely that steric and/or solvation barriers have to be involved to explain such behaviour and that these barriers are even more operative in non-aqueous media, in the relatively concentrated suspensions encountered in technological systems.

In addition to affecting dispersions of solids in liquids discussed rather briefly above, surface active agents play a part in other dispersion processes such as that of liquids in liquids (emulsions) and gases in liquids (foams). The theory behind these other dispersion processes is basically the same, in that the stabilising mechanisms already discussed operate in these systems also. Certain different conditions do apply, however, in that instead of flocculation through minor contacts (as with solids), the final state in an emulsion is the coalescence of two drops to form a larger drop which can lead to complete separation of the two phases, and in a foam complete collapse through continued ruptures of the various lamellae constituting such a foam. These phenomena are discussed below, under emulsion paint systems.

SELECTION OF SURFACE ACTIVE AGENTS

The bewildering variety of types and numbers of surface active materials which are available from the many manufacturers of such chemicals can be disheartening to anyone considering using such aids for the first time. In this section we examine the general ideas behind the selection and examination of these agents and indicate, where possible, the aids to selection which can be put forward with the present state of knowledge in this field. More detailed applications and uses of specific surface active compounds are described in later sections.

In most technological problems concerning the dispersion of solids in liquids there is an obvious, or more frequently less obvious, connection with surface properties, and hence with surface activity. It is important, in the first instance, to consider what possible relations exist between effect and cause. The effect can all too obviously be seen but the cause is often more difficult to determine. This is frequently due to the fact that many of the factors involved are interconnected and it is difficult to separate the variables. In such cases it is often advisable to carry out some simple tests, even if they are not apparently commercially or technically feasible, in order to define the problem. An application of the basic theories of surface chemistry with this problem in mind can often result in a solution or partial solution to the technological difficulty.

In aqueous systems it is usually fairly easy to decide the requirements for using an anionic, cationic or non-ionic surface active agent, and the general rule that adsorption must occur at the solid–liquid interface enables us to carry out some simple tests to select a short list of possible agents. In our testing procedures, of course, we must know whether the problem can be solved by improving the wetting of the pigment, or by increasing the stability of the final suspension under our particular application conditions. At this stage it is sometimes desirable to carry out further tests, perhaps of increasing severity, in order to select the preferred agent for our purpose. Economic considerations have also to be examined as many problems can be solved to the required extent by more than one type of surface active agent. It is very difficult indeed to generalise on the cost of surface active materials as so much depends on the ready availability of raw materials, but as a result of the large market in the detergent field and simpler processing from petroleum products, the anionic agents tend to be the cheapest, the non-ionics generally are about twice as expensive and the cationics four times the cost of the anionics, but considerable variations exist within this rough guide. Among the polymeric materials the β-naphthalenesulphonic acid/formaldehyde condensates and sulphite cellulose lye liquors in many modifications are widely used because of their low cost and very wide applicability to dispersing a great variety of solids from calcium carbonate to highly priced sophisticated organic pigments.

In aqueous media amphipathic adsorption is the most frequent type of adsorption process encountered; as this is by far the most predictable behaviour, this considerably aids in the selection of surface active compounds for particular uses, such as detergency, textile processing and the like, which have created a considerable demand for soluble surface active agents. In most cases a satisfactory solution to most technological problems in aqueous media can be obtained from the large selection of commercially available surface active compounds.

When we come to consider dispersions in non-aqueous media considerable differences are observed. In the first place the great majority of surface active agents used as dispersing agents in non-aqueous media are highly specific to the systems involved. There do not appear to be any agents of comparable ubiquity to the sulphite cellulose lye liquors or naphthalenesulphonic acid/formaldehyde condensates in aqueous media, for example. Secondly, as has been mentioned previously, many non-aqueous systems such as paint vehicles, some printing ink media and rubber, have fairly powerful deflocculating properties in themselves (due to surface active constituents arising from their manufacture), so that the addition of a special surface active compound is not necessary. An important corollary of this is that in certain cases where such a surface active component is required, considerable difficulties can be experienced because of the very severe competition (for the surface of the solid) between the added surface active material and the usually much larger amounts of surface active components in the media.

The problem of selecting surface active agents for use in non-aqueous media is made more difficult by this specificity and usually it is advantageous to exercise our chemical knowledge to select agents which might be chemisorbed on the surface of the solid or to modify, by chemical means, the surface of the pigment so that adsorption of surface active materials, already present in our medium, may be increased. In such cases, in order to differentiate between several agents, a combination of a practical test as devised by an experienced technologist and a simple test for wetting or adsorption should be applied. Tests for adsorption can be very simple; a row of glass vials containing (i) a solvent or dilute solution of the medium concerned, (ii) a certain weight of the solid to be dispersed, and (iii) various surface active agents at a convenient concentration can, by examination of the relative sedimentation rates and final volume of the sediment, provide considerable information easily and fairly rapidly. Similar tests can be devised for wetting, flow and other technical problems with simple apparatus and some ingenuity.

The desirability of having a convenient yard-stick to measure the surface activity of surface active materials and the relationship between this measurement and technological performance has occupied the minds of many surface chemists over the years. The first development of this kind which exhibited a broad generality was the concept of the HLB number introduced initially by Griffin[4] about 30 years ago and based on emulsion data and originally of an empirical nature.

The surface activity of a compound depends on the size and polarity of the hydrophilic and lipophilic groups in the molecule. This is the basis of

Griffin's HLB (Hydrophile–Lipophile Balance) system of classification[5] by which HLB numbers could be determined by a very laborious experimental method, but they could also be calculated approximately from the formula of the surface active agent.

Griffin pointed out[5] that this approach still lacked exactness in that different agents could have the same HLB value but be of different chemical type. This difference in chemical type could give considerably different technological performances: as an example Griffin cited the possibility of esters of a particular fatty acid providing better emulsification in a given system than any other fatty acid esters. He attributed such behaviour to the differences in the attraction of the lipophilic group in the agent for the lipophilic surface.

In the same paper[5] the required HLB values are listed for the preparation of water in oil emulsions (4–8), oil in water emulsions (9–17) and for the solubilisation of hydrophobic compounds in water (14–17). Davies[6] has extended the calculation of HLB values by devising a system of group numbers and has arrived at the formula

$$\mathrm{HLB} = \Sigma\ (\text{hydrophilic group numbers}) - n(\text{group number per } \mathrm{CH_2} \text{ group}) + 7.$$

Davies[6] has also shown that, at least under certain conditions, the experimental HLB values can be correlated with a theory based on the relative rates of coalescence of droplets of oil in water and water in oil.

The application of the HLB system in the selection of surface active agents has been extended to pigment coloration by Pascal and Reig[7] and more recently by Weidner[8]. The former authors found that the required HLB values for a number of organic and inorganic pigments were constant in both aqueous media (for emulsion paints) and in non-aqueous paint systems. The values found using three series of non-ionic surface active agents of different HLB values established that peak performance (colour value) occurred with the same HLB number in each series but that the different lipophilic groups exerted an influence on the level of the maximum colour effect. This will of course have a considerable application in the selection of surface active agents for use in universal tinters, as is pointed out by Pascal and Reig.[7] It is interesting to note that the required HLB values of the organic pigments range from 8–11 for the relatively easily dispersed toluidine pigments to 14–16 for the very much more difficult phthalocyanine blues. Inorganic pigments excluding lamp-black, which is considerably less polar than the others on the list, have required HLB values of from 13 to over 20. The degree of strength of the derived paints varied with the composition of latex, thickener and other additives but the required HLB for the pigment remained constant.

The use of HLB values has been examined further with regard to the compatibility of titanium dioxide with phthalocyanine green in an aqueous emulsion paint system[9] and for the selection of emulsifiers for emulsion polymerisation. Greth and Wilson[10] found that while certain emulsifiers and emulsifier blends affected the physical properties of polystyrene latices such as particle size, viscosity, and stability, the maximum stability and most rapid conversions were obtained with HLB values of 13 to 16. Testa and Vianello,[11] on the other hand, conclude, from an examination of the emulsion polymerisation of vinyl chloride, that the HLB value of the emulsifiers does not correlate with their performance and suggest that this is due to the relatively low solubility of the monomer in the polymer. Bondy[12] has given some broad generalisations on the required HLB values for emulsifiers for emulsion polymerisation systems. He points out that, not surprisingly, the more polar the monomer the higher is the required HLB; the more hydrophobic the disperse phase the more sensitive it is to deviations from the optimum HLB value; with systems containing two or more monomers the required value is intermediate to those required for the pure components. More interestingly, he observes that blends of chemically dissimilar types of agents give broader stability peaks than do blends of similar chemical types. This latter observation suggests that interactions between the different chemical types of agent in the surface layer may be contributing to the stability of the emulsion drops.

More recently Condorelli[13] has carried out an investigation into the HLB system for selecting surface active agents for organic pigment dispersions. As a result of examination of eight different pigments with four series of non-ionic agents of differing HLBs in water, a water-based paint, xylene and a gravure medium he concludes that there is an optimum HLB or HLB range for a chemically homogeneous surface active agent series. However, the pigment 'required HLB' varies according to the chemical type and optimum dispersion does not necessarily equate to optimum applicational performance.

The HLB system has had some considerable success but also some limitations in application to pigment dispersion. Shinoda and co-workers[14] were the first to point out that for non-ionic agents, whose solubility changes markedly with temperature, the HLB also changes with temperature. The phase inversion temperature (or HLB temperature) is therefore a very important characteristic property of surface active agent solutions for the understanding of emulsification and solubilisation. It must also be kept in mind that specific interactions can occur between surface active agents, solid surfaces and non-surface active constituents and that such interactions could swamp the concept of the HLB number, especially in non-aqueous systems.

DISPERSION OF PIGMENTS IN AQUEOUS MEDIA

The dispersion of solids in aqueous media is widespread in many industrial processes, from the washing of clothes to the preparation of dispersions of expensive pharmaceutical products. A vast accumulation of knowledge of the factors affecting dispersion, and the large number of special surface active materials which have been synthesised, indicate the tremendous range of interest in this branch of surface chemistry. In many processes where dispersed pigments are used, such as in the coloration of emulsion paints and the mass pigmentation of regenerated cellulose and synthetic fibres, it is important for the manufacturer to understand the surface chemistry involved in the manufacturing process, in which the pigment may be only a minor constituent.

In viscose pigmentation an aqueous dispersion of the pigment is stirred into cellulose xanthate in dilute caustic soda which is then de-aerated, to prevent the formation of small air bubbles which could cause weaknesses and unattractive defects in the finished yarn. This solution is then passed through a spinnerette, containing many very tiny holes, into a spinning bath containing sulphuric acid, water, zinc and possibly nickel sulphate and some dispersing agent. There are several important surface chemical effects involved in this process. Firstly, the pigment dispersion must be fine enough initially not to block the spinnerette holes, and it must be stable enough not to form flocculates which could block the holes. The initial pigment dispersion must be wetted out thoroughly in order to avoid the occurrence of bubbles in the finished fibre and this high degree of wetting must be preserved throughout the process in order to avoid the nucleation of gases (at the pigment–viscose interface) during processing. In general, dispersing agents of sodium salts of β-naphthalene-sulphonic acid/formaldehyde condensates have found considerable application in such systems. One of the advantages of these types of agent is that although they are well adsorbed at the pigment–water interface they do not decrease the surface tension at the air–water interface to the same extent as the conventional amphipathic surface active agents and this leads to better milling behaviour in that foaming and frothing are less likely to occur. Non-ionic agents are used as additives in the spinning bath to prevent crater formation (build-up of inorganic salts) on the spinnerettes and for producing special skin effects. Alkylphenol/ethylene oxide condensates and amine/ethylene oxide adducts are used for this purpose and some pigment pastes using alkylphenol/ethylene oxide adducts as dispersing agents have also been manufactured. These non-ionic agents are probably efficacious in that their dispersing and wetting powers are still operating to a great extent even in the very strong electrolyte solutions in the viscose spinning bath.

An interesting use of quaternary surface active agents has been patented by Hoechst,[15] who claim that by using pigments containing a feebly acidic group or groups which can become acidic in the presence of a strong base, adding a quaternary ammonium or phosphonium agent, a polyglycol, water and a heavy metal salt and heating to 80°C, the pigment goes into solution and remains in solution on cooling. These soluble products are said to be especially useful in viscose pigmentation but the rather large quantities of cationic surface active agent used would appear to be a financial difficulty compared to the more conventional surface active agents used at present.

The advantages of water-based paint systems over solvent-based systems are freedom from obnoxious vapour, very much reduced fire hazard in storage of large quantities of paint and, for the user, greater ease in cleaning brushes and paint kettles and mopping up spillages. The latter advantages are very attractive for the do-it-yourself market which has become considerable in the last fifteen to twenty years. Despite the disadvantages mentioned above, solvent-based paints have been developed over the years to give excellent coverage, adhesion and protection to a variety of substrates. The main problems associated with the introduction of water-based systems are how to achieve a sufficiently high performance/cost relationship compared to conventional solvent paints with regard to adhesion, protection and durability. Water-based emulsion paints containing fine latices of a variety of synthetic copolymers, pigments, thickeners and other additives were the first type to achieve commercial success, especially for interior decoration, and more recently paints containing water-soluble alkyd resins have also had a considerable impact on industrial finishes. These products, together with improved application methods such as the electro-coating technique, have had a considerable impact on surface coating technology and, with the increasing understanding of the surface chemistry and physics of these processes, seem certain to continue to show improvements.

In the field of emulsion paints success is largely due to surface active agents. The essence of all emulsion paints is the formation of a fine dispersion or latex of water-insoluble organic polymer which, after evaporation of the water, will coalesce into a film-forming binder. Such latices are formed by the emulsion polymerisation of a certain polymerisable monomer (or monomers) which is insoluble in water. The presence of surface active agents in the aqueous polymerisation of water-insoluble monomers (such as methyl methacrylate) increases the rate of polymerisation and allows it to go almost to completion.[16]

We have seen that amphipathic substances in aqueous solution exist as single molecules (or ions) and, above the critical concentration range, as aggregates or micelles. These micelles, because of their essentially hydrocarbon interior, are able to solubilise considerable quantities of hydrocarbons, so that the solubility of a hydrocarbon in a micelle-containing solution may be many times its solubility in water. The amount solubilised depends on the relative solubility of the hydrocarbon in water and in the micelle and on the capacity for the amphipathic compound to form micelles.

When a monomer is emulsified in an aqueous solution of an amphipathic compound, it is distributed in three loci: (i) in the emulsion droplets which comprise by far the largest fraction of the total monomer, (ii) in true solution in the aqueous medium and (iii) solubilised in the micelles. In emulsion polymerisation processes polymerisation takes place via a free radical mechanism and water-soluble initiators are used mainly. Polymerisation, according to the general theory as first propounded by Harkins,[17] takes place primarily almost entirely in the micelles where the solubilised polymer is converted to polymer. At a certain stage the polymer particle can no longer be said to be in the micelle. This is due to the monomer diffusing through the aqueous phase, swelling the polymer and continuing the polymerisation process. At this stage the polymer exists in the aqueous phase as a discrete particle stabilised by an adsorbed layer of the amphipathic emulsifying agent.

The monomer emulsion droplets then supply monomer to the aqueous solution to keep the dissolved monomer concentration constant and thus slowly disappear. The emulsifying agent in solution also decreases as the polymerisation progresses as this is used to stabilise the growing polymer dispersion, and the number of micelles therefore rapidly decreases until with their disappearance the formation of primary polymer particles virtually ceases. The polymerisation reaction continues to completion in the polymer–monomer particle, but since the initiation of new polymer particles has stopped, it is evident that the amount of emulsifying agent is a factor in determining the molecular weight of the final polymer. In practice, modifiers (chain transfer agents, inhibitors, and retarders) are also used to control the polymerisation. It has been suggested by several authors that micelles do not act as the locus of initiation of polymerisation. Roe,[18] in a paper containing several interesting calculations on particle generation and stabilisation, has argued that particle generation occurs always in solution and that the growing particle or oligomer is stabilised by the adsorption of surface active agent. It would appear that while this may be a true picture and would certainly explain several anomalies described by Roe in the same paper, in normal emulsion polymerisation recipes where the micellar concentration is relatively high, the rate of radical generation of about 10^{13}/s/ml would suggest that collision of the radical or growing low-molecular-weight oligomer with a micelle is far more probable than the chances of mutual termination. As such the most likely situation is that within a very short time of initiation the radical is solubilised within a micelle and this, together with the capacity for the micelles to solubilise the monomer, gives rise to the ideal conditions for rapid polymerisation. The very large number of micelles found at concentrations not much above the c.m.c. thus accounts for the large number of particles formed in emulsion polymerisations carried out in such surface active agent solutions.

McCoy[19] has reviewed the role of surface active agents in emulsion polymerisation and, in addition to the classical case outlined above, examines two other cases, (i) where the monomer is soluble in water but the polymer is insoluble and (ii) where the polymer is insoluble in and only slightly swollen by the monomer. In the first case, exemplified by vinyl acetate, colloidal macromolecules such as polyvinyl alcohol, carboxymethylcellulose and high-molecular-weight non-ionic surface active agents are used as stabilisers. These compounds do not form micelles and the main polymerisation route occurs initially in solution until the polymer becomes water-insoluble and precipitates. The water-soluble macromolecules stabilise this precipitating polymer and allow the polymerisation to continue to an extent governed by concentration of stabiliser and other modifiers. The nature of the emulsifier or protective colloid in this case can affect the final particle size of the latex. Micellar surface active agents can also be used to polymerise vinyl acetate but here we can encounter complications resulting from polymerisation taking place both in solution and in the micelles, so we get two routes to polymer particles.

The other case, where the polymer is insoluble in, and only slightly swollen by, its monomer, has been examined by Evans *et al.*[16] using vinylidene chloride as the monomer. A micellar surface active agent, sodium lauryl sulphate, was used as stabiliser. Initially polymerisation takes place in the micelles as in the case of styrene, etc., but, because the monomer cannot swell the polymer particles, there is a less rapid transfer of monomer from the emulsified monomer droplets than in the case of styrene polymerisation. This, coupled with the reduction in surface active agent available to stabilise these monomer droplets, leads to coalescence of monomer droplets and to polymer–monomer droplet collisions which overcome the slow-down in reaction rate governed by the diffusion of monomer through the solution.

More recently van der Hoff[20] has given a comprehensive review of emulsion polymerisation in which the role of surface active agents is examined with respect to initiation, particle formation and stability. A more detailed examination of all aspects of emulsion polymerisation has been given by Gardon[21] in which the effects of comonomers and surface active agents on particle size distribution are discussed. The effects of rate, time and order of addition of reactant and surface active agent in both batch and semi-continuous production with respect to influence on the final properties of the latices are considered also by Gardon in the same chapter.[21]

The stability of latices when polymerisation is completed depends on the amount and type of surface active agent present. The behaviour of the latex is most important in application, and stability to mechanical shear, freezing and the effect of the stabilisers on flow behaviour have all to be considered. At an early stage in the study of such systems it was discovered that the usable non-ionic surface active materials were more free from objectionable foaming, and latices incorporating such stabilisers were exceptionally stable, even to freezing and thawing. This is an important factor in the stability of aqueous paint systems but a non-ionic agent, when used as the sole emulsifier for monomers which are essentially water-soluble, gives latices of increased particle size. Combinations of non-ionic and anionic agents are used in practice as this enables greater control to be obtained over particle size and molecular weight.[22] Weidner[8] has examined some aspects of the interactions of anionic and non-ionic surface active agents in latex paints and finds that there are considerable differences in rheological behaviour depending on the anionic agent used, and that the hiding power of the paint dropped with increasing concentration of non-ionic agent in the paint. It seemed likely that many more complex interacting factors would arise when pigments were added to the latex system and when this pigmented material was formulated as an emulsion paint, and indeed this is the case.

In addition to pigments and the latex suspension other additives are present in an emulsion paint. Plasticisers to promote coagulation and give improved paint films, thickeners to improve can stability, evaporation, rheology, etc., and possibly defoamers are also present in the final formulated paint. These additions are made to improve scrubbing resistance, etc., but they can have possible adverse effects on pigment dispersion, and indeed their incorporation can present considerable difficulties depending on the adsorption–desorption characteristics and relative concentration of surface active agent in both the latex and the concentrated aqueous pigment paste.

Many of these effects have been described in an excellent paper by O'Neill[23] on problems associated with polyvinyl acetate emulsion paints. O'Neill points out that although flocculation of pigment can occur on mixing a white base paint with a coloured pigment leading to a reduction in tinting strength, not all losses of tinting strength can be ascribed to failures in the surface active agent stabilising mechanism. The slow loss in tinting strength of certain Benzidine Yellow pigments was shown to be due to a crystallisation of the yellow pigment in plasticised latex particles, and that this could be inhibited by the presence of certain colloid macromolecules and certain surface active agents which presumably acted by hindering contact of the pigment particles and plasticised latex particles. Such products are probably effective in preventing crystallisation because they form a more coherent and less labile adsorption layer around the pigment particles.

The adverse effect of excess surface active agents on wet-scrubbing resistance of the paint film and water vapour permeability has been emphasised by many authors. Excess surface active agent does not improve pigment or latex dispersion stability, and indeed can markedly alter the rheological behaviour of certain compositions. These unwanted effects of surface active agents can be overcome by using a latex formed by copolymerising the vinyl compound with a small quantity of a water-soluble vinyl derivative (such as acrylic acid), which on neutralisation imparts a charge stabilisation to the latex particle, and/or by a careful selection of surface active agent. Another possible method of overcoming the effect of excess surface active agent on the resistance of the paint film to wet-scrubbing has been claimed by Hoechst[24] who, by using an ethylene oxide adduct of *p*-hydroxybenzophenone as the emulsifier, found that exposure to light had the effect of causing the emulsifier to act as a cross-linking agent for the resin in the paint film. Other surface active agents have been examined in an attempt to overcome this difficulty and the following examples illustrate the use of high-molecular-weight non-ionic agents which are claimed to give products of low water sensitivity. Plastic milling of pigments with polyvinyl alcohol and polymeric *N*-vinylpyrrolidone[25] of formula

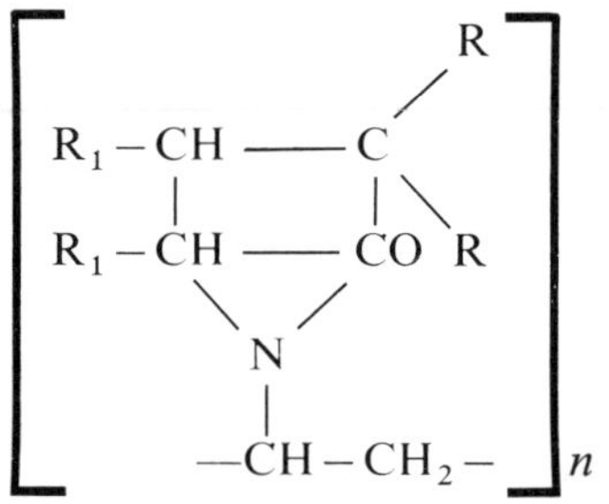

$$\left[\begin{array}{ccccc} & & & & R \\ & & & / & \\ R_1-CH & \text{——} & C & & \\ | & & | & \backslash & \\ R_1-CH & \text{——} & CO & & R \\ & \backslash & & / & \\ & & N & & \\ & & | & & \\ & & \text{—}CH-CH_2- & & \end{array}\right]_n$$

where R is hydrogen or methyl, R_1 is hydrogen, methyl or ethyl and the molecular weight is between 500 and 200 000, and mixing with styrene–butadiene latices is said to give paints of low water sensitivity, while Bowyer[26] claims that excellent gloss and water-resistant vinyl ester paints can be obtained by using mixtures of an anionic (sodium salt of sulphated methyloleate) and a non-ionic block copolymer of propylene oxide and ethylene oxide. Such block copolymers are surface active due to the fact that propylene oxide polymers are essentially hydrophobic. Propylene oxide can form adducts with alcohols, glycols or amines to give an almost infinitely adjustable hydrophobic block which can then be further reacted with ethylene oxide to incorporate one or more hydrophilic entities into the molecule. Other block copolymers prepared by treating polyamines or polyols with ethylene oxide and propylene oxide have been claimed[27] to give stable pigment dispersions in polyvinyl propionate-based emulsion paint containing cellulose ether as a thickening additive.

Combinations of non-ionic and anionic surface active agents of relatively high HLB values have been suggested for stabilising pigments and latices for emulsion paint systems. Ethylene oxide adducts of the readily available alcohols such as cetyl and lauryl alcohols and alkylphenols with alkylarylsulphonates have been used. Mixtures of alcohols reacted with ethylene oxide and carboxylic acid salts of the type

$$C_{16}H_{33}(OC_2H_4)_{12}OCH_2.C\begin{matrix} \nearrow\!\!\!\!\nearrow O \\ \searrow O^-Na^+ \end{matrix}$$

have also been suggested[28] and these probably owe their effectiveness as dispersing agents to the close packing which might be expected at the pigment–water and latex–water interfaces. Combinations of non-ionic surface active agents and polymeric carboxylic acids of not too high molecular weight have also been claimed[29] as pigment-dispersing aids which, by treatment with acid after dispersion, give high-solid easily dispersible pigment preparations for several types of emulsion paints.

Polymeric surface active compounds are being used more and more as pigment-stabilising agents. Williams[30] claims the use of olefin–maleic anhydride copolymers reacted with molar proportions of substituted amines to form the half-amide, which when used as the salts give good dispersion of many pigments in water with substantial shelf-life. Low-molecular-weight styrene–maleic anhydride copolymers[31] as their ammonium salts are claimed as pigment-dispersing agents giving improved resistance to scrubbing in various emulsion paint systems. Moriyama[32] has compared the stability of a range of inorganic pigments to a variety of polymeric and oligomeric surface active agents using sedimentation as a guide to dispersion stability. Vinyl copolymers, salts of polyacrylic acid, salts of styrene–maleic anhydride and C_8-olefin–maleic anhydride copolymers were examined alongside the β-naphthalenesulphonic acid/formaldehyde condensates.

Foaming

The problem of foam production which can arise from incorporation of a surface active agent to improve pigment dispersion in aqueous media has been mentioned previously. Foam formation in liquids arises because of the mechanical properties of the surface layer and, as large enough differences in properties between the surface layer and the bulk liquid can only be brought about by adsorption at the air–liquid interface, it is easy to see that the persistent foams are encountered generally in aqueous systems where amphipathic adsorption is greatest. The theoretical background to foams and foaming has been comprehensively reviewed by Cooper and Kitchener[33] and several monographs (listed in the bibliography) have been devoted to this subject.

Stabilisation of foams depends on the properties of the surfaces of the thin liquid films which separate the dispersed gas. The main factors have been known for a hundred years, i.e. since Gibbs showed that when a film containing an adsorbed surface active agent is stretched locally, the disturbance of the adsorption equilibrium leads to a decrease in the amount of surface active agent adsorbed per unit area. This gives rise to an elasticity effect, since the surface tension immediately rises and tends to counteract further expansion. Gibbs's arguments apply to static conditions and Marangoni showed that under dynamic conditions the effect of surface elasticity is greater than Gibbs's equations indicate. The other main factor in foam stability is surface viscosity, first postulated by Plateau, and now completely substantiated experimentally by many workers. The surface viscosity can be several orders of magnitude higher than the bulk viscosity of the liquid due to the concentration of surface active species at the interface and can be altered quite dramatically by specific interactions between two or more species giving considerable synergistic effects.

Our main interest is in the destruction of foams and the technical literature contains hundreds of examples of compositions which are claimed to inhibit or destroy foams. These substances must in general possess properties which are the opposite of those needed for foam stabilisation, i.e. they must eliminate or considerably diminish surface elasticity, increase the speed of drainage in the lamellae, decrease the stability of the interfacial films, wet out solid stabilisers into the liquid, and so on. Anti-foaming agents can therefore work in one or more ways, by displacing the foaming agent from the interface and thus reducing surface elasticity. In order to do this effectively the agent must have a low intrinsic surface tension, and must spread over the foam lamellae. A low solubility is also an obvious advantage as this increases the effectiveness by enabling small quantities of agent to be used. These factors are all present with the silicone anti-foamers and are also likely to be important in their effect on controlling flooding of pigments in non-aqueous paints. Care must always be exercised in the application of anti-foam agents, however, as their powerful surface active effects can lead to other problems if they are used incorrectly or at too high a concentration. In addition to anti-foam agents which lower the surface elasticity other agents are known which act only by increasing the rate of drainage, i.e. lowering the surface viscosity. Many cases are also known in which foams are stabilised by finely divided, partly wetted solids and addition of wetting agents which allow the liquid to completely wet the solid will break such a foam and usually hinder the formation of fresh foam.

Water-borne paints

With the advent of emulsion paints and the considerable growth in the do-it-yourself market it was an obvious extension to try to develop water-based air-drying gloss paints in the hope that the advantages of ease of application and perhaps cheapness would appeal in the market-place. Water-reducible coatings for industrial applications were also appealing, because of a reduction of fire and solvent fume hazard, and the resulting lowering of insurance costs, but for many years consumer reaction has been governed by performance and cost compared to the well-established paints using organic solvent-based coatings and most of the development work in this field has been concentrated on reconciling these requirements. Steady, but fairly slow, progress was made in water-borne paints mainly in the design and modification of resins, cross-linking and utilisation of mixed resin binder systems, until the mid 1960s. The introduction of Rule 66 in Los Angeles County to limit solvent, especially aromatics, branched chain ketones and olefin content, accelerated development considerably. This and subsequent regulations were drawn up primarily to help reduce smog formation and the main restrictions are on photochemically oxidisable chemical structures which give rise to smog-forming nuclei.

Water-borne coatings are produced with the resin in the form of emulsion polymer, a finer colloidal dispersion or as a water-soluble resin. Paints are made in all of these different forms using a variety of resin systems as either thermoplastic or thermosetting types. The various forms of water-borne resin systems have certain advantages and disadvantages compared to each other with respect to ease and variety of application, application speed, gloss, toughness, durability and cost.

Hunt[34] has examined the prospects of water-based air-drying gloss paints and came to several conclusions based on his investigations. If the resin is in the emulsified form it must wet and flow out over the pigment particles so that a non-chalky film is obtained. The emulsified particles are rather viscous, and to improve the flow plasticising components are added. Hunt reports that he attempted to incorporate the pigment into such plasticised emulsion resin bases using aqueous pigment dispersions, by triple-roll milling dry pigment into the alkyd then emulsifying. In every case, while the gloss obtained was good and wet-edge time and brushing behaviour satisfactory, the rheological properties of the brushed film were such that brush-marks did not flow out.

Several problems are also met using water-soluble resins. Normal ester-type alkyds tend to hydrolyse in the aqueous alkaline conditions required to keep them in solution and when this is overcome by the use of special hydrolysis resistance resins, the better resins are solubilised by the rather expensive method of using ionic carboxylic acid amine salts. The rheological properties of these moderately high-molecular-weight polyelectrolytes in water are such that, in order to obtain viscosities similar to those of good solvent-based paints, certain glycol ester water-soluble solvents must be added to reduce the viscosity. This, together with the necessity to adjust the rate of evaporation of the solvent system to achieve a good wet-edge retention, levelling and drying time, means that fairly substantial quantities of fairly expensive solvents have to be used. Hunt concludes that it is unlikely that water-based systems will ever be cheaper than organic solvent-based gloss paints because of the increased sophistication of the binders required, the necessity of using expensive glycol ether solvents, the use of amines to solubilise the resins and the possibility that more expensive surface-treated pigments may be required. While it seems possible that suitably surface-treated pigments might be obtained at little or no increase in cost over conventional products, the other difficulties are unlikely to be remedied easily.

Tasker and Taylor[35] have reviewed the situation of water-soluble resins with special reference to electrodeposition, a process which offers considerable advantages in the use of such systems. By the nature of the process such paints are used mainly as primer stoving finishes and are based either on emulsion systems, or on solutions of water-soluble modified alkyd, urea–formaldehyde, phenolic and acrylic resins. After stoving it is important that the film contains as small a concentration of water-soluble impurities as possible, otherwise the painted article shows poor water resistance. This obviously could create problems in the dispersion of pigments for such finishes, as most conventional surface active agents would lower the water resistance of the final film to a considerable degree. Fortunately the solubilised resins behave as dispersing agents and milling in conventional milling equipment gives satisfactory dispersions. Fry and Bunker[36] have examined the problem of wetting of various substrates and conclude that the paint has to have a surface tension below that of the lowest-energy surface concerned in order to obtain satisfactory wetting. In many cases the use of alcohols, to improve the solubility characteristics of the resin, has a beneficial action in this respect but they suggest that the use of surface active compounds which could be rendered hydrophobic on stoving might be beneficial in wetting. At the moment the continued improvement in water-soluble resins has increased the latitude for water-soluble impurities in pigments[37] and, in addition to the use of alcohols, white spirit and other solvents, small quantities of non-ionic surface active agents have been used as foam inhibitors in such systems.

Pigment dispersion does not seem to be regarded as too great a problem in electrodeposition systems at the moment although it does seem to have received some attention by Landon and Ashton,[38] who find that the addition of Calgon S in small quantities of the order of 0·15% on paint gives increased gloss in a titanium dioxide/acrylic resin system. The use of ammonium polyacrylate dispersing agents which give increased gloss and thicker films, is also mentioned by these authors. Tasker and Taylor[39] have examined the dispersion of a tinted titanium dioxide in two different water-soluble acrylic systems and find that flocculation is not markedly different from that obtained in the usually well deflocculated sprayed films. There is still some difficulty in relating adequate gloss and opacity to film thickness and here control of the pigment volume concentration may be important. The relationship between the pigment volume concentration in the bath and that in the deposited film has been examined by Robinson and Tear,[40] who found, not unexpectedly, that this depended on the relative electrophoretic mobilities of the water-soluble resin and the pigment particles. As electrophoretic mobility is affected by bath temperature, pH, solids content and applied voltage it is quite easy to appreciate that very good control of these factors is necessary in order to obtain deposited films of adequate gloss and appearance. Improvements in the control of adsorption on the pigment particles, possibly by a carefully selected surface active agent, could permit greater latitude in the bath conditions particularly if the surface charge density on the pigment could be brought to an optimum value for that of the resin system being employed.

Zinc oxide pigment has been used in paints, especially exterior paints, to improve resistance to ultraviolet light, controlled chalking, improved tint retention, increased durability and mildew resistance. Zinc oxide is a reactive pigment and forms zinc soaps with carboxylic acids. Morley-Smith[41] has examined the formation of such soaps using several fatty acids in a suitable solvent system and has reported on the viscosity behaviour with several zinc oxides including surface-modified grades. More recently, Princen[42] has examined interactions between zinc oxide and titanium dioxide in aqueous media. The behaviour here depends on the sign of the ionic charges on the surface of the pigments; if opposite signs exist at certain pH regions the pigments mutually flocculate. As might be expected there is a complex relationship between exchange of ions between the surface and solution and surface area, particle size and particle volume. Princen points out that non-ionic surface active agents are unable to overcome this problem because such agents are only physically adsorbed on the surface and are competing with the chemisorption process. The addition of phosphates is said to completely overcome this problem by chemisorption of the phosphate. Conventional anionic agents of the β-naphthalene-sulphonic acid/formaldehyde condensate type decreased the initial viscosity of the paints but increased the rate of interaction between the pigments.

An interesting solution to this problem and to the effect of surface active agents on water resistance has been suggested by Kubie *et al.*,[43] who formulated emulsion paints with and without zinc oxide present using linseed fatty acids or esters reacted with sorbitol or sorbitol/ethylene-oxide adducts. These agents are capable of reacting with film-forming binders in the paint and polymerising, thus increasing the resistance to water and in addition they give paints, containing zinc oxide pigment, which maintain a stable viscosity for over two years.

In recent years there has been considerable effort in preparing water-borne systems with improved properties and a great deal of work has been done by many companies, especially in the United States. Few references to any problems of pigment dispersion in such media have been noted in the literature and it must be assumed that there are no problems at this time.

Miscellaneous aqueous dispersions

In addition to their use in emulsion paints and viscose pigmentation, aqueous pigment dispersions are used for colouring paper both for newsprint and wallpaper. The difficulties associated with the dispersion of Prussian Blue are well known in the printing industry and we examine these later, but even in aqueous media it is found that large quantities of dispersing agents are required to obtain stable dispersions. Rogers and Todd[44] have found that the use of oxalic acid together with the sodium salt of a β-naphthalene-sulphonic acid/formaldehyde condensate enabled them to obtain satisfactory dispersions with low (0·3%) anionic surface active agent concentrations. Hoover and Sinkowitz[45] have also claimed high-solid, low-viscosity slurries of kaolin for use in paper coatings by using as little as 0·25% of the sodium salt of α-aminobenzyl-diphosphonate as dispersing agent.

DISPERSION OF PIGMENTS IN NON-AQUEOUS MEDIA

The use of surface active compounds to facilitate dispersion of pigments in non-aqueous media presents many problems to any reviewer. Some of these difficulties have been mentioned in the section dealing with the selection of surface active agents but the greatest difficulty by far is the lack of any predictive theory of dispersion stability in non-aqueous systems. The lack of understanding of all the factors involved in the dispersion of pigments in such media means that the worker in this field must rely to a very large extent on empirical sorting tests when seeking a dispersing agent for a particular system. In recent years, however, a number of useful papers have been published containing summaries of *ad hoc* information and these, together with several more fundamental papers on the nature of interactions at the pigment–vehicle interface, can form a useful starting point in this field.

Twenty years ago the number of fundamental papers on dispersion stability in non-aqueous media was small and the work tended to be rather fragmentary but since then several groups of workers and many individuals have made important contributions to this field, as can be seen by reference to Chapter 1. Notable among the centres where such work has been carried out are the National Printing Ink Research Institute at Lehigh University and the Paint Research Station of the Research Association of British Paint, Colour and Varnish Manufacturers. As a result of this and other work our understanding of interactions at the vehicle–solid interface has improved and, as adsorption is a pre-requisite for both wetting and dispersion, this has considerably improved our chances of being able to successfully apply surface active compounds to solve dispersion problems in non-aqueous media.

Various stabilising mechanisms have been suggested to explain the stability of dispersions in non-aqueous media. A very useful summary of the situation in this field has been given by Chessick,[46] who concludes that the repulsive forces set up by the interaction of electrical double layers round the particles confer the greatest degree of stabilisation. This may well be so with some systems, especially when the solvent is fairly polar, but in most commercial non-aqueous dispersions complete stabilisation is extremely rare and often undesirable, as it could lead to hard compact claying on storage, and with commercial non-aqueous dispersions control of the degree of flocculation tends to be the important factor. Controlled stabilisation by electrical repulsion alone tends to be rather difficult to apply over the variety of conditions to which the suspension might be subjected in use, and for this reason the author considers that steric barriers are more likely to offer practical solutions to dispersion stabilisation in non-aqueous media. Recently there has been considerable interest in attempts to quantify the physical barriers set up by adsorbed polymeric molecules and considerable progress has been made (see Chapter 1).

One of the major difficulties in working in non-aqueous media is in finding a dispersing agent which is adsorbed on the surface of the pigment. As mentioned previously, in many non-aqueous systems there are already present surface active materials which can compete for the adsorption sites on the pigment surface. A possible solution to this difficulty is to coat the pigment with the surface active compound before attempting to disperse it in the medium, and many successes have been achieved by this means. In the case of polymeric surface active agents such treatment probably owes its success to the occurrence of adsorption hysteresis. This arises because in many systems involving polymer adsorption, equilibrium is achieved very slowly, due to the large number of attachment points of the adsorbed molecule, and the polymer which is adsorbed initially is therefore likely to remain on the surface.

The surface active compound can be regarded as forming a bridge between the solid and the liquid in which it is dispersed. This is especially true when the polarity of the solid and the medium differs. In an interesting paper Trudgeon and Prihoda[47] illustrate this point among others; they obtained some evidence that a pigment–solvent series could be set up which indicated that in the sequence of hydrophilic to organophilic pigments the best dispersing solvents were in the order of the hydrogen-bonding parameters of the solvents as defined by Lieberman.[48] Trudgeon and Prihoda also show that a similar resin–pigment series can be set up and point out that according to their measurements, the largely used industrial solvents such as xylene and white spirit flocculate all the pigments examined.

Many papers have been published since that time on the use of solubility parameters stemming from Burrell's[49] original conception. The recent use of solubility parameters is based on the Hildebrand solubility parameter as modified for polymers, some fractional polarity parameter and a hydrogen-bonding parameter. This approach has had some very useful results in predicting solubility behaviour of resins (binders) in solvent and mixed solvent systems. Nelson *et al.*[50] describe a computer-based formulation procedure which in addition to using the above three contributing parameters allows simultaneous consideration of many additional ones such as Los Angeles Rule 66, viscosity, time to 90% weight of solvent evaporated, and cost. Attempts have been made to apply the concept to pigment dispersion by Hansen,[51] Eissler *et al.*,[52] Schreiber[53] and Sorensen.[54] Like the use of the required HLB for pigments, mentioned previously, this has met with limited success, largely because of the problems of defining specific interactions which take place between pigment surface, resin and solvent, and the changes that occur at these interfaces due to adsorbed moisture and surface active compounds (either as insoluble or slow-to-equilibriate surface-coating treatment, or as a soluble surface active additive). Sorensen,[55] in a more recent paper, deals with this by consideration of an acid–base concept in the interaction between pigments, binders and solvent.

As specific interactions appear all-important in materially altering (hopefully improving) dispersion of pigments in non-aqueous media it would seem useful to look at the various types of interaction that might be harnessed in the design of specific surface active compounds to aid pigment dispersion. Pigments can be classified as polar (many inorganic pigments including titanium dioxide, lead chrome) or non-polar (e.g. phthalocyanines, vats and quinacridones), with considerable variation arising from surface treatment of the pigment at the manufacturing stage. Non-aqueous solvents and solvent mixtures vary from low polarity (e.g. white spirit) through intermediate (xylene/butanol mixtures) to moderately polar (ethanol/methyl ethyl ketone mixtures). In dispersion of high pigment concentrations such as are found in commercial paints and printing inks, as has been mentioned previously, we consider that steric stabilisation offers the most likely stabilising barrier against flocculation. The problem therefore reduces to designing or obtaining a surface active agent with a suitable solvatable chain or chains for the solvent concerned and a group or groups which will adsorb tenaciously to the pigment surface.

There are several ways in which this problem can be tackled. Some obvious solutions are to employ some surface chemical or physical reaction involving the acid/base concept mentioned previously, block copolymers using blocks which have different solubility in the medium, multiple polar interacting groups in the same molecule to interact with polar pigment surfaces, large coloured flat nuclei such as copper phthalocyanine derivatives, or combinations of two or more of these mechanisms. Topham[56] has reviewed some of the patent literature which makes the above approach and he gives several examples of the types of stabilising chain and strongly adsorbing (anchor) groups.

Titanium dioxide is a widely used white pigment and there are numerous grades with different coatings and/or isolation techniques for particular purposes. It would be expected that these coatings materially affect adsorption behaviour at the pigment–solvent interface and thus that they can be regarded to a certain extent as surface active compounds. This problem is not examined in any detail here as it is dealt with in Chapter 9, but it is interesting to note[57] that treatment of titanium dioxide (and other pigments) with fairly simple organic compounds such as acetylacetone or phthalimide and drying at 120°C is said to give improved brilliance to the final paint and more rapid dispersion in an alkyd resin/white spirit medium. Howard and Ma[58] investigated the stability of titanium dioxide pigments, surface coated to give surfaces of different acidity, using methyl methacrylate polymers and copolymers as the stabilising agents. Adsorption measurements and relative settling rates of the differently surface-treated pigments were measured in several solvents of different solubility parameter and the effect of the introduction into the stabilising agent of small amounts of carboxylic acid of basic functionality was studied. The effect of water-wet solvents on adsorption and stability is included together with much additional information of which lack of space limits the fuller discussion merited. The authors conclude that acid–base interaction is the most important general feature which governs the adsorption of these polymers and that while adsorption is necessary it is not a sufficient condition for dispersion stability. While homopolymers give better stability of acidic pigments in toluene than in the more polar methyl ethyl ketone, copolymers with small amounts of anchoring groups are necessary for adsorption. The stability pattern is complex with carboxylic acid anchors but cationic anchor groups act best on acidic surface pigments.

The addition of surface active agents to the medium in general is less likely to improve dispersion stability than a surface treatment at the pigmentation stage in colour manufacture. Depending on the strength of the anchoring groups, this is due in part to competitive adsorption of other surface active film-forming binders and in part to adsorption hysteresis. In the latter case, especially when dealing with slow-to-equilibriate polymer adsorption, hysteresis means that the agent getting to the surface first is more likely to remain there. This does not imply, however, that the addition of simple surface active agents to the medium will have no effect in paint manufacture. The addition of cationic surface active agents is said to aid the wetting of pigments, probably in some cases for forming bonds of a chemisorption type;[59] this chemisorption probably plays a part in the improved adhesion of paints to damp surfaces which these cationic surface active agents are said to encourage. With agents which do not form chemisorption bonds, it is possible that because of their smaller molecules their rate of adsorption is high relative to the polymeric dispersing fraction of the resin, and although in time the simpler molecules are replaced on the surface by the polymeric material, the improved wetting obtained initially by the use of simple surface active agents increases the rate of attainment of fine particle size and thus lowers the grinding time in the paint mill. Kresse[60] encountered a wetting problem which resulted in the deposition of an iron oxide pigment at the juncture of the air–paint–container interface. Cationic surface active agents based on diamine oleates successfully overcame this difficulty. Another effect which is claimed to be explained by wetting is the subject of a patent by van Loo and Bitter,[61] who claim that dialkyl sulphosuccinates improve the gloss finish when paint is applied in humid conditions by maintaining a low (wetting) contact angle throughout the drying cycle and thus preserving the gloss.

The use of water-soluble surface active agents during aqueous milling to surface treat and/or finish pigments designed for use in non-aqueous media is recorded in several patents.[62] Here pigments are milled in the presence of either acid or alkali saponifiable surface active agents and when milling to a fine particle size is judged complete, the agent is hydrolysed, neutralised and washed free of electrolyte leaving the less water-soluble residue of the surface active agent coating the pigment. The pigment is thus somewhat hydrophobic and flocculates, which makes filtration easier; the surface active remnant possibly acts as a soft-texturing agent during evaporation of the water remaining in the filter cake.

The discovery of copper phthalocyanine as a potential blue pigment over 40 years ago, and its development to the present day where it is the acknowledged most versatile and universal blue pigment, has required the solution to several problems associated with its physical form and s[illegible]ristics. One of the largest applications for this pigment is i[illegible]ating field, in paints based on alkyd, nitrocellulose and [illegible], etc. In spite of the many good qualities of copper [illegible] such as its shade and intensity, three major defects have [illegible] with its use in paints, namely crystallisation, flocculation a[illegible]ted poor rheological properties of the pigmented mill b[illegible]

The high tinctorial strength of copper phthalocyanine leads to a considerable quantity of paints being prepared as reduced shades with white pigments such as titanium dioxide and it is in such reduced shades that the flocculation effect can be most easily seen. The white and blue particles flocculate to different extents depending on the shear rate at which the surface coating is applied and on the rate of drying of the solvent. The differences in polarity between the pigments have been suggested as the reason for such differing flocculation behaviour and certainly the paper by Trudgeon and Prihoda[47] supports this suggestion. Numerous attempts have been made to overcome these problems, many of them on lines similar to those used to overcome the problem of crystallising in aromatic solvents where it was found that partial chlorination of the phthalocyanine inhibited crystal growth in such solvents.

The use of conventional surface active agents has been conspicuously unsuccessful in curing the flocculation problem associated with phthalocyanine blues and even in cases where additives such as benzylcellulose have been used to improve the flocculation resistance in nitrocellulose lacquers,[63] quantities in excess of 100% by weight on pigment were used. Moser[64] claimed that aluminium tribenzoates prepared *in situ* gave useful non-flocculating properties but here again usage is high. Sulphonated phthalocyanine derivatives and mixtures of these with more conventional agents also gave non-flocculating pigments but the quantities of agent employed always led to rather weak powders often with reduced tinctorial yield.

Numerous other substituted phthalocyanines have been used in admixture with the pigment, usually at the pigment manufacturing stage, but no explanation of the surface chemistry involved in rendering these mixtures non-flocculating, except presumably the idea of altering the polarity of the molecule, was offered. As examples of this kind of treatment the claims of Siegel[65] on substituted sulphonamides and Hoelzle[66] on hydroxymethylphthalocyanine derivatives can be cited.

The use of salts of *o*-carboxybenzamidomethyl-substituted copper phthalocyanine has been claimed by McKellin *et al.*[67] and, in a very interesting paper,[68] Lacey *et al.* offer as an explanation of the deflocculating effect of these compounds a combination of a charge effect insufficient to stabilise the particles, and a hydrogen bonding of the solvent to the additive. Several patent specifications[69] have described the use of basic pigment derivatives as deflocculating agents for pigments in both aqueous and non-aqueous media. The mechanism of the deflocculating action of these basic pigment derivatives has been examined by Black *et al.*,[70] who have shown that the soluble compounds are well adsorbed on a variety of pigments from both aqueous acetic acid and white spirit solutions. Measurements of adsorption behaviour, deflocculation tests, infrared and cryoscopic studies show that the deflocculation of the soluble and insoluble basic pigment derivatives on finely divided pigments is due to hydrogen bonding between the basic groups and the acidic hydrogen atoms of the film-forming binders in the paint.

In the years since the publication of the above patents, many others, too numerous to mention here, have appeared claiming various substituted phthalocyanine derivatives as dispersing and crystal-stabilising agents.

In addition to flocculation, especially in reduced paints, two other paint defects associated with colloid chemical behaviour are flooding and floating. These two aspects have been described by Patton[71] in his excellent book as being due to the hydrodynamic behaviour of the pigment particles during evaporation of the paint solvent. Factors affecting hydrodynamic behaviour are particle size, density of the pigments and state of flocculation. These factors are interrelated and, as a result, more than one solution to such problems can be found. Crowl,[72] in a study of phthalocyanine blue and titanium dioxide pigmented paints, concludes that the finer the pigment the more prone it is to floating except where coflocculation occurs between the blue and the white pigments. Heselmeyer and Wahn,[73] on the other hand, suggest that floating is due to flocculation of one of the pigments in a paint. Addition of agents of long-chain amine salts of a polycarboxylic acid as suggested by Heselmeyer and Wahn may give controlled flocculation of the titanium dioxide or perhaps coflocculation between the blue and white pigments. Silicones, which have very low surface tensions and very high spreading pressures, would also be expected to level out the variations in hydrodynamic behaviour during solvent evaporation in paint films. Only small quantities of these agents are required to correct flooding and use of excess is to be avoided as this can lead to other defects.

Many papers have been published on the effect of conventional surface active agents on pigments in alkyd and other paint media. The effect of surface active agents on the flow properties of paint systems has been examined by Singer,[74] who finds that selectivity is very high between pigment-vehicle and agent. The improved rheological properties of his paints as measured by flow point could be due to improved wetting caused by specific interactions at the surface of the pigment or perhaps to removal by emulsification of the adsorbed water at the pigment surface. The effects of small quantities of water on the flocculation of pigments in non-aqueous media have been discussed and explained by Zettlemoyer.[75] The flocculating effects of water could be a possible explanation for the improved behaviour of surface active agent treated cadmium pigments reported by Cooper[76] as he found that the amount of water adsorbed on the finely divided treated pigment was reduced from more than 1 to about 0·05% by weight.

The various papers reported here are sufficient to indicate that surface active agents probably do not act by amphipathic adsorption in non-aqueous systems, and that for any satisfactory explanation of the effect of such agents in non-aqueous systems it is necessary to examine the possibilities of specific interactions between pigment and agent, and even agent and medium. Organic metal compounds which can have possible interactions with the pigment surface and/or certain components in the solvent have been used for a considerable time in the paint industry. Resin treatment of pigments with abietic acid and metallic rosin salts formed *in situ* have been used for many years without a great deal of understanding of the physical chemistry of the process. In a recent paper, Czajkowski and Jones[77] have examined the effects of rosination in the preparation of Lithol Red R and, with the aid of modern techniques, have thrown some light on the physical form changes which occur. Metal stearates also find uses in controlling rheological properties in the long-term storage of paints. It seems likely that the formation of multivalent salts of stearic acid can give rise to possible interactions, especially as many of these salts are not clearly defined tristearates but may contain distearates and monostearates. Recently organometallic salts of other metals such as titanium and zirconium have been examined. Sidlow[78] stated that butyl titanate has a useful effect in decreasing the time of milling of certain pigments in a long oil alkyd but that its use is limited to 0·5% by its antioxidant effect. This is another agent exhibiting considerable selectivity since the dispersion of Prussian Blue and copper phthalocyanine in both the long oil alkyd and a linseed stand oil is retarded by the presence of this additive. More complicated titanates such as di-*n*-octylene-glycol-stearyl-butyl titanate are claimed by Russell[79] to be good dispersing agents for free or colloidal carbon in organic solvents. Metal amino alcohol derivatives of zirconium such as diethyldi(triethanolamine) zirconate (*N*, *N*-distearate) and metal polyhydric alcohol derivatives of the same metal which could possibly be polymeric in character are also said by Koehler and Lamprey[80] to decrease the grinding time of pigments in alkyd media.

The major part of the preceding section has dealt with alkyd paints and the associated low dielectric white spirit solvent. Other paint systems based on melamine–formaldehyde and urea–formaldehyde have solvents of greater polarity and faster evaporation rate required for stoving finishes. The greater polarity of the solvent can lead to marked alteration in the behaviour of adsorbed surface active agents, and constituents in the solvent or solvent mixture can also exhibit competitive adsorption on the adsorption sites on the pigment surface. It is not surprising, therefore, to find that some agents which perform adequately in an alkyd/white spirit paint do not appear to have the same effect or, even when compared to other agents, the same order of efficiency in a particular stoving medium. The different film-forming binders and the dispersing effect of constituents in these binders also complicate the picture and hence make selection of a surface active agent to achieve a particular technical effect more difficult.

As has been mentioned previously, paint systems can be formulated in non-aqueous solvents of low to moderate polarity. It has already been shown that with solvents of greater polarity a marked alteration can be found in the behaviour of adsorbed surface active agents and that constituents in the solvent or solvent mixture can exhibit competitive adsorption on the adsorption sites on the pigment surface. In some cases with suitably designed anchor groups and solubilising chains this does not show up to a great extent. It is not surprising, however, to find that some agents which perform adequately in an alkyd/white spirit paint do not appear to have the same effect or even, when compared to other agents, the same order of efficiency, for example, in a stoving medium with differing solubility parameter(s).

In the more polar solvents and media surface active agents of a more polar nature such as salts of organic bases with phosphoric acid have been claimed to give stable pigment dispersions. Dimethylcyclohexylamine mono- or di-basic phosphates are said by Dreher and Rack[81] to be good dispersing agents in nitrocellulose and spirit lacquers at very low concentrations. Substituted *N*-methyl pyrrolidones[82] are also claimed to be useful in nitrocellulose lacquers and also, somewhat surprisingly, in alkyd paints and in lithographic inks.

It has been mentioned previously that in non-aqueous media amphipathic adsorption is not usually encountered and that certain difficulties can arise due to competitive adsorption processes. Waite,[83] however, has given an extremely interesting account of the development of graft copolymers which are amphipathic in an aliphatic hydrocarbon medium. By careful selection of the copolymers and control of the degree of grafting exercised by the chemistry of the disproportionation and combination reactions, amphipathic polymers have been obtained. Such products have been claimed to have been successfully used in dispersion polymerisations, in the dispersion of pigments and preformed polymer particles and for the stabilisation of emulsions in non-aqueous media. Oil-soluble polymeric dispersing agents for pigments have been claimed by Gardon *et al.*[84] from copolymers of long-chain methacrylates, methyl acrylate, maleic anhydride and/or itaconic acid. The problem of adsorption of polymeric dispersing agents at pigment surfaces has been overcome by several authors either by heating pigments having negative surface charges with fatty acid polysalts of polyethylene imine, which renders the particles hydrophobic,[85] by graft polymerisation of monomers on fresh pigment surfaces[86] or by cationic polymerisations on appropriately surface-treated pigments.[87] An interesting patent[88] illustrating the modification of pigment surface with resulting technological advantages claims that if a pigment which has been coated by a water-soluble dye containing sulphonic acid groups is treated with a long-chain fatty diamine acetate, for example, the pigment exhibits improved dispersibility, flow and resistance to loss of pigment strength when incorporated into decorative paints. More recently Kitahara *et al.*[89] have described the synthesis of an azo pigment in an organic mixture in the presence of both lauryl methacrylate–styrene and lauryl methacrylate–glycidyl methacrylate copolymers using a solvent-soluble diazotising agent. Infrared measurements showed that the latter copolymers bind chemically to the pigment through partial esterification between the glycidyl group and a carboxylic group in the pigment while the former copolymer could not so bind. The LMA–GMA copolymer inhibited crystal growth and gave stable dispersions in cyclohexane whereas the LMA–St copolymer did not.

The dispersion of pigments in printing inks is important for several reasons but the effect of dispersion on rheological behaviour is perhaps the major criterion. Because of the application methods, flow properties are all-important in inks and this is certainly the first condition which a printing ink must satisfy in order to be considered for potential use.

There is a wide variety of printing ink media in use in modern printing processes, from lithographic inks based on bodied linseed oil or synthetic resins dispersed in drying oils of high viscosity to gravure inks which are fairly thin inks. Pigments dispersed in a highly fluid vehicle which contains little or no drying oil give inks which dry mainly by adsorption and evaporation. The technical requirements for such inks vary considerably and many surface chemical effects are important in arriving at a satisfactory solution to the printing requirements. Many books have been written on the subject of printing inks and hundreds of articles published which have shed considerable light on the many complex processes which occur in the printing process and it is not proposed to enter into great detail here. It is, however, appropriate to observe that there are many different types of printing inks and that in general the problem of specificity of action of surface active agents which is present in non-aqueous paint systems can also be found in the printing ink field.

Variation in the viscosity of the ink vehicle and the methods of incorporating pigments into such vehicles have considerable effects on the shear which can be applied to pigment particles and agglomerates and therefore on the speed and fineness of the ultimate dispersion. The rheological properties are important in the application of inks; and associated phenomena, such as tack or stickiness of the ink, which is important in obtaining an even distribution on the press and proper transfer to the paper, are also controlled to some extent by the dispersion properties of the material. As in other media dispersion properties are controlled by particle size distribution, particle shape, degree of flocculation, etc., and various wetting and deflocculating agents have been used in such systems. As in paint systems where it has been shown that certain constituents in the medium have a deflocculating action, certain constituents in printing inks also show similar effects. It has been known for some time that with lithographic varnishes, increasing the acid value to about 5 improved the wetting and the dispersion, but the use of these agents is offset by attendant disadvantages such as livering and reduced drying properties. Less reactive surface active agents such as metallic soaps of naphthenic acid, sulphonated castor oil, sulphonated phenyl derivatives of fatty acids, and many other polar compounds have been suggested in addition to natural products such as lecithin and rosin. and its modifications. Cationic or amphoteric surface active agents are said to offer considerable advantages in improving pigment dispersion and pigment soft-texturing. This latter property refers to the improved texture and dispersibility in non-aqueous media which arises when the pigment is made water-repellent in aqueous media by specific adsorption with reversed orientation by some surface active agent in the aqueous phase just before drying the pigment. If the pigment is dried when it is water-wetted there is a high energy of adhesion of water to the surface and as evaporation proceeds the particles tend to be drawn together to give irreversible aggregation and a poor texture. The use of surface active agents which can give adsorption with reversed orientation, i.e. lipophilic part exposed to the water, improves the texture firstly by giving a flocculated structure in the water and hence reducing the shrinkage occurring on drying, and secondly by reducing the energy of adhesion to the water.

Carr[90] indicates that texture can be rated numerically by measurement of the surface area of the pigment in m^2/g divided by the oil absorption in g per 100 g of pigment. The paper gives a theoretical basis for this evaluation and a large number of results obtained for a variety of pigments in linseed oil. The use of stearyl *n*-propylenediamine dioleate to reduce the grinding time of pigments in printing inks and to give improved opacity and superior gloss has been described by de Vries.[91] He also said that a water-soluble alkylamine acetate was useful as a soft-texturing aid in pigment production and that the amphoteric analogue of this agent, an *N*-alkylaminobutyric acid, had had considerable success in the soft-texturing treatment of Prussian Blue. Topham[92] has explained the use of fatty amines to improve the dispersibility of azoaceto-acetarylamide pigments in gravure inks by a reaction of primary fatty amine with the keto groups in at least part of the surface of the dispersed particle during isolation. Topham *et al.*[93] have also claimed the reaction products of chloromethylated copper phthalocyanines with poly-12-hydroxystearic acid as dispersing agents for copper phthalocyanine pigment in hydrocarbon solvents which can be obtained in a finely divided deflocculated state suitable for use in gravure printing inks.

Despite the successes which have been achieved with cationic agents these still depend on specific adsorption for their effects and with many pigments this is absent. The rheological properties of some Benzidine Yellow pigments in printing inks have been consistently poor but the advantages of such pigments are such that they are used, albeit with some difficulty, in trichromatic printing inks. Carr[94] has described some experiments designed to examine the flow behaviour of Benzidine Yellow in printing ink. The effect of triethanolamine oleate and a non-ionic surface active agent added to the ink and present at the pigment coupling stage was investigated using a Ferranti–Shirley cone-plate viscometer. From the various flow-curves Carr concludes that the rheological properties are more improved when the surface active agent is present at the pigment coupling stage but with the most difficult Benzidine Yellow pigment no improvement in flow properties could be obtained with a wide variety of surface active agents.

The use of vinyl pyrrolidone as a dispersing agent for carbon black is attributed by Honak[95] to the establishment of a sphere of polar molecules round the particles. The use of titanium compounds as coating compounds for pigments to improve dispersion in lithographic varnish is the subject of a patent by Bernstein.[96] He claims that when tetrabutyl titanate is added to a pigment in hexane and subsequently hydrolysed at the pigment–hexane interface the hydrolysis product is $Ti(OH)_4$ and not the partial hydrolysis products obtained by alternative methods. The other hydrolysis product, butanol, is removed from the surface by solution in the hexane and the coated pigment can be filtered off and dried. Improved fluidity is claimed when such a treated pigment is incorporated into a printing ink and improved dispersion in rubber is also said to result.

The coloration of plastics and the problems associated with obtaining heavy depths of shade by dyeing man-made fibres has created an interest in the mass coloration of these products by pigments. In most cases it was found that the easiest way of obtaining suitable products for mass coloration was by preparing master batches of high pigment concentration, perhaps up to 70% pigment, by heating a proportion of the plastic material and mixing in the pigment in a heavy-duty mixer so that it was plastic milled into the product. Alternatively, master-batching the pigment with the plasticiser which could then be incorporated into the material at the appropriate loading has also been used. Variations on these techniques still find application and the use of milling with fine grinding media in water has been exemplified[97] using a pigment and a polyethylene wax to give a master batch suitable for use in various plastics such as polypropylene and polyvinyl chloride. Blackburn and Field[98] claim the use of low-melting finely divided powdered polyolefin by direct addition to the reaction stage used in the preparation of the pigment, stirring at a temperature below the melting point of the polyolefin and isolating the product which may then be used as master batches for plastics coloration. The use of surface active compounds is considerably more complicated in this field and is unlikely to be as successful as in paint and printing ink media due to the high temperatures which are reached in the melt-spinning or extrusion of many plastic materials such as nylon and Terylene. However, some surface active compounds are claimed to be useful aids in the mass pigmentation of nylon. Geiger and Geiger[99] claim that if a mixture of a pigment paste with caprolactam and an oleyl alcohol/ethylene oxide condensate is prepared, then added to excess caprolactam which is then polymerised, a fine dispersion of the pigment is obtained in the final polymerised material. In another patent[100] it is claimed that basic derivatives of β-naphthalene sulphonic acid condensed with formaldehyde, made by reacting the derived sulphonyl chloride with diamines such as dimethylaminopropylamine, are useful dispersing agents for carbon black when milled as their acetate salts in water and, moreover, such milled pastes when incorporated into nylon by autoclaving the paste with nylon 6·6 salt give finely divided pigment in the final melt-spun fibres. Granular pigment dispersions combining the use of low-molecular-weight polyethylene and 1-(2-hydroxyethyl)-2-heptadecyl-2-imidazoline used as the acetate salt and precipitated on the pigment surface have been claimed[101] to give granular carbon suitable for coloration of polyamides, polyvinyl chloride and acrylic/butadiene/styrene copolymer. Polymers have also been claimed[102] for treating carbon in suitable organic solvents in which the polymer dissolves, then removing the solvent by drum or spray drying to obtain a product for pigmenting plastics.

Monsanto[103] state that master-batching of polyacrylonitrile degrades the polymer owing to the high temperatures which are required in order to plastic mill the pigment. Addition of pigment or dyestuffs to the monomer inhibits polymerisation, and although many finely divided aqueous pastes are known, these are unsatisfactory in that if more than 3% water is incorporated in the spinning solution gelling occurs and this causes blockage of the spinnerettes and consequently gives poor results. Monsanto claim to have overcome these difficulties for the mass pigmentation of polyacrylonitrile and its copolymers by milling an aqueous pigment batch paste with polyethylene glycols of molecular weight between 1000 and 2000, then drying off the water at 50°C. A finely divided pigment suspension containing up to 70% pigment is obtained and this suspension is said to give good results in the mass coloration of polyacrylonitrile.

Other dispersants which have been used for pigments in non-aqueous media which, although interesting, do not fall conveniently into any of the previous categories have been exemplified by Maxey and Castle. Maxey[104] claims that the salts of *N*-cyclohexyl-*N*-palmitoyltaurine and the like are useful dispersing agents for pigments in the manufacture of wax crayons, candles and other waxes. The very large aliphatic portion of these molecules is presumably an important factor in the increased efficiency of these products over more conventional surface active agents. Castle[105] claims that very fine dispersions of pigments can be obtained in various fluoroalcohols and that such products can be used in the coloration of plastics, leather, etc. The perfluoro hydrophobic group, where the hydrogen atoms in an aliphatic hydrocarbon have been replaced by fluorine atoms, has a very low surface free energy and compounds containing such radicals have been examined by several workers[106] but the present cost would appear to limit their practical use as surface active compounds to very specialised applications. White,[107] in a comprehensive review of coloration of rubber and plastics, gives several references to surface treatment and pigment preparation.

Easily dispersed pigments

The use of cationic or amphoteric surface active agents to improve the soft-texturing of pigments dried down from aqueous media has been mentioned previously and, while this technique has proved to be successful in reducing the work required to overcome the deleterious effect of hydrophilic aggregation, the possibility of still further reducing the energy requirements for pigment dispersion in non-aqueous media has been a target for pigment manufacturers for many years. Recently considerable advances have been made in the production of easily dispersed pigments,[108] and several pigment manufacturers now offer ranges of these products which can be tailored for particular end uses such as paints, pigment inks and plastics.[109]

Moilliet and Plant[110] describe a method of forming easily dispersible pigments whereby hydrophilic aggregation is almost totally prevented by the formation of a stable, open structure of pigment particles and an oil-soluble resin or polymer. It is essential that a stable structure of sufficient strength be formed in order to withstand the hydrophilic forces arising due to evaporation of water, and several methods of doing this are indicated in the paper and the references therein. It would appear that the structure-forming 'cement' is required to be soluble in the medium into which it is desired to disperse the pigment in order to regenerate the fine particles, but evidence is produced that the structure-forming resins do not act as deflocculating agents in the hydrocarbon solvents nor is their action due to improvements in oil wettability. Their sole function appears to lie in the formation of an open matrix which adheres to the primary pigment particles and does not allow these to be drawn together by the adhesion of water or water-soluble surface active materials to their surface. Thus, while the pigment primaries are easily regenerated in a solvent which dissolves the matrix 'cement', other oil-soluble deflocculating agents are required to give good suspension stability in the solvent medium into which the pigment has been easily dispersed. Most paint systems and many printing inks have already dissolved in them suitable deflocculators for pigments and this has enabled easily dispersible pigments to gain a considerable foothold in the marketing of dry pigment preparations. A recent variation on the above method has been claimed by Marr *et al.*[111] using an alkaline water-soluble resin and aqueous pigment suspension whereby on acidification the resin is precipitated on the pigment surface. The product is isolated and brought into contact with a volatile amine in the vapour phase whereby the resin is converted to a water-soluble product. This product gives an easily dispersible pigment composition in aqueous coating systems.

UNIVERSAL TINTERS

In large-scale manufacture of paint, either aqueous or solvent-based, the coloration is carried out by preparing a tinter or stainer with a high pigment content which is then added to the bulk of the paint to give the finished coloured product. Recently, with the growth of do-it-yourself shops, there has been introduced the idea of increasing the selection of colours by producing paint to a specific colour by tinting in the shop. This can be done for both emulsion paints and gloss paints by means of universal tinters, which as their name implies, can be used satisfactorily to tint both water and solvent-based paints. Cole[112] in a review article has examined the basis of these products and concludes that the required compatibility with both water-based and solvent-based paints is generally achieved by the use of non-ionic surface active agents of appropriate HLB number and a solvent having certain solubility characteristics. The use of a co-solvent which is compatible with both water and drying oils, e.g. polyols or their ethers and salts of etherified alkylated ethylene oxide adducts of hydroxyacetic or propionic acid, has been claimed by Ritter *et al.*[113] and the use of a water-miscible non-ionic surface active agent with a dehydrated castor oil/soya lecithin mixture has been claimed as a useful medium for preparing pigments as universal tinters by Secker.[114] Kocian[115] employs another approach by plastic milling the aqueous pigment with oleic acid, separating off the water (flushing) and adding morpholine then a suitable co-solvent or solvent mixture such as ethylene glycol, diacetone alcohol and ethyl lactate.

All the products described so far are pastes of pigments in a liquid medium but Daubach *et al.*[116] have proposed that powdered universal tinters can be prepared by a suitable combination of an oil-soluble dispersing agent of HLB between 4 and 12 and a water-soluble dispersing agent. In an example quoted the oil-soluble dispersing agent is a condensation product of a sulphonated phenol, urea, and formaldehyde which was then further condensed with phenol and formaldehyde and mixed with the triethanolamine salt of an alkylbenzenesulphonic acid. The aqueous pigment paste is plastic milled with the above components and can then be dried to obtain a powder containing the pigment in finely divided form.

As can be seen from the above examples there are various ways of obtaining pigment dispersions which are compatible in both aqueous and non-aqueous media. It is unlikely, however, that tinters prepared by several of the above methods will prove as efficient as tinters in different alkyd paint media. It would appear that the particular tinter system could be optimised for perhaps only one emulsion paint formulation and one type of gloss paint, although with the almost continuous variations which are possible in such systems this might be fairly difficult to detect.

PIGMENT FLUSHING

In many pigment manufacturing processes the production of the final pigmentary form (i.e. particle shape, size and polymorphic form) is carried out by a chemical or physical reaction in an aqueous medium. In many cases this gives a flocculated suspension due to the presence of electrolytes, and filtration and washing is required before a batch paste is obtained. If the batch paste is then dried a product of a poor texture is obtained and the aggregation of primary particles which occurs during the evaporation of the water has to be overcome by some mechanical milling process to obtain a satisfactory dispersion in a non-aqueous medium. The use of surface active agents as soft-texturing aids and the more recent surface-treated easily dispersed pigments in the reduction and almost elimination of hydrophilic aggregation have been mentioned above. The very much older method of overcoming this aggregation is to flush the pigment from the aqueous phase into a non-aqueous water-immiscible vehicle and thus avoid drying the pigment altogether. The easiest way to carry out this process is to make the pigment surface oleophilic by adsorption of a surface active compound with reversed orientation. The manufacture of white-lead paints was carried out using this method more than a century ago[117] and it seems likely that the success of this depended on the formation of insoluble lead soaps on the surface of the pigment particles. Water-soluble cationic surface active agents which also form interfacial layers with reversed orientation on pigments in aqueous media are also advocated for promoting flushing and are probably the most widely used surface active agents for this purpose. Bass[118] points out that cationic agents have the additional advantage that they can be applied under acid conditions and in the presence of heavy metal ions which are often present in pigment slurries. The flushing process is usually carried out in a heavy duty mixer of the Werner–Pfleiderer type and the separated aqueous phase is decanted. It is important that the surface active agent does not act as an emulsifying agent, otherwise the flushing process is impaired. The formation of an oleophilic surface layer round the pigment particle leads to flocculation of the pigment in the aqueous phase and this aids the flushing process, but as a result flushing tends not to be 100% efficient and some water is carried into the oil phase, this is usually removed by drying under vacuum. In addition certain imidazolines[119] have been claimed to promote the flushing of iron blue pigments. The combination of 1-hydroxyethyl-2-heptadodecenylimidazoline and oleyl alcohol ethylene oxide adduct was claimed to give a smooth readily mixed flushed pigment of increased solids content.

The fundamentals of the pigment flushing process have been examined by Gomm *et al.*[120] who, by considering the thermodynamics of the process together with the use of a technique devised to estimate the wettability of small particles, postulate that the equilibrium water wettability of the pigment is increased when precipitated anionic surface active agents are used. These authors suggest that in spite of this increase in water wettability, the surface treatment promotes flushing due to the free energy changes arising at the pigment surface during the dissolution of the precipitated oil-soluble agent. This flushing overcomes the increase in water wettability and thus carries the pigment into the oil phase, albeit with some associated water.

The disadvantages of flushing as a method of preparing non-aqueous dispersions are that unless there is a close relation between the pigment manufacturer and the pigment user, the user has to be content with the vehicle that is incorporated in the flushed pigment or have such considerable business that he can afford to purchase batch paste and carry out his own flushing process into his preferred medium.

PIGMENT–RESIN PRINTING OF TEXTILES

The production of textile prints using finely divided pigments together with resins which when heated cross-link and bind the pigment particles to the fabric has been carried out for a number of years. Instead of the conventional water-soluble gums and modified starches used as thickeners in the dyestuff printing trade, pigment printing of textiles uses either oil-in-water (O/W) or water-in-oil (W/O) emulsion thickenings. The thickening effect depends on the rheological properties of the emulsions, which being dispersions of liquid in liquid have flow properties somewhat similar to the flow properties of solids in liquids, being influenced by concentration, particle size and flocculation of the disperse phase.

In recent years most of the improvements in pigment printing have been by modification of the basic chemicals to give improved handle and fastness properties which, together with considerable unpatented 'know-how' in the very complex formulations, has succeeded in reducing the complexity of the process. Ratcliffe,[121] in a recent review, makes the above comments but in addition points out that much work has been done in getting away from the use of solvent to achieve all-aqueous pigment printing. The development of synthetic thickeners has enabled the pigment printer to overcome the harsh handle of the earlier formulations. Probably the most critical feature of such systems is the rheological properties of synthetic thickeners, which are based mainly on polycarboxylic acids, necessitating the use of non-ionic surface active agents for pigment dispersion. If anionic surface active agents are used the electrolyte effect on the thickener could be such that the rheology of the print paste is affected over the differing shear rates found during processing. Ratcliffe points out that the disadvantages of white spirit both from cost, hazard and ecological impact are such that the all-aqueous systems are likely to be the method of pigment printing in the future. One example of the different types of surface active additives which may be added to print pastes is that claimed by Helfert *et al.*,[122] who advocate the use of tridecyl citrate to give deeper, more brilliant colours on polyester–cotton printed with a vat pigment in a synthetic resin print paste.

PROBLEMS AND PROSPECTS

In the preceding sections a fairly wide range of problems connected with the use of surface active agents in the dispersion of pigments in liquids has been covered. As can now be seen the field is indeed very wide and in this review the terms of reference of surface active compounds, pigments and indeed liquids have been interpreted very freely indeed. The limited depth of coverage is perhaps regrettable in some ways, but it is essential in that there have been literally thousands of references to pigment-treating in the literature. There seems little doubt that over the past 10 to 20 years considerable advances have been made in surface treatment and dispersion of pigments, many of them rather empirical, but some as a result of the considerable increase in our understanding of the pigment–medium interface. The greater sophistication of many of the pigment-using industries, changes in methods (some brought about by the increased cost of petroleum-based raw materials) and the impact of environmental pressures are all continuing to play a significant part in demand for better and different dispersion processes. Technological requirements, perceived or yet to be recognised, will continue to be the spur and carrot for further improvements, to achieve the improvements in ease of mixing and increased speed of dispersion required for the future.

What has been achieved in recent years? Techniques of producing products such as easily dispersible pigments which depend on a controlled surface treatment of the pigment at the manufacturing stage, and the development of special coatings of chemisorbed surface active or surface activating materials together with 'tailored' polymeric deflocculants have shown considerable promise in overcoming the problems of dispersion in non-aqueous media. Although many of these developments have been technology-led they have to a considerable extent depended on our increased knowledge of stabilising mechanisms and the effect of differing solvent/surface active agent interactions. Until perhaps six to ten years ago the major scientific challenge was to develop non-aqueous dispersions to the techno-commercial level of aqueous dispersions but now, with the changing emphasis on aqueous systems, further challenges may appear. It appears self-evident that it would be unlikely that one particular product can be made to satisfy the highest technological requirements in a variety of media. Products which at present fulfil such requirements are arrived at by a judicial selection of properties to achieve a satisfactory compromise of the various conflicting requirements. This may be an acceptable solution in the present structure of technical requirements and cost, but from a purely scientific outlook the best results can be obtained only by not compromising too much.

The surface treatment of pigments has been carried out for hundreds of years but only recently has the scientific understanding of the colloid chemistry involved been harnessed to help in the selection and synthesis of more effective products. Technological change will continue and, while puzzling results and problems will be better understood and possibly explained, more problems and newer puzzles will arise. Some of these will be capable of explanation, at least to some degree, if we continue to ask the question attributed to J. Clerk-Maxwell—'What's the go of that?' The challenge of technology will always be there demanding explanations and further refinements of our scientific understanding.

BIBLIOGRAPHY

General

E. JUNGERMAN, *Cationic Surfactants*, Marcel Dekker Inc., New York, 1970.

W. M. LINFIELD (Ed.), *Anionic Surfactants*, Marcel Dekker Inc., New York, 1976 Parts 1 and 2.

J. L. MOILLIET, B. COLLIE and W. BLACK, *Surface Activity*, 2nd edition, Spon, London 1961.

R. SAPPOK and B. HONIGMANN in *Characterisation of Powder Surfaces*, Ed. G. D. Parfitt and K. S. W. Sing, Academic Press, London, 1976, p. 231.

M. SCHICK, *Non-Ionic Surfactants*, Edward Arnold, London, 1967.

N. SCHONFELDT, *Surface-Active Ethylene Oxide Adducts*, Pergamon, Oxford, 1969.

H. WARSON, *The Application of Synthetic Resin Emulsions*, Benn, London, 1972.

Foams

J. J. BICKERMAN, *Foams, Theory and Industrial Applications*, Reinhold, New York, 1953.

R. BUSCALL and R. H. OTTEWILL in *Colloid Science*, Specialist Periodical Reports Vol. 2, Chemical Society, London, 1976, p. 191.

E. MANEGOLD, *Schaum*, Strassenbau, Heidelberg, 1953.

Flushed Pigments

T. A. LANGSTROTH in *Pigment Handbook*, Ed. T. C. Patton, Wiley, New York, 1973, Vol. 3, p. 447.

REFERENCES

1. D. L. Anderson, *J. Am. Oil Chem. Soc.*, **34** (1957) 188.
2. H. Arai and S. Horin, *J. Colloid Interface Sci.*, **30** (1969) 372; S. Saito, *Kolloid Z.*, **216** (1968) 10; S. Saito and T. Taniguchi, *J. Colloid Interface Sci.*, **44** (1973) 114; Th. F. Tadros, *J. Colloid Interface Sci.*, **46** (1974) 528; E. D. Goddard, R. B. Hannan and G. H. Matteson, *J. Colloid Interface Sci.*, **60** (1977) 214.
3. B. V. Derjaguin and A. Titijevskaya, *Proc. 2nd Int. Congress on Surface Activity* (*London*), **1** (1957) 211.
4. W. Griffin, *J. Soc. Cosmetic Chem.*, **1** (1949) 311.
5. W. Griffin, *J. Soc. Cosmetic Chem.*, **5** (1954) 249.
6. J. T. Davies, *Proc. 2nd Int. Congress on Surface Activity* (*London*), **1** (1957) 426.
7. R. H. Pascal and F. L. Reig, *Off. Dig. Federation Soc. Paint Technol.*, **36** (1964) 839.
8. G. L. Weidner, *Off. Dig. Federation Soc. Paint Technol.*, **37** (1965) 1351.
9. J. Rapach, *Am. Paint J.*, **51** (1967) No. 47, 92.
10. G. G. Greth and J. E. Wilson, *J. Appl. Polym. Sci.*, **5** (1961) 135.
11. F. Testa and G. Vianello, *J. Polym. Sci.*, **C27** (1969) 69.
12. C. Bondy, *J. Oil Colour Chem. Assoc.*, **49** (1966) 1045.
13. G. Condorelli, *XI FATIPEC Congress* (1972) 631.
14. See K. Shinoda and H. Kuneida, in *Microemulsions*, Ed. L. M. Prince, Academic Press, London, 1977, pp. 58–59 and references therein.
15. Farb. Hoechst, FP 1,396,714.
16. J. H. Baxendale, M. G. Evans and J. H. Kilham, *Trans. Faraday Soc.*, **42** (1946) 668; J. H. Baxendale and M. G. Evans, *Trans. Faraday Soc.*, **43** (1947) 210.
17. W. D. Harkins, *J. Am. Chem. Soc.*, **69** (1947) 1428.
18. C. P. Roe, *Ind. Eng. Chem.*, **60** (1968) No. 9, 20.
19. C. E. McCoy, Jr., *Off. Dig. Federation Soc. Paint Technol.*, **35** (1963) 327.
20. B. M. E. van der Hoff, in *Solvent Properties of Surfactant Solutions*, Ed., K. Shinoda, Edward Arnold, London, 1967, p. 285.
21. J. L. Gardon, in *Polymerisation Processes*, Eds. C. E. Schildknecht and I. Skeist, Wiley, New York, 1977, p. 143 ff.
22. M. E. Woods, J. S. Dodge, I. M. Krieger and P. E. Pierce, *J. Paint Technol.* **40** (1968) 541.
23. L. A. O'Neill, *J. Oil Colour Chem. Assoc.*, **41** (1958) 780.
24. Farb. Hoechst, Dutch Patent 67.14978.
25. General Aniline and Film Corp., UKP 835,637.
26. W. Bowyer, UKP 885,604 to Imperial Chemical Industries Ltd.
27. H. Belde, K. Oppenlaender and E. Daubach, USP 3,841,888.
28. W. Ritter, K. Hofer, K. U. Steiner and E. Hess, FP 1,286,980.
29. J. B. Blackburn and A. W. Field, Ger. P. 2,547,539 to Ciba–Geigy AG.
30. C. R. Williams, USP 3,235,526 to Monsanto.
31. Anon., *Pigment Resin Tech.*, **6** (1977) No. 5, 7.
32. N. Moriyama, *Colloid Polym. Sci.*, **254** (1976) No. 8, 726.
33. G. Cooper and J. A. Kitchener, *Quart. Rev.* (*London*), **13** (1959) 71.

34. T. Hunt, *J. Oil Colour Chem. Assoc.*, **53** (1970) 380.
35. L. Tasker and J. R. Taylor, *J. Oil Colour Chem. Assoc.*, **48** (1965) 122.
36. E. S. J. Fry and E. B. Bunker, *J. Oil Colour Chem. Assoc.*, **43** (1960) 640.
37. A. Strickland, *Paint Varnish Prod.*, **53** (1963) Nov., 61.
38. G. Landon and I. H. Ashton, *J. Oil Colour Chem. Assoc.*, **49** (1966) 202.
39. L. Tasker and J. R. Taylor, *J. Oil Colour Chem. Assoc.*, **49** (1966) 640.
40. F. D. Robinson and B. J. Tear, *J. Oil Colour Chem. Assoc.*, **53** (1970) 265.
41. C. T. Morley-Smith, *J. Oil Colour Chem. Assoc.*, **41** (1958) 85.
42. L. H. Princen, *Off. Dig. Federation Soc. Paint Technol.*, **37** (1965) 766.
43. W. L. Kubie, J. L. O'Donnell, H. M. Tester and J. C. Cowan, *J. Am. Oil Chem. Soc.*, **40** (1963) 105. See also USP 3,140,191.
44. L. R. Rogers and W. Todd, UKP 973,428 to Imperial Chemical Industries Ltd.
45. M. F. Hoover and G. D. Sinkowitz, USP 3,713,859 to Calgon Corp.
46. J. J. Chessick, *Am. Ink Maker*, **40** (1962) Jan., 28.
47. L. Trudgeon and H. Prihoda, *Off. Dig. Federation Soc. Paint Technol.*, **35** (1963) 211.
48. E. P. Lieberman, *Off. Dig. Federation Soc. Paint Technol.*, **34** (1961) 30.
49. H. Burrell, *Off. Dig. Federation Soc. Paint Technol.*, **27** (1955) 726 and 1069.
50. R. C. Nelson, V. F. Figurelli, J. G. Walsham and G. D. Edwards, *J. Paint Technol.*, **42** (1970) 644.
51. C. M. Hansen, *J. Paint Technol.*, **39** (1967) 104.
52. R. L. Eissler, R. Zgol and J. A. Stolp, *J. Paint Technol.*, **42** (1970) 483.
53. H. P. Schreiber, *J. Paint Technol.*, **46** (1974) 35.
54. P. Sorensen, *J. Oil Colour Chem. Assoc.*, **50** (1967) 226.
55. P. Sorensen, *J. Paint Technol.*, **47** (1975) 31.
56. A. Topham, *Prog. Org. Coatings*, **5** (1977) 237.
57. Laporte Titanium Ltd., FP 1,437,065.
58. G. J. Howard and C. C. Ma, *J. Coat. Technol.*, **51** (1979) No. 651, 47.
59. M. K. Schwitzer, *Off. Dig. Federation Soc. Paint Technol.*, **33** (1961) 1111.
60. P. Kresse, *A. Paint. J.*, **54** (1969) No. 5, 28.
61. M. van Loo and V. W. Bitter, USP 2,886,456 to Sherwin-Williams Co.
62. E. Spietschka and M. Urban, UKP 1,414,116 to Farb. Hoechst; K. H. Wolf, R. Hoernle, G. Popp, K. Nonn and J. Spille, Ger. Offen. 2,160,208 to Bayer; D. R. Harris and L. E. Rowley, Ger. Offen. 2,646,211 to Pigment Manufacturers of Australia.
63. B. T. Stephens, USP 2,851,371, to Pittsburgh Plate Glass Co.
64. F. H. Moser, USP 2,965,662 to Standard Ultramarine and Colour Co.
65. A. Siegel, USP 2,861,005 to E. I. du Pont de Nemours.
66. K. Hoelzle, UKP 893,165 to Ciba.
67. W. H. McKellin, H. T. Lacey and V. A. Giambalvo, USP 2,855,403 to American Cyanamid Co.
68. H. T. Lacey, G. L. Roberts and V. A. Giambalvo, *Paint Varnish Prod.*, **48** (1958) Apr., 33.
69. A. Schoellig, R. Schroedel and H. J. Hasse, UKP 949,739 and 985,620 to Badische Anilin u. Soda Fabrik; G. Barron, W. Black and A. Topham, UKP 972,805 to Imperial Chemical Industries Ltd.
70. W. Black, F. T. Hesselink and A. Topham, *Kolloid Z.*, **213** (1966) 150.

71. T. C. Patton, *Paint Flow and Pigment Dispersion*, Wiley-Interscience, New York, (1979) pp. 592–7.
72. V. T. Crowl, *J. Oil Colour Chem. Assoc.*, **50** (1967) 1023.
73. F. Heselmeyer and W. H. Wahn, *Paint Manuf.*, **38** (1968) No. 8, 16.
74. F. Singer, *Off. Dig. Federation Soc. Paint Technol.*, **32** (1960) 762.
75. A. C. Zettlemoyer, *Off. Dig. Federation Soc. Paint Technol.*, **29** (1957) 1238.
76. E. C. Cooper, *Paint Manuf.*, **33** (1963) No. 2, 57.
77. W. Czajkowski and F. Jones, *J. Soc. Dyers Colour.*, **93** (1977) 313.
78. R. Sidlow, *J. Oil Colour Chem. Assoc.*, **41** (1958) 577.
79. C. A. Russell, USP 2,913,469 to National Lead Co.
80. J. D. Koehler and H. Lamprey, UKP 885,679 to Union Carbide Corp.
81. E. Dreher and F. Rack, UKP 884,147.
82. F. J. Prescott, *Paint Varnish Prod.*, **50** (1960) No. 12, 31.
83. F. A. Waite, *J. Oil Colour Chem. Assoc.*, **54** (1971) 342.
84. J. L. Gardon, M. Kalandiak and La Verne N. Bauer, USP 3,413,255 to Rohm and Haas.
85. P. F. Pascoe, *Farbe u Lack*, **74** (1968) No. 3, 245.
86. A. B. Taubman, G. S. Blyskosh and L. P. Yanova, *Lakokras. Mat. Prim.*, **3** (1966) 10 (from *Chemical Abstracts*, **65** (1966) 9167g).
87. R. Kroker and K. Hamann, *Angew. Makromol. Chem.*, **13** (1970) 1.
88. G. F. Bradley, D. Price and A. Hamilton, UKP 1,356,254 to Ciba–Geigy.
89. A. Kitahara, S. Kowasaki and K. Kon-no, *J. Colloid Interface Sci.*, **63** (1978) 170.
90. W. Carr, *J. Oil Colour Chem. Assoc.*, **49** (1966) 831.
91. R. J. de Vries, *Paint Manuf.*, **29** (1959) Feb., 59.
92. A. Topham, *J. Appl. Chem.*, **18** (1968) 233.
93. H. P. D. Paget, L. R. Rogers, A. Topham, J. K. D. Royle and J. F. Stansfield, UKP 1,343,606 to Imperial Chemical Industries Limited.
94. W. Carr, *J. Oil Colour Chem. Assoc.*, **45** (1962) 28.
95. E. R. Honak, *Plaste Kautchuk*, **11** (1964) No. 6, 372 (from *Chemical Abstracts*, **63** (1965) 16611b).
96. I. M. Bernstein, USP 3,025,173.
97. Ciba, UKP 1,042,906.
98. J. B. Blackburn and A. W. Field, UKP 1,518,094 to Ciba–Geigy.
99. G. Geiger and A. Geiger, FP 1,266,096 to Sandoz.
100. Imperial Chemical Industries Limited, Belg. P. 675,964.
101. Hercules, USP 3,755,244.
102. P. Werle, H. Gräf and E. Walter, UKP 1,519,320 to Deutsche Goldund Silber-Scheideanstalt.
103. Monsanto, UKP 990,122.
104. W. J. Maxey, USP 2,919,993 to General Aniline and Film Corp.
105. J. E. Castle, USP 3,129,053 to E. I. du Pont de Nemours.
106. J. L. Moilliet, B. Collie and W. Black, *Surface Activity*, 2nd edition, Spon, London, 1961, pp. 442–3.
107. H. G. White, *Rev. Prog. Coloration*, **8** (1977) 79–80.
108. Pigment Dispersions Supplement, *Paint Technol.*, **34** (1970) No. 9, i–xx.
109. J. E. Todd, *Paint Oil Colour J.*, **159** (1971) 138.
110. J. L. Moilliet and D. A. Plant, *J. Oil Colour Chem. Assoc.*, **52** (1968) 289.

111. P. W. Marr, D. W. Beattie and J. D. Easton, USP 4,166,811 to Dominion Colour Company Limited.
112. R. J. Cole, *Rev. Curr. Lit. Res. Br. Colour Varn. Mfrs.*, (1964) 635.
113. W. Ritter, K. Hafer, K. U. Steiner and E. Hess, FP 1,286,980 to Sandoz.
114. C. W. Secker, Jr., USP 2,996,397 to E. I. du Pont de Nemours.
115. L. Kocian, FP 1,329,489 to Sandoz.
116. E. Daubach, W. Fischer and L. Setzer, UKP 918,516 to Badische Anilin u. Soda Fabrik.
117. See T. A. Langstroth, *Color Eng.*, **6** (1968) No. 4, 40.
118. D. Bass, *Paint Manuf.*, **27** (1957) 5.
119. J. J. Einerhand, L. J. H. Leonardus, R. J. Laenen, J. W. J. Hoofs, Dutch Patent 76.07496 to Ten Horn (from *Chemical Abstracts* **89** (1978) 112591n).
120. A. S. Gomm, G. Hull and J. L. Moilliet, *J. Oil Colour Chem. Assoc.*, **51** (1968) 143.
121. W. A. Ratcliffe, *Rev. of Prog. Coloration*, **7** (1976) 43.
122. H. Helfert, G. Uhl, K. Dachs, R. Fikentscher and H. G. Sharpenberg, Ger Offen 2,361,950 to Badische Anilin u. Soda Fabrik.

CHAPTER 5

PRINCIPLES OF PRECIPITATION OF FINE PARTICLES*

H. FÜREDI-MILHOFER
Ruder Bošković Institute, Zagreb, Yugoslavia
and
A. G. WALTON
Case Western Reserve University, Cleveland, USA

INTRODUCTION

The precipitation of solids from liquids is of importance in many diverse areas of science, including oceanography, metallurgy, physiology and chemistry. In industry, paint formulation, polymerisation and plastic manufacture, saline water conversion (via ice), photographic chemicals and pigment manufacture, and many other processes, involve the principles of precipitation. In some cases precipitation is a process which is to be avoided (sedimentation and scaling); in others it is to be promoted (separation of trace elements). A colossal amount of data relating to one form or another of precipitation process has accrued in the literature. The majority of the observations reported have, however, been qualitative in nature and it is only within the last three decades or so that more quantitative information has become available.

Traditionally, precipitation phenomena have lain within the realm of activity of the analytical chemist who is concerned with separation of the various components which annoyingly get together in nature. Large commercial enterprises are founded upon the separation of materials by precipitation. This chapter is not, however, directed toward the analytical

* This chapter is devoted to the memory of Professor B. Težak, who was teacher to one of us (H.F.M.) and a mutual friend.

aspects of precipitation and separation, but is devoted to outlining the current status of knowledge in the area of the mechanism of precipitate formation. Even in this more limited area the literature is extensive and studies attempting to unravel the complexities of precipitate formation have been carried out for at least 80 years.

Perhaps one of the more important earlier observations was Ostwald's recognition that the normal solubility of a substance could be exceeded without precipitation occurring.[1] Other workers, of whom von Weimarn[2] is most frequently recognised, have shown that the number, size and shape of precipitated particles are a function of the concentration of solute in excess of solubility. In other words, the relation between the reactant concentrations and the solute solubility determines many of the physical (and chemical) characteristics of the precipitate. In the following paragraphs these parameters will be explored, predominantly for precipitation of inorganic and organic crystals from solution.

BASIC CONCEPTS

Precipitation is the formation of a new phase from a homogeneous parent phase, such as the formation of droplets from vapour, bubbles from liquids, and solids from melts and solutions. In the following treatment we will be concerned with the precipitation of solids from solutions with particular attention to electrolyte precipitation. Although many of the basic principles originate in the thermodynamic treatment of vapour condensation, precipitation from electrolyte solutions is much more complicated and it will be our task to present in a systematic way the most important factors which have to be taken into account.

Supersaturation

Classical nucleation theory shows that for the formation of stable nuclei of a new phase an energy barrier must be exceeded, which expresses itself physically as a critical supersaturation. Ostwald's observation of stable solute concentrations in excess of the normal solubility is easily explained by this consideration. The supersaturation, S, is defined in terms of the difference in chemical potentials of a supersaturated and a stable saturated solution

$$S = \exp[(\mu(a) - \mu(a_s))/RT] \qquad (5.1)$$

where μ is the chemical potential, and a and a_s are the activities of the solute in a supersaturated solution and at equilibrium, respectively.

In the simplest case, for constant temperature and pressure, eqn. (5.1) reduces to

$$S=(a/a_s)_{T,P} \tag{5.2}$$

since $\mu(a)=RT\ln a$.

For a binary electrolyte

$$S=[(a_A a_B/K_{s_0})^{1/2}]_{T,P} \tag{5.3}$$

where a_A and a_B are ionic activities and K_{s_0} is the solubility product.

We shall see that many of the properties (mainly thermodynamic) of an electrolyte solution and the resulting precipitate(s) depend on the supersaturation; others (kinetic and chemical properties), however, depend on the reactant concentration ratio as well. Therefore, in order to get reliable information on the precipitation of electrolytes from solutions, a wide region of reactant concentrations should be considered.

Metastable limit

Following Ostwald's observation, numerous experiments have been carried out with the purpose of establishing the maximum stable supersaturation. This supersaturation has been termed the metastable limit, or critical supersaturation. Actually, the maximum stable supersaturation is a function of the time and tool of observation, which should be stated when comparing different systems. The metastable limit might be reached isothermally, by increasing concentrations either continuously or discretely (by direct mixing of reactants), or it might be attained at constant composition by supercooling. The first question that arises is what determines this metastable limit and how can it be related to the molecular chemistry of precipitation?

It is now known that the critical degree of supersaturation is established both by the nature of the precipitating phase and by the presence of impurities. Removal of more and more of the impurity particles by successive filtration raises the metastable limit, but whereas it is simple to achieve 3000% supersaturation for barium sulphate, it is impossible to exceed 300% for silver chloride. An ample demonstration of the catalysing action of impurities is shown in Table 5.1, where the supercooling required for solidification of water is shown as a function of the model impurities added.

We must conceive then that the birth or nucleation of a precipitate

TABLE 5.1
CRITICAL DEGREES OF WATER SUPERCOOLING IN THE PRESENCE OF POWDERED SUBSTRATES

Substrate	*Critical supercooling* (°*C*)
Teflon	>16
Benzophenone	>16
Thallium iodide	> 6.2
Lead iodide	4·1
Silver iodide	2·5
Silver chloride	4·5
Mercuric sulphide	5·6
Cadmium sulphide	6·5

involves the accretion of ions or molecules on an impurity surface; the consequent clustering and subsequent growth eventually leads to the formation of the new phase. This type of initiation is known as *heterogeneous nucleation* and predominantly initiates precipitation at low levels of supersaturation. At high levels of supersaturation heterogeneous nucleation is overtaken by *homogeneous nucleation*, which becomes the dominant process responsible for the formation of the new phase.

Precipitation processes

The physical reason for the existence of a critical supersaturation may be found in the fact that small particles (as nuclei) have a higher solubility than larger ones.[3] Consequently nuclei, once formed, must grow to reach a critical size, at which their solubility is at least as low as the concentration of solute in the mother liquor. Further development of the new phase proceeds by interplay of several precipitation processes, *crystal growth* being just one of them. An equally important role is to be ascribed to *aggregation* which, depending on the experimental conditions, may become dominant in the very early stages of precipitate formation[4–6,38], or at a later stage. Further transformation of the primary dispersion or precipitate is achieved by *ageing*, either in the sense of Ostwald ripening, or by chemical and structural transformation of metastable phases into thermodynamically more stable ones.

The properties of precipitates depend on the rates and mechanisms of the prevalent precipitation processes, but how and when one or the other process will become rate controlling depends on the experimental conditions under which precipitation is carried out. We will see that the initial reactant concentrations (supersaturation and reactant concentration ratio) are of utmost importance in determining the rates and mechanisms of precipitation processes and consequently the properties of precipitates.

THEORETICAL CONSIDERATIONS

Nucleation

Nucleation has been treated theoretically by thermodynamical, statistical, mechanical, and kinetic approaches. The thermodynamic approach, although 40 years old, has been the basis of most of the experimental work concerned with nucleation of crystals from solutions. It will therefore be given the most attention in this review. For additional information the reader is referred to refs. 7–10.

Homogeneous nucleation

In classical thermodynamic nucleation theory the homogeneous nucleus is conceived as an aggregate of critical size, which is in unstable equilibrium with the parent phase. At concentrations below the critical level the cluster grows or dissociates reversibly, while at the critical concentration upon accretion of an additional ion or atom irreversible growth ensues.

$$n\mathrm{X} \rightleftharpoons \mathrm{X}_n + \mathrm{X} \rightarrow \tag{5.4}$$

It is furthermore assumed that macroscopic properties may be assigned to such a nucleus. In the case of a solid–liquid transformation the nucleus would then be a piece of solid phase large enough for its solubility, i.e. the activity of the solution in equilibrium with the nucleus, to be given by the Gibbs–Kelvin relation

$$\frac{\rho}{M} RT \ln S = \frac{\Delta\mu}{v} = \Delta G_{\mathrm{v}} \tag{5.5}$$

where ρ is the density of the solid phase and M is the molecular weight, S is the supersaturation (eqn. 5.1), $\Delta\mu$ is the difference between the molecular chemical potential of solid and solution phases, $\bar{v}$ is the molecular volume ($M/\rho N_0$) and ΔG_{v} the volume (Gibbs) free energy. N_0 is the Avogadro number.

In nucleation theory the creation of particle surface and its associated energy of formation relates to the energetics of cluster formation. The free energy $\Delta G^{0\prime}_{AB}$ for the formation of a critical metastable cluster is

$$\Delta G^{0\prime}_{AB} = \Delta G'_v + \Delta G_s \tag{5.6}$$

where $\Delta G'_v$ is related to the heat of crystallisation and to the formation of *bonds* in the cluster, and ΔG_s is the free energy associated with the creation of the cluster surface. Equation (5.6) is an approximation since it equates a standard free energy with the other free energy terms, but a more rigorous statistical mechanical derivation yields a similar equation. From eqn. (5.5) the volume free energy is given by

$$\Delta G'_v = V_c \Delta G_v = -n\, RT \ln S \tag{5.7}$$

where n is the number of monomers in the cluster and V_c the cluster volume.

The free energy associated with the formation of the surface of a cluster is difficult to define exactly because of the presence of edges, corners and defects. For the present purpose it is (as usual) assumed that the contributions of these and of changes in surface energy with cluster size may be neglected and the free energy is given by

$$\Delta G_s = A\gamma_{S/L} \tag{5.8}$$

where A and γ are the surface area and solid–liquid interfacial free energy, respectively. For a faceted cluster a similar relation may be proposed

$$\Delta G_s = \sum_{i=1} A_i \gamma_i \tag{5.9}$$

where the area of the ith facet A_i is governed by the Gibbs–Wulff condition $\Sigma_{i=1} A_i \gamma_i$ = minimum for a given volume. Insertion of eqns. (5.7) and (5.8) into (5.6) gives for a spherical nucleus

$$\Delta G^{0\prime}_{AB} = 4/3\pi r^3 \Delta G_v + 4\pi r^2 \gamma_{S/L} \tag{5.10}$$

where r is the radius of the cluster.

We can now see that the energy barrier to homogeneous nucleation, $\Delta G^{0\prime}_{AB}$, is simply related to the excess of the cluster free energy over that of the same number of entities in the equilibrium solubility system, with the addition of a surface term. Maximisation of eqn. (5.10) with respect to r gives for the critical radius

$$r^* = -\frac{2\gamma_{S/L}}{\Delta G_v} = \frac{-2\gamma_{S/L}\bar{v}}{kT \ln S} \tag{5.11}$$

and for the activation energy

$$^{*}\Delta G^{0\prime}_{AB} = \frac{16\pi\gamma^3_{S/L}}{3\Delta G^2_v} = \frac{16\pi\gamma^3_{S/L} v^2}{3k^2 T^2 \ln^2 S} \qquad (5.12)$$

where k is the Boltzmann constant. Equation (5.11) is known as the Gibbs–Thompson relation and eqn. (5.12) was first derived by Gibbs[10] to describe the condensation of (large) droplets from vapour.

The rate of homogeneous nucleation J is usually regarded as being controlled by the rate of addition of the postcritical ion to the critical nucleus and, for stationary conditions, may be expressed by

$$\frac{dN}{dt} \cdot \frac{1}{V} = J = A_k \exp\left(\frac{-^{*}\Delta G^{0\prime}_{AB}}{kT}\right) \qquad (5.13)$$

where N is the number of nuclei produced per unit time t and volume V. A_k is a composite term including a diffusional energy barrier and the concentration of solution species and is generally taken as $\sim 10^{25}$.

Inserting eqn. (5.12)

$$J = A_k \exp\left(\frac{-16\pi\gamma^3_{S/L}\bar{v}^2}{3k^3 T^3 \ln^2 S}\right) \qquad (5.14)$$

The derivation of eqn. (5.14) involves the assumption that the embryo partition functions can be evaluated and that the formation of clusters does not significantly deplete the overall ionic concentration.

A plot of eqn. (5.14) showing J as a function of the supersaturation is given in Fig. 5.1. It can be seen that the rate of nucleation remains negligibly small until the critical supersaturation S^* is reached above which nucleation becomes extremely fast. Ideally, if no impurities were present in the system, the critical supersaturation, as defined by eqn. (5.14), would correspond to the breakdown of the metastable state. In real systems, however, precipitation is usually initiated at much lower supersaturations by heterogeneous nucleation. It is nevertheless possible to determine the critical supersaturation for homogeneous nucleation by experiment. Quantitative interpretation of such experiments is possible if the critical nucleation rate is assigned a value, which is usually assumed to be one nucleus per second per unit volume. In this case eqn. (5.14) may be written as

$$\ln S^* = \left(\frac{16\pi\gamma^3_{S/L}\bar{v}^2}{3k^3 T^3 \ln A_k}\right)^{1/2} \qquad (5.15)$$

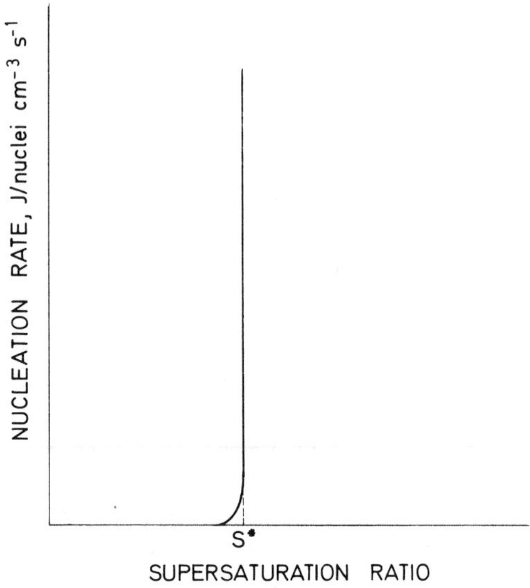

Fig. 5.1. The rate of homogeneous nucleation J as a function of the supersaturation ratio, defined according to eqn. (5.2). S, critical supersaturation.*

If S^* can be determined all quantities are known except for $\gamma_{S/L}$ and consequently the results of an experimental investigation are usually expressed in terms of $\gamma_{S/L}$. This is a useful parameter for relating nucleation to solid–solvent interaction and would normally have values of 15–500 ergs/cm^2. The low values would indicate the interaction between a soft inorganic crystal and a non–interacting solvent or between a strongly interacting high-lattice-energy crystal and a strongly polar solvent. Weak interaction between a high-lattice-energy crystal and solvent leads to high values for $\gamma_{S/L}$.

Heterogeneous nucleation

Since most precipitation reactions involve impurities as heterogeneous nuclei it is desirable to pursue a similar theoretical approach for this situation. Heterogeneous nucleation may be conceived as a sequence of diffusion of ions or molecules through the bulk solution, adsorption on the impurity, diffusion on the substrate surface and two-dimensional clustering at an active site.[11]

We will show later that under certain conditions diffusion through the bulk solution becomes rate controlling, but for the present purpose we will consider the surface reactions to be the rate-determining steps. Avoiding for the moment any assumption of the size or shape of the nucleus the energetics of surface nucleation may be assessed by considering the following parameters:

interaction energy E_{AA} between ions or molecules in the adlayer;
interaction energy E_{AS} between adions or molecules and substrate;
lattice parameters of the nucleating phase and substrate l_A and l_S, respectively.

Considering the results of vapour condensation on solid substrates, Voorhoeve[12] has pictured the following three possibilities.

(i) $E_{AA} \gg E_{AS}$. This situation resembles homogeneous nucleation and the difference between l_A and l_S (lattice mismatch) is of little significance. Low surface coverages but high critical supersaturations are expected.

(ii) E_{AA} and E_{AS} are of the same order of magnitude but l_A and l_S are very different. In this case a large amount of the new phase may be adsorbed before nucleation takes place. As a consequence high critical surface coverages are expected and the surface of the substrate may be modified before the nucleation event. The critical supersaturation will then depend on the properties of the modified surface. A similar situation seems to exist in the condensation of cadmium on tungsten[13] and on germanium.[14]

(iii) E_{AA} and E_{AS} are comparable and the distances l_A and l_S are reasonably close. This is a case of coherent nucleation where all molecules can participate in the formation of a nucleus and the critical supersaturation is lowered because of strong interaction of admolecules with the substrate. Consequently low surface coverages and low critical supersaturations are expected. With large E_{AS} no nucleation barrier would exist at all and epitaxial growth could proceed at any supersaturation.

For coherent nucleation of crystals from solutions Upretti and Walton[11] have given the following quantitative relation:

$$\ln S^* = 4\bar{d}^2\gamma_e^2/\beta kT(2Q_{ads} - Q_D + BkT) \tag{5.16}$$

where $\bar{d}$ is the average ion diameter, γ_e the cluster–solution edge free energy, Q_{ads} the adsorption energy, Q_D the energy barrier to surface diffusion and B a kinetic factor. β is the ratio of lattice dimensions of solute crystal and substrate.

Presumably for precipitation processes many different impurities will be involved and $\beta \sim 1$ is an acceptable approximation. In terms of predicting the degree of supersaturation in precipitating systems two terms predominate, namely γ_e and Q_{ads}. If the adsorption energy is high, i.e. the solute is strongly adsorbed by the impurity, then a relatively low supersaturation can be achieved. In general it might be expected that ionic crystalline fragments would be the best nuclei. We can also see that the level of critical supersaturation of ionic salts on an ionic substrate should be less than for organic crystals on similar substrates. This is in accord with the common laboratory experience that organic materials are not readily nucleated by foreign materials. Secondly, γ_e, the edge energy, which is related to the lattice energy of a crystal, plays an important part in nucleation. Materials of high lattice energy would normally have a high γ_e and would consequently support higher supersaturation. Thus barium, lead, strontium and calcium sulphates and carbonates can be highly supersaturated whereas monovalent materials sodium chloride, silver chloride, etc., generally do not supersaturate to any extent. Finally, fairly soluble materials will normally have a high kinetic constant (B) and will thus support less supersaturation.

Atomistic treatment of nucleation

One of the most serious limitations of classical nucleation theory is the use of macroscopic thermodynamic parameters to describe the properties of small aggregates. This becomes particularly serious at high supersaturations where the critical nucleus is so small that macroscopic parameters lose their significance (see eqn. (5.11)).

To avoid these limitations an atomistic theory was developed for the purpose of describing nucleation in microscopic or atomistic terms. (An outline of this theory is given in ref. 15.) This theory retains the basic concept of a critical nucleus which is formed by addition of monomers to subcritical entities (eqn. (5.4)). As in classical nucleation theory the nucleation rate equation pertains to an equilibrium situation (stationary state) but the final equation (eqn. (5.14)) is modified by the Zeldovich factor which takes into account the decomposition of nuclei and their depletion by growth.

Kinetic treatments

In a consideration of non-stationary effects in nucleation Toshev and Gutzov[16] have considered the significance of the induction period (the time which elapses before a stationary state is reached) for phase changes in different media. By combining kinetic with thermodynamic reasoning the induction period (pertaining only to nucleation) may be defined as

$$\tau = \frac{1 \cdot 6\gamma}{kTz\ z'\ln^2 S} \tag{5.17}$$

i.e. the induction period is proportional to γ but inversely proportional to $\ln^2 S$. In eqn. (5.17) z is the rate of addition of monomers to a critical nucleus.

$$z = \frac{\alpha kT}{\delta^5 \eta} \tag{5.18}$$

(where δ is the average molecular (or ionic) distance, η is the viscosity and α is a steric factor which for complex molecules may assume values $< 10^{-2}$); z' is a probability factor which has been included to assess the effectiveness of collisions of monomers with aggregates. Apparently this factor becomes important when the formation of crystalline aggregates is considered, because the rate of formation of an ordered phase is necessarily smaller than for isotropic matter. The magnitude of z' ($0 \leqq z' \leqq 1$) depends on the nucleation mechanism and has been estimated to vary between 10^{-2} and 10^{-6}.[16]

In an attempt to assess the significance of τ for the formation of crystals in electrolyte solutions we have calculated τ values for two extreme values of $\gamma_{S/L}$ and z' (Table 5.2). The values of the other factors are fixed at: $\delta = 10^{-7}$ cm, $\eta = 1$, $\alpha = 10^{-1}$, T = 295 K $S = 10$ and $k = 1{\cdot}38 \times 10^{-16}$. Apparently the contribution of the nucleation time lag to an experimentally observed induction period in precipitation kinetics is rather insignificant, or it becomes significant only at very high interfacial energies and/or low probability factors z'.

TABLE 5.2

VALUES OF THE INDUCTION PERIOD FOR EXTREME VALUES OF THE INTERFACIAL ENERGY AND THE PROBABILITY FACTOR z'

$\gamma_{S/L}$(*erg/cm*^{2})	z'	*Induction period* (*s*)
15	10^{-2}	$2{\cdot}7 \times 10^{-5}$
15	10^{-6}	$2{\cdot}7 \times 10^{-1}$
500	10^{-2}	9×10^{-4}
500	10^{-6}	9

With the development of computer simulation techniques new possibilities for treating the kinetics of nucleation without making detailed assumptions about the size and shape of a critical nucleus arose. Such treatments provide a better insight into the dynamics of the early events leading to a phase transition. A detailed analysis of the corresponding literature is beyond the scope of this chapter but some important conclusions, stemming from a simulation of vapour deposition on bare surfaces,[17] will be mentioned. In this work the development of size distributions (monomers, subcritical embryos, supercritical aggregates) during deposition was computed. It is shown that in open systems the concentration of monomers first rises to a maximum then begins to decrease. The supersaturation and the nucleation rate show similar behaviour. Due to the depletion of monomers by growing aggregates eventually a saturation of clusters is reached, i.e. the nucleation rate approaches zero. At the same time an 'Ostwald ripening' situation ensues in which small, previously supercritical clusters become unstable and dissociate. (We shall see at a later stage that a similar conclusion may be reached by purely thermodynamic reasoning.) Thereafter, until coalescence becomes important, the size distribution modes translate to larger sizes and become lower and broader.

The number of precipitate particles

The theories of homogeneous and heterogeneous nucleation allow us to make predictions about the number of particles generated at a certain supersaturation. After the metastable limit has been exceeded the formation of particles is initiated by heteronuclei, i.e. by different impurities in solution. The number of particles thus formed depends on the number of impurities present and their catalysing activity. It is thus either constant or approaching a maximum value (usually 10^6–10^7 particles/cm^3). At a certain critical supersaturation, characteristic of homogeneous nucleation, the rate of nucleation dramatically increases (eqn. (5.14) and Fig. 5.1) having as a consequence an increase in the number of nuclei by several orders of magnitude. The above reasoning is illustrated in Fig. 5.2.

Nucleation of polymers

One of the commercially important precipitation processes involves the precipitation of polymeric particles prior to the formulation of plastics. In the polymerisation process we may regard an *n*-mer as

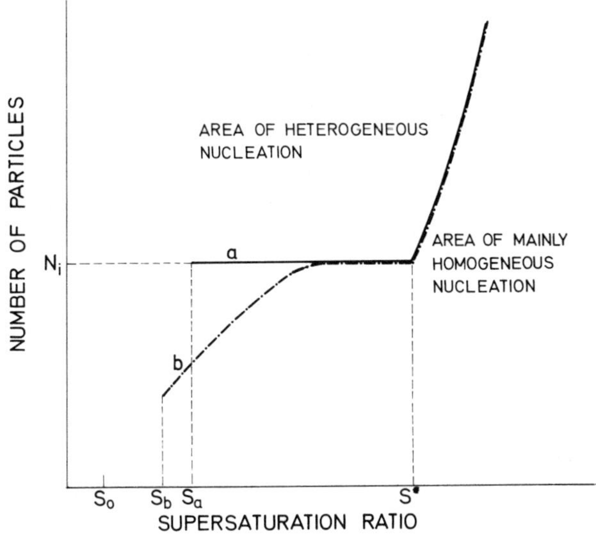

Fig. 5.2. The number of precipitate particles as a function of the supersaturation ratio, defined according to eqn. (5.2). Curve (a) is for a system of N_i impurity particles, all equally efficient in catalysing heterogeneous nucleation. Curve (b) corresponds to the precipitation on to N_i impurities of varying nucleation efficiency. S_0, solubility; S_a and S_b, metastability limits for curves (a) and (b) respectively; S^*, critical supersaturation for homogeneous nucleation.

composed of n monomer units, thus

$$nX \rightarrow X_n \tag{5.19}$$

Although it is tempting to think of such a process as nucleation of an insoluble particle this is not so, at least not in terms of the classical concept of nucleation. Reaction (5.19) is not reversible and cannot, therefore, be treated by equilibrium thermodynamics. The step involved in the nucleation of polymers from solution is concerned with configurational changes in the molecule. Thus the nucleation of a crystalline polymer might be described in terms of chain folding, the driving force for chain folding, nucleation and crystallisation being the supersaturation in terms of the molecules of specific length. Non-crystalline polymeric particles probably undergo similar but more random configurational changes on nucleation.

The nucleation of polymerising particles from solution has some unusual features. Since the rate of addition of monomer units to the *n*-mer is usually much more rapid than the diffusion of the *n*-mer to a suitable nucleating impurity, the system is able to undergo spontaneous homogeneous nucleation. The homogeneous nucleation process is typified by the large numbers of particles which can be produced in such a system (up to $10^{15}/cm^3$ and higher). Polymeric products from solution polymerisation processes are hence usually agglomerates and the control of agglomeration often controls the quality of plastics formulated from such polymers.

Secondary or ancillary nucleation

It is known that if a seed crystal of ice is introduced into supercooled water, many ice particles are precipitated prior to complete solidification. Similarly, in the seeding of clouds, the introduction of relatively few silver iodide particles often leads to the precipitation of a vast number of rain drops. This proliferation of particles is brought about by secondary or ancillary nucleation. Examples more akin to the central theme of precipitation from solution are those of the seeding of supersaturated solutions with crystals of the solute phase. The introduction of one organic crystal may induce the formation of many secondary crystals and there is experimental evidence indicating that the same phenomenon occurs when inorganic precipitates are seeded under certain conditions.[18] It is believed that nuclei generated on the surface of the seed crystals may be carried into the bulk solution by convection and then act as ideal heterogeneous nuclei. Jackson and co-workers have shown in a series of remarkable micrographs[19] that for ammonium chloride crystallisation, it is the formation of small dendrites, which fragment and fall into bulk solution, that causes ancillary nucleation. Actually, of course, this is not nucleation in normal usage, but is rather ancillary crystal growth.

Crystal growth

Crystal growth is visualised as the result of a succession of events, i.e. transport of ions or molecules through the solution, adsorption at the solid–solution interface, surface diffusion, reactions at the interface (dehydration, nucleation) and incorporation of the reaction products into the crystal lattice. The rate of crystal growth is controlled by the slowest of these processes and consequently crystal growth kinetics are important in the evaluation of the rate-controlling mechanism. As for nucleation, the principles of crystal growth have been defined some 30 years ago (see for instance ref. 20). The theories are based primarily on considerations of crystal growth from vapour or melt, but with some caution they may be applied to the growth of crystals from solutions[21-29].

Considering first crystals with perfectly flat surfaces we conceive that they cannot grow without nucleation. Once a nucleus has been formed it will spread over the crystal surface to fill up a monolayer. For small crystals growing at low supersaturation the time needed to cover the surface is much shorter than the time needed for nucleation. Consequently growth is nucleation controlled and the basic principles of two-dimensional nucleation, as outlined in the previous section, apply. This mechanism has been named[21,22] mononuclear growth (I).

For larger crystals and higher supersaturations the time required to fill up a monolayer may become comparable to or longer than the time of nucleation. A polynuclear mechanism (II) then applies, i.e. the surface is at all times covered by surface nuclei, spreading themselves and intergrowing.[21,22,24-27]

In both cases there exists a nucleation barrier which expresses itself in a critical supersaturation. It has, however, been observed that crystals grow readily at supersaturations well below the critical one. This problem was solved by Frank[28] who proposed a growth pattern initiated by a screw dislocation (III). The development of the screw dislocation occurs either by addition of material directly from solution, or by addition of molecules diffusing across the crystal surface. Since both mechanisms lead to a constant rate of advancement of the step across the crystal surface, it can be seen that the steps will move through a greater arc near the centre than at the edge, where it will lag behind. Consequently, the screw dislocation winds itself into a spiral as shown in Fig. 5.3.

Quantitative treatments of mechanisms I and II[21,24] are based on the classical theory for two-dimensional growth, i.e. heterogeneous nucleation, and take into account surface migration of ions or molecules to and their capture at the nucleation site. A rigorous treatment given by Lewis[24] leads to the following equations:

$$F_{\mathrm{I}}=l_{\mathrm{e}}^2\, C\,(S'-1)\,R_0 x_{\mathrm{s}}^2 \exp\,(-{}^*\Delta G/kT) \tag{5.20}$$

and

$$F_{\mathrm{II}}=(4C)^{1/3}\,(S'-1)\,R_0 x_{\mathrm{s}}^{4/3} \exp\,(-{}^*\Delta G/3kT) \tag{5.21}$$

F_{I} and F_{II} are the growth rates of new crystal planes, l_{e} is the edge length of the crystal plane, x_{s} is the surface diffusion distance in atomic units, C is a composite term including the activation energy ${}^*\Delta G$, the

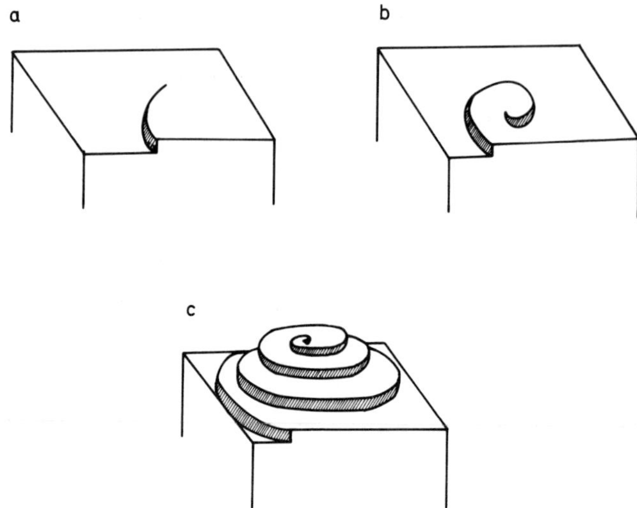

Fig. 5.3. The development of a growth spiral from a screw dislocation. Although deposition of new material at the screw dislocation edge occurs at a uniform rate, the anchor dislocation centre causes the relative rate of 'winding' to be larger at the origin.

supersaturation S', and a capture factor. $C \cong 1$ is an order of magnitude approximation. S' is the supersaturation defined as

$$S' = R/R_e \tag{5.22}$$

and

$$R_0 = \zeta R_e \tag{5.23}$$

where R and R_e are the actual and equilibrium incident fluxes, and ζ is a retardation factor ($\zeta < 1$). We may then put

$$(S' - 1)\ R_0 = F_\infty \tag{5.24}$$

where F_∞ is the maximum possible growth rate at a given supersaturation.

The activation energy $^*\Delta G$ is given by

$$^*\Delta G = \beta'^2/4kT \ln S' \tag{5.25}$$

where β'^2 is an edge energy parameter.

The validity condition for eqn. (5.20) is $F_1/F_\infty < 4x_s/l_e$, while at

$F_1/F_\infty > 4x_s/l_e$ eqn. (5.21) applies. Both eqns. (5.20) and (5.21) show the corresponding crystal growth rate approaching F_∞ at finite supersaturations, S'_{I} and S'_{II} respectively, which depend on the material parameters of the crystal, β' and x_s. Above these supersaturations growth is inhibited by the corresponding mechanism.

With $C=1$ in the first approximation, eqns. (5.20) and (5.21) become

$$F_{\mathrm{I}}=l_e^2 x_s^2 F_\infty \exp\left(\frac{-\beta'^2}{4k^2T^2 \ln S'}\right) \tag{5.26}$$

and

$$F_{\mathrm{II}}=1{\cdot}59 x_s^{4/3} F_\infty \exp\left(\frac{-\beta'^2}{12\, k^2T^2 \ln S'}\right) \tag{5.27}$$

Putting the pre-exponential factors in eqns. (5.26) and (5.27) equal to $K_{\mathrm{I}} F_\infty$ and $K_{\mathrm{II}}F_\infty$ respectively and taking logarithms:

$$\log F_{\mathrm{I}}=\log K_{\mathrm{I}} F_\infty - \frac{0{\cdot}05\ \beta'^2}{k^2T^2 \log S'} \tag{5.28}$$

and

$$\log F_{\mathrm{II}}=\log K_{\mathrm{II}} F_\infty - \frac{0{\cdot}015\beta'^2}{k^2T^2 \log S'} \tag{5.29}$$

According to eqns. (5.28) and (5.29) plots of the rate versus the inverse supersaturation on a log/log scale should in both cases give straight lines, the slopes of which are proportional to β' in a first approximation. (Since β' appears also in the pre-exponential term C, eqns. (5.20) and (5.21), through the activation energy, a true value of the edge energy parameter can only be obtained by an iterative procedure).[24]

The theoretical treatment of screw dislocation controlled growth is given by Burton, Cabrera and Frank.[29] Using the same symbols as above one obtains[24] for growth from a single dislocation:

$$\begin{aligned} F_{\mathrm{III}} &= (S'-1)R_0 \frac{x_s}{2\pi r^{*\prime}} \tanh\left(\frac{2\pi r^{*\prime}}{x_s}\right) \\ &= (S'-1)R_0 \frac{\ln S'}{\ln S'_{\mathrm{III}}} \tanh\left(\frac{\ln S'_{\mathrm{III}}}{\ln S'}\right) \end{aligned} \tag{5.30}$$

where

$$\ln S'_{\mathrm{III}} = \frac{\pi^{1/2}\beta}{kTx_s} \tag{5.31}$$

and $r^{*\prime}$ is the critical radius of the two-dimensional cluster.

Since tanh $x \to 1$ when $x > 1$, and ln $S' \to S' - 1$ when $S' < 2$ it follows that

$$F_{\mathrm{III}} \to \frac{(S'-1)^2 R_0}{\ln S'_{\mathrm{III}}} \text{ when } 2 > S' > S'_{\mathrm{III}}$$

and since tanh $x \to x$ when $x < 1$, it follows that $F_{\mathrm{III}} \to (S'-1)R_0$ when $S' > S'_{\mathrm{III}}$. Thus the growth rate varies quadratically with supersaturation if $S' < S'_{\mathrm{III}}$ and linearly if $S' > S'_{\mathrm{III}}$.

Verification of the above theories is possible under strictly controlled experimental conditions[24] but may not be feasible for the precipitation of slightly soluble salts from electrolyte solutions. For such a situation Nielsen[21,22] proposed an alternative approach. Assuming a constant number of spherical (or cubic) crystals which all have the same size, he included the particle radius r as a parameter and defined the dependence of the linear growth rate on the supersaturation and on crystal size as follows.

In mononuclear growth the rate of growth is related to the surface area and the solute concentration as

$$\frac{\mathrm{d}r}{\mathrm{d}t} = K_{\mathrm{M}} r^2 c^m \tag{5.32}$$

where K_{M} is a constant and m is the number of units (ions, molecules) in the two-dimensional nucleus.

If the rate of growth is controlled by a polynuclear mechanism, it is independent of the crystal surface, since every nucleation event involves only a small part of the total surface area. Consequently

$$\frac{\mathrm{d}r}{\mathrm{d}t} = K_{\mathrm{P}} c^p \tag{5.33}$$

where K_{P} is a constant and p is the order of the reaction.

For screw dislocation controlled growth at $c \approx s$ the author defines

$$\frac{\mathrm{d}r}{\mathrm{d}t} \approx \frac{0{\cdot}05\ DkT}{\gamma} (c-s)^2 \tag{5.34}$$

and at $c \gg s$

$$\frac{\mathrm{d}r}{\mathrm{d}t} \approx kc \tag{5.35}$$

where D is the diffusion coefficient, c is the solute concentration and s is the solubility. Equations (5.20), (5.21), (5.30) and (5.32)–(5.35) imply that above a certain supersaturation crystal growth becomes uninhibited by either of the above mechanisms. However, crystal growth from solution may be impeded by yet another 'bottleneck' mechanism, i.e. the transport of ions or molecules to the crystal. In the simplest case this is achieved by diffusion and it may be shown[21,22] that for spherical particles the linear rate of growth is given by

$$\frac{dr}{dt}=\frac{D\bar{V}(c-s)}{r} \tag{5.36}$$

where $\bar{V}$ is the molar volume of the precipitating substance. In case of convection the transport of matter is faster than in purely diffusion controlled growth.

The above considerations show that the principal parameters determining the rate and mechanism of crystal growth are the supersaturation, the crystal size, the available surface area (eqns. (5.32)–(5.36)) and material parameters of the crystal faces, i.e. the edge free energy parameter β' and the diffusion distance x_s (eqns. (5.20), (5.21)).

Figure 5.4 shows how the rate controlling mechanism of crystal growth depends on the solute concentration (the curves are plots of eqns. (5.32)–(5.36)).

A powerful tool to test theoretical models of crystal growth kinetics and to facilitate the understanding and illustration of growth mechanisms is given by the recent development of computer simulation techniques. The possibilities of these methods are far from exhausted. So far results have been obtained on simple lattices and one-component crystals, and applications to growth of simple cubic lattices from vapour have been successful. However, in a transition to n-component crystals and solid–solution interfaces additional problems arise, which have yet to be accounted for. A detailed analysis is beyond the scope of this chapter, but the reader is referred to some recent reviews on the subject.[23,30–32]

Flocculation and stabilisation of precipitates

General considerations

The principles of flocculation (aggregation and stabilisation) as defined in Chapter 1 apply to newly formed precipitates as well as to dispersions. The Smoluchowski equation[33] (see Chapter 1) applies to the flocculation of originally monodispersed precipitates when the movement of particles is

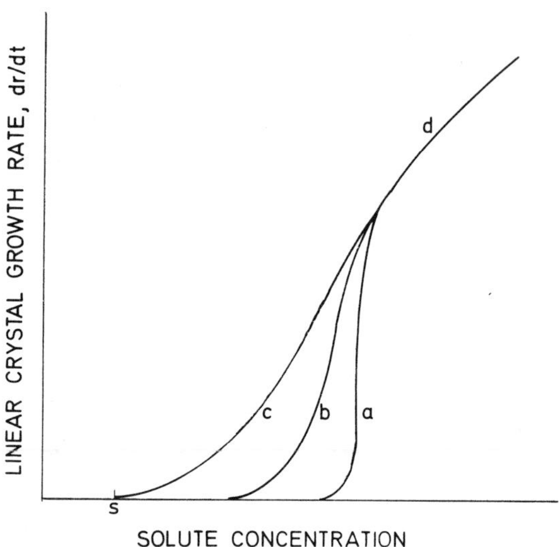

Fig. 5.4. The linear rate of crystal growth as a function of the solute concentration for different rate-determining mechanisms. Curve (a), mononuclear growth, eqn. (5.32); curve (b), polynuclear growth, eqn. (5.33); curve (c), growth controlled by screw dislocation, eqn. (5.34); curve (d), growth controlled by bulk diffusion, eqn. (5.36); s denotes the solubility. (After A. E. Nielsen.[22]*)*

due to Brownian motion. A dispersion of newly formed primary particles (nuclei) may be expected to conform to these requirements and we may thus estimate under what conditions Smoluchowski-type flocculation is likely to occur.

According to this theory, for aqueous dispersions at 25°C with initially N^0 particles per cm^3, the time required to reduce the number of particles by half is

$$t_{1/2} \approx 2 \times 10^{11} / N^0 \text{ s} \tag{5.37}$$

We have seen that, for heterogeneous nucleation, the maximum number of impurity particles $N^0 \sim 10^6$–10^7 particles per cm^3. Thus $t_{1/2} \sim 2 \times 10^4$–$2 \times 10^5$ s or 5·5–55 h.

Apparently in such systems flocculation is not significant in the early stages of precipitate formation but will be preceded by crystal growth. However, since all impurities may not catalyse nucleation equally well, after a certain time the precipitate will have become polydisperse and the Smoluchowski equation no longer applies. Swift and Friedländer[34] have considered the flocculation of a heterogeneous hydrosol with a continuous size spectrum. The kinetics of flocculation due to Brownian motion are described as

$$\frac{\mathrm{d}N}{\mathrm{d}t} = -\frac{N^2 A_1 kT}{3\eta} \tag{5.38}$$

where N is the number of particles, η is the fluid viscosity and A_1 is an integral function including the size distribution of particles. The derivation of eqn. (5.38) makes use of the self-preservation hypothesis, i.e. it is assumed that particular asymptotic forms of the size spectrum do not vary with time. For homogeneous hydrosols $A_1 = 4$ and eqn. (5.38) reduces to the Smoluchowski equation. For all heterogeneous hydrosols $A_1 > 4$ and depends on the particular size distribution. Consequently, as crystals grow they will reach a certain size distribution at which the rate of flocculation becomes significant. The effect is autocatalytic and becomes more important under forced convection (shaking or mixing). Evidently, by flocculation the surface area available for growth is reduced which causes a reduction in the growth rate above that due to decreasing supersaturation. Eventually the rate of crystal growth becomes insignificant while, as a consequence of flocculation, an asymptotic value of N and a corresponding size distribution are approached.

If precipitation is initiated by homogeneous nucleation the number of particles generated is larger by several orders of magnitude. The maximum number of particles so far detected by experiment is $\sim 10^{12}$ particles/cm^3,[35] and consequently the half-time of flocculation according to the Smoluchowski equation would be $t_{1/2} \sim 0{\cdot}1$ s. If the number of primary particles (or clusters) were only 10 times larger, $t_{1/2}$ would be $\sim 0{\cdot}01$ s, which is within the time scale predicted for induction periods in homogeneous nucleation (Table 5.2). It may then be conceived that flocculation of nuclei or primary particles generated by homogeneous nucleation occurs parallel to or immediately after nucleation. Homogeneous nucleation thus leads to heavy aggregation unless particles are stabilised by repulsive forces (coulombic, or those arising from solvation, adsorbed layers, etc.; see Chapter 1). Aggregated precipitates are also commonly observed in polymerisation systems where the monomer–polymer reaction generates large numbers of primary particles.

According to the concepts of Odén[36] and Težak[6] aggregation is the dominant mechanism in precipitation. Težak envisaged embryos, nuclei and primary particles as being formed by aggregation of complex solution constituents (ion pairs, mononuclear and polynuclear complexes, etc.). Primary particles then aggregate into secondary structures (spherulitic aggregates, mosaic crystals, etc.). It would seem that this picture very well describes the formation of precipitates initiated by prevalently homogeneous nucleation. The hypothesis is supported by experiments pertaining to the formation of amorphous calcium phosphate.[37,38] Other experimental data, pertaining to the formation, ageing and dissolution of monodispersed barium sulphate particles,[39] indicate that the above mechanism might be important in some heterogeneously nucleating systems as well.

The flocculation of primary particles and secondary precipitates is of course a colloidal phenomenon described in detail in other parts of this book.

In situ studies of the conditions leading to the flocculation and stabilisation of hydrophobic sols have been made by Težak and collaborators. Their extensive work in this area will not be reviewed in detail here (for partial reviews of the early work of this group see refs. 40 and 41, and Chapters 5 and 6 in ref. 9). However, it is germane to point out that their data, which are qualitatively in accord with the usual concepts of the flocculating action of electrolytes and of reduced stability in solvents of low dielectric constant,[42] show that the intimate nature of the interaction at the solid–liquid boundary is extremely important in characterising the stability of the system.

Chemical aspects

Težak's model,[6] apart from stressing the importance of aggregation in the initial stages of solid phase formation, emphasises also the decisive role of the chemical composition of the mother liquor in determining the composition and morphology of precipitates. This point has been proven by exhaustive experimental evidence. It is of interest to point out the work of Matijević and collaborators (for reviews see ref. 43), who succeeded in preparing well defined monodispersed sols of metal hydrous oxides (iron, aluminium, titanium, chromium, manganese, copper, etc.) by ageing the respective salt solutions at elevated temperatures. Their results show the decisive role which certain anions (Cl^-, ClO_4^-, SO_4^{2-}, etc.), and also polymers have in determining the size and shape of the precipitate particles.

Another important aspect which emerges from the work of Težak and collaborators[6,40,41,44] is the effect of specific ionic interactions on colloid stability. An example is the ability of citrate ions to coagulate and again restabilise positive silver bromide sols.[40,41] The latter effect is apparently due to specific adsorption and consequently reversal of charge at the solid–liquid interface. Later work using radiotracer techniques has shown[44] that the solid–liquid interface of silver iodide behaves like an ion exchanger with regard to di- and tri-valent cations. This behaviour was correlated to the coagulating ability of the respective cations.

The important role of specific chemical interactions at the solid–liquid interface becomes even more obvious if the surface chemistry of hydrophilic colloids (hydrous oxides, salts of weak acids and strong bases and vice versa) is considered. It has been shown in Chapter 3 of this book that for hydrous oxides hydrogen and hydroxide ions may be considered potential-determining. The surface charge of salts depends on the concentrations of all ionic species in solution, which again are dependent on the pH. For instance the surface charge of calcium carbonate depends on the charge balance

$$\Gamma_+ - \Gamma_- = (Ca^{2+}) + (CaOH^+) + (CaHCO_3^+) - (HCO_3^-) + (CO_3^{2-}) \quad (5.39)$$

where parentheses denote ionic activities. Consequently the following equilibria must be considered:

$$(CO_2)_g \rightleftharpoons (CO_2)_{sol.} \quad (5.40)$$

$$(CO_2)_{sol.} + H_2O \rightleftharpoons H_2CO_3 \quad (5.41)$$

$$H_2CO_3 + OH^- \rightleftharpoons HCO_3^- + H_2O \quad (5.42)$$

$$HCO_3^- + OH^- \rightleftharpoons CO_3^{2-} + H_2O \quad (5.43)$$

$$Ca^{2+} + HCO_3^- \rightleftharpoons CaHCO_3^+ \quad (5.44)$$

$$Ca^{2+} + OH^- \rightleftharpoons CaOH^+ \quad (5.45)$$

$$CaOH^+ + OH^- \rightleftharpoons Ca(OH)_2 \quad (5.46)$$

Using the respective stability constants for the above equilibria[45] one may find the isoelectric point (i.e. p.) in solution, which corresponds to the point of zero charge (p.z.c.) at the solid–liquid interface, provided that the surface has not been significantly altered by specific adsorption. It is thus obvious from equilibria (5.40)–(5.46) that the surface charge of salts of strong bases/weak acids and of weak bases/strong acids also depends on the pH. The pH at the p.z.c. may of course be determined by microelectrophoresis, surface titrations or other means (Chapter 3 and ref. 46).

Probably the best understood of hydrophilic solid–solution interfaces are those of hydrous oxide dispersions. It has been shown[47] that at least one layer of water molecules is tightly held at such interfaces and there is evidence that adsorbed metal ions retain their primary hydration sheaths.

There is general consensus in the literature,[43,46–50] that the Gouy–Chapman and Stern adsorption isotherms, including the modification by Graham (see Chapter 3), cannot fully explain the adsorption of ions at hydrous oxide surfaces. According to James and Healy[47] the change in standard free energy of adsorption of a species i is

$$\Delta G^0_{\mathrm{ads_i}} = \Delta G^0_{\mathrm{coul_i}} + \Delta G^0_{\mathrm{solv_i}} + \Delta G^0_{\mathrm{chem_i}} \tag{5.47}$$

i.e. it consists of a coulombic, a solvation and a 'chemical' free energy term. The last term takes care of specific chemical interactions at the interface and cannot always be independently evaluated. The nature of the chemical interactions at the interface is not quite clear. A surface complex formation model has been proposed,[46,49] in which the hydrous oxide groups at the surface are treated, similarly to amphoteric functional groups in polyelectrolytes, as complex-forming species. The authors have defined 'intrinsic' stability constants for such complexes.

Adsorption of various solution species at the oxide–solution interface may cause flocculation, but also restabilisation by charge reversal. It has indeed been observed that many ions, particularly those which hydrolyse strongly, cause two or three charge reversals at different pH values.[47,51] Such phenomena are interpreted as consequence of the precipitation of the adsorbate species as a new surface phase, i.e. the formation of surface coatings. Surface coatings have a wide application in the pigment industry.[51,52] In a quite different area, dental medicine, surface coatings are of paramount importance in protecting intact tooth enamel as well as preventing carious lesions from further erosion.[53]

Ageing

After a precipitate has been formed it undergoes numerous physical and chemical changes when in contact with the mother liquor, but also after liquid has been removed. Some of these changes (crystal growth, aggregation) have already been discussed. All other changes are commonly described with the term 'ageing'.

Ostwald ripening

Any two-phase system consisting of a polydisperse precipitate in contact with its mother liquor will be thermodynamically unstable because of its large interfacial area, which is a source of free energy. A tendency then exists to minimise the free energy by coarsening of the particles. This is achieved by dissolution of small particles and growth of large ones until eventually, after infinite time, only one large crystal remains. Under more practical circumstances, it is the parts of precipitate crystallites with high energy (edges, corners, dendrite arms) which will dissolve preferentially, the excess solute then being redeposited at surface positions of lower energy (steps, dislocations). The phenomenon is usually called Ostwald ripening since it has first been explained by Ostwald[54] who adapted the Gibbs–Kelvin equation. This equation shows that the vapour pressure p_i of a liquid droplet of radius r is greater than the saturation vapour pressure of the liquid p and is related to the surface tension γ_l and the droplet radius r by

$$kT \ln\left(\frac{p_i}{p}\right)=\frac{2\gamma_l V_i}{r} \qquad (5.48)$$

where V_i is the molecular volume.

We have seen earlier that computer simulations predict some sort of Ostwald ripening to take place immediately after nucleation. This point can also be made on the basis of purely thermodynamic arguments.[3,55,56]

Equation (5.11) shows the critical radius of a nucleus being inversely proportional to the supersaturation. If, by direct mixing, the supersaturation is built up to exceed the critical level, nucleation will commence and nuclei will start growing. As a result the supersaturation declines, while the critical radius increases with time. Since the rate of growth is a function of the supersaturation (eqns. (5.20), (5.21), (5.32)–(5.36)) it steadily decreases.

Thus the smallest crystals may not grow fast enough to match the increase of the critical radius and thus become unstable and dissolve. Consequently, Kahlweit[3,56] sees precipitation of a polydisperse system as a continuous race between the critical size r^* and the size r of individual particles.

Ostwald ripening has been theoretically treated by several authors.[55–58] In all treatments aggregation was assumed to be negligible and the ageing rate and development of the size distribution with time was evaluated. While according to some of the theories[55,57,58] the ageing rate and the size distribution asymptotically approach constant values, the derivation by Kahlweit[3,56] shows the ageing rate reaching a maximum during the early stages of precipitation and then decreasing again to approach zero as $t \rightarrow \infty$.

In any case there appears to be consensus on the point that Ostwald ripening is most significant in the early stages of nucleation and growth. This then is another factor (besides aggregation) which influences the number of particles formed in a precipitation experiment.

Various ions accelerate and some surfactants inhibit Ostwald ripening. In collaboration with aggregation, ripening has yet another important effect, that of locking adsorbed impurities into the internal region of a consolidated crystal. The methods, parameters and problems of *co-precipitation* are outside the scope of this chapter, but in principle any foreign ions present in solution are adsorbable to a greater or lesser degree upon the precipitate surface. The greater the surface area the more impurity is adsorbed and the more is eventually precipitated. As particles aggregate the adsorbed material is trapped on the inner surface of the precipitated crystal. Thus processes which will require a highly pure product must avoid the large precipitate surface areas produced by homogeneous nucleation and/or dendritic growth.

Formation and transformation of metastable phases

In a precipitation process in which the formation of several phases is possible, the least stable phase with the highest solubility will generally precipitate first. This phenomenon has already been observed in the early 1800s and was later recognised as a rule and formulated by Ostwald[59] as his 'law of stages'. According to Stranski and Totomanov[60] the relative rates of nucleation and crystal growth for the stable and unstable forms will determine which form separates first from solution. The following reasoning then applies.

According to eqn. (5.11) the critical radius r^* is proportional to the interfacial energy $\gamma_{S/L}$. Consequently, at a given supersaturation the phase with lower interfacial energy requires a smaller accretion of ions to form a thermodynamically stable nucleus and is thus energetically favoured. The same is clear from eqn. (5.14) which shows that the stationary nucleation rate is

$$J \sim \frac{1}{\exp \gamma_{S/L}^3} \qquad (5.49)$$

It may thus be expected that the phase with the lowest interfacial energy nucleates first from a supersaturated solution.

Many commonly occurring crystals may exist as polymorphic forms. A well known example is calcium carbonate, which has at least five modifications, of which aragonite and calcite are the most commonly found.

The formation of different hydrates of a solid is also frequently observed in industrial practice as well as in biological systems. Of these the one with the lowest solubility at a given temperature, pressure and vapour pressure will be the thermodynamically most stable one, but will usually not be the first to precipitate.

At high supersaturations, where homogeneous nucleation prevails, the formation of highly hydrated, amorphous or poorly crystalline precursors (amorphous calcium phosphate, gels of hydrous oxides) is not uncommon.

The formation of metastable higher hydrates is easily understood from the previous arguments (eqns. (5.11) and (5.49)). In addition it should be kept in mind that the energy required for desolvation of strongly hydrated cations might be quite large and hydration forces may not be easily overcome by electrostatic interactions.

At high supersaturations the critical radius may be rather small (eqn. (5.11)) and for crystals of large unit cells it may become even smaller than one unit cell.[61,62] It has been shown earlier that under such conditions particles enlarge by aggregation rather than by crystal growth. Consequently, if precipitation is initiated predominantly by homogeneous nucleation, highly hydrated cations are likely to form poorly ordered structures which retain considerable amounts of molecular water (adsorbed and coordinated).

Ageing in contact with the mother liquor leading to the transformation of metastable phases, frequently involves concomitant dissolution of the less stable and reprecipitation of the thermodynamically next more stable form.[63–67] Nucleation of the secondary precipitate is usually heterogeneous with the precursor serving as the nucleating phase.[64,66,68] Amorphous precipitates may have to undergo internal rearrangements (dehydration, crystal ordering) before effectively nucleating the secondary precipitate.[69,70] Progressive ordering processes in the solid phase, accompanied by the liberation of water and resulting in a more crystalline precipitate, have been demonstrated in many hydrous oxide gels.[67] Specifically, condensation polymerisation and aggregation–cementation processes have been observed.

Small ions, polymers and other additives may have a profound influence on the properties, stability and rate of transformation of metastable phases,[63,69,71] as well as on the properties (size, shape and composition) of the products of ageing.[63,66,69] Under certain conditions the formation of new surface phases has also been observed.[63]

MORPHOLOGY OF PRECIPITATES

Crystals formed by spontaneous precipitation

The morphology of precipitates is determined by the mechanisms and relative rates of all precipitation processes which have so far been discussed. Obviously the mechanism of nucleation is of overriding importance in determining the course of subsequent events. We have seen that heterogeneous nucleation is usually followed by a period of crystal growth and subsequent aggregation. In systems in which precipitation is initiated by prevalently homogeneous nucleation, aggregation and/or stabilisation of primary particles are dominant in the early stages of precipitate formation. Hydrophobic precipitates then form colloid dispersions. Hydrophilic precipitates under such conditions tend to form highly hydrated precursors; even the formation of amorphous, gel-like precipitates is feasible.

Regardless of the course of previous events, subsequent ageing processes may change the composition, structure and morphology of the precipitates. It has also been shown[6,9,43] that the morphology of precipitates is extremely sensitive to the chemical composition of the mother liquor, i.e. to the concentrations of all ions, molecules, macromolecules and other constituents present. Consequently a precise definition of all experimental conditions is necessary before any predictions about precipitate morphology can be made.

We shall now explore the influence of precipitation processes on the characteristics of precipitates (number, size and shape of precipitate particles) in more detail.

Since it has been shown that the rates and mechanisms of nucleation and crystal growth depend on the supersaturation, this parameter will be given particular attention.

Number of particles

We have seen (Fig. 5.2) that the number of particles generated in a certain system depends on the mechanism of nucleation. However, the number of particles which are detected by experiment may be much smaller, particularly in the supersaturation region, where homogeneous nucleation prevails.

For every experimental tool there exists a certain lower limit for the size of particles which can be observed. In the homogeneous nucleation region this size is reached by aggregation, rather than by crystal growth.

Thus the observed particles may be aggregates of many more primary particles and consequently the original number of particles may be larger by orders of magnitude than the number of particles actually counted. The effect has indeed been observed in the laboratory of one of the authors,[37] when amorphous calcium phosphate was precipitated with and without gelatin as a peptising agent (Fig. 5.5).

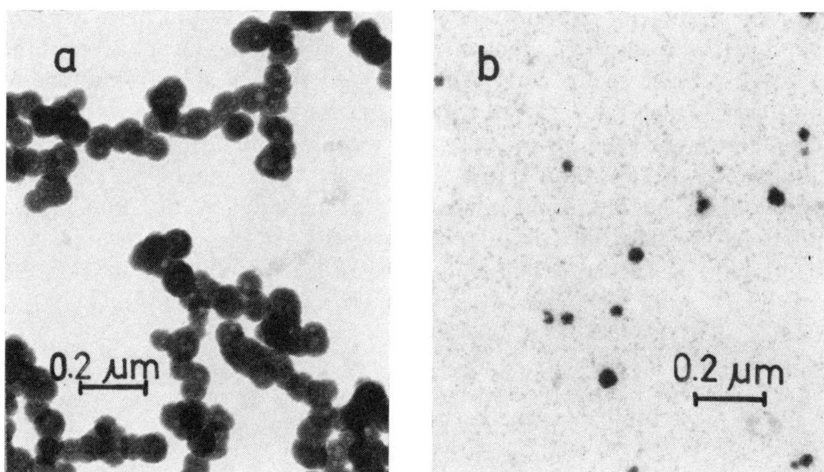

Fig. 5.5. Electron micrographs showing the morphology of amorphous calcium phosphate precipitated (a) without additives, and (b) in the presence of 0·4 mg/cm³ of gelatin. Gelatin acts as a peptising agent preventing aggregation in the early stages of precipitate formation. (After Füredi-Milhofer et al.[37])

Ostwald ripening is another process which commences immediately upon nucleation and may therefore lead to erroneous particle counts.[56] For this case also the discrepancy between the number of particles nucleated and those actually observed will increase with increasing supersaturation, whilst the effect may be neglected at low supersaturations only.[56]

Extreme caution is therefore essential if particle counts are utilised for the determination of nucleation rates.

Size and shape of particles

The size distribution of particles observed in a precipitation experiment is again influenced by the mechanisms and relative rates of all precipitation processes, and is basically a function of the supersaturation and time.

From nucleation theory a maximum of particle sizes is expected to appear just before the onset of homogeneous nucleation. This is easily explained by considering the number versus supersaturation curve in Fig. 5.2. In the heterogeneous nucleation region where the number of particles is constant or approaching a constant value, particles are growing with increasing supersaturation. With the vast increase in particle number due to the onset of homogeneous nucleation a significant drop in the average particle size occurs. We have seen that even if an increase in particle size does occur in this region, it is due to aggregation rather than crystal growth. The critical supersaturation for homogeneous nucleation and hence the position of the particle size maximum is of course a function of the interfacial energy of the material in question. These relations are illustrated in Figs. 5.6 and 5.7 for low, intermediate and high interfacial energy materials (curves (*a*), (*b*) and (*c*) in Fig. 5.6

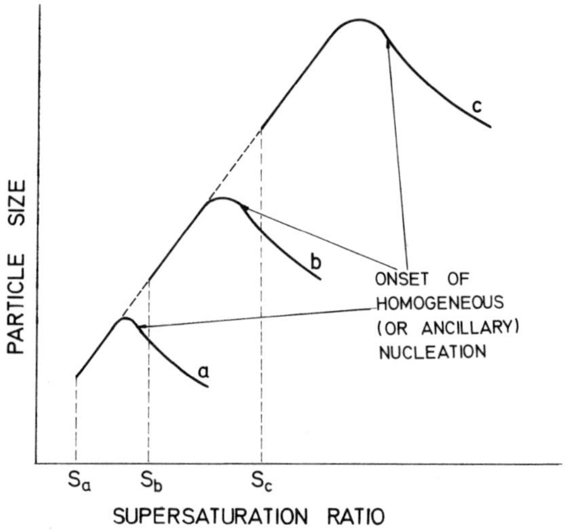

Fig. 5.6. Precipitate particle size (after say one hour) as a function of the initial supersaturation ratio. Curve (a) corresponds to low, (b) to intermediate and (c) to high interfacial energy materials. S_a, S_b and S_c denote the metastability limits for curves (a), (b) and (c) respectively. In the area of heterogeneous nucleation the number of initiating impurity nuclei is assumed constant.

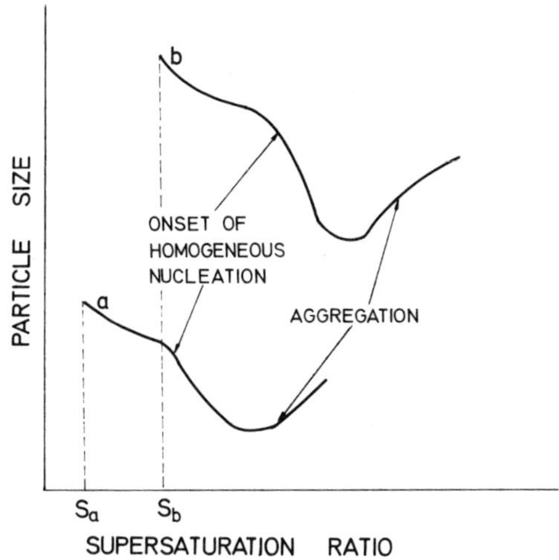

Fig. 5.7. The effect of a wide spectrum of initiating impurity nucleators on the final average precipitate particle size. Curve (a) is for low, and (b) for high, interfacial energy materials. S_a and S_b are metastability limits for curves (a) and (b) respectively.

correspond to interfacial energies of 50, 100 and 200 ergs/cm^2). Figure 5.6 corresponds to curve (*a*) and Fig. 5.7 to curve (*b*) in Fig. 5.2.

Actually many precipitates are known to give size versus supersaturation curves corresponding to those in Figs. 5.6 or 5.7. Von Weimarn,[2] for example, reported many systems showing the size maximum early in this century. It has also been shown[72] that the crystallisation maximum described by Težak[73] corresponds to the same phenomenon. Evidently some control over the particulate size can be maintained by manipulation of appropriate degrees of supersaturation.

Before giving a more detailed analysis of the dependence of size distributions upon supersaturation in the heterogeneous nucleation region a short discussion of the influence of crystal growth on the size and shape of particles is in order.

As in nucleation, one of the dominant parameters in crystal growth is the degree of supersaturation. At very low degrees of supersaturation, small crystals may be expected to grow with a shape close to that which corresponds to the minimum interfacial free energy, i.e. the Gibbs–Wulf criterion $A_i\gamma_i = \text{minimum}$, should be satisfied.

Confining our attention initially to the equilibrium form of crystal, it might be assumed that the lowest energy configuration would correspond to flat planar surfaces, thus minimising the effective surface area. Actually, according to the Gibbs concept, the interface is diffuse, and Mutaftschiev[74] has calculated that for the crystal–liquid interface the diffuse layer can be represented by defects which lie at a depth of one molecule or ion either side of the surface of tension. Thus, at normal temperatures we may conceive that the interface undergoes thermal fluctuations which render it rough in terms of molecular dimensions. There is also some evidence that thermal instability causes surface steps to form. This point would seem to be important in relation to the mechanism of crystal growth since it is assumed that surface steps are necessary for growth propagation.

Although the equilibrium form of crystal surfaces involves roughening at the molecular level, the gross structure of pure crystals involves macroscopically flat surfaces. The equilibrium shape will thus conform to a configuration involving low energy surfaces. The surface energy may be calculated in some cases for the idealised crystal–vacuum interface, and from such calculations Stranski and others[75,76] have deduced the equilibrium shape of some simple crystals. For example, the (100) planes of alkali halide crystals are of lowest surface energy and the equilibrium shape is thus cubic.

However, it does not necessarily follow that the equilibrium shape of crystals grown from solution is the same as that calculated for crystal–vapour equilibrium. If the relative energetics of the crystal surface–environment are unchanged, as they would be if the solvent was inert, crystals would indeed have the same equilibrium form as that calculated for the vapour environment. It is rare that solvents are inert and unreactive towards inorganic crystals, because the solute would not be soluble in such a solvent. In organic systems hydrocarbon solvents might be regarded as inert and unreactive towards non-polar materials, but here again we may expect that in general solvent–solid interactions will modify surface energetics and will hence modify the shape of crystals produced from a specific solvent.

The energetics of a crystal–liquid interface may be considered to be composed of two major terms, the solid–vapour surface free energy ($\gamma_{S/V}$) and a solid–liquid interaction term ψ_{12}. These parameters may be related by

$$\gamma_{S/L}=\gamma_{S/V}+\gamma_{L/V}+\frac{\psi_{12}}{A} \tag{5.50}$$

where $\gamma_{L/V}$ is the solvent–vapour interfacial free energy (the surface tension) and A the surface area of solid in contact with the liquid. The interfacial energy of a particular crystal facet is then governed by the specific surface free energy and solvent interaction for the facet. For sodium chloride (100) planes the appropriate relation would be

$$\gamma_{S/L}(100) = \gamma_{S/V}(100) + \gamma_{L/V} + \frac{\psi_{12}(100)}{A} \tag{5.51}$$

It can be seen that the larger the solid–liquid interaction, the lower is the solid–liquid interfacial free energy (ψ_{12} is negative by convention) and hence that face is more energetically favourable and more predominant. It turns out that since the (100), (010) and (001) planes of the crystal sodium chloride structure have the same surface configuration of ions, they will also have the same (solid–vapour) surface free energy and solid–liquid interaction. The balance of these forces allows the equilibrium shape of NaCl-type crystals, grown from solution, to be of cubic habit unless some other planes are energetically more favourable.

If we consider the (111) planes of a sodium chloride crystal, we find that the planes are occupied by ions of one type only, and hence, the surfaces composed of such planes will carry a net electrostatic charge. If this surface is exposed to a polar solvent, strong interaction will occur, and hence the ψ_{12} term of eqn. (5.50) will be large, and it might be expected that (111) planes will be stabilised by polar solvents. Undoubtedly (111) planes are stabilised by polar solvents, but in all known cases $\gamma_{S/V}(111) \gg \gamma_{S/V}(100)$ and this effect renders insignificant the solvent–solid interaction. It is possible, therefore, to justify the fact that the equilibrium form of NaCl-type crystals in solution is usually cubic.

The situation for non-cubic crystals is obviously much more complicated. In this case the arrangement of ions in the crystal faces is not the same. If the solvent does not interact discriminately with any one face, the crystal, in solution, should take on the theoretical low surface free energy ($\gamma_{S/V}$) form. The low surface energy form often reflects the underlying crystal structure and precipitate crystals can often be identified by relation of shape to crystal structure (cubic, hexagonal).

The principles of organic crystal habit are also fairly clear. For example, organic molecules which possess both polar and non-polar segments will have a habit in which polar planes are predominant if grown from a polar solvent and non-polar (or less polar) planes predominate if grown from a non-polar solvent. A schematic diagram is shown in Fig. 5.8. Wells[77] has been instrumental in pointing out the habit-modifying influence of several organic solvents. Examples taken from Wells's work are: anthranilic acid crystals are bipyramidal when grown from ethanol, and prismatic with pyramidal faces when grown from acetic acid; iodoform forms bipyramidal crystals when grown from aniline and prismatic crystals when grown from cyclohexane.

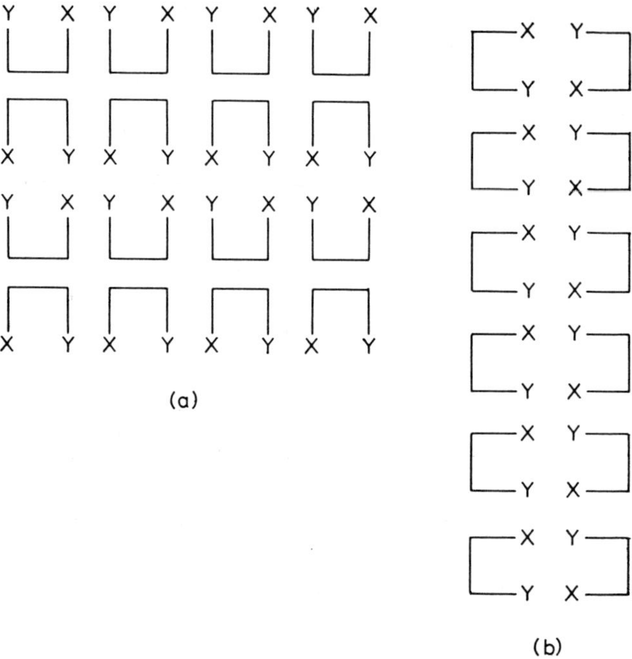

Fig. 5.8. *Schematic, two-dimensional crystal grown (a) from a polar solvent and (b) from a non-polar solvent.*

For a two-dimensional crystal of equilibrium shape, it is easy to show that

$$\gamma_{S/L}(1)l_1 = \gamma_{S/L}(2)l_2 \tag{5.52}$$

where $\gamma_{S/L}(1)$ is the interfacial free energy of side length l_1, etc. By analogy we might expect that

$$\gamma_{S/L}(1)A(1) = \gamma_{S/L}(2)A(2) = \gamma_{S/L}(3)A(3) \ldots \tag{5.53}$$

i.e. the product of the interfacial free energy and the area of that surface is a constant.

It might be argued that since each plane in a crystal has a discrete surface, and hence interfacial free energy, it should be represented in the equilibrium form of a crystal. Such a conclusion might indeed have some validity if true equilibrium could be obtained. However, this is very unlikely in real precipitation systems and even a pseudo-equilibrium shape is obtained only if extreme care is taken to ensure optimum conditions for growth (for instance in single-crystal growth experiments). Clearly, under conditions when crystal growth is the rate-determining mechanism of precipitation, the size and shape of crystals may be deduced from growth kinetics rather than from equilibrium considerations.

It has been shown (Fig. 5.4) that the rate-determining mechanism of crystal growth changes with supersaturation. At low supersaturations growth on surface defects (with the possibility of the formation of screw dislocations) and/or nucleation controlled growth are feasible. Beautiful examples of screw dislocations formed on polymer[78] and inorganic crystals[78,79] have been published. If screw dislocation is the rate-determining mechanism the formation of whiskers is possible,[80] although not usual for solution-grown inorganic crystals.

Apart from the supersaturation, the rate equations corresponding to the above mentioned mechanisms (eqns. (5.20), (5.21), (5.30)) contain material parameters characterising the crystal faces, i.e. the edge energy parameter β' and the surface diffusion distance x_s. The inclusion of the edge energy term in these equations explains the preferential growth of high energy planes in a crystal.[24]

When, due to variations of β' between planes, $S' \gg S'_{II}$ and S'_{III} on some planes and $S' < S'_{II}$ and S'_{III} on others, crystal growth is anisotropic. As a consequence, low symmetry crystals form needles or platelets while cubic crystals have only the slowest growing low energy planes.

The disappearance of a high energy crystal plane by preferential growth is visualised in Fig. 5.9.

At supersaturations, at which crystal growth is uninhibited by surface processes (still below the critical supersaturation for homogeneous nucleation), transport of ions through the solution becomes the rate controlling mechanism (Fig. 5.4). Under such conditions the development of dendritic crystals becomes feasible. A simple explanation for this phenomenon is as follows.

Growth of ionic precipitates from solution may be regarded as a two-

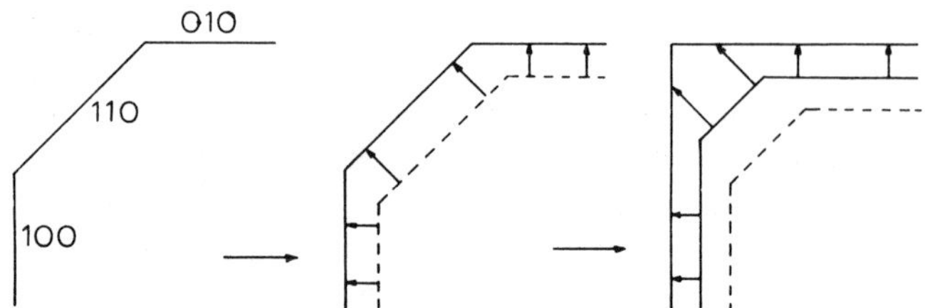

Fig. 5.9. Schematic diagram of the elimination of a high energy (110) crystal face. Growth occurs most rapidly on high energy faces such that eventually they become eliminated (for slow growth).

way process; solvated ions diffuse to the surface and subsequently, at some stage, desolvate. The released solvent molecules diffuse away from the surface causing a counter current. At high degrees of supersaturation the steady state concentration of solvent molecules in the interfacial region reaches a maximum, and the concentration contour resembles that in Fig. 5.10.

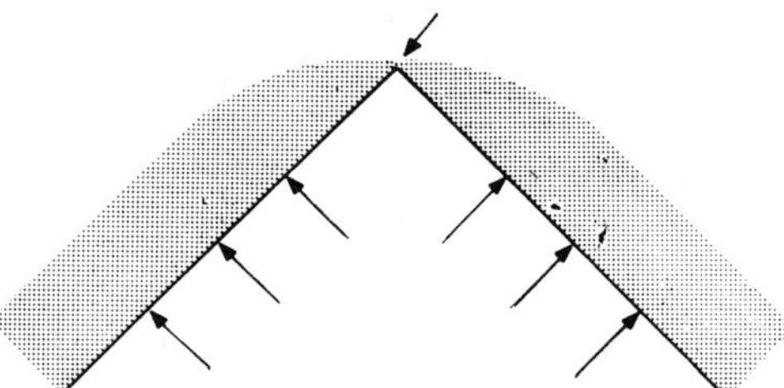

Fig. 5.10. Schematic representation of solvent molecules released from the forming solid hindering growth on plane surfaces.

The limitation on the growth rate imposed by this boundary layer probably does not extend to crystal corners which are not blocked and hence growth at the corners becomes predominant as shown in Fig. 5.11(*a*) and (*b*). Thus, the development of dendritic crystals ensues with the dendritic arms extending in well defined crystallographic orientation to the parent crystal.

A feature of dendritic growth which is of some importance in an understanding of precipitation is that of dendrite fragmentation. Jackson and co-workers[19] have shown that under rapid growth conditions, dendrite arms separate from the parent crystal and thus generate secon-

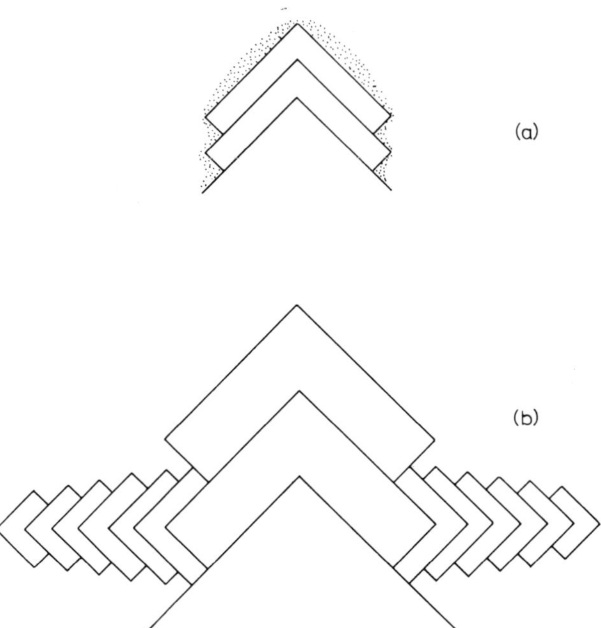

Fig. 5.11. Representation of the development of a crystal dendrite. The corners not blocked by solvent diffusing from the surface grow preferentially.

dary nuclei. This is not a result of mechanical fracture, but is rather a crystallisation/dissolution phenomenon.

Jackson's observation seems to tie together studies of so-called secondary or ancillary nucleation processes in widely diverse fields. For example, the seeding of clouds with silver iodide or some other suitable particles would clearly be of no commercial value if it required one particle for each rain-drop which was produced. In fact, up to 10 000 drops are often produced for each seed particle. The explanation would now seem to be that dendritic ice crystals are nucleated on the seed particles and subsequent fragmentation causes a chain nucleation process. In the liquid phase, seeding supercooled water with ice also produces many secondary crystals.

The foregoing considerations imply that apart from the maximum preceding the onset of homogeneous nucleation, the particle size versus supersaturation curve exhibits a discontinuity at a medium supersaturation, where the formation of compact crystals (or needles, platelets, etc.) changes to dendritic crystal growth. Such a discontinuity has indeed been observed when calcium oxalate was precipitated at different initial supersaturations.[72] A corresponding change in the slope of the particle number versus supersaturation curve indicates an increase in N, probably due to fragmentation of dendrites. These relations are shown in Fig. 5.12, while typical crystal morphologies, as observed 24 h after

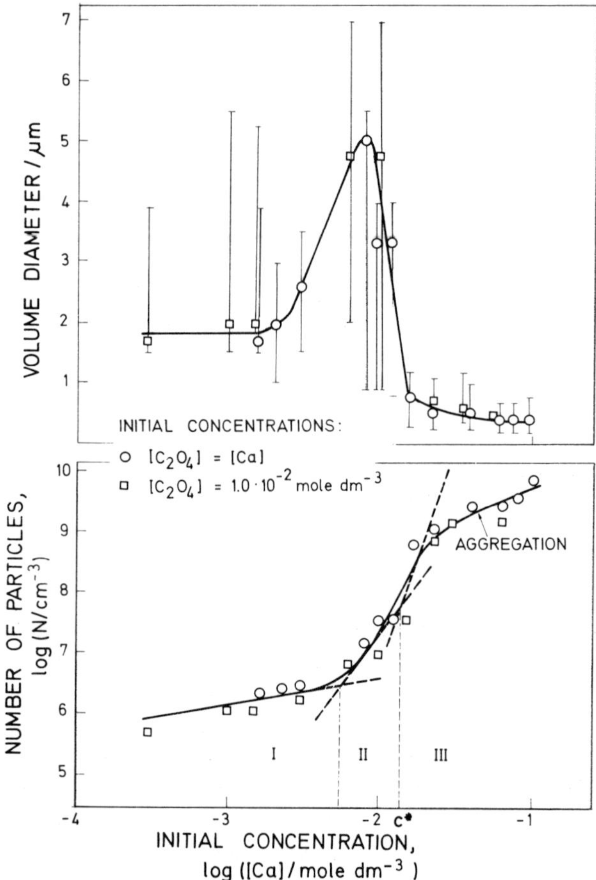

Fig. 5.12. The number of particles (>0·5 μm) per cm³ of calcium oxalate precipitates (lower diagram) and the corresponding populated range of particle sizes (upper diagram) as a function of the initial reactant concentrations (initial pH 6·5, 0·3 mol/dm³ NaCl, 25°C, observation time 10 min, experiments performed by Coulter counter). Points in the upper diagram represent population maxima. I, II and III are concentration regions within which different morphological types of precipitates prevail (see Fig. 5.13); c denotes the critical concentration for homogeneous nucleation. (After Füredi-Milhofer et al.[72])*

sample preparation in corresponding concentration regions,[81] are shown in Fig. 5.13.

The sequence, compact crystals–dendrites–microcrystalline aggregates has indeed been found in the precipitation of many slightly soluble precipitates, typical examples being the sulphates of barium[21] and lead.[82]

The kinetics of the formation and transformation of metastable phases is another factor which greatly influences the size and shape of precipitate particles. As an example, the development with time of a size distribution of calcium oxalate, precipitated by prevalently hetero-

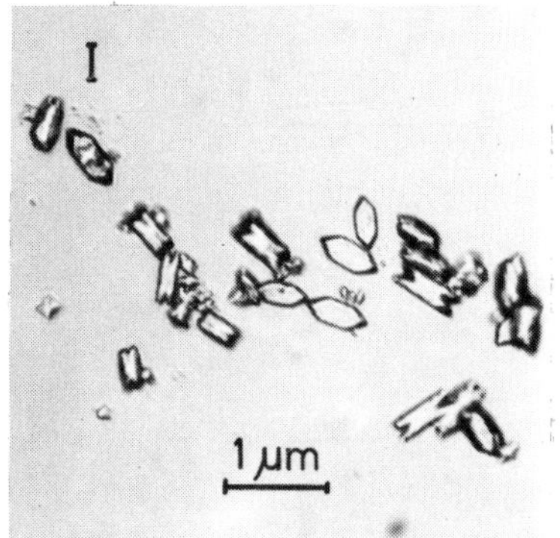

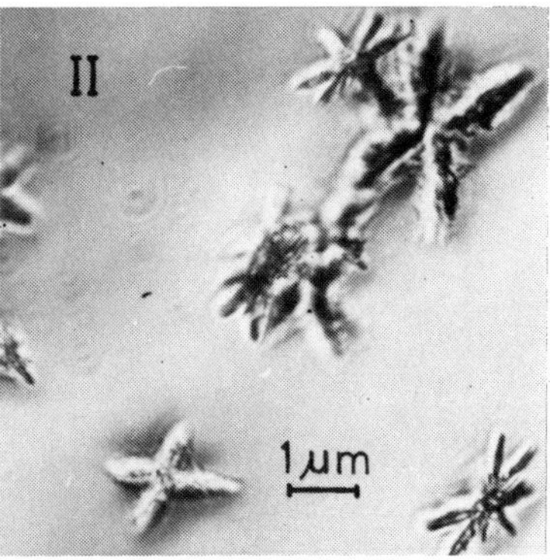

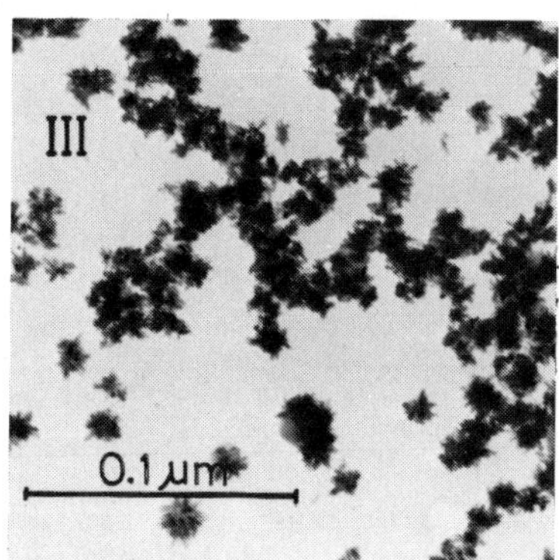

Fig. 5.13. Micrographs (I, II) and electron micrograph (III) showing morphological types of calcium oxalate typically found 24 h after sample preparation in concentration regions I, II and III, shown in Fig. 5.12. I, platelets of calcium oxalate monohydrate ($CaC_2O_4{\cdot}H_2O$, whewellite); II, dendrites; III, microcrystalline aggregates. (After Babić et al.[81])

geneous nucleation is shown in Fig. 5.14.[83] The appearance of several maxima on the number distribution curves indicates the presence of different solid phases (i.e. different hydrates of calcium oxalate) which transform and grow at different rates. The large number percentage of small particle sizes, evident from the first three curves (from above) shows that many particles have sizes below the detection limit of the instrument (Coulter counter).

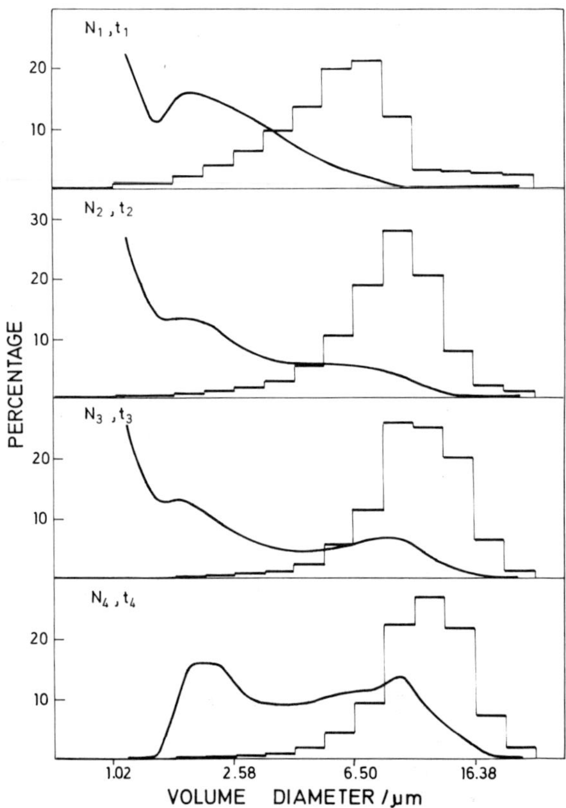

Fig. 5.14. Schematic volume distribution histograms and number distribution curves showing the development of a size distribution of calcium oxalate with time; $t_1<t_2<t_3<t_4$*. Precipitation was initiated from 0·3 mole/dm³ sodium chloride solutions by heterogeneous nucleation upon impurity nucleators. The appearance of several maxima on the number distribution curves indicates the presence of different hydrates of calcium oxalate which transform and grow at different rates.* (*After Füredi-Milhofer et al.*[83])

Influence of impurities

In a broad sense the effect of impurities on crystal morphology comprises changes of the equilibrium shape of crystals, promotion or inhibition of nucleation, growth, dissolution and transformation of metastable phases and stabilisation and flocculation of colloids. The last mentioned effect has been discussed in detail (see Chapters 1 and 2 and earlier in this Chapter). All other effects may be explained by considerations of surface energetics and nucleation and growth kinetics.

First we have to conceive that impurities adsorb at specific sites of the crystal–solution interface, i.e. faces with high surface free energy are preferred. As a result the surface free energy (or edge free energy) of such faces decreases, which expresses itself in a change in the anisotropic growth kinetics. For example, surface active agents generally adsorb on the (111) planes of NaCl-type crystals. The situation is then the opposite to the normal growth sequence since growth now virtually ceases perpendicular to the (111) surface and the crystal grows with octahedral habit, the (111) surface and the crystal grows with octahedral habit, the (100) planes having grown out of existence. Many modifications of the preceding type are known, the modification of sodium chloride by urea being a much quoted example.

Low symmetry crystals which usually form platelets may, in the presence of the impurities, appear as needles. This was observed[84] when calcium hydrogen phosphate dihydrate was precipitated in the presence of citrate ions; an example is shown in Fig. 5.15.

In the case of solution growth of ionic crystals constituent ions which are in excess may specifically adsorb at certain faces and thus cause habit modifications.[85] This is manifested in the case of the growth of calcite, where the excess in solution of Ca^{2+} or $CO_3{}^{2-}$ ions gives rise to elongated or flat crystals respectively.[86]

If crystals are small and compact and contain a considerable quantity of surface defects, impurities may adsorb equivalently on all faces. In this case habit modification may not occur, but inhibition of growth and dissolution, or for metastable phases, inhibition of phase transformation may be observed. Nucleation, growth and dissolution are inhibited when the adsorption kinetics are slower than the molecular exchange between the crystal and the mother liquor.[87] On the other hand, for weak adsorption, when communication between adsorbate and solution is faster than between substrate molecules and the environment, foreign ions do not actually block the surface but lower the surface free energy of nucleus or crystal face. As a consequence the activation energy for nucleation or growth should decrease and the rate of the respective process increase on foreign ion adsorption.[85]

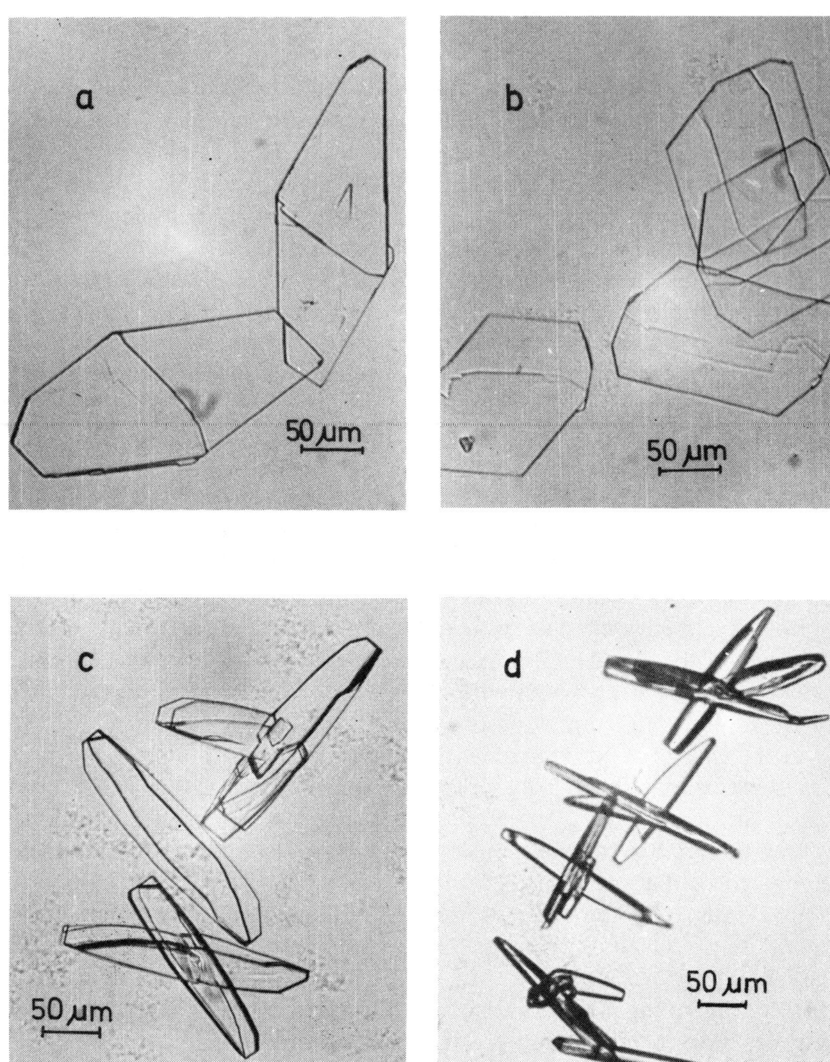

Fig. 5.15. Micrographs showing crystals of calcium hydrogen phosphate dihydrate ($CaHPO_4 \cdot 2H_2O$) precipitated without additives (a), and in the presence of increasing concentrations of citrate ions ($b < c < d$). (After Brečević and Füredi-Milhofer.[84])

Some examples of enhancement of nucleation[88] and mononuclear crystal growth[89] have indeed been reported (see also Fig. 5.20, later). However, many more examples of the inhibition of crystal growth by foreign substances have been described. Some crystallisation inhibitors (pyrophosphate, polyphosphonates) have been extensively used in medical science as well as in industrial practice. The problem of scaling is only one of the problems in industry which might be solved by the use of effective inhibitors of crystallisation.

Addition of traces of surfactants to relatively slowly growing crystals often improves their crystal form by further slowing down their growth rate. This effect is commonly used in crystal growing plants. On the other hand, at higher degrees of supersaturation and faster growth rates, the same surfactants may promote dendrite formation, probably by the traffic jam mechanism. It is then not surprising that control of crystal morphology by additives has become something of an industrial art.

The growth of crystals on substrates

So far we have considered the morphology of crystals which were nucleated either by heterogeneous nucleation upon non-specific impurities or by predominantly homogeneous nucleation. However, the role of the nucleating substrate may be of decisive importance in determining growth kinetics and the properties of the resulting precipitates. In order to explore these interactions, the growth of crystals on well defined model substrates has been studied by many authors. One of the results emerging from such studies is the description of epitaxy.

If the model for heterogeneous nucleation is correct, then the ions or molecules of a solute phase must take up some special configuration relative to the substrate in the most efficient energetic process. Macroscopically, such a process manifests itself as epitaxy or oriented crystallisation. Epitaxy of crystals on planar substrates has been known for many years and hundreds of examples are known. In Fig. 5.16 is shown the epitaxial arrangement of RbI crystals on a mica substrate. It can be seen that the rubidium iodide crystals, which have the sodium chloride cubic lattice array, grow with their (111) planes adjacent to the (001) mica plane. Figure 5.17 shows a schematic diagram of the arrangement of the RbI ions relative to those in mica. The most efficient substrates for epitaxy are those in which the lattice parameters match those in the depositing phase, as supposed in eqn. (5.16). Usually crystals do not grow epitaxially on substrates when there is a mismatch of $>15\%$, though isolated examples are known where there is no lattice matching

Fig. 5.16. Oriented overgrowth of rubidium iodide crystals on (muscovite) mica. The crystals grow with their (111) planes adjacent to the (001) plane of mica.

criterion. One interesting example of this phenomenon is the epitaxy of polymers on ionic crystals. Willems and others[90,91] observed that polyethylene crystals deposited from solution on to the (001) face of sodium chloride crystals grew epitaxially along the (110) plane of the substrate. Similar arrangements have also been found for other polymers (see Fig. 5.18). The initial suggestion was that the chain folds matched the lattice requirements of the substrate. However, substantial evidence[92-94] indicates that a range of polymers (polyethylene, polypropylene, polyoxymethylene, poly(ethylene oxide), polyacrylonitrile, nylon, isotactic polystyrene) will all grow epitaxially on a range of alkali halides, quartz, graphite, etc., completely independently of the lattice matching requirements. The mechanism and energetics of polymer nucleation have been studied in the laboratory by one of the authors with the conclusion that although the matching between the folded chain structure of poly-

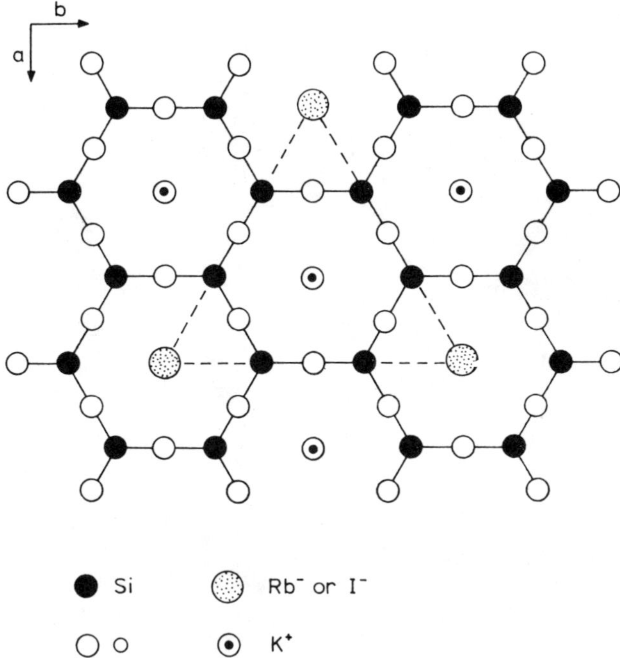

Fig. 5.17. The surface structure of mica is such that the heterogeneous nucleation of crystals of close lattice match is directly controlled by ions (or ion holes) in the mica surface. An arrangement of a few ions in the sub-critical cluster is shown. In practice surface nucleation probably occurs in the region of a surface defect (probably caused by the inclusion of one or more foreign ions which distort the surface structure).

ethylene and the substrate plays a small part, it is predominantly the effect of induced dipoles which orientates the molecules.[92–94]

Many examples of the growth of crystals on substrates are known in industry, scaling being just one of them. The use of solid crystalline additives to control plastic crystallinity has also been developed. Physiology seems to be another area where epitaxy is of extreme importance. An example is the formation of renal calculi, which may involve epitaxial growth of urinary salts upon one another.[95,96] A good lattice match facilitates this process, but a certain amount of lattice mismatch can be tolerated.[95]

Of course, the best lattice match is obtained when crystal growth is induced upon seeds of the same salt. This technique, originally in-

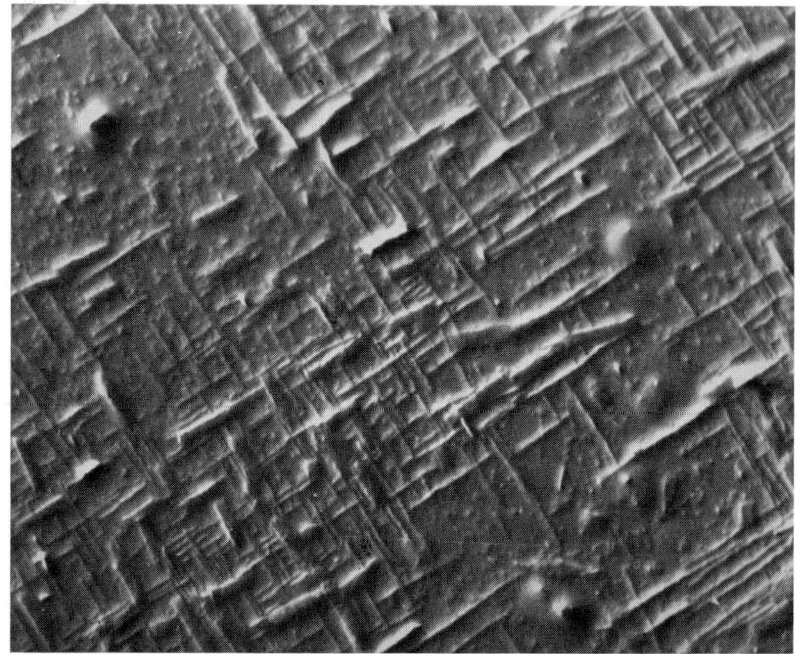

Fig. 5.18. Epitaxial growth of poly-O-acetyl-L-hydroxy-proline on sodium chloride showing orientation of the crystallites in the <110> direction.

troduced by Marc,[97] and Davies and Jones,[98] was extensively used by Nancollas and co-workers[99-101] to obtain reproducible crystal growth kinetics. This work will be described in more detail below.

EXPERIMENTAL APPROACH

Given the theoretical background outlined in the foregoing sections it remains to be seen which are the experimental parameters determining the properties of precipitates. Both thermodynamic and kinetic considerations point to the great influence of the supersaturation but chemical considerations show that the ratio of ionic activities of the reactants and the activities of foreign ions and molecules in solution are equally important. In relation to the above the ionic strength and the pH of the electrolyte solution must be strictly defined. Other important parameters are the temperature and the reaction time.

In this section we shall describe some of the experimental approaches which are frequently used with the purpose of providing

(i) qualitative and quantitative information on specific, usually complex, precipitation systems, and
(ii) quantitative information pertaining to the mechanisms of precipitation processes, specifically nucleation and crystal growth.

Stationary state experiments

Complex precipitation systems

Experiments designed to evaluate the influence of solution composition on the properties of the resulting precipitates in complex precipitation systems involve the investigation of a large range of concentrations (activities) of reactants or additives. The reaction time is usually kept constant and long enough to allow the establishment of a stationary state in the system.

A useful first approach to a new system is the construction of solubility diagrams using solubility and complex stability constants[45] of slightly soluble and soluble species which are likely to form in the system. Such diagrams show which of the solid phases are stable within a given range of pH and reactant concentrations and may also indicate the existence of potential metastable precursors.[102] For descriptions of the construction and use of solubility diagrams the reader is referred to the relevant literature (see for instance refs. 46 and 103).

Further information on the experimental system is obtained by systematic variations of the concentrations (activities) of the reactants (or additives) and observing some characteristic feature of the precipitate (turbidity, morphology or other). The concentrations are changed in discrete steps; this is achieved by simply mixing together stable solutions containing different, known reactant concentrations. The chosen experimental parameter is then observed at a predetermined time after sample preparation. Evidently the manner of mixing will play an important part in the rate of formation of the precipitate but if the components are mixed rapidly it may be assumed that a fairly uniform solution can be achieved in a reasonably short time (compared with the time of observation). The data obtained by this method may be represented by precipitation curves, plotting the measured quantity against the concentration (activity) of one component (A) at constant concentration activity of the other (B). A series of such curves with different constant concentrations of (B) may be used for the construction of a precipitation diagram,[6,104,105] i.e.

a plane showing contours of equal values of the chosen experimental parameter as a function of the initial concentrations (activities) of components (A) and (B). These may be replaced by boundaries showing concentration regions within which precipitates with similar characteristic properties prevail. Obviously, two-component systems may be completely characterised by one precipitation diagram while for the characterisation of multicomponent systems two or more such diagrams may be required.[102,105]

Precipitation diagrams are extremely useful as they yield information on the influence of the solution composition

- on the properties of precipitates,[6,102,104–106]
- on the prevailing solid/solution equilibria,[6,104–106]
- on the mechanisms of some precipitation processes (nucleation and crystal growth),[72,102] and
- on material parameters characterising the homogeneous nucleus of the nucleating solid phase.[72,102]

A schematic presentation of a precipitation diagram of a two-component system is given in Fig. 5.19. For simplicity it was assumed that complex formation in this system is negligible and that the chemical composition of the precipitate is independent of the reactant concentration ratio. It can be easily shown[105] that in such a diagram (provided it is plotted in terms of ionic activities on a log/log scale) the solubility boundary is a straight line perpendicular to the line indicating equimolar ratios of the reactants. Again, for simplicity reasons, we have plotted the boundaries of the precipitation regions parallel to the solubility limit considering thus only the influence of the supersaturation on the properties of the precipitates. This influence has been discussed in detail earlier, and Fig. 5.19 may be seen as a summary of these considerations. In addition it shows some prominent features of precipitation diagrams, i.e. the precipitation boundary and the boundary between the regions of heterogeneous and prevalently homogeneous nucleation.

The precipitation boundary is defined as the boundary between clear solutions and detectable precipitates and its position is of course dependent on the time and tool of observation. At long observation times it may coincide with the solubility boundary and may conveniently yield information on the prevailing solids and soluble species including the respective solubility and complex stability constants. This approach was first demonstrated by Kratohvil, Težak and Vouk[106] and has been described in detail in subsequent review articles.[6,104,105]

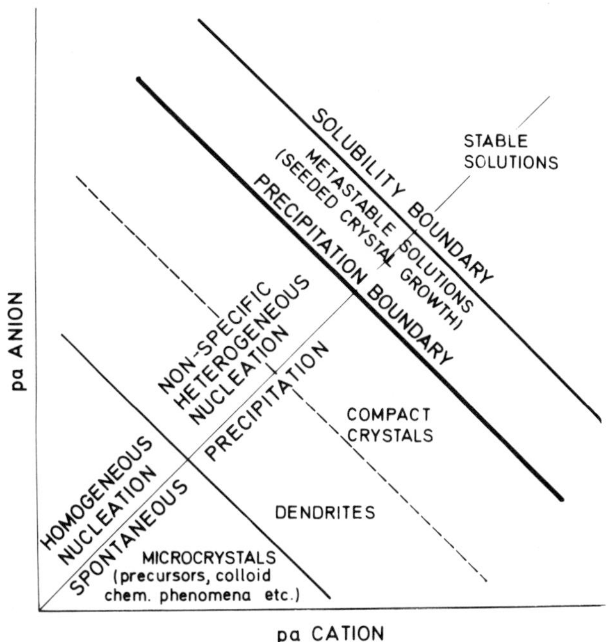

Fig. 5.19. Schematic presentation of a precipitation diagram of a two-component system constructed on the assumption that complex formation is negligible and that the chemical composition of the precipitate is independent of the reactant concentration ratio. The coordinates are negative logarithms of initial ionic activities of the constituent ions. The mechanism of nucleation and the morphology of the precipitates are indicated within the proper range of reactant concentrations.

Under stationary state conditions (metastable equilibria) the precipitation boundary is shifted towards higher supersaturations, dividing thus the region of metastable solutions (where crystal growth can be induced by specific substrates) from the region of spontaneous precipitation.

In terms of nucleation and crystal growth kinetics the precipitation boundary corresponds to a constant induction time, which is the time of observation. In other words, precipitation boundaries are 'isochrones' showing the dependence of the induction period on solution composition.[89] Since the times of observation are usually long compared with induction periods characteristic of nucleation (Table 5.2 and ref. 16) the information thus obtained most likely pertains to mononuclear crystal growth kinetics (ref. 21 and Fig. 5.4).

A set of experimental precipitation boundaries, observed at different time intervals (10 min – 1 month) is shown in Fig. 5.20. Obviously, such boundaries are not necessarily parallel to the solubility limit. The results shown in Fig. 5.20 were interpreted as discussed above and it was

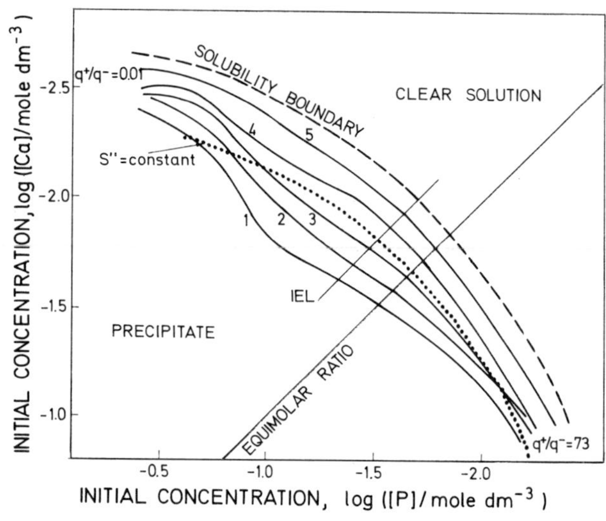

Fig. 5.20. Schematic presentation of precipitation boundaries of the system calcium chloride – sodium phosphate – 0·15 mole/dm³ sodium chloride, pH = 5·0 ± 0·1, 25°C. In all systems the solid phase precipitated was calcium hydrogen phosphate dihydrate ($CaHPO_4.2H_2O$). Curves 1–5 are boundaries obtained at different observation times, increasing from 10 min (curve 1) to 1 month (curve 5). The calculated solubility boundary and a curve at constant supersaturation ($S'' = AP/K_{s0} = [Ca^{2+}][HPO_4^{2-}]/[Ca^{2+}]_s[HPO_4^{2-}]_s = \text{constant}$) are indicated. IEL, calculated isoelectric line; q^+/q^-, calculated ratio of positive and negative charges in solution. (After Kosar-Grašić et al.[89])

concluded that the rate of mononuclear crystal growth in the investigated system depends not only on the supersaturation, but on the reactant concentration ratio as well and that it is indeed enhanced when positive or negative ions are in excess in the mother liquor.[89] This observation ties in with the consideration that nucleation and crystal growth may be enhanced by surface adsorption.[85]

The boundary between the regions where precipitation is initiated by heterogeneous and prevalently homogeneous nucleation respectively in most precipitation diagrams appears as a discontinuity which may be recognised by turbidimetry, morphological observations or other means. The reason for this effect may be found in the profound influence which the mechanism of nucleation has on the number, size and morphology of precipitate particles. Therefore the heterogeneous–homogeneous nucleation boundary may be used to determine the critical supersaturation for homogeneous nucleation, as an average value over a wide range of reactant concentrations.[72] An example is shown in Fig. 5.21. The value for S^* thus obtained may be used to calculate the material parameters characterising the homogeneous nucleus of the nucleating phase. i.e. the interfacial energy (eqn. (5.15)) and the critical radius (eqn. (5.11)) as well as the critical activation energy for homogeneous nucleation (eqn. (5.12)).

Experimental tests of nucleation theory

Ever since the thermodynamic nucleation theory was proposed, experimentalists have tried to verify eqn. (5.15) by determining the critical supersaturation for homogeneous nucleation and calculating the interfacial energy of the critical nucleus. Most early experiments, however, in which the supersaturation was increased continuously, met with the difficult task of preventing heterogeneous nucleation. Attempts to achieve this end by removing impurities proved more or less impossible in practice.

The miniaturisation technique, originally proposed by Vonnegut,[107] and further developed by White and Frost[108] and other authors[109–112] has been more successful. The principle of this technique is to break up the solution into more droplets than there are impurities. Hence, some droplets will contain no impurities and nucleation, if any, must be homogeneous. In order to obtain large populations of non-communicating droplets of narrow volume range and to prevent their evaporation during the experiment it is usual to make the dispersions in some inert medium (e.g. an oil). The size of the droplets should be small enough to minimise the heterogeneous effect (droplets should be smaller than 5×10^{-3} cm in diameter) and yet contain enough solute so that upon crystallisation the particles are sufficiently large to be detected. A suitable detection device for such a system might for example incorporate filming the microscopic field, illuminated by polarised light. Crystallisation then leads to the appearance of twinkling particles in the drops. An essential requirement of the droplet technique is that supersaturation be established after drop formation. This is achieved either by supercooling of saturated solutions[108–110] or by homogeneous gene-

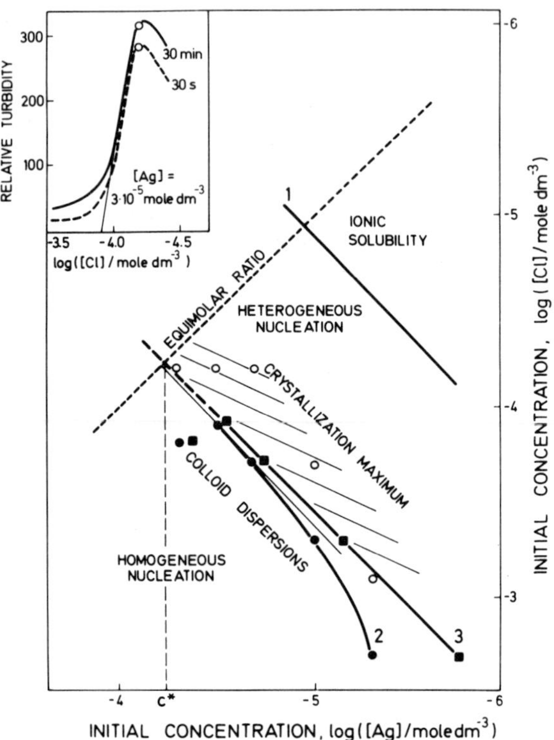

Fig. 5.21. The determination of the critical supersaturation for homogeneous nucleation ($S^* = (a^{*2}/K_{s0}^{1/2})$) from a precipitation diagram of silver chloride, constructed from literature data.[73,106] As expected from theory (Fig. 5.6) in this system the change from heterogeneous to prevalently homogeneous nucleation is evidenced by turbidity maxima (inset), termed by Težak[73] 'crystallisation maxima'. The concentration products at the heterogeneous–homogeneous nucleation boundary (full circles, line 2) were obtained by drawing tangents to the steepest parts of turbidity curves (inset). Line 3, full squares, shows the corresponding (recalculated) activity products. Open circles denote crystallisation maxima; line 1 is the precipitation boundary;[106] c is the critical concentration for homogeneous nucleation. (After Füredi-Milhofer et al.[72])*

ration of ions at constant temperature.[111,112] The first type of experiment yields the critical temperature which, for ionic clusters, is related to the critical supersaturation by

$$nRT\ln S^* = \frac{\Delta H \Delta T_c}{T} \tag{5.54}$$

where ΔH is the heat of solution, ΔT_c the critical supercooling for the homogeneous nucleation of clusters, and n is the number of ions in a neutral molecule.

Thus eqn. (5.15) becomes

$$\Delta T_c = \left(\frac{16\pi\gamma_{S/L}^3 \bar{V}^2 n^2 T N_0^3}{3R\Delta H^2 \ln A_k} \right)^{1/2} \tag{5.55}$$

where $\bar{V}$ is the molar volume.

Equation (5.55) is generally used in studies of homogeneous nucleation by the supercooled drop method. The other type of experiment (homogeneous generation of ions) yields number/time data which are difficult to interpret because the supersaturation (and thus the driving force) varies with time, and the kinetics of the supersaturation-creating reaction is known only by comparison with a macroscopic (beaker) experiment.[112] Moreover, the interpretation of particle number versus time data in terms of nucleation kinetics is difficult because of the interplay of ageing and aggregation in the early stages of precipitate formation.

Apart from the difficulties with technique which arise in both approaches, the problems arising from sources of spurious nucleation in the droplet method are considerable. For example, although the medium itself may be inert, it may contain many impurities which may catalyse nucleation at the oil–solution interface. Furthermore, since this interface is likely to be diffuse at the molecular level, there exists the distinct possibility that local fluctuations in solute density could cause spurious nucleation in the interfacial region. If effects of the preceding type are present, calculated interfacial energies would be too low if homogeneous nucleation were assumed to be the predominant nucleation mechanism.

The experimental set-up becomes much simpler if the supersaturation is not increased continuously but in discrete steps, for instance as described in the previous subsection. This way it is possible to obtain curves of the type shown in Fig. 5.2 in which, upon onset of homogeneous nucleation, the number of particles increases by orders of magnitude with a corresponding decrease in particle size. The critical reactant concentration corresponding to homogeneous nucleation is thus easily detected (see lower diagram in Fig. 5.12) and may be used to calculate S^* and the interfacial energy and radius of the critical nucleus (eqns. (5.11) and (5.15)). Nielsen[113] was the first to identify a curve of this type for barium sulphate precipitation and used it to calculate the interfacial energy of the barium sulphate–water interface. Subsequently other systems have been studied by these means[35,72,114,115] or, more recently, by evaluating the heterogeneous–homogeneous nucleation boundary from precipitation diagrams[72] (see previous section).

An alternative treatment of data, proposed by Nielsen,[35] uses the classical nucleation equation (eqn. (5.14)) and includes the assumption that crystal growth after homogeneous nucleation is diffusion controlled. As a result equations linking the number of particles and the induction period to the supersaturation have been obtained. For experimental verification a 'rapid flow' technique was employed to measure short 'induction periods' and peptising agents were used in some cases to prevent excessive aggregation.[35,115] The inherent difficulties of this approach, i.e. concomitant Ostwald ripening and aggregation, could probably not be completely overcome, but it was nevertheless possible to verify the proposed equations for some 2,2 electrolytes[35,115] while for 1,1 electrolytes considerable discrepancies between theory and experiment have been obtained.[115]

A compilation of a great number of interfacial energies, calculated from critical supersaturations, is given in ref. 115. Using these data the authors demonstrated an inverse linear relationship between the logarithms of the solubilities and the interfacial energies of slightly soluble electrolytes, i.e. the smaller the solubility of an electrolyte, the higher is the interfacial tension at the crystal—solution interface. This result has been supported by some more recent data obtained by evaluating precipitation diagrams.[72]

While tests of homogeneous nucleation theory require examinations of spontaneous precipitation at high supersaturations, heterogeneous nucleation theory has been tested by determining the critical supersaturation ratio for the nucleation of crystals on well defined substrates. Most popular among substrates has been mica, which can be easily prepared and is remarkably free from surface defects. Newkirk and Turnbull[116] examined the nucleation of ammonium iodide on to the surface of various micas and determined that, as expected, lattice matching between substrate and deposit plays a part in the energetics of heterogeneous (epitaxial) nucleation. The critical supersaturation increases with increase in lattice mismatch (25°C ambient temperature) as is shown in Table 5.3. Later work[11] has shown that the lattice matching requirement is borne out for the nucleation of a series of alkali halides on mica, and furthermore the critical supersaturation decreases with increase in operating temperature as required by eqn. (5.16). The alkali halides grow with their (111) planes adjacent to a mica (001) plane and thus form a rather limited class of epitaxial arrangement.

TABLE 5.3
CRITICAL SUPERSATURATION RATIO S^* FOR NUCLEATION OF AMMONIUM IODIDE ON VARIOUS MICAS AS A FUNCTION OF LATTICE MATCH
(after Newkirk and Turnbull[116])

Mismatch (%)	*ln* (*supersaturation ratio*)
1·13	0·61 ± 0·01
1·53	0·62(5) ± 0·01
2·73	0·81 ± 0·04
3·40	1·00 ± 0·09
3·60	0·96 ± 0·14

The theoretical predictions for two-dimensional heterogeneous nucleation cannot be expected to apply exactly because we are dealing with a situation in which macroscopic thermodynamic considerations have been applied to clusters of relatively few ions. Particularly serious is the assumption that γ_e is independent of cluster size. It has indeed been shown[117] that not only does the edge energy of clusters change with size, but the gross changes in the energetics of cluster formation caused by modified edge energetics can cause major morphological effects.

The nucleation of non-polar organic crystals on to defined substrates is difficult to examine because of the low interaction energy between admolecules and substrates. In these cases the critical supersaturations are extremely large and therefore inconvenient to study. These conclusions are, of course, in line with common experience relating to the difficulty in nucleating many types of organic crystals from supersaturated solution. Experiments on the nucleation of succinic acid on mica, for which it was possible to obtain reasonable (though not precisely reproducible) results, again show that the critical supersaturation decreases with increase in temperature (Table 5.4).

Precipitation kinetics

It has become obvious from this presentation that kinetic factors play an equally important role in determining the properties of precipitates to that of thermodynamic and chemical factors. Obviously then, following the kinetics of precipitate formation has been the method of choice for many authors. In principle, kinetic studies show which of the precipitation processes are rate-determining under a certain set of experimental conditions and may also yield information on the rates and mechanisms

TABLE 5.4
NUCLEATION OF SUCCINIC ACID ON MICA

Saturation temperature (°*C*)	*Critical supersaturation ratio, S**
20	1·628
25	1·60
30	1·57
35	1·55

of those processes. Most attention has so far been given to the kinetics of crystal growth and dissolution, but aggregation and ageing processes have also been successfully studied using the theoretical background outlined in other parts of this chapter. Rather than giving a complete review of the relevant literature we shall confine our attention to basic principles and discuss the implications of some more frequently used methods and approaches.

Experimental methods

The basis of a kinetic analysis is the progress curve, i.e. a curve showing changes of the solute concentration or the amount of solid phase formed as a function of time. Some of the methods most frequently used to obtain such curves are listed and commented on in Table 5.5, which also contains references to literature on the more recently developed techniques.

Usually it will be advantageous to combine solution analysis with information on the properties of precipitates, which may be obtained by particle size analysis.[83,118,119] Furthermore, interpretation of the results obtained from progress curves will be greatly facilitated by additional chemical and physical chemical characterisation of the precipitates formed. For this purpose optical and electron microscopy, X-ray and electron diffraction, thermogravimetric analysis, i.r. spectroscopy, determination of specific surface area, conventional chemical analysis and other methods might be used.

The choice of methods will of course depend on the particular experimental system and on the kind of information required. It must, however, be stressed that the complex reaction sequences which are found in most experimental systems make it necessary to use as many methods as possible to follow the course of formation and transformation of precipitates.

Kinetic studies of spontaneous precipitation

Such studies have been typically performed in ranges of low-to-medium supersaturations where precipitates are formed by heterogeneous nucleation upon non-specific impurities (Fig. 5.19). The attention of most investigators has been centered around the rate and mechanism of crystal growth.[120-123] A much exploited model system in early investigations has been barium sulphate. Turnbull[124] was able to demonstrate that the fraction of barium sulphate precipitated with time, shows a sigmoidal relation which is dependent upon the initial concentration and mode of reactant mixing. Evidently the curve is related to the appearance of solid surface and is autocatalytic in nature. The kinetic relation could be described by the equation

$$\frac{d\alpha}{dt} = k_1 \alpha^{2/3} (1-\alpha)^p \tag{5.56}$$

Turnbull showed that the precipitation curves could be superimposed by varying k_1, thus indicating the uniformity of the process. In eqn. (5.56) α is the fraction of barium sulphate precipitated, i.e.

$$\alpha = (c_0 - c)/(c_0 - s) \tag{5.57}$$

where c_0 and s are the initial and equilibrium solute concentrations respectively. $\alpha^{2/3} = \mathrm{A}$ is the surface area of the solid and

$$1 - \alpha = k_2 (c - s) \tag{5.58}$$

where $k_2 = 1/(c_0 - s)$ is constant at constant initial solute concentration. Eqn. (5.56) may then be written in a general form as

$$R/\mathrm{A} = k_2^p (c - s)^p \tag{5.59}$$

where $R = d\alpha/dt$ is the precipitation rate.

If the number of particles is constant and the precipitate has a sufficiently narrow size distribution that an average volume radius r may be defined, eqn. (5.56) takes the form of eqn. (5.33) which was proposed for crystal growth controlled by a surface process through a polynuclear mechanism.[21,22]

In fact, relations of the form of eqn. (5.59) have been found applicable to a number of different systems and have been interpreted in terms of

TABLE 5.5
A COMPARISON OF COMMONLY USED EXPERIMENTAL METHODS FOR STUDIES OF PRECIPITATION KINETICS

Property monitored	*Method*	*Advantages*	*Restrictions*
Solution concentration[a]	Chemical and radio-tracer analysis	Wide application	Discontinuous; previous separation of precipitate necessary
	Conductimetry	Sensitive; fast; continuous	Restricted to low ionic strength
	Ion selective electrode potentiometry[83]	Selective; continuous; usage possible at any ionic strength	Restrictions depend on the kind of electrode used. Electrodes used for divalent ions have restricted sensitivity and relatively slow response
Concentrations of ions needed to keep constant hydrogen ion activity or solution concentration[a]	pH-stat potentiometry[38,118,119]	Direct quantitative relationship between the amount of precipitate formed and the amount of OH^- or H^+ ions used to keep constant pH. Changes in the system due to pH changes (dissolution, recrystallisation) are avoided.	Restricted to systems in which the pH changes as a consequence of precipitation (anions of weak acids or cations of weak bases are components of the solid phases)
	Constant composition[101c]	Constant chemical potential of all constituent ions, i.e. constant driving force, particularly useful in seeded crystal growth experiments	Applicable only if a suitable specific ion electrode is available

	Turbidimetry	Fast; simple; continuous; optimal range of particle sizes 5–200 nm	Quantitative interpretation difficult
Precipitate properties[b]	Particle number and size analysis[120,121] (Coulter counter)	Fast; sensitive; yields quantitative information on rates of crystal growth and aggregation; particularly useful in media of high ionic strength	Ionic strength > 0·05 required; detects only particles of volume diameter > 0·50 μm; discontinuous
	Isotope exchange[122,123]	Yields direct information on ageing processes	Discontinuous; previous separation of precipitate necessary. Restricted to systems for which suitable isotopes are available

[a] Information obtainable on crystal growth kinetics only.
[b] Information obtainable on crystal growth, aggregation and ageing; see specific comments.
[c] The activities of all ionic species are maintained constant by the simultaneous addition of corresponding reagent solutions under the control of a specific ion electrode.

crystal growth controlled by a surface process. Different models have been proposed by different authors[98,125–127] with the purpose of predicting the value of p and explaining its meaning (for a review see ref. 100). It seems that, for very low supersaturations, a parabolic rate law applies, as also follows from the BCF treatment[29] of screw dislocation controlled growth (eqns. (5.30) and (5.34)).

For higher supersaturations higher values of p (3–4) have been found[35,38,83,125] and it has moreover been demonstrated[35] that the mechanism of crystal growth can change from diffusion to surface reaction controlled as the supersaturation decreases during precipitation.

It is then in any case of interest to plot the results of a kinetic experiment in the form of eqn. (5.56) or (5.59). A plot in terms of eqn. (5.56) is preferable because it allows direct comparison of solution and particle size analysis, since α, defined as in eqn. (5.57), may also be expressed in terms of precipitate volumes, i.e. $\alpha = V/V_{max}$, where V_{max} is the maximum obtainable volume for any experiment (depending on the initial solute concentration).[83]

If surface reaction controlled crystal growth is rate determining one should, in a log/log diagram, obtain a straight line with the slope equal to p.

Figure 5.22 shows a schematic plot of a progress curve (upper diagram) and the corresponding rate versus supersaturation curve (lower diagram) of a system in which precipitation was initiated by heterogeneous nucleation. Curves of this type have been repeatedly obtained in the laboratory of one of the authors.[38,83,118] (Although induction periods have been frequently observed these are not shown in the Figure because they are irrelevant to the present discussion.)

As follows from the sigmoidal shape of the concentration versus time curves, the corresponding rate versus supersaturation curves typically exhibit maxima (section a in Fig. 5.22). The initial slopes of the rate curves are so steep that the time corresponding to the maximum rate almost coincides with the time of the onset of precipitation. This part of the curves of course relates to nucleation and initial growth of the particles. Further sequences of events depend on the initial reactant concentrations. In a certain range of initial supersaturations the maximum of the rate curve is followed by a linear part (sections b in Fig. 5.22) and during the corresponding time period the number of particles is usually constant. This section may be described by any of the usual crystal growth equations (such as eqns. (5.56) or (5.59)) and has been used to find the order of the crystal growth process, p.[38,83] The last part

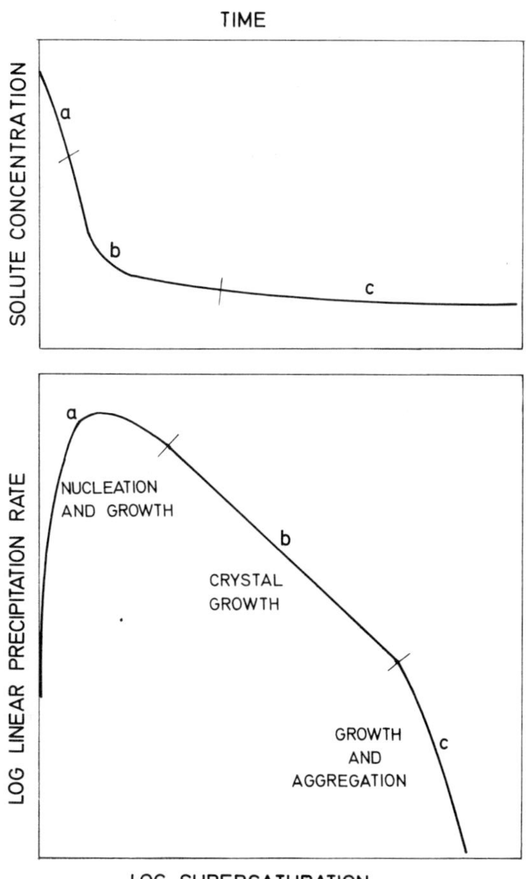

Fig. 5.22. Schematic presentation of a typical kinetic precipitation curve, showing the depletion of the solute concentration as a function of time (upper diagram). Lower diagram: The corresponding linear rate of precipitation (R/A) as a function of the supersaturation, defined as in eqn. (5.56) or (5.59). Sections (a), (b) and (c) in both diagrams correspond in time. Section (a) signifies nucleation and fast initial crystal growth; in section (b) crystal growth is predominant; the change in slope from (b) to (c) may be due to concomitant growth, aggregation and/or recrystallisation.

of the rate versus supersaturation curve (section c in Fig. 5.22) deviates towards higher supersaturations, thus indicating that the precipitate rate decreases faster than expected by the crystal growth equation. Such deviations show that in addition to crystal growth other processes (aggregation and/or recrystallisation) have become significant. Of course, the type of the rate versus supersaturation curve shown in Fig. 5.22. is not necessarily obtained at all supersaturations. At very low supersaturations section c may not appear, while at relatively high supersaturations no linear dependence of the rate on the supersaturation may be observed at all because crystal growth occurs concomitantly with aggregation and/or recrystallisation during the whole precipitation period. To obtain further information about the deviation shown in section c, Fig. 5.22, additional experiments (varying precipitation conditions and using different tools of observation) are required. Useful information might be obtained by counting, isotope exchange, observation and characterisation of the precipitates, etc.

If phase changes occur during precipitation, discontinuous progress curves of the form shown in Fig. 5.23 may be expected. The pH and

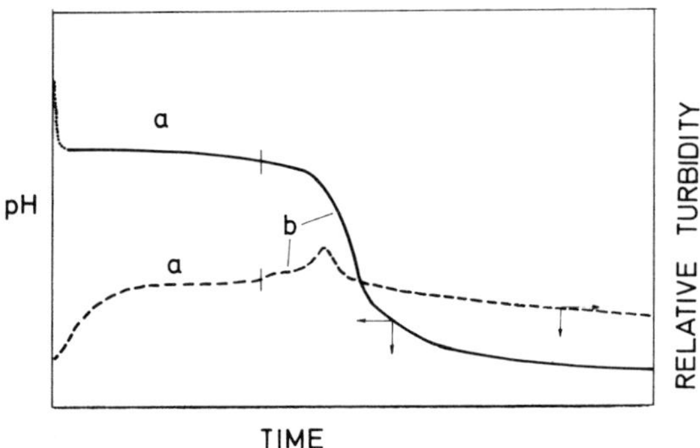

Fig. 5.23. Schematic presentation or typical pH and turbidity versus time curve showing the formation and period of metastability of a metastable precursor (amorphous calcium phosphate, sections (a)) and its transformation by secondary precipitation of the next more stable phase (octacalcium phosphate, sections (b)). Note that in the precipitation of calcium phosphates, changes in pH indicate solid phase formation. (After Brečević and Füredi-Milhofer.[68])

turbidity curves shown in this particular example pertain to the formation and transformation of amorphous calcium phosphate (similar curves have also been obtained when changes in solution calcium and phosphate concentrations were recorded[123]). The shape of the curves indicates a first precipitation step (formation of the metastable phase) followed by a period of metastability (a) and subsequent precipitation of the next more stable phase (b). Quantitative information may be obtained by separate kinetic analysis of each time period (see for instance refs. 38 and 128). Progress curves of the type shown in Fig. 5.23 have also been reported to characterise the formation and transformation of several metastable hydrates.[64,100,129]

The above examples show that in studying precipitation kinetics it is essential to examine a large enough range of initial supersaturations and to use as many complementary physicochemical methods as possible.

Crystal growth from low degrees of supersaturation

The control of precipitate morphology by causing growth from low degrees of supersaturation has been an active area of research. Of the possible methods of performing such a manipulation (evaporation, supercooling and precipitation from homogeneous solution), precipitation from homogeneous solution (PFHS) has met with the most consistent success.[130-133] PFHS methods involve the chemical generation of either anion or cation (but usually anion) in the presence of the excess of the other ion. For example, generation of sulphate ions into strontium nitrate solution causes the precipitation of well formed strontium sulphate crystals. A summary of some of the precipitates which are formed as coarse crystalline materials by PFHS methods is given in Table 5.6.

The sequence of steps involved in PFHS is shown in Fig. 5.24. It is the buildup of supersaturation, a burst of heterogeneous nucleation and then collapse of supersaturation as the precipitate particles grow. Since the crystals produced by PFHS methods are very regular in shape we may assume that aggregation is usually absent.

Efforts to determine the kinetics of precipitate formation have been attempted by several workers. La Mer and Dinegar[131] found that the growth rate of sulphur precipitates was diffusion controlled. The same was also found for the precipitation of barium sulphate[132] (generation of sulphate by $S_2O_6^{2-}$, $S_2O_3^{2-}$ reaction) and for the precipitation of AgI from ethanol[133] (generation of iodide by C_2H_5I hydrolysis). The above analyses were based upon a rate law of the form

$$\frac{dm}{dt} = k_3 A^{1/2} S \tag{5.60}$$

where m is the mass of precipitate and k_3 the rate constant.

However, analysis of precipitation kinetics under the preceding cir-

TABLE 5.6[a]

FORMATION OF COARSE CRYSTALLINE PRECIPITATES BY PRECIPITATION FROM HOMOGENEOUS SOLUTIONS

Precipitate	*Method*	*Reagent*
Sulphates (Ba, Pb, Sr, Ca, etc.)	Generation of anion	Dimethylsulphate hydrolysis, sulphamic acid hydrolysis, thiosulphate–persulphate reaction.
Sulphides (Sb, Bi, Mg, Cu, As, Cd, Pb, Sn, Hg)	Generation of anion	Thioacetamide hydrolysis
Phosphates (Zr, Hf)	Generation of anion	Hydrolysis of triethyl phosphate, tetraethyl pyrophosphate, metaphosphoric acid.
Chlorides (Ag, Tl)	Generation of anion	Hydrolysis of allyl chloride.
Phosphates (Ca)	Change of pH	Hydrolysis of urea or cyanate.
Oxalates (Mg, Ca, Zn, La)	Generation of anion	Hydrolysis of dimethyl or diethyl oxalate.
Oxalates (Mg, Ca, Zn)	Change of pH	Urea + bioxalate.
Calcium salts (fluoride, sulphate)	Generation of cation	Hydrolysis of ethylene chlorohydrin in presence of Ca–EDTA complex.
Carbonates (rare earths, Ca, etc.)	Generation of CO_2	Hydrolysis of trichloroacetates, cyanate, etc.
Chromates (Pb, Ag)	Change of pH	Hydrolysis of urea or cyanate.

[a]From *Formation and Properties of Precipitates* by A. G. Walton, John Wiley and Sons, New York, 1967, p. 152.

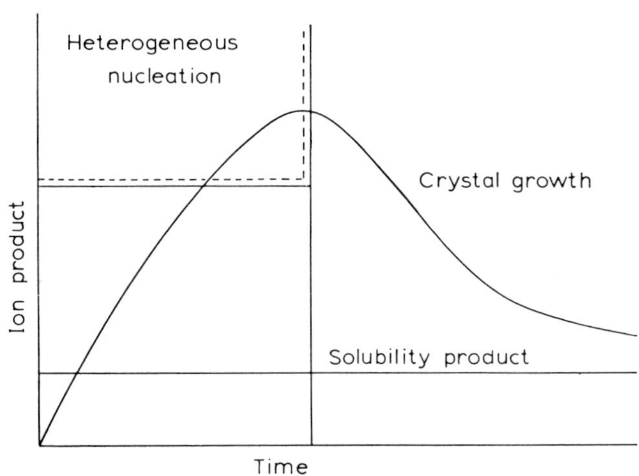

Fig. 5.24. The level of concentration of the solute in solution is shown as a function of time for a slow generating reaction (PFHS). The reaction might, for example, be the hydrolysis of allyl chloride into excess silver ions; the ion product $[Ag^+][Cl^-]$ would then be represented on the ordinate.

cumstances is complicated by the fact that the amount of solute remaining in solution is not only controlled by the precipitation reaction itself, but is counterbalanced by the generation of new material. Thus, if the rate of precipitation is followed by a technique which measures the amount of solute remaining in solution (e.g. conductivity), the rate of disappearance is slower than it would be in the absence of a generating reaction, i.e.

$$R = f(\mathrm{A}, S) - f'(R') \tag{5.61}$$

Conversely, if the rate of precipitation is followed by a technique which detects the appearance and size of the precipitate, the rate is faster than it would be in the absence of a generating reaction. Under these circumstances, the resolution of eqn. (5.61) in its integral form is extremely difficult.

Walton and Klein[134] have examined the kinetics of AgCl formation by PFHS (hydrolysis of allyl chloride) in the region for which the generating reaction just balances the precipitation rate. Under these circumstances the supersaturation is constant and the rate was found to be proportional to A, which is normally typical of an interface controlled reaction. The reaction order could not be determined.

Another method to study the kinetics of crystal growth is by the use of seed crystals.[97–101] Such studies are performed at extremely low supersaturations, between the solubility limit and precipitation boundary (see Fig. 5.19). Reactant solutions which are metastable for an extended period of time are inoculated with well defined, pre-aged crystals of the mineral of interest and changes in solution concentration are then followed by a suitable technique (see Table 5.5). Although this method is applicable to a restricted set of experimental conditions and is not strictly analogous to precipitation *ab initio*, the advantages of growth studies on such systems are many. By avoiding the complicating effects of spontaneous nucleation, ageing and aggregation, it is possible to obtain highly reproducible results on growth kinetics which of course pertain only to very low supersaturations.

The parabolic rate law ($p = 2$ in eqn. (5.59)) has been found applicable for growth of a number of salts on their seeds. This is in accordance with the predictions of the BCF theory[29] (eqns. (5.30), (5.34)) and some other models which assume a reaction at the crystal surface[98,126,127] (dehydration or other) to be the rate-determining step. However, no distinction between the two possibilities (screw dislocation or surface reaction control) could so far be made.[100]

More recently the method has been employed for studies of the growth of different solid phases upon one another[63,135] and of the transformation of metastable phases.[64] The recently developed constant composition method (ref. 101 and Table 5.5) allows monitoring of the growth process at constant supersaturation and consequently constant driving force, which is particularly important in studies of systems where the formation of several metastable phases is feasible.

ACKNOWLEDGEMENT

Thanks are due to Mr D. Škrtić for a comprehensive literature search and the preparation of the line drawings.

REFERENCES

1. W. Ostwald, *Z. Physik. Chem.*, **22** (1897) 284; **23** (1897) 365; **34** (1900) 444.
2. P. P. von Weimarn, *Chem. Rev.*, **2** (1925) 217.
3. M. Kahlweit, in *Physical Chemistry*, Academic Press, New York, 1970, Vol. 10, p. 719.
4. D. Balarev, *Der Disperse Bau der festen Systeme*, T. Steinkopff, Dresden, 1939.
5. J. Traube and W. von Behren, *Z. Physik. Chem.*, **138** (1928) 85.
6. B. Težak, *Disc. Faraday Soc.*, **42** (1966) 175.
7. J. P. Hirth and G. M. Pound, *Condensation and Evaporation*, Pergamon Press, Oxford, 1963.
8. A. G. Walton, in *Nucleation*, Ed. A. C. Zettlemoyer, Marcel Dekker, New York, 1969, p. 225.
9. A. G. Walton, *The Formation and Properties of Precipitates*, John Wiley and Sons, New York, 1967, p. 1.
10. J. W. Gibbs, *Collected Works*, Vol. 1. In *Thermodynamics*, Yale University Press, New York, 1948.
11. M. C. Upretti and A. G. Walton, *J. Chem. Phys.*, **44** (1966) 1936.
12. R. J. H. Voorhoeve, *Surf. Sci.*, **28** (1971) 145.
13. J. B. Hudson and J. S. Sandejas, *Surf. Sci.*, **15** (1969) 27.
14. R. J. H. Voorhoeve, R. S. Wagner and J. N. Carides, *J. Cryst. Growth*, **13/14** (1972) 167.
15. R. P. Andres, in *Nucleation*, Ed. A. C. Zettlemoyer, Marcel Dekker, New York, 1969, p. 69.
16. S. Toshev and I. Gutzov, *Kristall und Technik*, **7** (1972) 43.
17. D. Robertson and G. M. Pound, *J. Cryst. Growth*, **19** (1973) 269.
18. T. P. Melia and W. P. Moffit, *Ind. Eng. Chem. (Fund.)*, **3** (1964) 317.
19. K. A. Jackson and J. D. Hunt, *Acta Metall.*, **13** (1965) 1212.
20. 'Crystal Growth', *Disc. Faraday Soc.*, **5**, 1949.
21. A. E. Nielsen, *Kinetics of Precipitation*, Pergamon Press, Oxford, 1964.
22. A. E. Nielsen, *Croat. Chem. Acta*, **42** (1970) 319.

23. P. Bennema, *J. Cryst. Growth*, **24/25** (1974) 76.
24. B. Lewis, *J. Cryst. Growth*, **21** (1974) 29; 40.
25. N. Cabrera and W. K. Burton, ref. 20, p. 40.
26. W. B. Hillig, *Acta Metall.*, **14** (1966) 1968.
27. F. L. Binsbergen, *Kolloid Z. Polymere*, **237** (1970) 289.
28. F. C. Frank, ref. 20, p. 48.
29. W. K. Burton, N. Cabrera and F. C. Frank, *Nature*, **163** (1949) 398; *Phil. Trans. Roy. Soc. London*, **A 243** (1951) 72.
30. H. Müller-Krumbhaar, 'Monte Carlo simulation of crystal growth', in *Monte Carlo Methods in Statistical Physics*, Ed. K. Binder, Springer, Berlin, 1979, pp. 261.
31. H. Müller-Krumbhaar, 'Surface dynamics of growing crystals', in *Festkörperprobleme*, Advances in Solid State Physics, Ed. J. Trensch, Vieweg, Braunschweig, 1979, vol. 19, p. 1.
32. K. A. Jackson and G. H. Gilmer, *Faraday Discuss. Chem. Soc.*, **61** (1976) 53.
33. M. von Smoluchowski, *Z. Physik. Chem.*, **92** (1917) 129.
34. D. L. Swift and S. K. Friedländer, *J. Colloid Sci.*, **19** (1964) 621.
35. A. E. Nielsen, *Kristall und Technik*, **4** (1969) 17.
36. S. Odén, *Kolloid Z.*, **26** (1920) 100.
37. H. Füredi-Milhofer, Lj. Brečević, E. Oljica, B. Purgarić, Z. Gass and G. Perović, in *Particle Growth in Suspension*, Ed. A. L. Smith, Academic Press, London, 1973, p. 109.
38. H. Füredi-Milhofer, Lj. Brečević and B. Purgarić, *Faraday Discuss. Chem. Soc.*, **61** (1976) 184.
39. J. J. Petres, Gj. Deželić and B. Težak, *Croat. Chem. Acta*, **40** (1968) 213.
40. B. Težak, E. Matijević, K. F. Schulz, J. Kratohvil, M. Mirnik and V. B. Vouk, *Disc. Faraday Soc.*, **18** (1954) 63.
41. E. Matijević, *Chimia*, **9** (1955) 287.
42. W. Ostwald, H. Kokkoros and K. Hoffmann, *Kolloid Z.*, **81** (1937) 48.
43. E. Matijević, *J. Colloid Interface Sci.*, **43** (1973) 217; **58** (1977) 374.
44. M. J. Herak and M. Mirnik, *Kolloid Z.*, **179** (1961) 130.
45. L. G. Sillén, *Stability Constants of Metal Ion Complexes*, The Chemical Society, London 1961, 1971.
46. W. Stumm and J. J. Morgan, *Aquatic Chemistry*, Wiley-Interscience, New York, 1970, p. 474.
47. R. O. James and T. W. Healy, *J. Colloid Interface Sci.*, **40** (1972) 65.
48. E. Matijević, M. B. Abramson, K. F. Schulz and M. Kerker, *J. Phys. Chem.*, **64** (1960) 1157.
49. W. Stumm, C. P. Huang and S. R. Jenkins, *Croat. Chem. Acta*, **42** (1970) 223.
50. J. A. Davies, R. O. James and J. O. Leckie, *J. Colloid Interface Sci.*, **63** (1978) 480.
51. G. D. Parfitt, *Croat. Chem. Acta*, **45** (1973) 189.
52. D. N. Furlong, K. S. W. Sing and G. D. Parfitt, *J. Colloid Interface Sci.*, **69** (1979) 409.
53. J. A. Gray and M. D. Francis, in *Mechanisms of Hard Tissue Destruction*, Publ. No 75 of the American Association for the Advancement of Science, Washington D. C., 1963, p. 213.

54. W. Ostwald, *Z. Physik. Chem.*, **34** (1900) 295.
55. W. J. Dunning, in *Particle Growth in Suspensions*, Ed. A. L. Smith, Academic Press, London, 1973, p. 3.
56. M. Kahlweit, *Adv. Colloid Interface Sci.*, **5** (1975) 1.
57. J. M. Lifshitz and V. V. Slezov, *J. Phys. Chem. Solids*, **19** (1961) 35.
58. C. Wagner, *Ber. Bunsenges. Phys. Chem.*, **65** (1961) 581.
59. W. Ostwald, *Lehrbuch für allgemeine Chemie*, W. Engelmann, Leipzig, 1896–1902, Vol. 2, p. 444; quoted in ref. 55.
60. J. N. Stranski and D. Totomanov, *Naturwissenschaften*, **20** (1933) 905; quoted in ref. 55.
61. Ref. 9, p. 36.
62. G. W. Sears. In *Physics and Chemistry of Ceramics*, Ed. C. Klingsberg, Gordon and Breach, New York, 1963.
63. G. H. Nancollas and B. Tomažić, *J. Colloid Interface Sci.*, **50** (1975) 451.
64. B. Tomažić and G. H. Nancollas, *Invest. Urol.*, **16** (1979) 329.
65. B. Subotić, D. Škrtić, I. Šmit and L. Sekovanić, *J. Cryst. Growth*, **50** (1980).
66. T. Sugimoto and E. Matijević, *J. Colloid Interface Sci.*, **74** (1980) 227.
67. G. C. Bye and K. S. W. Sing, in *Particle Growth in Suspensions*, Ed. A. L. Smith, Academic Press, London 1973, p. 29.
68. Lj. Brečević and H. Füredi-Milhofer, *Calc. Tiss Res.*, **10** (1972) 82.
69. H. Füredi-Milhofer, H. Bilinski, Lj. Brečević, R. Despotović, N. Filipović-Vinceković, E. Oljica and B. Purgarić, in *Physico-Chimie et Cristallographie des Apatites d'Intérět Biologique*, Colloques internationaux CNRS No. 230, Centre National de la Recherche Scientifique, Paris 1975, p. 303.
70. Lj. Brečević and H. Füredi-Milhofer, in *Industrial Crystallization 1976*, Ed. J. W. Mullin, Plenum Publ. Co., New York, p. 277.
71. J. D. Termine, R. A. Peckauskas and A. S. Posner, *Arch, Biochem. Biophys.*, **140** (1970) 318.
72. H. Füredi-Milhofer, M, Marković, Lj. Komunjer, B. Purgarić and V. Babić-Ivančić, *Croat. Chem. Acta*, **50** (1977) 139.
73. B. Težak, *Z. Physik. Chem.*, **192** (1943) 101.
74. M. B. Mutaftschiev, *Compt. Rend.*, **259** (1964) 572.
75. I. N. Stranski, *Z, Krist.*, **105** (1943) 287.
76. I. N. Stranski, ref. 20, p. 13; K. Moliere, W. Rathje and I. N. Stranski, ref. 20, p. 21.
77. A. F. Wells, ref. 20, p. 197.
78. Ref. 9, p. 49.
79. Z. Šolc and O. Söhnel, *Kristall Technik*, **8** (1973) 811.
80. Ref. 9, p. 165.
81. V. Babić, B. Purgarić, Z. Despotović and H. Füredi-Milhofer, in *Urolithiasis Research* (*1976*), H. Fleisch, W. G. Robertson, Eds. L. H. Smith and W. Vahlensieck, Plenum Press, New York, 1976.
82. I. M. Kolthoff and B. von't Riet, *J. Phys. Chem.*, **63** (1959) 817.
83. H. Füredi-Milhofer, D. Škrtić, M. Marković and Lj. Komunjer, in *Urolithiasis: Clinical and Basic Research*, Ed. L. H. Smith, Plenum Publ. Co., New York, in press.
84. Lj. Brečević and H. Füredi-Milhofer, *Calc. Tiss. Int.*, **28** (1979) 131.
85. B. Mutaftschiev, *CRC Critical Reviews in Solid State Science*, **1976**, 157.

86. G. K. Kirov, I. Vesselinov and Z. Chernova, *Krist. Tech.*, **7** (1972) 497.
87. R. Kaischew and B. Mutaftschiev, *Bull. Chem. Inst. Bulg. Acad. Sci.*, **7** (1959) 145, quoted in ref. 84.
88. A. A. Khalaf and W. R. Wilcox, *J. Cryst. Growth*, **20** (1973) 227.
89. B. Kosar-Grašić, B. Purgarić and H. Füredi-Milhofer, *J. Inorg. Nucl. Chem.*, **40** (1978) 1877.
90. J. Willems, *Disc. Faraday Soc.*, **25** (1954) 111.
91. E. W. Fisher, *Disc. Faraday Soc.*, **25** (1954) 204.
92. J. A. Koutsky, A. G. Walton and E. Baer, *J. Polym. Sci.*, **4** (1966) 611.
93. J. A. Koutsky, A. G. Walton and E. Baer, *Polym. Lett.*, **5** (1967) 177.
94. K. A. Mauritz, E. Baer and A. J. Hopfinger, *J. Polym. Sci: Macromolecular Rev.*, **13** (1978) 1; S. E. Rickert, C. M. Balik and A. J. Hopfinger, *Adv. Colloid Interface Sci.*, **11** (1979) 149.
95. K. Lonsdale, *Nature*, **217** (1968) 168.
96. F. L. Coe, *Nephrolithiasis*, Churchill Livingstone, New York, 1980.
97. Marc, *Z. Phys. Chem.*, **61** (1908) 385; **67** (1909) 470; quoted in ref. 92.
98. C. W. Davies and A. L. Jones, ref. 20, p. 50; *Trans. Faraday Soc.*, **51** (1955) 812.
99. G. H. Nancollas and N. Purdie, *Quarterly Rev.*, **18** (1964) 1.
100. G. H. Nancollas, *Adv. Colloid Interface Sci.*, **10** (1979) 215.
101. M. B. Tomson and G. H. Nancollas, *Science*, **200** (1978) 1059; P. Koutsoukos, Z. Amjad, M. B. Tomson and G. H. Nancollas, *J. Am. Chem. Soc.*, **102** (1980) 1553.
102. H. Füredi-Milhofer, *Croat. Chem. Acta*, **53** (1980) 243.
103. W. E. Brown, *Soil Sci.*, **90** (1960) 51; E. C. Moreno, T. M. Gregory and W. E. Brown, *J. Res. Nat. Bur. Stand.*, **72A** (1968) 773, **74A** (1970) 461.
104. B. Težak, *Croat. Chem. Acta*, **40** (1968) 63.
105. H. Füredi-Milhofer, in ref. 9, p. 188.
106. J. Kratohvil, B. Težak and V. B. Vouk, *Arhiv kem.*, **26** (1954) 191.
107. B. Vonnegut, *J. Colloid Sci.*, **3** (1948) 563.
108. M. L. White and A. A. Frost, *J. Colloid Sci.*, **14** (1959) 247.
109. T. P. Melia and W. P. Moffitt, *Nature*, **201** (1964) 1024.
110. G. R. Wood and A. G. Walton, *J. Appl. Phys.*, **41** (1970) 3027.
111. O. E. Hileman, Jr., *Talanta*, **14** (1967) 139.
112. R. E. Massey and O. E. Hileman, Jr., *Can. J. Chem.*, **55** (1977) 1285.
113. A. E. Nielsen, *Acta Chem. Scand.*, **15** (1961) 441.
114. A. G. Walton, *Microchim. Acta*, **3** (1963) 422.
115. A. E. Nielsen and Ö. Söhnel, *J. Cryst. Growth*, **11** (1971) 233.
116. J. B. Newkirk and D. Turnbull, *J. Appl. Phys.*, **26** (1955) 579.
117. E. Hauser, L. Sogor and A. G. Walton, *J. Chem. Phys.*, **45** (1966) 1071.
118. B. Purgarić and H. Füredi-Milhofer, in *Proc. Int. Conf. Colloid and Surface Sci.*, Vol. 1., Ed. E. Wolfram, Akadémiai Kiadó, Budapest, 1975, p. 273.
119. B. Purgarić and Z. Tutek, *Anal. Chim. Acta*, **112** (1979) 193.
120. M. Marković and Lj. Komunjer, *J. Cryst. Growth*, **46** (1979) 701.
121. M. Marković, Lj. Komunjer and H. Füredi-Milhofer, In *Industrial Crystallization 78*, Eds. E. D. Jong and S. J. Jančić, North Holland, Amsterdam, 1979, p. 65.
122. R. Despotović, *Disc, Faraday Soc.*, **42** (1966) 175.

123. R. Despotović, N. Filipović-Vinceković and H. Füredi-Milhofer, *Calc. Tiss. Res.*, 18 (1975) 13.
124. D. Turnbull, *Acta Metall.*, **1** (1954) 684.
125. R. H. Doremus, *J. Phys. Chem.*, **62** (1958) 1068.
126. R. von Reich and M. Kahlweit, *Ber. Bunsenges Phys. Chem.*, **72** (1968) 66.
127. A. G. Walton, *J. Phys. Chem.*, **67** (1963) 1920.
128. E. D. Eanes and A. S. Posner, *Trans. N. Y. Acad. Sci., Ser. II*, **28** (1965) 233.
129. G. Gardner, *J. Colloid Interface Sci.*, **54** (1976) 298.
130. L. Gordon, M. L. Salutsky and H. H. Willard, *Precipitation from Homogeneous Solution*, John Wiley and Sons, New York, 1959.
131. V. K. La Mer and R. H. Dinegar, *J. Am. Chem. Soc.*, **72** (1950) 4847.
132. F. C. Collins and J. P. Leineweber, *J. Phys. Chem.*, **60** (1956) 389.
133. M. J. Jaycock and G. D. Parfitt, *Trans. Faraday Soc.*, **57** (1961) 791.
134. A. G. Walton and D. H. Klein, *Kolloid Z.*, **189** (1963) 141.
135. G. H. Nancollas and B. Tomažić, *J. Phys. Chem.*, **78** (1974) 2218.

CHAPTER 6

FUNDAMENTALS OF THE BREAKDOWN OF SOLIDS

G. C. Lowrison
Consulting Chemical Engineer, Bradford, UK

ENERGY AND FORCE

Energy and mass, although themselves truly interconvertible, together appear to be the fundamental and indestructible components of the universe. In practical terms, if energy is gained in one spatial location, it must have been lost from elsewhere.

Force is manifested where mass experiences a spatial energy gradient. Energy may be expressed as $ML^2/2T^2$, where M, L and T represent the dimensions of mass, length and time respectively. Thus

$$\frac{\partial}{\partial L}\frac{(ML^2T^{-2})}{2}=\frac{ML}{T^2} \tag{6.1}$$

which has the dimensions of force.

Where the energy field is uniform, even though it is great, no force will be experienced in it while it remains uniform. Conversely, the force might be considerable if the energy change overwhelms the distance over which the change takes place. Thus the energy change may be very small and yet the force be very large. The effectiveness of a force must be measured by the product of the force and the distance over which it acts, i.e.

$$\text{Energy} = \int \text{Force dL}$$

When a body enters a uniform field and produces an energy gradient either from the nature of its own energy content or the way in which it modifies the energy in the space of the field it has occupied, then that body will experience a force, the direction and magnitude of which would appear to be determinable by the methods of vector algebra though, as will be seen, the vectors must be localised. If the energy gradient at a point is the same in all directions, or is such that the resultant is zero or ceases to exist, there will effectively be no force at that point.

Examples of such effects are found in the common daily experience of gravitation or are commonly shown in elementary demonstrations of the properties of magnetic fields.

An energy gradient may also be created on a macroscopic scale by bodies of different energy contents coming into contact, e.g. a so-called *moving* body with a so-called *stationary* one as in mechanical impact of a hammer on an anvil, or the entry of a piece of metal into the hot combustion gases in a blacksmith's forge.

Since force is the result of spatial conditions, there can be no doubt as to the importance of the magnitude, direction and location of the gradient on the body which is experiencing it at that instant. The concept of a rigid body transmitting over its entire volume a force which does not change, is independent of its point of application, and causes no change in the location of energy, can only be useful when the differences, due to and the consequence of change of location of energy, are small enough to be ignored for particular purposes: clearly, such a concept has little application in considerations of the nature of fracture.

In the case of a force applied through a field, the differences due to position of the parts of a mass are often small enough to be ignored, but not if the volume of the mass is large enough to correspond with large differences of energy levels in the field. A piece of lead originally of any uniform cross-section, 5 m long and suspended so that its length is vertical, will, very soon after it has been suspended, break across the cross-section at the place of suspension. This could not happen if the gravitational field were uniform.

In the case of energy gradients conferred by one form of matter on another, the surfaces of contact between the bodies determine the location of the application of the force.

The immersion of a solid in a liquid to which pressure is applied presents a condition close to an imposed uniform energy field, the so-called *hydrostatic condition*, though of course there is an energy gradient with change of depth of liquid, just as there is in the gaseous atmospheric fluid and, at every section in the suspended body perpendicular to the gravitational force, there is a force acting equal to the weight of the part of the body below that section minus the upward force due to the pressure of the liquid on that part of the body.

It is seen, therefore, that there are forces which act on the bulk of the material and there are those which act in a more discriminating way on particular locations of the body, and that forces only exist while there is an energy gradient, so as soon as energy is evened out forces cease to exist.

A mass of material may acquire energy, i.e. be the subject of an energy gradient, in a number of ways:

(i) by impact through moving relative to another body;
(ii) by being caused to move against other bodies (friction);
(iii) by participating in a chemical reaction with other bodies;
(iv) by residing in a magnetic or gravitational field;
(v) by conducting an electric current or acquiring an electric charge; and
(vi) by receiving radiation from a hot or luminous body.

A force, i.e. an energy gradient, may be (a) static, i.e. of constant value with time, or (b) fluctuating, i.e. varying with time either rhythmically or randomly changing, and changing either suddenly or gradually. Figure 6.1 illustrates these different kinds of forces.

SOLIDS AND ENERGY

The energy in a solid has many locations:

(i) in the atoms,
(ii) that expressed by the spatial relationships between the atoms, e.g. as in the regular lattices of a crystal structure or in the irregular random form of a glass,
(iii) in the vibrations of the atoms,
(iv) that implicit in the imperfections in the exact lattice reproduction, e.g. pores where the atoms appear to be highly mobile and where impurities gather,
(v) residing between the surfaces of polycrystals, and
(vi) on the surfaces of the body adjacent to the atmosphere.

The different locations of energy give rise to many energy gradients.

The single crystal

The equilibrium position of an atom or ion in a crystal is the result of the equality of the force of attraction and the force of repulsion, or

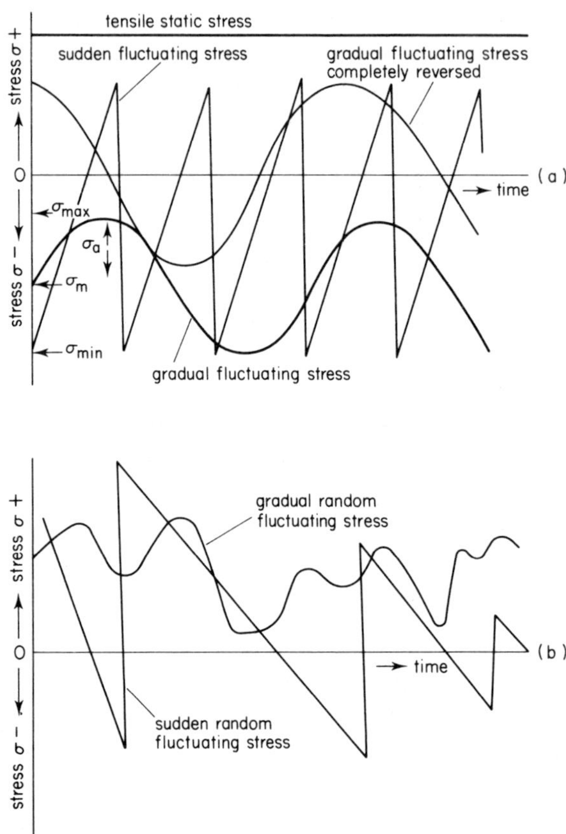

Fig. 6.1. Examples of stresses. (a) Rhythmically fluctuating stresses. (b) Random fluctuating stresses.

perhaps of propulsion, between the atoms or ions, i.e. the equilibrium position corresponds with a position of zero energy gradient.

The force of attraction corresponds with gravitational energy, though it appears to be much stronger, and the force of repulsion with the kinetic energy of the thermal vibration which would tend to move the atom away from its location in the lattice. Thus one force is largely independent of temperature, and the other may be said to be the temperature or the so-called *intrinsic pressure* which induces the solid to become liquid, and liquid to become gas, when the kinetic energy has reached a level sufficient to overcome the energy gradient of attraction.

The energy required for the spatial arrangement of the atoms in the lattice may be expressed by

$$W = \frac{A}{d^n} - \frac{B}{d} \tag{6.2}$$

where d is the distance between the centres of two adjacent atoms (or ions) and A, B and n are constants, n being considerably larger than unity. Thus to decrease d, the first component requires energy to be added to the system, whereas the second component will release energy, i.e. will do work or generate heat. The force between the atoms for a certain value of d is given by the energy gradient

$$\frac{\partial W}{\partial d} = -\frac{nA}{d^{n+1}} + \frac{B}{d^2} \tag{6.3}$$

For this gradient to be zero, the force of repulsion must equal the force of attraction

$$\frac{nA}{d^{n+1}} = \frac{B}{d^2} \text{ or } A = \frac{Bd^{n-1}}{n} \tag{6.4}$$

The energy of one atom relative to another in this state of equilibrium and at a distance apart d_0 is given by

$$W_0 = \frac{B}{d_0}\left(\frac{1}{n} - 1\right) \tag{6.5}$$

and relative to closer or more distant configuration (since n is greater than unity) will be negative, i.e. require less energy to keep it in that position. Energy is therefore required to be added to the system either to bring the atoms closer together or to take them further apart. The situation may be represented graphically by Fig. 6.2.

The exponent n in the equation becomes more influential as the distance between the atoms is decreased, i.e. as the atoms are compressed together. Born and Brody[1] have determined the value for n for sodium and potassium halides from compressibility data and arrive at an average value of 8·80. Values of this order can also be deduced from wave mechanical considerations.

Calculations on the energy required to separate a crystal lattice between lattice planes, taking account of the energy changes consequent upon the exposed atoms (or ions) being influenced by the interior of the crystal, lead to values for surface energy. The value of

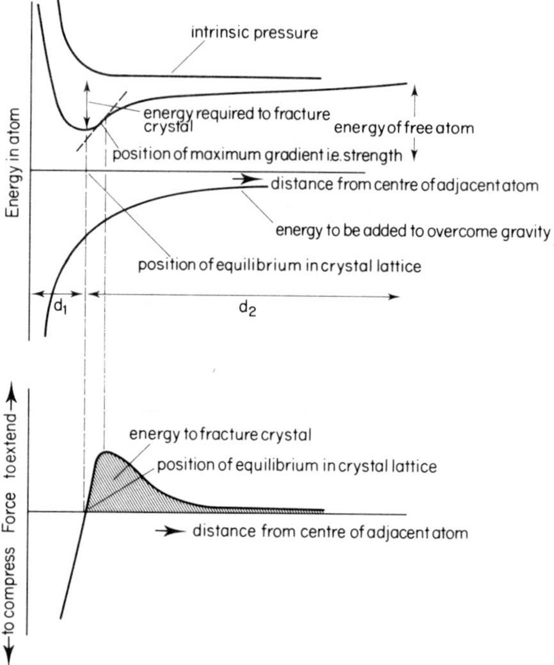

Fig. 6.2. Energy and force in the crystal lattice.

this energy of separation of the lattice also leads to the evaluation of the strength of the crystal. The strength of the material will correspond with the maximum value of the force between the atoms because strength is defined in terms of force. Now the force between the atoms (eqn. 6.3) will be a maximum at a distance (d_s) between the atoms corresponding with

$$\frac{\partial^2 W}{\partial d^2}=0=\frac{-2B}{d_s^3}+\frac{n(n+1)A}{d_s^{n+2}} \tag{6.6}$$

or when

$$\frac{2B}{d_s^3}=\frac{n(n+1)A}{d_s^{n+2}}$$

and

$$d_s^{n-1}=\frac{n(n+1)A}{2B}.$$

Now the equilibrium distance between the atoms (d_0) may be obtained from the above equation in the condition when $\partial W/\partial d=0$, i.e. when the force between the atoms is zero or when

$$d_0^{n-1}=\frac{nA}{B} \tag{6.7}$$

Thus

$$\frac{d_s}{d_0}=\left[\frac{n+1}{2}\right]^{1/n-1}$$

and if $n=9$ as determined by the methods indicated above,

$$\frac{d_s}{d_0}=5^{0\cdot125}=1\cdot22$$

and the conventional (or ordinary) strain corresponding with the ultimate strength of the crystal would be

$$\text{strain}=\frac{\text{extension}}{\text{original length}}=\frac{1\cdot22-1}{1}=0\cdot22$$

a figure of the same order as those obtained on the macro scale for many materials (see Table 6.1), particularly metals, but much less than those usually obtained with single crystals.

The condition in which the atoms are sufficiently energised to separate, i.e. for the solid to rupture, corresponds with the atoms being in a free

TABLE 6.1
ULTIMATE STRAINS OF COMMON MATERIALS

Material	*Ultimate strain*
Wrought iron	0·30
Steel (SAE1020)	0·30
Ni–Cr steel (SAE8620)	0·21
Wrought aluminium alloy (3S–0)	0·30
Hastalloy B	0·44
Cast iron	0·01
Malleable cast iron	0·10
Cast copper	0·02
Cast aluminium	0·02
Polystyrene	0·02
Titanium	0·25
Polyethylene	0·45
Brass (yellow)	0·20
Cellulose nitrate	0·28

state, i.e. to be liquefied, and it may be argued that the energy required to rupture a solid should be equal to the energy required to liquefy it. Various relationships have been proposed[2] along these lines. Thus if W_m is the energy at the melting point

$$W_m = 3kT_m \tag{6.8}$$

where k is Boltzmann's constant and T_m is the melting point (in degrees Kelvin), and it has been suggested that the rupture energy (W_r) at temperature T is

$$W_r = 3k(T_m - T) \tag{6.9}$$

Again

$$W_r = k\lambda/3$$

where λ is the latent heat of evaporation to which the factor $\frac{1}{3}$ is applied because the atoms move only in one dimension. If the strain at rupture is ϵ_u and if the coefficient of linear expansion is $\mathrm{d}l/\mathrm{d}T = \alpha$ and $\mathrm{d}H/\mathrm{d}T = C$ is the specific heat, then

$$\frac{C}{\alpha} = \frac{\mathrm{d}H}{\mathrm{d}l} \tag{6.10}$$

W_r, as will be seen later, may be approximated by $\sigma_u \epsilon_u / \rho$

and

$$W_r = \epsilon_u \frac{\mathrm{d}H}{\mathrm{d}l} = \frac{\epsilon_u C}{\alpha} \approx \frac{\sigma_u \epsilon_u}{\rho}$$

or

$$\sigma_u \propto \frac{\rho C}{\alpha}$$

in which σ_u is the tensile ultimate strength and ρ is the density. Table 6.2

TABLE 6.2
PROPERTIES OF MISCELLANEOUS MATERIALS AND ALLEGED CONSTANTS

Material	$\dfrac{(T_m - T)^2 \rho}{\sigma_u \epsilon_u}$	$\dfrac{2\lambda\rho}{\sigma_u \epsilon_u} \times 10^{-3}$	$\dfrac{\rho C}{\alpha \sigma_u} \times 10^{-4}$
Aluminium	0·384	0·240	0·119
Brass	0·216	0·041	0·070
Copper	0·236	0·046	0·094
Crown glass	2·27	0·742	0·113
Wrought iron	0·130	0·034	0·075
Marble	8·56	3·581	0·594
Quartz	6·46	1·043	2·02
Zinc	0·102	0·025	0·040

gives a few substances, their properties and values for some of these relationships; it is seen that on the macro scale they do not hold very accurately. Some of the reasons for this failure are indicated below.

Knowledge of the shape of the energy curve (Fig. 6.2) would allow other properties of the crystal to be deduced. The depth of the trough (W_r) is a measure of the energy which has to be added to the atom to get it out of its location in the lattice, and therefore of the energy required to fracture the crystal.

The maximum gradients on either side of the trough are measures of the maximum compressive (on the left hand side) and tensile (on the right hand side) forces which the atom will tolerate, i.e. the ultimate compressive and tensile strengths of the atom in the lattice. The width of the curve on either side of the trough (d_1 and d_2) measures the compressibility and extensibility of the atom depicted. According to this argument, compression can proceed until the atoms coalesce; in fact, as will appear later, compression leads to the generation of shear and tensile forces with respect to which the material has less strength than in true compression. Common sense suggests compression is a consolidating, not a breaking, action.

Referring to the force–distance curve in Fig. 6.2 it will be seen that a considerable part of the curve on either side of the null point is almost a straight line, i.e. in these regions the force required is approximately directly proportional to the displacement of the atoms from the null point. This relationship was suggested by Robert Hooke in 1678, not, of course, in reference to the atomic scale in crystalline structures, but to the macroscopic scale where it applies to many kinds of solids.

Hooke's law becomes most useful when force is expressed per unit of area over which it acts, i.e. as stress (or pressure), and when deformation is expressed as change in dimension per unit of original dimension, i.e. as strain. Hooke's law then becomes

$$\frac{\text{stress}}{\text{strain}} = \text{constant} \qquad (6.11)$$

The constant is usually known as Young's modulus (E) if the stress refers to tensile or compressive forces. (The constant for shear forces is known as the modulus of rigidity (G).)

So far as is known, there is no necessary connection between Hooke's law and the phenomenon known as elasticity, although Hooke himself said his law applied to materials which showed 'springyness'. Indeed, strictly there cannot be a connection, because Hooke's law is found not to be obeyed if very refined measurement is used; moreover elasticity is displayed beyond the limits of proportionality between stress and strain, albeit to only a slight extent, to what is called the *yield point*. Nevertheless elasticity, or the ability of a deformed body to return to its original shape when stress is removed, is always shown, again so far as is known, over the range of stress in which Hooke's law holds and in any body which displays this property, i.e. if a body obeys Hooke's law in one dimension, the law is obeyed in all dimensions simultaneously. What is also clear is that the phenomena of stress–strain proportionality and elasticity are shown only over the smaller stresses.

The energy possessed by a particular atom will depend upon its history and the energy of the atoms in a particular crystal may be expected to follow a statistical law. In a crystal of an element, i.e. in which the atoms are all alike, the energies of atoms might be expressed by the proportion f having energy W at an average temperature T, thus:

$$f = A \exp(-W/BT) \tag{6.12}$$

where A and B are constants. Some atoms possess enough energy to flow from the lattice. When the crystal has been strained, there will be more atoms with such energy than when the crystal is unstrained. The phenomenon of flow of solids is known as *plasticity* or, particularly in the case of metals, as *ductility*.

Many substances (some would claim all substances, given sufficient time) if subjected to such a constant stress that during practically short times they would display elasticity if the stress were removed, do, after a sufficiently long time, increase in strain, i.e. flow or creep, or if kept under constant strain, gradually dissipate, or relax, the stress which was originally applied to them and have little or no restoring force when released from clamps or presses which maintain their deformity (see Fig. 6.3).

It is seen that, even in the restricted structure of a crystal, both plastic and elastic behaviour must be expected to be shown simultaneously. The reciprocal of equation (6.12),

$$\frac{1}{A\exp(-W/BT)} = C \exp(W/BT) \quad \text{(say)} \tag{6.13}$$

is a well known form of expression for empirical relationships between viscosity (η) and temperature, e.g. the Andrade–Guzman equation[3]

$$\eta = C \exp(D/T) \tag{6.14}$$

in which C and D are constants. On the same basis of reasoning as

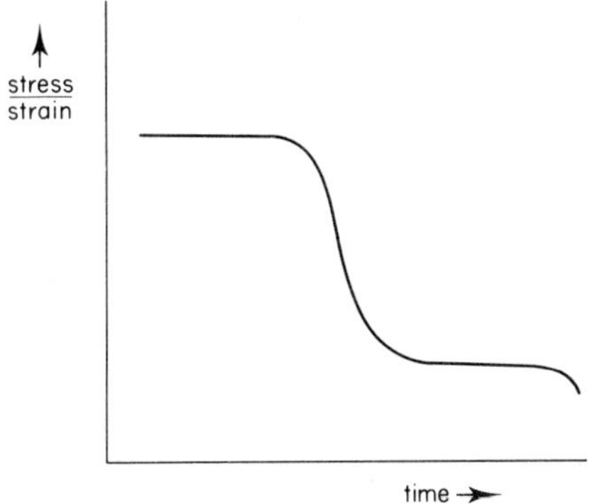

Fig. 6.3. Creep and relaxation: graph relating stress/strain to time.

above, D may be expected to be related to the latent heat of fusion for the substance.

All solids may therefore be described as being visco-elastic; some show negligible elasticity, others show negligible plasticity.

In crystalline materials, those simply related to the above considerations are single crystals of the elements, i.e. regular arrays of atoms of the same kind. However, even in such crystals, the planes in the crystals may contain different packings of atoms, according to the structure of the unit cell, and these different faces may have different values for given properties; indeed it is found that the highest value of a property in a crystal plane may be as large as ten times the lowest value in the same crystal, but in different planar directions, e.g. the work required to grind away equal masses from equal areas of surface.[4]

Crystals of compounds contain the complication of heterogeneous atoms.

The strengths of the forces between atoms in a crystal lattice may differ. Thus, they may be (i) three-dimensional, in that all forces are relatively strong, (ii) two-dimensional, in that forces between layers are weaker than those between atoms in the layers, e.g. in the cases of mica and graphite, or (iii) one-dimensional, in that forces between lines of atoms are weaker than those between atoms in the lines, e.g. in the cases of asbestos and of silk.

Most crystals met in practice contain imperfections. A position in the lattice may be unoccupied or a foreign atom may have been substituted. An atom may be displaced, e.g. occupy an interstitial position between normal layers. Layers may be out of normal sequence.

Directional differences in strength between atoms may also manifest themselves in slipping, or gliding between layers, in the single crystal whenever an applied force has a component in the direction of the 'weak' layers sufficiently strong to make them move relative to parallel layers. This gliding will also be classed as ductility. It is this gliding in single crystals that produces the enormous extensions which contribute to the anomalies between practice and theory in the above relationships.

Amorphous solids

Amorphous solids are composed of molecules which, at one extreme, are packed in quite random arrangement and range, through to degrees of associations in regular arrangement approaching the crystalline state at the other extreme. They represent, in their various forms, the transition from liquid to crystalline solid; this freezing process can be imagined to progress from relatively few associations of an indefinite number of molecules in the truly liquid state to the highly regular exact arrangement in the single crystal.

Polymerised substances are particularly prone to amorphous solidity, since they are often composed in the liquid state of molecules with a wide range of degrees of polymerisation and are heterogeneous, in this sense, before solidification.

As might be expected, amorphous materials undergo a much more gradual transition from solid to liquid properties, e.g. with change of temperature, than is the case with crystalline solids. This is illustrated in Fig. 6.4.

Very large molecules of amorphous substances will tend to conform, i.e. fold up on themselves; thus there is also the tendency for the molecules to become coiled and entangled. Stretching such substances may lead to the alignment of uni-dimensional crystals, at one extreme, or to brittle fracture, if the molecules are not amenable to disentanglement, at the other. Those molecules which can be unfolded by stretching and then show strains of several hundred per cent, i.e. vastly greater than three dimensional crystals, are called *elastomers* (rubber is a very familiar example).

Amorphous materials have as a rule a transition temperature known as the *glass transition temperature* (T_g). This is the temperature above which the material behaves plastically and below which it is brittle and elastic. Some amorphous materials, particularly fused silica, increase in stiffness (i.e. Young's modulus) with increase in temperature. A characteristic of amorphous material is the recovery of elastic strain in two parts, the first

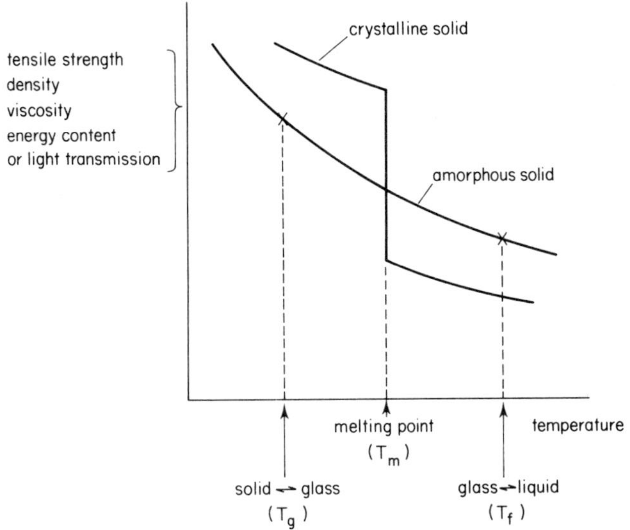

Fig. 6.4. Contrast between behaviour of crystalline and amorphous solids.

immediate as in most crystalline materials, and the second after some time.

Mixed crystalline and amorphous solids

Some materials consist of crystalline substances bonded together with amorphous substances, and uni- and bi-dimensional crystals are particularly prone to this kind of configuration with amorphous substances forming bridges between fibres and plates. Many ceramic materials are of this form. Such mixed solids tend to behave like amorphous substances.

Solids on the macro scale

Unless special care is taken in their production, solids met in practice are not single crystals. If they crystallise homogeneously, they consist of an assemblage of small crystals meeting together at various angles so that a polished section looks like a patchwork. The size of the small crystals or grains depends upon the circumstances of freezing and of treatment subsequent to freezing. The macro effect of these polycrystals is to cancel out distinctions due to the atomic array in single crystals, so that in general polycrystalline materials do not show the anisotropic properties of single crystals, e.g. the ductility due to gliding on specific planes (see Fig. 6.5, comparing the behaviour of a single crystal of zinc with a polycrystalline specimen of zinc) or the differences of hardness displayed in certain directions. The properties of polycrystalline material are the statistical averages of the directional properties of the single-crystalline material.

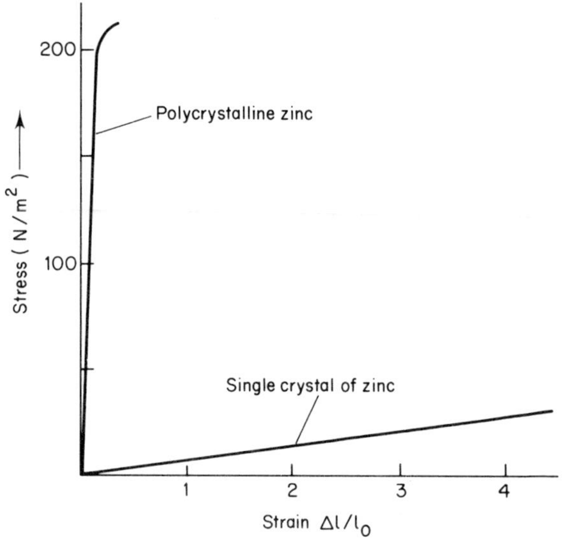

Fig. 6.5. Comparison of the behaviour of a polycrystalline specimen with a single crystal of zinc (after Polonyi and Schmid[5]).

However, rolling, hammering or drawing of polycrystalline material, in which there may be some flow of crystals, may result in some specific orientation of direction and the anisotropic properties may be exhibited. Drawing may result in only the core, nearest the axis of drawing, showing this selectivity. Again, from processes like electrodeposition, electrophoresis or biological activity, selection of directional selectivity may be shown by the solid. When recrystallisation takes place below the melting point, previous orientation of the crystals by such processes as those mentioned may be retained in the recrystallised product. Copper is strongly retentive, aluminium hardly at all.

The formation of polycrystals from the melt usually proceeds by way of dendrites, which change into equiaxial forms, until the distance between the crystals is insufficient for ordered growth and dissimilarities of array prevent the assimilation of one grain in the other, whereupon the remaining liquid freezes mostly into an amorphous form.

For a given material, the smaller the grain size the greater the mechanical strength (ultimately, the body would become a single crystal) though the smaller the grain size the greater the diversity of crystal orientation and the lower the ductility.

Solids met in practice are often mixtures of compounds and even if there are major constituents of the mixture which are crystalline, amorphous materials (impurities) may collect at the crystal boundaries. The nature of a solid mixture ranges through the substitution of the alloy (in which foreign atoms occupy normal lattice positions), the interstitial mixture in the lattice, the eutectic, the agglomerate (e.g. of gravel, sand and cement in concrete), and the composite (e.g. wood, composed of cellulose fibres embedded in lignin; apatite fibres embedded in collagen, as in bone; steel bars embedded in concrete, as in reinforced concrete).

Often, the mixture is stronger than the pure constituents because the properties of the constituents augment each other. Thus concrete is strong in compression, but weak in tension. Steel is more expensive to produce than concrete, but is much stronger in tension than is concrete. By immersing steel bars in a closely adhering matrix of concrete and choosing cross-sectional areas so that the elongation of the components are equal for a given force on the composite, a product results that is as strong in tension as it is in compression and that may therefore be used for beams which require this equality.

Again, vulcanisation of rubber, i.e. providing cross-links of sulphur between rubber molecular chains, increases the tensile strength of rubber, but it also increases Young's Modulus, i.e. makes it 'stiffer'. Carbon black will provide links between the rubber chains, but it confers less strength and stiffness than sulphur. By adjusting the ratio of sulphur to carbon black links, a compromise between strength and stiffness can be achieved.

Yet another example of the usefulness of composites is the reinforcement of organic plastics with fibres which have been so carefully grown that they are free from flaws and therefore 100 to 1000 times stronger than the everyday forms of the substances composing the fibres. Typical substances used as fibres are carbon, aluminium oxide and glass. By immersion in the plastics, the fibres are orientated and put into a handleable form. The ductility of the plastics permits the forces to be redistributed in the event of some fibres being severed.

As has been stated above, forces may be applied to solids in many ways. Perhaps one of the simplest examples of the application of a force to a solid is for one end of a wire or uniform fibre of a material to be gripped in a chuck fastened to a substantial frame and for the wire to drop to a chuck from which a pan may be suspended (see Fig. 6.6(*a*)). The pan is gradually and very lightly loaded with weights until the wire is taut and the length of the wire from chuck to chuck may be carefully measured. The diameter of the wire is also measured at several places along its length; if the wire (or fibre) has been carefully drawn during its manufacture, the diameter will be very nearly the same throughout its length to a high order of accuracy.

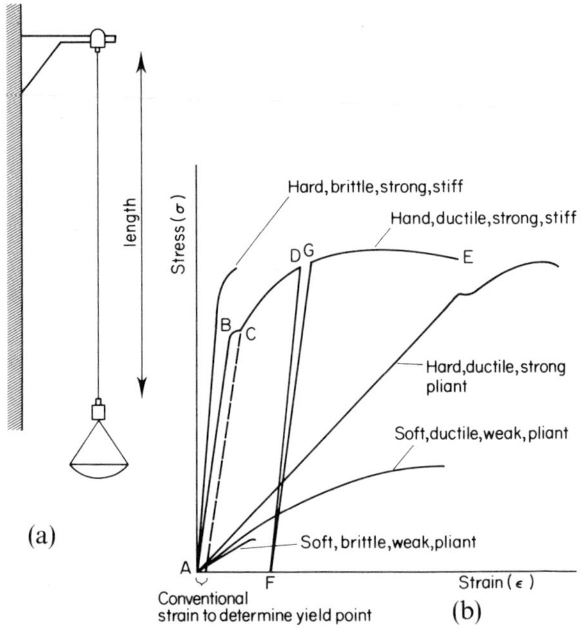

Fig. 6.6. (*a*) *Diagram of an arrangement to determine the extensibility of a wire or fibre.* (*b*) *Examples of stress/strain diagrams and properties they could indicate.*

If the pan is carefully loaded with weights, so as not to cause any impact or very fast change in length of the wire, and, at intervals of time, the length and diameters are carefully measured, it will be found that the wire extends and the diameter contracts, but that while the weight on the pan is unchanged, and if the period of the experiment is of the order of a few hours, the length and diameter remain constant. If the weight on the pan is plotted against the extension of the wire, a graph like one of the forms in Fig. 6.6(*b*) is obtained.

The graph has several important features which may be identified by the letters on the graph. There is a length AB which is very nearly a straight line passing through the origin and represents the conditions to which Hooke's Law applies, i.e. where extension is directly proportional to force. The point B for this reason is usually known as the proportionality limit. After B, the line deviates very slightly from a straight line until C is reached. When the force which has been applied to reach the point C, is slowly and carefully released, the whole of the extension and the diametrical contraction disappear. The point C is therefore the elastic limit or yield point. Because point C is both difficult to identify and important for constructional purposes, usually a conventional extension is defined in terms of the excess of the extension which would have been obtained if Hooke's law had continued to apply up to that point; Fig. 6.6(*b*) will perhaps make this clearer.

The application of additional force after that corresponding with the point C, causes an ever increasing departure from linearity between force and extension, and removal of the force leaves the wire permanently distorted. If the addition of force is continued, the wire eventually fractures corresponding to the point E on the graph. The region from the point C (the yield point) to the point E is the plastic or ductile region.

If, at some stage in the application of force past the point C, the force is carefully removed, the wire is found to be permanently extended so as to correspond roughly with a line with a slightly steeper gradient than AB and meeting the strain axis at a point F. Reloading the wire causes the extension to increase so that the curve is roughly parallel to AB and meets the previous graph at a point G, i.e. with a slightly enhanced yield point; this effect is known as *strain hardening*. The whole hysteresis effect is named after its discoverer Bauschinger. Thus the range of elasticity of a body may be increased by extending it beyond its yield point and then carefully releasing the force causing the extension.

Solids may be distinguished by the form of the force–extension graph that is obtained from them. To make comparisons possible between materials tested in different sizes, force is usually expressed as force per unit area over which it acts or as stress, and is often represented by the symbol σ. Deformation is usually expressed in the form of change of dimension per unit of original dimension or as strain, and is often given the symbol ϵ.

The length AC on Fig. 6.6(*b*) is used to express loosely the general property of elasticity. If the length AC is large, whatever its slope, then one would say loosely that the material represented was highly elastic. The slope of the line AB is expressed as stiffness and is usually expressed numerically as Young's Modulus. Thus rubber is highly elastic, but not very stiff. High tensile steel is highly elastic and very stiff.

The length CE is loosely used to express plasticity or ductility. If the length CE is small, the material is said to be brittle.

The bigger the area under the curve, whatever its shape, the greater the energy required to fracture unit volume of the material since deformation energy $=\int_0^{\epsilon_u}\sigma\, d\epsilon$, in which ϵ_u is the strain at the breaking point (the ultimate strain). The value of this energy is called the *toughness* of the material, often it is given the approximate value of $\sigma_u\epsilon_u$ in which σ_u is the stress at breaking point (the ultimate stress). The area under the curve from zero strain to the strain at the yield point represents the elastic energy of the material and is called the *resilience* of the material, i.e. the energy which the material can absorb and still remain elastic and the energy which is completely recoverable as work, if the stress is released slowly and carefully. Thus resilience $=\int_0^{\epsilon_{yp}}\epsilon d\epsilon$ in which ϵ_{yp} is the strain at the yield point. As the yield point is usually very nearly the same as the limit of proportionality and the curve up to the limit of proportionality is a straight line with the equation $\sigma=E\epsilon$, resilience is approximately

$$\int_0^{\epsilon_{yp}}\sigma d\epsilon=\int_0^{\epsilon_{yp}}E\epsilon d\epsilon=\left[\frac{E\epsilon^2}{2}\right]_0^{\epsilon_{yp}}=\frac{E\epsilon_{yp}{}^2}{2} \qquad (6.15)$$

a result easily obtained by inspection of the curve since it is the area of a triangle.

Many bodies have one dimension larger than any other, and this is called the length of the body; forces which act parallel to the length are called *axial* forces and those which act at right angles to the length are called *transverse* forces. Axial or transverse forces which act so as to extend the dimension in which they act are said to be *tensile* and the body relative to them is in tension. Axial or transverse forces which act so as to contract the dimension in which they act are said to be *compressive* and the body is said to be in compression relative to them. Tensile and compressive forces act so as to tend to fracture bodies at surfaces perpendicular to the direction of the forces. Forces which act so as to separate surfaces parallel to the direction in which they act are said to be *shear* forces and the bodies relative to them are in a state of shear.

Forces can produce a mixture of effects; states of tension, compression and shear can act simultaneously in the same body. For example, if a beam is supported at its ends and is subjected to a force acting downwards on it at a point between the supports (see Fig. 6.7(*a*)) the lower fibres parallel to the axis are in tension, the upper fibres are in compression, the fibres midway between are without stress, whereas the effect of the transverse forces, acting upwards at the supports and downwards from the imposed force, is to cause the cross-sectional surfaces of the beam to shear. Twisting a body at right angles to its axis produces a state of shear, but is usually called *torsion.*

When a body is extended in one direction, it is found that it contracts

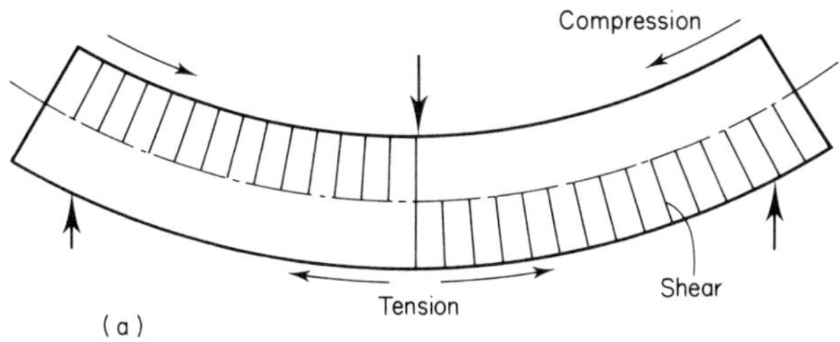

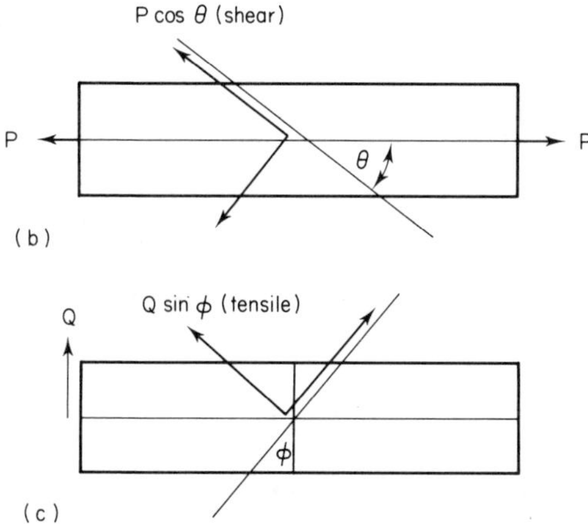

Fig. 6.7. (a) Mixed forces in a beam. (b) Tensile force (P) giving rise to shear (P cos θ). (c) Shear force (Q) giving rise to tension (Q sin φ).

in a direction at right angles to the extension. Similarly, when a body is contracted in one direction, it extends in a direction at right angles to the contraction. Within elastic limits the ratio

$$\frac{\text{lateral change}}{\text{longitudinal change}}$$

is usually constant and is called Poisson's ratio (μ) and is often the same value in both the dimensions at right angles to the direction of the inducing force. After the material has entered the plastic region, the body remains approximately constant in volume until it fractures. In the elastic range typical values for μ are in the range 0·25–0·30. Since the volume of a unit cube when placed under tensile strain would change to

$$(1+\epsilon)(1-\mu\epsilon)^2 \approx 1+\epsilon(1-2\mu)$$

another way of stating the condition in the plastic range is to say that μ changes to 0·50. Some values of Poisson's ratio, in the range of elasticity, for miscellaneous substances, are given in Table 6.3.

TABLE 6.3
POISSON'S RATIO FOR SOME MATERIALS

Steel	0·30	Concrete	0·10
Cast iron	0·25	Rubber	0·50
Marble	0·28	Quartz	0·07
Granite	0·22	Cork	0·11
Glass	0·23	Lead	0·45

Since there is lateral change accompanying any axial change, the cross-section of the material alters from the original area as the force is increased and the true stress (S) is the force divided by the new cross-section; the true strain (K) is the sum of all the minute extensions compared with the lengths existing just prior to the minute extension. Thus

$$K = \int_{l_0}^{l_f} \frac{dl}{l} \tag{6.16}$$

in which l_0 and l_f are the original and the final length respectively, and

$$K = \ln\left(\frac{l_f}{l_0}\right) \tag{6.17}$$

But the nominal strain is

$$\epsilon = \frac{l_f - l_0}{l_0} = \frac{l_f}{l_0} - 1 = e^K - 1 \tag{6.18}$$

If the force P is applied over an original area A_0, the nominal stress σ is P/A_0, but the application of the force P will cause a lateral change of dimensions so that A_0 becomes $A_0(1+\mu K)^2$ and the true stress

$$S = \frac{F}{A_0(1+\mu K)^2} \tag{6.19}$$

Thus
$$\frac{\text{true stress}}{\text{nominal stress}} = \frac{S}{\sigma} = \frac{1}{(1+\mu K)^2} \tag{6.20}$$

and within the proportional limit, in which $\sigma = E\epsilon$,

$$S(1+\mu K)^2 = E\epsilon = E(e^K - 1) \text{ or} \tag{6.21}$$

$$S = \frac{E(e^K - 1)}{(1+\mu K)^2}$$

In very few practical cases do the true stress and strain differ sufficiently from the nominal stress and strain to warrant the extra trouble of calculation. For example, suppose $\epsilon = 0{\cdot}05$, $\sigma = 230\ \text{MN/m}^2$, and $\mu = 0{\cdot}30$; then $K = 0{\cdot}04879$, and $S = 223{\cdot}4\ \text{MN/m}^2$.

A tensile force such as P shown in Fig. 6.7(*b*) will have a component which is a shear force on any plane not at right angles to the direction of the tensile force and if the shear strength of the material, in the plane containing the shear force, is weaker than the tensile strength of the plane at right angles to the tensile force P, the material will break in shear before it breaks in tension, although the applied force would be described as tensile. But unless a material has no strength at all in shear (as in a liquid) it cannot break in shear along a plane due to a force acting at right angles to the plane. Both strength and force are more conveniently measured as stress and if the area of the body at right angles to the force P is A, the shear stress on the plane at an angle θ to the direction of P is given by

$$\text{stress} = \frac{P \cos\theta \sin\theta}{A} = \frac{P \sin^2\theta}{2A} \tag{6.22}$$

and is therefore a maximum when $\theta = \pi/4$.

Conversely, a force in shear, Q, will have a component at any angle except at right angles to its direction, which will cause tension or compression (see Fig. 6.7(*c*)). Again, such a stress component will be maximum when $\phi = \pi/4$.

Thus shear forces can result from tensile or compressive forces and vice versa. The problem of finding which forces are important to the strength of the material is essentially one of finding what and where are the maximum tensile (or compressive) force and the maximum shear force, since in many materials strengths in tension, compression and shear are different. The problem may be resolved in the following way.[6]

A force at a point is difficult to define since most practical concepts are associated with stress, i.e. force per unit area. If an area ΔA can be circumscribed about a point over which the stress may be considered constant, then the force acting on that area may be calculated by the product of the stress and the area.

All forces may be resolved into axial (σ) and shear (τ) forces. Thus any body may be represented as a composition of rectangular units (as in Fig. 6.8), the size of which depends upon the uniformity of the stresses. All of these stresses may be resolved on to three orthogonal axes and considered to act from a point. The three orthogonal axes may be turned until only axial forces are present (clearly this can be done because shear stresses can be transformed to axial stresses in different planes). When

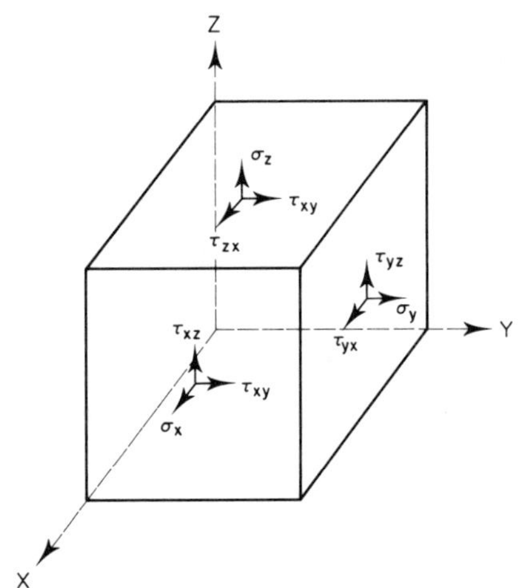

Fig. 6.8. Elementary volume and its stresses.

this condition has been reached, one of the axial stresses must represent the maximum axial stress present in the body because any plane inclined to the plane over which it acts would produce components which would include a shear force and a corresponding smaller axial stress. The three stresses which include the maximum axial stress and act along the three orthogonal axes are called *principal stresses.*

At another angle ($\pi/4$ to other axes if no viscous forces are present) only shear stresses would be acting and the largest of these would be the maximum shear stress suffered by the body.

The values of the principal stresses of the system shown in Fig. 6.8 are the three roots of the cubic equation[6]

$$S^3-(\sigma_x+\sigma_y+\sigma_z)S^2+(\sigma_x\sigma_y+\sigma_y\sigma_z+\sigma_x\sigma_z-\tau_{yz}^2-\tau_{xz}^2-\tau_{xy}^2)S \\ -(\sigma_x\sigma_y\sigma_z+2\tau_{yz}\tau_{xz}\tau_{xy}-\sigma_x\tau_{yz}^2-\sigma_y\tau_{xz}^2-\sigma_z\tau_{xy}^2)=0 \qquad (6.23)$$

This equation may be solved for particular values by, for example, the method of Cardan,[7] but the general expressions for the roots are too cumbersome to be useful. The values for the principal shear forces are much simpler and are $\pm\frac{1}{2}(\sigma_y-\sigma_x)$, $\pm\frac{1}{2}(\sigma_x-\sigma_z)$ and $\pm\frac{1}{2}(\sigma_x-\sigma_y)$.

The relationship between compressibility and 'space' in a body

The space in a body depends upon such factors as the volumes of the atoms composing the body, the energy of the atoms and the position of the null point between them, the size of the grains in the polycrystals and their degree of crystallinity, their homogeneity, their dihedral angles and their orientation, the form and nature of foreign material in those angles, and the structure of the agglomerate.

Compressibility is measured as change in volume per unit of original volume divided by the stress. The compressibility of elements is found to vary with the atomic volume, i.e. the larger the atomic volume, the larger the compressibility. Interatomic spaces in solids manifest themselves in such phenomena as occlusion of gases, clathrate formation and molecular sieving. Intercrystalline space is exhibited by porosity of metals to gases and its dependence upon temperature.

Spaces result from the mode of formation of the agglomerate, e.g. by sedimentation, electrochemical deposition and biological growth. The application of energy so as to compress the body will eliminate some of these spaces and often brings about the elimination of air pockets. Often metals become more dense when they are stamped in dies or less dense when they are drawn through orifices. Some materials show far more readiness to be compressed than to be extended, or vice versa: for example, cork may be compressed to about one-tenth of its bulk, and rubber may be extended to about ten times its length.

But after such pockets or spaces have been eliminated, the above considerations suggest that short of coalescence of the atoms, increasing compression becomes more and more difficult, and Bridgman[8] has shown that compressibility of solids becomes less as pressure is increased. Kendall[9] suggests that there are three ways (see Fig. 6.9) in which

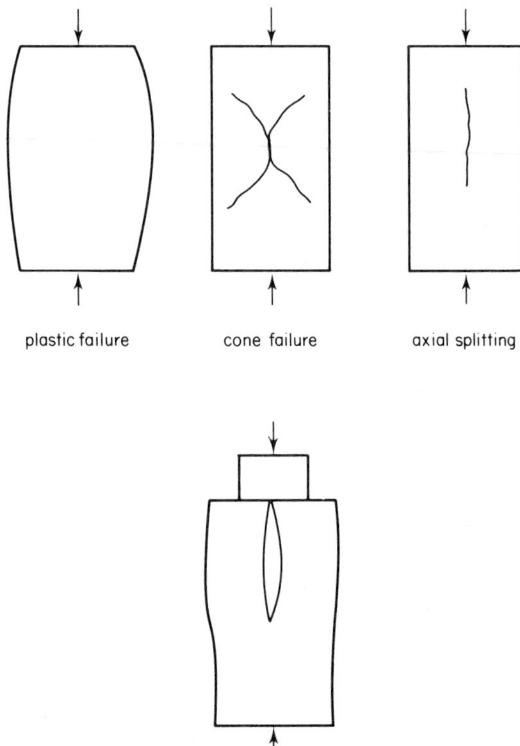

Fig. 6.9. Types of compressive failure. Axially split particle acting as two struts. (After Kendall.[9])

compression can manifest itself:

(i) bulging out at the sides at right angles to the direction of the compressive force,

(ii) conical cracking, i.e. the formation, at approximately 45° to the direction of the compressive force, of cracks which may be single or several and criss-crossed to form pyrimidal pieces, and

(iii) splitting along a plane parallel to the direction of the compressive force.

Form (ii) is the manifestation of the shear force components of the compressive force to which reference was made above.

The form of fracture (iii), the splitting along a plane parallel to the direction of the compressive force, might occur as a result of deformation in the form of (i). Kendall shows that if the particle is below a certain size and the piece of equipment through which the force is applied bears a certain relationship to the particle size, then plastic deformation rather than fracture will occur. This work will be considered further in the section on the limits to size reduction below.

The areas of bodies through which energy is exchanged

The underlying assumptions of the above considerations have included (i) the existence of uniform distribution of stress over the whole cross-section in stress, and (ii) the slow application of energy changes so that all energy change is in the form of strain energy up to the yield point of the material. These circumstances do not necessarily hold in practice and the effect of deviations from them will now be considered.

In many cases energy is transferred over areas which project from the main part of the body, donating or accepting energy as for example in the impact of a flat bar with an irregular lump of material, in the collision of two spheres, and in the application of the edge of a triangular rib to a plate of material. In such cases, the stresses might not be expected to be transmitted through the bodies so that they are distributed uniformly over the cross-sectional area reached at each stage in the progress of the strain.

This subject was first investigated by Hertz,[10] who showed that, in general, the area of contact between two bodies could be considered to be elliptical[6] and the coordinate of the stress perpendicular to the ellipse of contact could be considered with the ellipse to form a semi-ellipsoid. The maximum stress is at the centre of the ellipse of contact and has a value $1\frac{1}{2}$ times the average stress on the area. The maximum shearing stress is at a certain depth (depending upon the radii of curvature of the contact points) below the surface of the material. In a radial direction from the centre of the ellipse and at the periphery of the ellipse a tensile stress acts; at right angles to this tensile stress an equal compressive stress acts; these two stresses compound into a shear stress acting at an angle of $\pi/4$ to each of the other stresses.

In the case of two spheres, radii R_1 and R_2, in contact with a compressive force P, the area of contact is a circle of radius a, given by

$$a=\left(\frac{3\pi P(K_1+K_2)R_1R_2}{4(R_1+R_2)}\right)^{\frac{1}{3}} \tag{6.24}$$

in which $K=(1-\mu^2)/\pi E$ for the respective spheres. The maximum pressure (σ_{max}) at the centre of the circle of contact is given by

$$\frac{3P}{2\pi a^2}=0{\cdot}270\left(\frac{P(R_1+R_2)^2}{(K_1+K_2)^2\,R_1^2R_2^2}\right)^{\frac{1}{3}} \tag{6.25}$$

The shear stress at the boundary of the circle is $(1-2\mu)\sigma_{max}/3$. If one of the impactors is a flat plate, i.e. R_2 is infinite, then

$$\sigma_{max}=0{\cdot}270\left(\frac{P}{R_1^2(K_1+K_2)^2}\right)^{\frac{1}{3}} \tag{6.26}$$

Effect of rate of application of force

It has been mentioned that both elastic and plastic (viscous) behaviour could be expected in some degree from all solids, mainly as a result of the spread of kinetic energy of the atoms composing the solid. Thus a given stress (σ) imposed upon a body may result in

(i) elastic deformation (ϵ) which is related to the stress by the equation $\sigma=E\epsilon$, or if σ is shear stress by $\sigma=G\epsilon$ in which G is the modulus of rigidity, and

(ii) viscous deformation given by Newton's equation

$$\sigma=\eta\frac{\mathrm{d}v}{\mathrm{d}y}=\eta\frac{\mathrm{d}^2x}{\mathrm{d}y\mathrm{d}t} \tag{6.27}$$

in which v is velocity, y is the coordinate perpendicular to the direction of the velocity, x is the coordinate parallel to the direction of the velocity, η is viscosity and t is time.

Thus total deformation (γ) resulting from the imposition of a shear stress σ is

$$\gamma=\epsilon+\frac{\mathrm{d}x}{\mathrm{d}y} \tag{6.28}$$

and
$$\frac{d\gamma}{dt} = \frac{1}{G}\frac{d\sigma}{dt} + \frac{\sigma}{\eta} \tag{6.29}$$

which is known as Maxwell's equation. Thus it is seen that deformation and stress in a body are functions of time. Fig. 6.10 indicates the effect of strain rate on the ultimate strength of mild steel and copper, and on the duration of stress on the ultimate strength of glass. It is seen that the effect of speed is to increase the strength of the material.

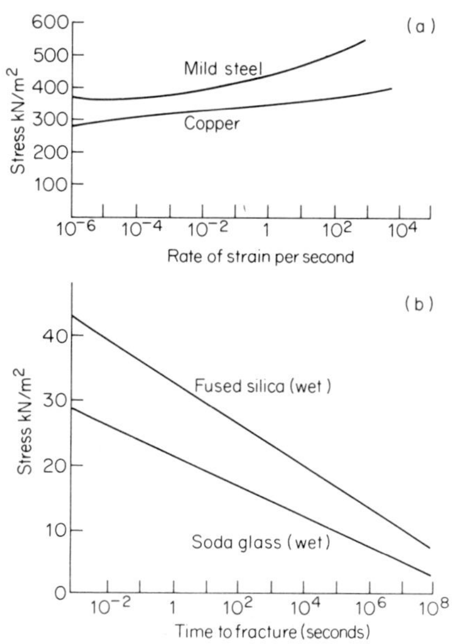

Fig. 6.10. Time effects on strength. (a) Effect of strain on ultimate strength. (b) Effect of duration of stress.

When two bodies impact together elastic waves are created in both materials and these waves appear to be of three kinds: (i) spherical dilational (longitudinal) waves in which material particles move to and fro parallel to the direction of the force, (ii) spherical distortional (shear) waves in which particle movement is transverse to the direction of the force, and (iii) surface (Rayleigh) waves which only have significant amplitudes near the surface of the interface. The velocity of the longitudinal waves (c_L) is given by

$$c_L = \left[\frac{3K(1-\mu)}{(1+\mu)}\right]^{\frac{1}{2}} \tag{6.30}$$

and of the shear waves (c_S) by

$$c_S = \left[\frac{G}{\rho}\right]^{\frac{1}{2}} \tag{6.31}$$

in which K is the bulk modulus ($V\sigma/\Delta V$, V being the volume and ΔV the change in volume caused by the stress σ). As $K = 2G(1+\mu)/3(1-2\mu)$, c_L is always larger than c_S. Rayleigh waves have velocities smaller than, but very close to, those of shear waves.

At an interface there is both reflection and transmission of the wave; the stresses (σ_R and σ_T respectively) compared with the stress reaching the interface, are given by

$$\sigma_R = \frac{\rho_2 c_2 - \rho_1 c_1}{\rho_2 c_2 + \rho_1 c_1}\sigma \tag{6.32}$$

$$\sigma_T = \frac{2\rho_2 c_2}{\rho_2 c_2 + \rho_1 c_1}\sigma \tag{6.33}$$

in which the suffixes indicate the properties of the material before and after the interface. The sign of the expression indicates in this case (contrary to the usual convention) + for compression and − for tension. The partition of stress between reflection and transmission is therefore

$$\sigma_R/\sigma_T = \tfrac{1}{2} - \frac{\rho_1 c_1}{2\rho_2 c_2} \tag{6.34}$$

For gases under adiabatic conditions $K = \gamma P$ in which γ is the ratio of the specific heats at constant pressure and volume ($\gamma = 1{\cdot}41$ for air), P is the pressure of the gas; μ is of course zero for gases.

The spherical waves entering a solid body produce (i) compression or tension, which so far as the axes of the hemispherical waves are concerned, is a maximum at the centre of the front, and (ii) hoop stresses from the semicircular front, which produce a sideways bursting or collapsing thrust. If the reflected wave is tensile, the body may fail in tension: this phenomenon of an impact apparently causing compression in a bar, but actually causing fracture at the end farther away from the place of impact, by the compression wave being reflected at the material–

air interface, is known as *Hopkinson spalling*. Reflections can also occur from the sides of the material (see Fig. 6.11).

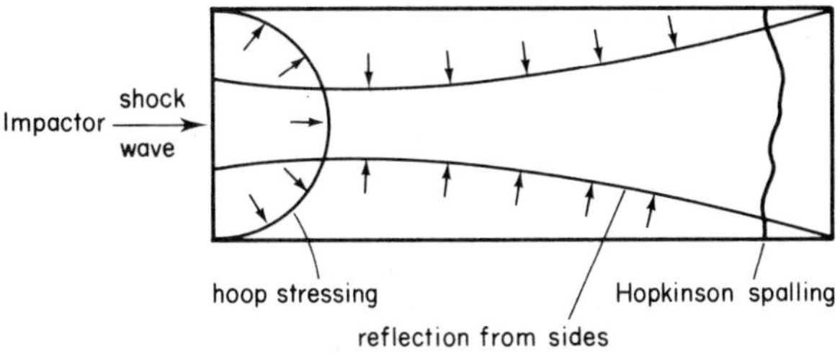

Fig. 6.11. Shock waves and Hopkinson spalling.

The effect of velocity of impact, besides depending fundamentally on the quantity of energy involved and the properties of the materials concerned, also depends upon the duration of the impact. At comparatively low velocities, when the impactors do not possess enough energy to exceed their elastic limits, the wave motion takes only a very small fraction of the energy, the remainder being absorbed as strain energy and in the corresponding plastic deformation discussed at the beginning of this section. The wave motion created on impact is sufficiently fast, compared with the time of contact, for a vast number of reflections of waves to occur and for the conditions to be very close to static stressing.

For somewhat higher velocities, the yield point is exceeded and either brittle fracture or plastic flow ensues (according to the properties of the materials concerned). If the latter, one impactor may penetrate the other, according to the relative geometries, or one or both may become a flat disc on the other.

Futher increase in velocity provides enough energy on deceleration for one or both impactors to become liquid and stress to become measured in the hydrodynamic form ρV^2, where V is the relative velocity of impact.

As the velocity approaches that of sound, the energy of the waves becomes very considerable and the effects of reflection and hoop stresses, described above, become apparent. Crack velocities are half the dilational velocities and, with reflection, waves can be formed which, so to speak, nullify the stresses of crack formation before they can be exercised. However, with all the other complications of hoop stressing and reflection from the sides, fracture still becomes highly probable.

At supersonic velocities, energies become available which can vaporise the impactors and the effect becomes explosive.

Hopkins[11] gives the figures set out in Table 6.4 for several different metals impinging on the same metal, the projectile being a flat-ended cylinder, and the target a large flat plate: V_1 is the maximum velocity at which the impact is elastic, plastic deformation occurs at velocities from V_1 up to the limit V_2, V_3 is the limit for the region in which hydrodynamic behaviour prevails, from V_3 to about $3V_3$ is the transonic regime of shock waves, and beyond $3V_3$ is the explosive regime.

TABLE 6.4
VELOCITY REGIMES FOR VARIOUS METALS
(after Hopkins[11])

Metal	*Velocity (m/s)*		
	V_1	V_2	V_3
Steel	46	360	4 600
Duralumin	65	430	5 300
Aluminium	13	190	5 300
Copper	8	130	3 750
Lead	2	40	2 100

An interesting illustration of the effect of velocity of application of force, is in electro-hydraulic crushing in which a shock wave is generated by an electrical discharge in water in which material to be comminuted is suspended. Shock wave velocities of the order of 100 000 m/s are created by the electrical discharge and material is shattered.[12] Examples are shown in Table 6.5.

TABLE 6.5
ELECTRO-HYDRAULIC CRUSHING
(after Carley-Macauly[12])

Material	*Material size (mm)*		*Spark energy*	*Yield*
	Feed	*Product (80% passing)*	*(J)*	*(kg/J)*
Alumina	10	0·2	32	$1{\cdot}8 \times 10^7$
Quartzite	200	30	1 250	$3{\cdot}6 \times 10^9$
Cornish granite	30	5	500	$1{\cdot}8 \times 10^9$

Radiation, if sufficiently energetic, e.g. neutrons greater than 1 MeV and travelling at say 3900 m/s, will dislodge atoms from a crystal lattice into interstitial positions or so as to collect together into platelets which cause the material to expand. If a polycrystalline material is irradiated, the effect on the grains with their different crystal orientations is to produce strain at the grain boundaries and this can lead to crack growth and even disintegration to powder.[13]

Effect of fluctuating stresses

Associated with fluctuating stresses is a phenomenon known as *fatigue.* The distinctive effect is that materials submitted to fluctuating loads behave as if they were much weaker than they are under static loads. The cross-section over which rupture has taken place, often has a characteristic appearance. In metals there are usually two zones, one of which is smooth and velvety and the other, the same as that which is obtained with static fracture. In the former there are concentric 'tide' marks appearing to originate from certain points usually on the surface of the material, from which radial ledges also emanate. It is found that the distance between the tide marks is related to the amplitude of the stress fluctuation.

The ratio between the areas of the fatigue and static zones is related to the magnitude of the fluctuating stresses compared with the static ultimate strength of the material.

A rhythmical force which fluctuates evenly on either side of zero, i.e. is varying from a given tensile force to a compressive force of the same magnitude and which may be considered in its simplest form as sinusoidal, is called a completely reversed force (see Fig. 6.1). If the maximum value of this force, calculated as a stress, is plotted against the number of cycles to failure (i.e. rupture), two characteristic curve shapes are obtained depending upon whether the material is essentially a ferrous metal or not. The curve shapes are indicated in Fig. 6.12. In the case of the ferrous material, the shape starts with a horizontal line from 0 to about 100 cycles, i.e. the failure stress cannot be distinguished from the ultimate strength; thereafter the line drops at an angle to about half the ultimate strength at about 10^7 cycles and then proceeds as a horizontal line indefinitely. Thus one may speak of an infinite life for a ferrous material under a completely reversed fluctuating stress, the value of which stress is known as the *fatigue endurance limit.*

For a non-ferrous material the shape is less simple. Again, it is usually difficult to detect any change of strength up to 10^2 cycles, but thereafter the strength drops steadily and appears to approach asymptotically the axis representing number of cycles. Thus for non-ferrous material there is

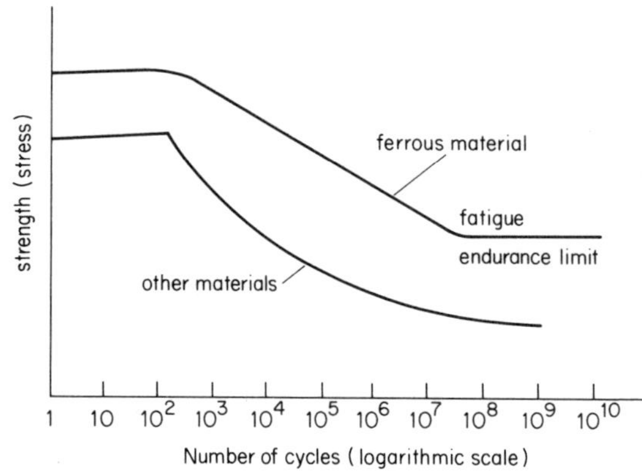

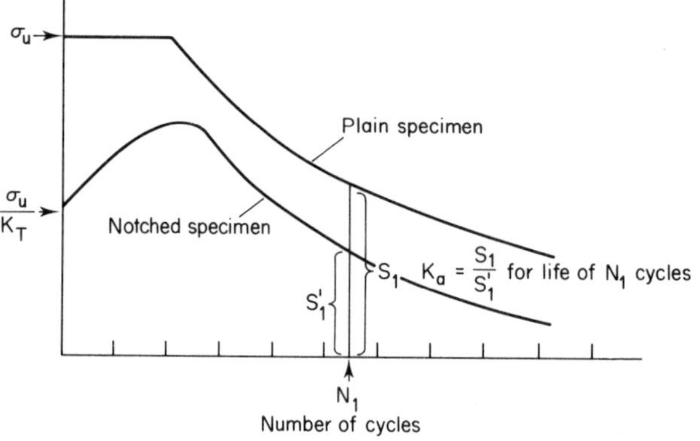

Fig. 6.12. Stress cycle (S–N) fatigue diagrams.

no such thing as infinite life: life (i.e. the number of cycles) has to be related to the maximum stress.

For rhythmically fluctuating stresses which are not completely reversed, a technique has been developed, principally by Goodman[14] and by Soderberg,[15] of considering such stresses to be made up of a static component (σ_m) about which a stress of a given amplitude (σ_a) oscillates. The maximum stress of the fluctuation is σ_{max}, and the minimum stress is σ_{min}; for purposes of identification, a tensile stress may be considered positive and a compressive stress negative, and a more positive stress considered larger than a less positive or a more negative stress.

Now

$$\sigma_{max} = \sigma_m + \sigma_a \text{ and } \sigma_{min} = \sigma_m - \sigma_a$$

Hence $\sigma_m = \frac{1}{2}(\text{maximum} + \text{minimum stresses})$

and $\sigma_a = \frac{1}{2}(\text{maximum} - \text{minimum stresses})$

An empirical result which seems to be applicable is that for a given life (number of cycles):

$$\frac{\sigma_m}{\sigma_u} + \frac{\sigma_a}{\sigma_f} = 1 \tag{6.35}$$

in which σ_u is the ultimate strength of the material (if both tensile and compressive stresses are present, the weaker of the tensile and compressive ultimate strengths is used, whereas if only one or the other kind of stress is present, then the appropriate value is used; a safer value to use is the yield strength), and σ_f the fatigue strength of the material for completely reversed loads and for the number of cycles for which the item is being designed or will have to withstand. Thus for given values of σ_u(or σ_{yp}) and σ_f, the static stress which may accompany a given amplitudinal stress (or vice versa) may be calculated.

Shape of body

In a previous section the effect of the areas over which energy was exchanged between bodies, particularly the behaviour of the impact of bodies of different relative sizes, was discussed. When a small projectile hits a large target, stresses are produced in the target which are by no means uniform over the area of the target. But even if care is taken to distribute the force over the cross-section of the body when it is first applied, if the body changes in cross-section perpendicular to the direction of the force, the force will cease to be uniformly spread and a phenomenon known as *stress concentration* becomes manifest.

The phenomenon is most clearly revealed through the use of a material like celluloid (glass may be used, but it is difficult to shape and to stress), phenol–formaldehyde resin, methyl methacrylate, cellulose acetate or epoxy resin. Brewster, many years ago,[16] found that polarised white light produced coloured patterns in stressed glass and proposed that this method could be used to measure stress in structures. The idea was developed, particularly by Coker.[17] A model of the part is made in one of the materials mentioned above. The model is stressed in the way that the part is to be stressed, but using fractions of the loads to suit the weaker material of the model. Polarised monochromatic light is used rather than polarised white light, to obtain differences in black and white rather than in colours. It is passed through the model perpendicular to the plane to be investigated and the emergent light is passed through another polarised set so that it will pass only that light vibrating at right angles to the polarisation plane of the light incident to the model. If the stress is uniform over the model in that plane, the light leaving the second polariser will be uniformly dark or light, but, if there are differences of stress, then the light will have different velocities in the areas of different stress and the emergent light will accordingly consist of rays in different phases. The image of this light will consist of light and dark areas. Examples of such images are sketched in Fig. 6.13.

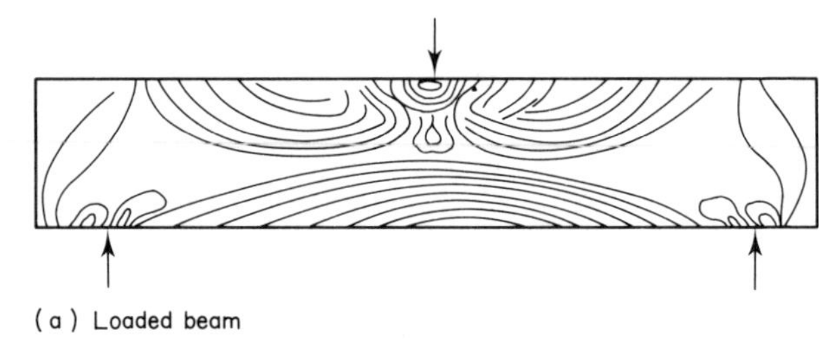

(a) Loaded beam

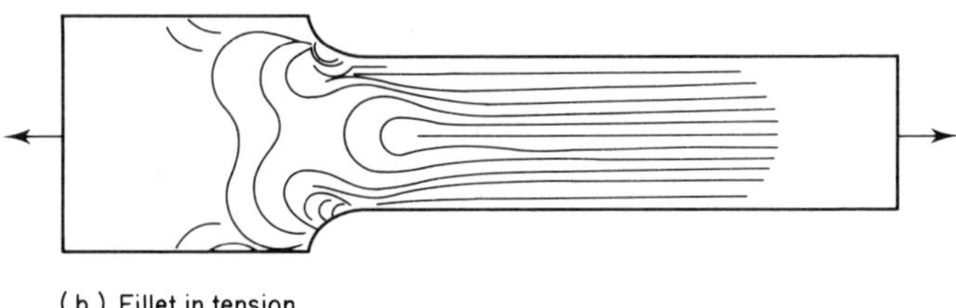

(b) Fillet in tension

Fig. 6.13. Sketches of photoelasticity. (After Timoschenko and Goodier.[6]*)*

The phase difference (Δ) is given by

$$\Delta = \frac{2\pi K t_0(\sigma_x - \sigma_y)}{\lambda} \tag{6.36}$$

in which K is known as the stress optical coefficient and is a constant, but only for a given wavelength λ of monochromatic light and at a given temperature, t_0 is the thickness of the model (assumed to be uniform) and σ_x and σ_y are the principal stresses (see eqn. (6.23)) at the particular points considered. The amplitude of the emergent light (which determines its intensity) is given by $a\sin 2\alpha \sin \Delta/2$, in which a is a constant, and α the angle of the plane of the incident light to the principal stress σ_x. Thus intensity of light will be zero where $(\sigma_x - \sigma_y)$ corresponds with integral multiples of some initial value making $\Delta/2 = 2\pi n$, where n is an integer (because $\sin 2\pi n = 0$) and a series of light and dark fringes will be produced in the image corresponding with changes of stress and the distance between the dark lines corresponds with the rate of change of stress with distance; close lines are seen where stress is changing rapidly with distance.

If a series of photographs of such images are examined, it will be seen that dark lines come closer where the body under stress changes in cross-sectional area; where the section is uniform the dark lines are parallel and set further apart. Thus stress is not uniformly distributed over cross-sections which are changing rapidly with distance, and the shape of the body suffering stress is important. Lumps of material met in nature, quarried from the ground or produced in some typical commercial breaking equipment, are not of simple geometrical shape or of uniform cross-section. The application of a uniform stress throughout such bodies is impossible.

Values for the stress concentration factors (the actual stress divided by the nominal stress, i.e. force divided by area) can be calculated only for the simplest cases such as the ellipse (and the special case of an ellipse, the circle). Even these calculations before the days of the electronic computer were most laborious.

Inglis[18] calculated the relationship for an ellipse

$$\sigma_{max} = \sigma\left(1 + \frac{2a}{b}\right) \tag{6.37}$$

in which $2a$ is the length of the cross-sectional axis perpendicular to the line of stress, $2b$ the length of the cross-sectional axis parallel to the line of stress, and σ the nominal stress corresponding to the applied force divided by the area of the material either side of the ellipse (see Fig. 6.14). Thus for a circle $(a = b)$, σ_{max} is three times the nominal stress; for an ellipse in which a is very much larger than b, as in a suitably orientated crack, σ_{max} is enormous, and this would suggest why cracks are propagated in a stressed body. It might be argued that if b were extremely minute, σ_{max} approaches infinity; or when there is no hole, the stress is infinite, but as explained earlier in this chapter, as b approaches the lattice dimensions there is no such thing as a crack, any distances of this

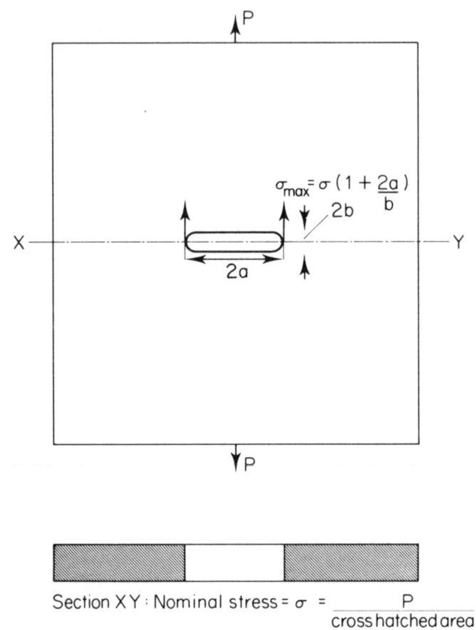

Fig. 6.14. Stress concentration at an elliptical hole.

order of magnitude being accommodated by atomic displacement. It should also be noted that eqn. (6.37) applies to similar ellipses (i.e. constant a/b ratio) and not to their absolute size.

If a body contains a notch of some kind, there is the possibility that it will be weaker than a body of uniform cross-section even when reduced by the area occupied by the notch. Broadly the smaller the radius of curvature of the notch, the greater the factor of stress concentration.

The stress concentration for static stresses is usually given the symbol K_T, or K_T = maximum stress/nominal stress. These factors are so large in many cases that the effective elastic limit of the material can be brought down to a fraction of that of the unnotched form.

On the macro scale, the importance of surface finish is realised from the above considerations. Rough surfaces are those with many grooves, and grooves are enhancers of stress (see Fig. 6.15). There is, however, the counter-acting influence of work-hardening in many materials when they are rough-turned or planed and this may be sufficient, in some cases, to nullify the stress-concentrating effect to weaken the material. A fine finish is often of greater value than its purely aesthetic importance.

Griffith[19] pointed out that the equation of Inglis (eqn. (6.37)) suggests

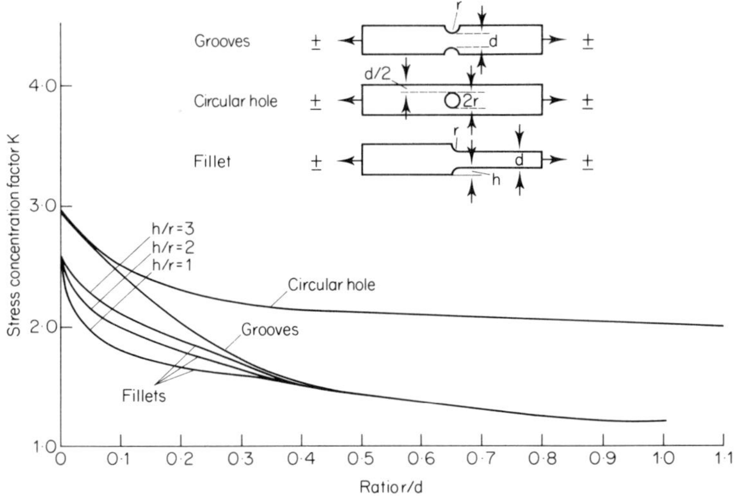

Fig. 6.15. Variation of stress concentration factor with r/d ratio.

that similar ellipses have the same stress concentration factors, and when applied to scratches on a body, the absolute width has no effect. But Griffith had shown that in fatigue tests particularly, the bigger the scratches, the weaker the body; moreover the weakening due to scratches was much less than the high factors predicted by the equation.

If in eqn. (6.37), for a (the major semi-axis of the ellipse) is substituted c (the half length of the crack), and since the radius of curvature (ρ) of the apex of such an ellipse is b^2/a (or b^2/c with the new term), $\sqrt{\rho c}$ may be substituted for b. Thus Inglis's equation becomes

$$\sigma_{\max}=\sigma\left(1+\frac{2c}{\sqrt{\rho c}}\right)=\sigma\left(1+2\sqrt{\frac{c}{\rho}}\right) \tag{6.38}$$

and if c is very much larger than ρ

$$\sigma_{\max}\approx 2\sigma\sqrt{\frac{c}{\rho}}$$

If the energy required to open a crack, just outside the range of molecular attraction is equated with the new surface energy (γ per unit area) produced on either side of the crack (compare Rittinger's hypothesis eqn. (6.45)) and this is the energy associated with the last atomic lattice separated on either side of the crack, of atomic spacing d_0, assuming that this is accomplished in a brittle elastic solid (i.e. one having no ability to deform plastically) then

$$\frac{\sigma_{max}^2}{2E}\,d_0^3=2\gamma d_0^2 \text{ or } \sigma_{max}=2\sqrt{\frac{E\gamma}{d_0}} \tag{6.39}$$

Now σ_{max} is the maximum mechanical strength that the crystalline material can have and is seen to be composed of constants for the material from which σ_{max} may be calculated.

Above it was shown that

$$\sigma_{max}\approx 2\sigma\sqrt{\frac{c}{\rho}} \tag{6.40}$$

in which σ was the nominal stress; and for growth of the very smallest cracks, the extremity has a width of about twice the lattice distance and a radius of curvature (ρ) of about the lattice distance (d_0), which remains constant although the rest of the crack may be widening. If d_0 is substituted for ρ

$$\sigma_{max}\approx 2\sigma\sqrt{\frac{c}{d_0}}\approx 2\sqrt{\frac{E\gamma}{d_0}} \tag{6.41}$$

and the nominal stress which will cause a crack of half length c to grow in a material with Young's modulus E and surface energy per unit of surface γ is

$$\sigma=\sqrt{\frac{E\gamma}{c}} \tag{6.42}$$

Now as c increases, σ decreases, i.e. as the crack grows, the required nominal stress to make it grow farther decreases. Thus once the crack starts to grow, it will continue until it reaches the boundary of the body.

Returning to the point made by Griffith that the stress concentration at a scratch seemed to depend upon the absolute width of the scratch, and not upon the length to width ratio as in the Inglis equation, it is seen that if the radii of curvature of cracks in the body are of the order of the lattice distance of the crystals, the stress concentration, σ_{max}/σ, depends mainly on the square root of the crack length; and in a body composed of many cracks of such small width, small scratches, though large compared with the cracks, may be imposed upon the body without much altering the nominal stress the body will tolerate. However, there will be a size of scratch which will be wide enough to impose a stress concentration in excess of the effect of these very small cracks, eventually, of course, being sufficient to decrease significantly the cross section bearing the stress as well as coming within the macroscopic province of Inglis's equation.

Various attempts have been made by Orowan[20] and others to modify Griffith's theory, which applies to brittle materials, so that it applies to plastic materials.

It is seen that the strength of a material, completely free from cracks and other notches, would be enormously greater than one with these defects. The work also suggests the reasons for the discrepancies between the various energy equalities which have been proposed (mentioned above) and the energy of fracture which mechanical behaviour adduces.

Griffith[19] suggested that fatigue was the result of crystal surfaces being caused, by the fluctuating forces, to rub together, of the debris produced by the rubbing being amorphous and, as is so often the case, of amorphous material being denser than the crystalline material and of holes thereby being created in the mass becoming sources of stress concentration. The theory would suggest that a ferrous metal requires a certain minimum stress (fatigue endurance limit) before the crystals will rub together, whereas other materials do not have this limit, and that given enough cycles the material will fracture at any stress down to the asymptotic limit. The theory could still hold if the amorphous rubbings were of greater volume than their crystalline source, because the larger volume could set up stresses in new directions and thereby cause fracture. It would not, of course, explain fatigue in materials with the amorphous form equal in density to the crystalline.

However, the phenomenon of fatigue is experienced in single crystals as well as in polycrystals. It appears that fracture begins in slip bands, that these are volumes in which a large number of dislocations exist and are in a direction of crystallographic weakness, in which the fluctuating forces have strong enough shear stress components to cause slipping. The to-and-fro movements of the bands will to some extent tend to cancel out each other's effects, but some effects will persist and be augmented (e.g. screw dislocations which have twist sense) over a very large number of cycles; this is, of course, characteristic of fatigue processes and would explain why a smaller number of fluctuations have little effect.[21] Slip bands appear in many directions in a polycrystalline material, but crack formation, in one grain, will confer enhanced stress in that direction to other suitably orientated grains. Over a large number of cycles, cracks in one grain will join up with cracks in other grains and a band of weakness will be induced which will be sufficiently large to cause fracture, even in unfavourably (from the point of view of fracture) orientated grains in its path. Finally even the most resistant grains will be fractured by the residual forces they have been left to support. This theory explains why fatigue strength of a polycrystalline material increases with decreased grain size and also, a common concomitant of fatigue, why minute slivers of material appear on the surface of affected areas.

Fluctuating stress with stress concentration

When stress versus number of cycle (*S–N*) diagrams are constructed for a material with a given notch, a graph of somewhat peculiar shape is obtained (see Fig. 6.12). For cycles up to 10^3, the figures are indefinite and a line is even more difficult to define than usual (for fatigue diagrams generally the lines have to be the result of a very large number of tests and of considerations of statistical probability), but there is often an improvement in the stress which can be tolerated both in static conditions (σ_u/K_T) and for those with a small number of cycles. After 10^3 cycles the graph falls always below the corresponding graph for an unnotched (plain) specimen, but not usually parallel with it. The stress concentration factor for a fluctuating stress over a given number of cycles is the quotient

$$\frac{\text{maximum (completely reversed) stress at } N \text{ cycles for notched specimen}}{\text{maximum (completely reversed) stress at } N \text{ cycles for plain specimen}}$$

and is usually designated K_a.

Thus for a notched specimen, the equations of Goodman and Soderberg mentioned above become

$$K_T\frac{\sigma_m}{\sigma_u}+K_a\frac{\sigma_a}{\sigma_f}=1 \quad \text{(Goodman}^{14}\text{)} \tag{6.43}$$

and

$$K_T\frac{\sigma_m}{\sigma_{yp}}+K_a\frac{\sigma_a}{\sigma_f}=1 \quad \text{(Soderberg}^{15}\text{)} \tag{6.44}$$

Random fluctuations of stress

Many conditions experienced in industry do not involve rhythmically (i.e. regular) fluctuating stresses, but the fluctuations are quite irregular. Consider, for example, the wing of an aeroplane and the buffeting it receives in a long distance flight. In such a case an attempt is made to analyse the number of cycles which are related to small bands of loads over the whole range of loads which the body has to suffer over a representative period. A histogram is thereby constructed. The loads are converted into trial stress (by assuming a cross-sectional area). The greatest stress (say σ_1) is then set against the maximum number of cycles which the material would bear in rhythmical conditions at that stress (say N_1 cycles). The life of the part is then said to have been fractionally consumed to the extent n_1/N_1, where n_1 is the actual number of cycles at a force corresponding with the stress σ_1. This procedure is followed for each calculated stress, in descending order of stress. In this way, eventually a number of fractions n_1/N_1, n_2/N_2, n_3/N_3 ... n_x/N_x are obtained and the life remaining for the body is said to be

$$1-\sum_{i=1}^{i=x} \frac{n_i}{N_i}$$

This method known as Miner's method[22] has been found to overestimate the life by a factor which may be as high as five. There are methods which seem to give more accurate results[23] but which require much more detailed analysis.

In a vibration mill, which uses fluctuating stresses (e.g. 25Hz) the power consumption is claimed to be of the order of 10% of a tumbling mill, which confers a much lower number of cycles of fluctuating stress in unit time.

BREAKDOWN OF SOLIDS

Energy of fracture

From what has been discussed above, it may be deduced that when a body contains a certain quantity of strain energy numerically equal to its toughness × its volume, the body will break. Thus if σ_u is ultimate strength, ϵ_u is ultimate strain and V is volume, true strain energy per unit volume $=\int_0^{\epsilon_u}\sigma d\epsilon$. For brittle material $\int_0^{\epsilon_{yp}}\sigma d\epsilon \approx \sigma_{yp}^2/2E$, and for plastic material total strain energy $\approx \sigma_u \epsilon_u V$. In the comminution of materials on an industrial scale, it is rare for the materials to be pure simple chemicals or for the size of the feed or of the product to be uniform. Again, except in some special cases, neither are there economically acceptable industrial size-reduction processes which treat even closely graded material as individual pieces or supply energy to them so as to break them only once; for unless energy at least equal to the toughness is absorbed by the material and the pieces break, removal of the circumstances causing the energy transfer will permit the material, without breaking, to release much of the energy by contracting or expanding and doing useless work.

Industrial comminution equipment receives feed material within a range of sizes, and performs certain deforming processes on it, such as nipping it between surfaces, impacting it with surfaces which collide with it, tumbling it in a vessel with or without foreign units, cutting and attriting it. In such processes some pieces of material will not receive sufficient strain energy to cause fracture, others will receive multiple fracture. Moreover the material of the equipment will be strained, though, hopefully, not broken. Refined calculation of the energy required for comminution in such industrial processes does not seem to be possible. Nevertheless, attempts have been made to give some indication of the energy required in crushing and grinding processes.

As described earlier Griffith[19] related strain energy for brittle materials at fracture with the surface energy provided for crack formation; in other words, new surface was related to the strain energy necessary to incur fracture. This particular relationship was anticipated by Rittinger[24] in 1867.

Rittinger thought that in a comminuting process, in which particles were produced of approximately similar shape to that of their parent, the energy required was directly proportional to the new surface produced. Thus if D_1 and D_2 are the initial and final corresponding dimensions of two similarly shaped particles, W is the energy needed for the size reduction of unit weight of material and K is a constant

$W \propto$ (final surface area − initial surface area)

i.e. $W \propto$ (final number of particles × surface area of representative final particle − original number of particles × surface area of representative initial particle)

$$= K'\left(\frac{D_2^2}{D_2^3} - \frac{D_1^2}{D_1^3}\right) = K\left(\frac{1}{D_2} - \frac{1}{D_1}\right) \qquad (6.45)$$

Various experiments[25,26] have shown that this relationship is approximately true, particularly for small particles of brittle material.

Kick[27] developed an equation based upon the idea that comminution took place in a number of stages and that at each stage the size reduction was constant. Thus the relationship between the feed size and the product size determined the number of stages. At each stage the strain energy in the particles had to be the toughness of the material and was therefore proportional to the volume of the material processed. Thus for geometrically similar particles

Energy $= K_1 \times$ number of stages of size reduction $\times$ volume of material processed

Suppose, therefore, that corresponding dimensions on the feed and product particle are given by D_1 and D_{n+1} respectively, and that it is imagined there are n stages of size reduction so that if r is the size reduction ratio

$$\frac{D_1}{D_2} = \frac{D_2}{D_3} = \frac{D_3}{D_4} \ldots = \frac{D_n}{D_{n+1}} = r \tag{6.46}$$

or

$$\frac{D_1}{D_{n+1}} = r^n$$

Therefore $n = \log\left(\dfrac{D_1}{D_{n+1}}\right) \cdot \dfrac{1}{\log r} = K_2 \log\left(\dfrac{D_1}{D_{n+1}}\right)$ where $K_2 = \dfrac{1}{\log r}$

Now the total volume is constant, K_3, for each size reduction process Therefore,

$$\text{Energy required} = W = K_1 \ K_2 \ K_3 \ \log\left(\frac{D_1}{D_{n+1}}\right)$$

$$= K \ \log\left[\frac{D_{\text{Feed}}}{D_{\text{Product}}}\right] \tag{6.47}$$

Kick's reasoning would include both brittle and ductile materials since strain energy of both kinds of materials would be proportional to the volume, though there is not much evidence to support Kick's law.

Both the equation of Rittinger and that of Kick may be reduced to the differential equation[28]

$$\frac{dW}{dD} = \frac{K}{D^n} \tag{6.48}$$

in which n is 2 in the Rittinger equation and 1 in that of Kick. Bond[29] has proposed an equation which in the differential form gives $n = 3/2$, i.e. halfway between the exponent for the Rittinger and the Kick equations. Bond is supported to some extent by the work of Kihlstedt,[30] who showed that the surface area of comminuted material was proportional to

$1/\sqrt{D}$: therefore Bond's equation might be considered a correction of Rittinger's equation. Bond argued that as surface area per unit volume was proportional to $1/D$, crack length was proportional to $1/\sqrt{D}$.

In an industrial comminution operation, the reduced pieces are far from being uniform in size and there is usually quite a large ratio between the largest and the smallest particle size in both the feed and in the product. Moreover, some of even the largest feed sometimes gets through unchanged and if the criterion were the size of sieve through which all of the material passes, there would often be no size reduction evident at all. A common convention is to give D the value of the sieve size through which 80 per cent of the material passes having subtracted, from the quantity of feed and from the quantity of product, that quantity in the feed which was of smaller size than the sieve size designated.

Bond's equation with such a convention for particle size leads to a work index useful when trying to pitch the energy requirement for a comminution process. He suggests that it is possible to assign energy to a given size of material by assuming it has gained this energy by the comminution of a particle of infinite size whereupon $W=K/\sqrt{D}$ becomes that energy in which D is the size of the present material. Thus to find the energy which may be put into material of size D_1 to make it of size D_2, all that is necessary is to subtract the energy at present in the material of size D_1 from that which is required to be in material of size D_2 or

$$W=K\left(\frac{1}{\sqrt{D_2}}-\frac{1}{\sqrt{D_1}}\right) \tag{6.49}$$

A selection of values for K determined by Bond[29] are given in Table 6.6. (Bond's figures have been converted to SI units and the value K may be inserted directly in the above equation to give W in terms of MJ/kg.) The figures give orders of magnitude and are useful in determining the relative power requirements, on a given piece of equipment, in transferring from one duty to another.

Holmes[31] claims that the exponent of the particle size varies with the material and proposes the more general exponent r.

$$W=K(D_2^{-r}-D_1^{-r}) \tag{6.50}$$

Materials differ from the same quarry and even the same vat; their crack population, polycrystalline or amorphous condition and shape, and the way energy is applied, lead to such a mass of variables that even single specimens tested in the laboratory give results better expressed in

TABLE 6.6
A SELECTION OF MODIFIED BOND INDICES K

Material	*Density* $10^{-3} \times (kg/m^3)$	$K(MJ/kg)$	*Material*	*Density* $10^{-3} \times (kg/m^3)$	$K(MJ/kg)$
Bauxite	2·38	37·5	Kyanite	3·23	74·9
Cement clinker	3·09	53·6	Limestone	2·69	46·1
Chrome ore	4·06	38·1	Magnesium oxide	5·22	66·7
Coal	1·63	45·1	Phosphate rock	2·66	40·2
Dolimite	2·82	44·9	Potash	2·37	35·3
Feldspar	2·59	46·3	Quartz	2·64	50·7
Fluorspar	2·98	38·7	Rutile	2·84	48·1
Galena	5·39	40·5	Granite	2·68	57·1
Glass	2·58	12·2	Ilmenite	4·27	52·0
Graphite	1·75	178·8			
Haematite	3·76	50·3			

statistical form than a simple chain of results. The above equations are, at best, a rough guide.

Effect of environment on energy consumption

A phenomenon with which size reduction of solids may be concerned is called *corrosion fatigue.* This consists of the interaction of the environment with fluctuating loads so that the effect on a body is to weaken it far more than if it had to suffer only one of the conditions described above. The phenomenon is particularly well known in metals. Thus brass is little affected by sodium chloride solution even if the brass is stressed statically, but if a fluctuating, as opposed to static, load is imposed, the fatigue endurance limit is much decreased. Even fresh water will affect steel in fatigue conditions and the exclusion of air will enhance the fatigue tolerance of copper and brass.

In comminution operations, the various factors are difficult to identify. For example, it is well known that wet grinding is less energy-consuming than dry grinding and the Bond indices included in Table 6.6 may be divided by 1·3 to give the corresponding wet figures. Many materials when added in very small concentration (sometimes of the order of 0·01 per cent) appear to decrease energy requirements by large fractions (e.g. of the order of one-third). This is called the Rehbinder effect[32] after its discoverer. There is some confusion as to whether such an effect is really the result of the dispersion of the particles or the prevention of their agglomeration rather than of corrosion fatigue. A list of examples has been given elsewhere.[33]

The effect of water and other liquids may be due to the adherence, which they encourage, of the feed material to the grinding media and therefore of fixing targets for impact. Meloy[34] found that the improvement in grinding corresponded with higher surface tension of the liquid, provided the liquid wetted both the material to be ground and the material of the equipment. If the effect is due to the surface tension holding material to the equipment parts, the force due to surface tension (value γ) is a maximum over the semicircular circumference of the particle (assumed to be spherical),[38] and for the particle (density ρ) to be held against gravity, the force must equal the weight of the particle: to hold a particle of maximum radius r

$$\tfrac{4}{3}\pi r^3 \rho = \pi \gamma r$$

$$r = \sqrt{\frac{3\gamma}{4\rho}} \tag{6.51}$$

Meloy used water ($\gamma = 72{\cdot}7\,\mathrm{mJ/m^2}$) and n-hexane ($\gamma = 18{\cdot}4\,\mathrm{mJ/m^2}$) with the feed material quartz ($\rho = 2650\,\mathrm{kg/m^3}$) and magnetite ($\rho = 5050\,\mathrm{kg/m^3}$). If these values are substituted in the above equation they lead to the suggestion that the radii shown in Table 6.7 would be favourably comminuted in equipment which was wetted by water and by hexane. Feed between 3 and 4 Tyler (6·7 and 4·7 mm) placed in a ball mill produced fractions less than No. 10 Tyler (1·65 mm) as shown in Table 6.7. It is seen that the degree of grinding measured by the percentage

TABLE 6.7

COMPARISON OF MAXIMUM RADII WHICH WOULD ADHERE TO EQUIPMENT WITH DEGREE OF GRINDING OF FEED MATERIAL BETWEEN 6·7 AND 4·7

	Water		*n-Hexane*	
	Quartz	*Magnetite*	*Quartz*	*Magnetite*
Maximum radius (mm)	4·5	3·3	2·3	1·7
Percentage passing 1·65 mm screen	62	57	53	46

passing a 1·65 mm screen is related to the maximum radius of the particle which would adhere to the equipment.

Size and shape of broken particles

Crushing and grinding of materials are major operations in many factories and have the purpose of reducing the size of materials in order to (a) decrease at least one dimension to assist such operations as laying controlled thicknesses of the material or achieving greater dispersions, (b) increase surface area to improve access of liquids or gases to the solid, and (c) release components of agglomerates so that they may be separated, e.g. galena from quartz in run of mine product, chocolate from fondu in the recovery of constituents from imperfectly formed chocolates.

It is often highly desirable to form pieces of material of sizes within a closely defined range or, at least, to be aware of the range of sizes, and the relative quantities within defined size ranges, which are likely to be produced from a particular process.

Evidentally, solid materials can be formed into carefully defined shapes by the skilful use of chisels, saws, abrasive stones or by melting and casting, but the indefinitely aimed hammer, the application of pincers or the use of projectiles against a material, which are the common features of industrial comminuting equipment, yields a product of which chaos is almost the definition. The surprising thing is that if a large number of pieces of material is fed at reasonably constant rate to a machine using such crude breaking methods, then the size distribution of the product follows a broadly predictable form.

Typically, the size distribution may be ascertained by subjecting the product to a carefully controlled, and therefore reproducible, sieving procedure consisting of passing the product from each of a series of sieves to the next of decreased mesh size and weighing the residue on each sieve. A graph of percentage (P) through size x, plotted against size x is usually a relatively smooth curve. The graph becomes closer to a straight line if log P is plotted against log x and as usual, a still better straight line is obtained if log log P is plotted against log log x.

Various equations have been proposed to conform with these results and the equations are usually given the names of their proposers. Thus

Gaudin:[35] $$P=\left(\frac{x}{D}\right)^m \text{ or } \log P=m\log\left(\frac{x}{D}\right) \qquad (6.52)$$

Schuhman:[36]

$$100-P=\exp\left(\frac{x}{\bar{x}}\right)^n \text{ or } \log\log(100-P)=n\log\left(\frac{\mathrm{x}}{\bar{\mathrm{x}}}\right) \tag{6.53}$$

Rosin and Rammler:[37]

$$100-P=100\ \exp(-bx^n)$$

$$\text{or } \log\log\left(\frac{100}{100-P}\right)=n\log x+\log b \tag{6.54}$$

in which D is the sieve size for the largest particle, $\bar{x}$ is the arithmetical mean size of all the sieve sizes occupied, and m, n and b are constants.

All of these equations are redolent of probabalistic forms. They are useful (a) for putting the range of sizes produced by equipment in a quantitative and comparative form (the exponent becomes the slope of the graph in the logarithmic forms; a steep line (large exponent) indicates a narrow spread of sizes), (b) for extrapolating distributions from a few sieve results, and (c) for predicting the number of stages of comminution required.

More recently, stemming from work by such as that of Epstein,[38] Brown,[39] Broadbent and Callcott[40] and Austin,[41] the process of breakage has been analysed into (i) a selection function, $S(y)$, for the equipment which is the proportion by weight of the particles of a given size (y) presented to the equipment and broken in some specified duty by that equipment, e.g. cycles of a machine part, time in the equipment, and (ii) a breakage function $B(x, y)$ which is the proportion by weight of the particles broken from size y, and less than a particular size x. Thus from the supply of unit weight of particles of size y to a given comminuting machine, $S(y)$. $B(x, y)$ will be less than size x after a defined unit of duty, e.g. 1 s in the equipment.

The weight of material from all sources which is less than size x in unit time is

$$P(x)=\sum_{y_j=x}^{y_j=y_{\max}} S(y_j)\,B(x,y_j)+P(y_j=x) \tag{6.55}$$

Now, if $P(y,t)$ represents the function $P(y)$ at time t, and $S(y)$ is the selection function of that same size y, the quantity of material of size y at any time t is

$$\frac{\partial P(y,t)\mathrm{d}y}{\partial y}$$

and the above equation may be put in the differential form

$$P(x)=P(y=x)+\int_{y=x}^{y_{\max}} S(y).\ B(x,y)\frac{\partial(P(y,t))}{\partial y}\,\mathrm{d}y \qquad (6.56)$$

If $P(y, t)$ represents a surface of the relationship between $P(y)$, y and t,

$$P(x,\tau)=P(y=x,0)+\int_0^{\tau}\int^{y=y_{\max}} S(y)\ B(x,y)\frac{\partial(P(y,t))}{\partial y}\,\mathrm{d}y\mathrm{d}t \qquad (6.57)$$

represents the proportion of product less than size x at time τ.

Of course, this leaves the determination of $S(y)$, $B(x, y)$ and $P(y, t)$ as functions of both y and t to be ascertained for each piece of equipment. Various assumptions have been made about these functions and values have been experimenally determined for particular cases which have permitted distribution to be calculated and which have agreed very well with practical distributions.

Interesting results which link up with other work have been obtained by Austin *et al.*[42] As an example, if $S(y)\,B(x, y)$ is assumed to be a function of x, i.e. ax^a, then

$$\frac{\partial P(x)}{\partial t}=ax^a\int_{y=x}^{y_{\max}} \mathrm{d}(P(y,t))=ax^a\Big[P(y_{\max}t)-P(x,t)\Big] \qquad (6.58)$$

Now $P(y_{\max}t)$ must equal 1

or

$$\int_0^t \frac{\mathrm{d}(P(x))}{1-P(x,t)}=\int_0^t ax^a\mathrm{d}t$$

and

$$\Big[-\log(1-P(x,t))\Big]_0^t=ax^at \qquad (6.59)$$

whence

$$\frac{1-P(x,0)}{1-P(x,t)}=\exp(ax^at)$$

and if $P(x, 0)=0$, then at a time t

$$1-P(x)=\exp(-ax^a)$$

which is the Rosin–Rammler equation (eqn. (6.54)).

Nature abounds with recognisably regular shapes, and crystalline forms are common amongst mineral substances. Crystalline material tends to break along crystal planes whether the crystals be three-, two- or one-dimensional. Three-dimensional polycrystalline material tends to break into the grains which compose it, i.e. into the natural grain size. Two-dimensional crystals break into leaves or plates, e.g. graphite, mica.

One-dimensional crystals, if they are aligned to be parallel, tend to break into fibres, e.g. asbestos.

Even polycrystalline or amorphous material can be found with at least one dimension which is regular, e.g. in sedimentary or metamorphic strata, solid material can be removed with approximately flat parallel sides; clays form thin layers on top of each other and these can be easily split apart, for example to form slates.

Less frequently a mass of molten rock has cooled so steadily that the stresses of contraction have been evenly distributed in two horizontal dimensions and the rock has separated into approximately equal hexagons, as in the case of basalt in the Giant's Causeway in Northern Ireland.

Even a lump of material broken haphazardly is not without form; the majority of the pieces have sections with five or six sides, a condition which can be explained topologically.[43]

Certain comminuting equipment has a disposition or can be adjusted to give cubic rather than say tetragonal shapes, e.g. gyratory crushers give more cubic material than jaw crushers, impact crushers can be fitted with plates which aid cubic shaping of the product.

Limits to the size reduction of solids

There does not appear to be any fundamental limit to the upper size of the feed to comminuting equipment, save the skills of mechanical design and execution.

With regard to the lower limits of size reduction, the question arises as to when an assemblage of molecules ceases to be a gas and becomes a liquid or solid. Zwicky[44] concluded that there were, in crystals, natural blocks of 10 nm in diameter and the 'mosaic' structure was a source of weakness revealed as crack lines by X-rays.

Some materials grind to a minimum size, expressed by maximum surface area, and with further grinding decrease in surface area, presumably by agglomeration. Yet other material, such as very pure graphite (99·97% carbon), grinds to particles containing 7000 atoms without any sign of reaching a limit except for the lack of a device to grind it further. Such a particle of graphite is approaching Zwicky's limit.

Recently, Kendall[9] has examined the nature of compression failure. Compression failure appears to be of three forms, not necessarily followed by any particular kind of material: (a) plastic yielding in which the specimen bellies out; (b) cone failure in which cracks roughly at 45° form either singly on in twos, crossing each other; (c) axial splitting (see Fig.

6.9). These forms depend upon the arrangement of the compressing platens and one material may show all three forms of failure depending upon the circumstances of the test.

Kendall considers a particle compressed between two plates (Fig. 6.9). The particle has had a crack formed down the axis parallel to the direction of the compressive force, so that two struts now support the upper plate and bend due to the eccentric form of the plate. The stress in the struts is calculated by means of the beam equation,

$$\frac{\sigma}{y}=\frac{M}{I} \tag{6.60}$$

in which σ is the stress at a distance y from the neutral axis of the beam, M is the bending moment and I is the moment of inertia about the axis perpendicular to the neutral axis and about which the beam is bending. From the determination of σ, the total strain energy may be calculated by integrating the strain energy in each elementary volume at its corresponding distance from the neutral axis. Then using the method of Griffith[19] by which the surface energy of the crack is considered to be derived from the energy of the system including the strain energy, and also using the principle of the conservation of energy, by which the energy remains unchanged and particularly the rate of change of energy with the change of crack length is zero, Kendall deduces that the force required to split the particle is proportional to

$$\text{breadth of particle}(\text{surface energy per unit area } (\gamma) \times E \times \text{width of particle})^{1/2}$$

$$\left(1-\frac{\text{width of platen}}{\text{width of particle}}\right) \tag{6.61}$$

Now if breadth of particle = width of particle = d and the ratio of width of platen to width of particle is α, then

$$\text{force}=\frac{Kd^{3/2}\sqrt{\gamma E}}{(1-\alpha)} \tag{6.62}$$

The stress to cause splitting is therefore proportional to

$$\frac{\text{force}}{\text{area}}=\frac{K\sqrt{\gamma E}}{d^{1/2}(1-\alpha)} \tag{6.63}$$

or the stress required to cause splitting increases as the particle becomes smaller for a given (size of platen)/(size of particle) ratio.

If the platen size becomes equal to the particle size the stress for splitting would become infinite, but when the stress becomes equal to the compressive yield stress, the material collapses in plastic flow and then becomes resistant to further breaking. Thus there appears to be a lower limit to the size of material which can be broken by compressive forces, because beyond that limit the material responds by flowing.

Strain energy is proportional to the square of the stress σ; for brittle material it is approximately $\sigma^2/2E$. Strain energy is therefore, according to Kendall, approximately proportional to

$$\frac{\gamma E}{d(1+\alpha)^2}\cdot\frac{1}{E}=\frac{\gamma}{d(1+\alpha)^2} \tag{6.64}$$

which suggests that, for a given material, strain energy is inversely proportional to the size of the particle, which is in accord with Rittinger's law (eqn. (6.45)).

REFERENCES

1. M. Born, and E. Brody, *Z. Physik.*, **7** (1922) 217.
2. R. Houwink, and H. K. De Decker, (eds.) *Elasticity, Plasticity and Structure of Matter*, Cambridge University Press, 1971.
3. E. N. da C. Andrade, *Nature*, **125** (1930) 309; A. Guzman, *Ann. Fis. Quim.*, **11** (1913) 353.
4. W. D. Kusnezow, *Z. Physik.*, **42** (1927) 302; **44** (1927) 228.
5. M. Polonyi, and E. Schmid, *Z. Phys.*, **32** (1925) 684.
6. S. P. Timoschenko, and J. N. Goodier, *Theory of Elasticity*, McGraw Hill, New York, 1970.
7. H. S. Hall and S. R. Knight, *Higher Algebra*, 4th edition, Macmillan, London, 1940.
8. P. W. Bridgman, *The Physics of High Pressures*, G. Bell & Sons, London, 1949.
9. K. Kendall, *Proc. Roy. Soc.* (*London*), **A361** (1978) 245.
10. H. Hertz, *J. Math.* (*Crelle's J.*), **92** (1881) 156.
11. H. G. Hopkins, *Memoirs Manchester Lit. Phil. Soc.*, **116** (1973/74) 77.
12. K. W. Carley-Macauly, *Chem. Proc. Eng. 49*, Sept. 1968, 87.
13. F. J. P. Clarke, *Progress in Nuclear Energy Series IV*, **5** (1963) 221.
14. J. Goodman, *Mechanics Applied to Engineering*, 9th edition, Longmans, London, 1930, Vol. 1.
15. C. R. Soderberg, *Trans. ASME*, **52**, APM 52–2 (1930) 13.
16. D. Brewster, *Trans. Roy. Soc.* (*London*), **106** (1816) 156.
17. E. G. Coker, *Phil. Mag.*, **20** (1910) 740.
18. C. E. Inglis, *Trans. Inst. Naval Arch.*, **55** (1913) 219.
19. A. A. Griffith, *Phil. Trans. Roy. Soc.* (*London*), **A221** (1920/21) 163.

20. E. Orowan, *Rep. Prog. Phys.*, **12** (1948) 185.
21. N. F. Mott, *Phys. Soc. B*, **64** (1951) 729.
22. M. A. Miner, *Trans. ASME*, **67** (1945) A159.
23. J. Marin, *Mechanical Behaviour of Engineering Materials*, Prentice-Hall Inc., New Jersey, 1962.
24. P. R. Von Rittinger, *Lehrbuch der Aufbereitungs Kunde*, Ernst and Korn, Berlin, 1867.
25. G. Martin, E. A. Bowes, and F. B. Turner, *Trans. Ceram. Soc.*, **25** (1926) 63.
26. J. N. S. Kwong, J. T. Adams, J. F. Johnson and E. L. Piret, *Chem. Eng. Prog.*, **45** (1949) 508.
27. F. Kick, *Das Gesetz der Proportionalem Widerstand und Seine Anwendung*, Felix, Leipzig, 1885.
28. W. A. Walker, W. K. Lewis, W. H. McAdams and E. K. Gilliland, *Principles of Chemical Engineering*, McGraw Hill, New York, 1937.
29. F. C. Bond, *Trans AIMME*, **193** (1952) 484.
30. P. G. Kihlstedt, *Dechema Zerkleinern Symposion*, 1962, 218.
31. J. A. Holmes, *Trans. Inst. Chem. Eng.*, **35** (1957) 125.
32. P. A. Rehbinder, *Bull. Akad. Sci. USSR, Chem. Series*, (1936) 639.
33. G. C. Lowrison, *Crushing and Grinding*, Butterworths, London, 1975.
34. T. P. Meloy, *Dechema–Monogr.*, **57** (1966) 405.
35. A. M. Gaudin, *Trans. AIMME*, **73** (1926) 253.
36. R. Schuhman, Principles of Comminution, 1 Size Distribution and Surface Calculations; *AIME Tech. Publ. No* 1189 (1940).
37. P. Rosin, and E. Rammler, *J. Inst. Fuel* (*London*), **7** (1933) 29.
38. B. Epstein, *Ind. Eng. Chem., Ind. Edn.*, **40** (1948) 2289.
39. R. L. Brown, *J. Inst. Fuel* (*London*), **14** (1941) 129.
40. S. R. Broadbent, and T. G. Callcott, *J. Inst. Fuel* (*London*), **29** (1956) 191.
41. L. G. Austin, *Powder Tech.*, **5** (1971/2) 1.
42. L. G. Austin, P. T. Luckie and R. R. Klimpel, *Soc. Min. Eng. AIME*, **252** (1972) 87.
43. R. W. Smith, *Min. Eng.*, **13** (4) (1961) 390.
44. F. Zwicky, *Phys. Z.*, **24** (1923) 131.

CHAPTER 7

TECHNICAL ASPECTS OF DISPERSION

D. A. WHEELER
MBW Limited, Stone, Staffordshire, UK

INTRODUCTION

The objective of dispersion and dispersion equipment is to incorporate homogeneously particles into a liquid system. This simple objective is expanded or modified to suit the many industries involved, to the extent that dispersion means quite different things in different industries. In this context dispersion may incorporate the processes of reducing the size of particle agglomerates, of melting or fusing of solids to form a liquid phase prior to the dispersing of solids, and possibly of deaeration to produce an acceptable homogeneous mass within the dispersion equipment.

In the 10 years or so since the production of the first edition of this book there has developed a much better understanding of the processes involved. The fundamental principles of surface chemistry and rheology are better understood and applied to new materials and preparations by product formulators. Similarly the equipment manufacturers have continued to improve conventional machinery and develop new equipment and processes.

This chapter aims to provide an approach to the selection of dispersion equipment by considering the dispersion itself, and the performance or use of various types of dispersion machines. Examples are drawn from several industries to indicate the problems involved and the type of equipment used. In one chapter it is not possible to illustrate effectively all dispersion problems or to give justice to the many equipment manufacturers serving these industries. Tables 7.1 and 7.2 indicate the

TABLE 7.1
INDUSTRIES USING DISPERSION EQUIPMENT

Basic industry	*Types of feed products and process*			
	Solid only	*Solid/ liquid*	*High temperature*	*De-aeration*
Adhesive				
Water and solvent base		Typical		Some
Hot melt and pressure sensitive	Typical	Some	Frequently	
Inks				
Newsprint		Typical		
Litho and gravures		Typical		
Toners	Typical	Some	Usually	
Surface finishes				
Emulsion paints and enamels		Typical		
Powder coatings	Typical		Usually	
Pharmaceutical				
Ointments and creams		Typical	Seldom	Some
Toothpastes		Typical		Some
Plastics and rubber				
Colour master batches	Typical	Some	Usually	Usually
Plastisols		Typical		Some
and many others, including	Paper coatings, Dyestuffs, Foodstuffs, Ceramics			

variety of industries involved and an oversimplification of the types of equipment available. However, the omission even by reference of important dispersion processes and dispersion equipment is unavoidable.

The dispersion concept

The concept of dispersion incorporates a variety of ideas depending upon the experience and industrial background of those involved. When considering dispersion problems, words like mixing, kneading, homogenising, grinding and milling are only occasionally used selectively. Here the idea of dispersion is approached by considering mixing on the macro and micro scales. Macro mixing, also referred to as distributive mixing, is the distribution of the mix ingredients on a large scale. The mix ingredients may be in the correct proportions and uniformly distributed but if a small sample is examined the distribution

TABLE 7.2
EQUIPMENT TYPES AVAILABLE

Machine type	*Variant*
Tank with overdriven agitator	Propeller
	Impeller
	Vibratory
Tank with underdriven agitator	Stator/rotor
Planetary and change-can	Single arm
	Double arm
	Intermeshing
Universal	Z blades
	Dispersion blades
	Heavy duty
Ball and pebble mills	Horizontal
	Vibratory
	Planetary
Bead mills	Vertical
	Horizontal
Roll mills	
Colloid mills	Single surface
	Multi-surface

and many others suitable for batch or continuous operation in a wide variety of construction materials, including combinations of the machine types

may be unsatisfactory. Micro mixing, or dispersive mixing, is the distribution of the mix ingredients at the scale of the small sample. The ideal concept that dispersion is the ultrafine distribution of all ingredients irrespective of the application or the state of aggregation of solids is difficult to live with in practice.

In considering the product quality or performance characteristics of a dispersion process it is assumed that the dispersion be uniform in quality and reproducible. Uniformity and reproducibility are only possible when the qualities and proportions of the raw materials affecting uniformity are maintained and when the materials are processed under controlled conditions. This does not imply that only one controlled formulation processed in one type of equipment will produce the required product. In practice many formulators must adjust their recipes to suit the equipment or raw materials available, and consider the required total performance of the dispersion rather than, for example, only one characteristic of the product, such as particle size distribution of the dispersed solid, or colour, or viscosity.

The dispersion equipment exerts forces on the materials usually in two ways, first on a macro scale, frequently by some form of gross displacement mixing, and secondly by subjecting the macro mixed material to localised high intensity or micro scale mixing. The need for and intensity of each type of mixing action will depend upon the product ingredients, and the wide range of equipment available reflects this necessity to vary the time, intensity and degree of the mixing, grinding or kneading action in the macro and micro mixing scales.

The interrelationship between the product and the dispersion equipment must be appreciated and in many industries individual company success is largely dependent upon understanding these factors. Product quality is the key.

Product quality

There are many methods of quality assessment and test techniques available. In considering reduction of agglomerate size as a measure of quality of the dispersion, quite different results will apparently be achieved if varying testing methods are employed. Figure 7.1 illustrates this graphically.[1] The appropriate testing method must be related to the rate and extent of quality improvement required over the macro mixed

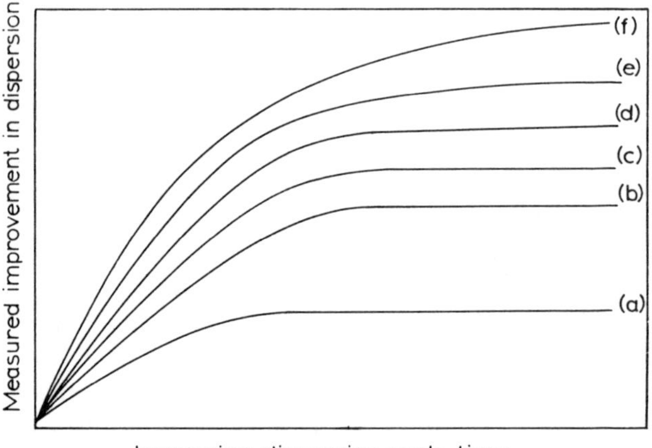

Fig. 7.1. Comparison of testing methods: (a) Hegman seals; (b) photometric sedimentation; (c) optical microscopy; (d) centrifugal sedimentation; (e) colour of finished product; (f) rheology.

material. Further, from time to time it is necessary to repeat an assessment of quality, and the effect of sample ageing or of its being affected by a previous test should be considered. Figure 7.2 shows the phenomenon of shear thinning. The curves were plotted from three separate sets of readings taken on the same sample in the order shown.[1] The changes in rheology are due entirely to the additional shear applied for each determination and non-repeatable data are gathered.

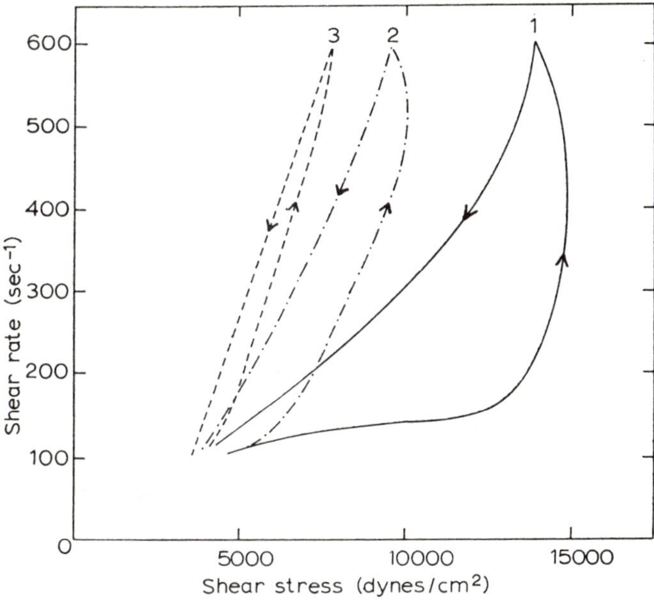

Fig. 7.2. The effect of shear thinning with a set of results obtained using the Ferranti–Shirley viscometer. The curves were plotted from three separate sets of readings taken on the same sample, in the order shown. The changes in rheology are due entirely to the additional shear applied for each determination.

It is tempting to rely on the new instrumentation which provides fast results and can be used by less experienced technicians. For example, the newer electrophotometers can provide data on the quantity and quality of solvents, resins and pigments as well as their relative proportions.[2] Results are reliable but their use may not be appropriate or desirable unless the data provided can be otherwise matched to acceptable duplicate samples. Computer-aided colour matching approaches are established, but the experience of traditional colour matchers is invaluable to supplement these data, particularly in areas of lightfastness processability, etc.

The definition of product quality must be carefully established, and its characteristics and measurement clearly defined, in order to establish the objectives of the dispersion process leading to the selection of the dispersing equipment. The two simple examples of the effect of test method and non-repeatable testing perhaps indicate the difficulties surrounding this requirement.

THE DISPERSION PROCESS

The production of a dispersion of a solid and liquid can be considered as a stepwise operation. The example of incorporating a pigment agglomerate in a viscous liquid vehicle illustrates the process involved in many dispersion applications. Initially the pigment agglomerate must be brought beneath the surface of the vehicle. This first contact usually involves problems associated with the pigment floating on the surface of the vehicle, and even when it is fully submerged entrained air needs to be removed. Pigment particles have air entrained in their clusters and moisture on their surface, which if not fully removed tend to form attractive sites for other pigment particles, and the pigment flocculates. The water may be removed by the conventional flushing process operating on a minute scale.[3] The dispersion process recognises that entrained air and water may need to be removed during the process at a later stage to ensure the production of a stable dispersed product.

In many applications it is desirable to reduce the agglomerates to their primary particle size. The breakdown of the agglomerates increases the surface area to be wetted requiring more vehicle, and so the product viscosity increases. It is generally accepted that primary particles of pigment are not significantly reduced during conventional dispersion processes.

Figure 7.3 illustrates the basic approach and consideration of each step will assist in the selection of a particular formulation, process and process equipment. Most systems will have more than two components but the basic principles remain.

The first step of contacting the solid and liquid together is primarily a material handling problem combined with the need for adequate metering or proportioning of the materials. The pre-blending of dry ingredients may reduce dusting problems and decrease the amount of

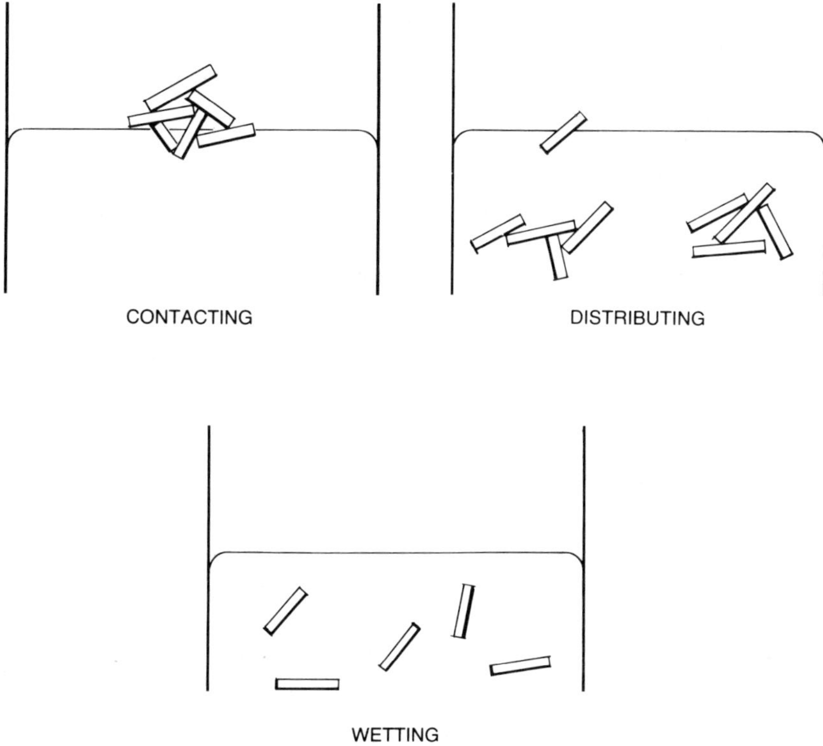

Fig. 7.3. Basic steps to complete wetting and dispersion.

entrained air. The use of agglomerated or compacted materials such as beaded carbon black will improve handling and contacting characteristics, but may present more difficult wetting out and other problems. For example, the use of soft beaded blacks for conventional newsprint inks in place of fluffy blacks does require the dispersing equipment to break up the occasional hard beaded black that is present in these grades.

The distribution step is straightforward and the initial wetting and deaeration of the agglomerates and also partial breakdown of the agglomerates can take place with the initial contacting in a variety of equipment. This pre-mix or mill base is then further processed to complete the wetting out stage to produce the desired dispersion.

As stated previously, advances in surface chemistry (Chapter 1) with the resulting availability and use of different coatings continue to improve the dispersibility of solids in liquids. Many economically attractive dispersions can now be produced in equipment originally conceived and designed to produce mill base and pre-mix type products. This trend is expected to continue to influence the selection and suitability of dispersion equipment.

The last step of breaking down the agglomerates to their primary particles and distributing them uniformly is carried out in a wide variety of equipment. The terms *milling* or *grinding* are applied to this step of the dispersion process, particularly in the paint and ink industries.

In addition to the formulation approach, comminution of the agglomerates can be achieved by one of three basic techniques: (a) impact, (b) shear, and (c) extension, and all equipment available for these processes uses one or more of these techniques.

Dispersion by impact

The destruction of agglomerates and their dispersion by direct impact of the particles is carried out either by directing the mill base as a jet at a static surface, or alternatively by rotating a fast moving rotor through the more slowly moving mill base. Dispersion in high speed impingement mills, frequently referred to as kinetic mills, depends partly upon the particular machine configuration, but primarily on product viscosity.

A typical high speed rotor–stator arrangement is illustrated in Fig. 7.4.

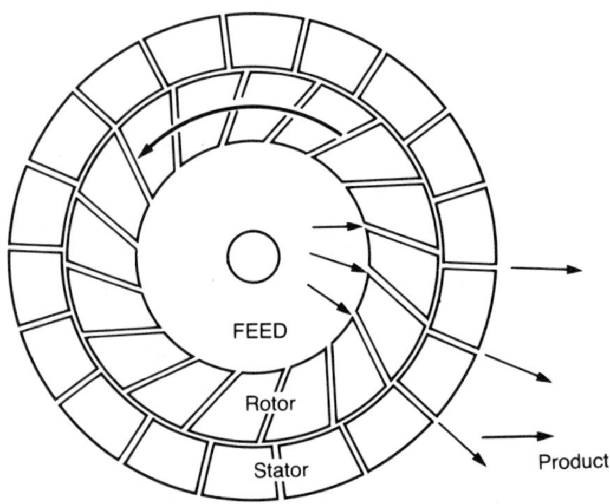

Fig. 7.4. Typical kinetic mill dispersion head assembly.

The feed material enters the centre of a rotor having a series of radial slots or holes. Forces generated by the high speed rotor cause the feed to accelerate down the slots to impinge against the close fitting stator. The stator is also provided with radial slots to give passageway for the feed material to enter the main product stream. Destruction of agglomerates will also occur between the faces of the rotor and stator by shear forces. The impact forces will be substantially reduced as product viscosity increases, due to the viscous drag forces created as the feed material travels along the rotor slots thus reducing particle impact velocities.

Dispersion by impact lends itself, therefore, to low viscosity systems and because the rotor–stator assemblies are usually mounted in tanks, batch systems have developed.

Dispersion by shear

Breakdown of agglomerates by shear depends upon stresses being exerted on the agglomerates by the vehicle as the vehicle flows in a laminar or viscous manner. Conventional laminar flow is illustrated in Fig. 7.5(a) and shown to be generated by the relative movement of parallel planes containing the fluid. In practice regions of laminar flow can be generated without the need for the shearing surfaces to be parallel or moving at constant velocity. The nip of a roll mill and the volume beneath a disc rotor are well-known regions of laminar flow.

The idealised view of the breakdown of an agglomerate by shear is illustrated in Fig. 7.5(b) and (c), and shows the tendency for the agglomerates to form particle streaks. These streaks in turn need to be distributed as more discrete particles. When using shear to break down agglomerates it is usually necessary to combine the high intensity shear zone with a macro mixing facility. Considerable attention to the subject of dispersion in shear zones, particularly in plastic polymer systems, has produced much experimental and theoretical data.[4,5]

Comminution of agglomerates by shear can most readily be achieved in higher viscosity systems.

Dispersion by extension

The knowledge that elongational flow can give a rapid mixing and dispersive action has been known in the confectionery industry for many years. The use of intermeshing double arm ‘pullers’ to disperse colour and flavourings into boiled sugar based confectionery is still widespread. Recent theoretical studies[6] and analysis of conditions in practical mixers[7] indicate that these dispersion techniques have great significance.

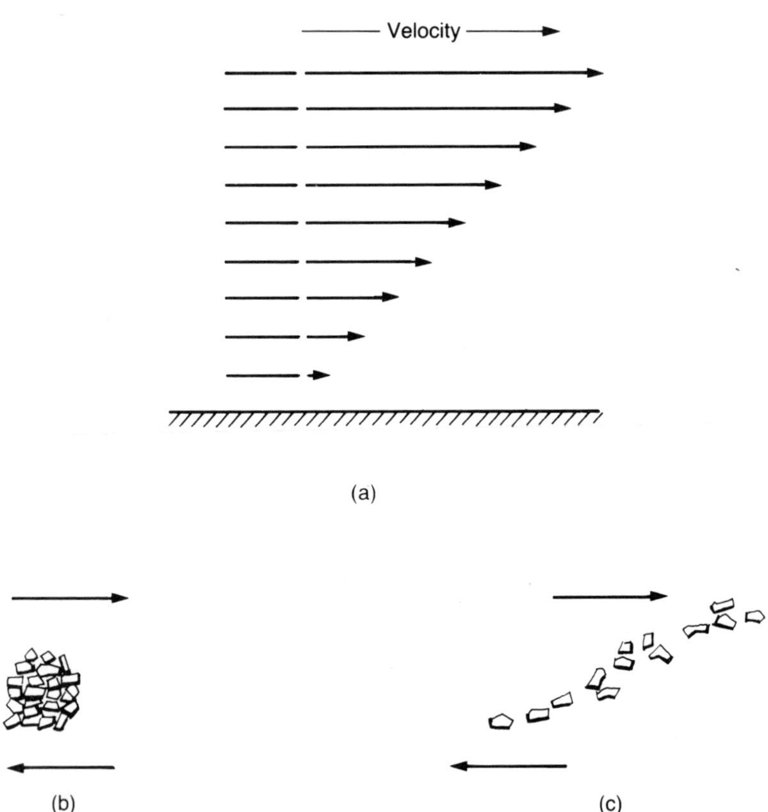

Fig. 7.5. (*a*) *Idealised laminar flow velocity distribution in viscous system;* (*b*) *initial agglomerate in moving viscous system;* (*c*) *partial breakdown of agglomerate before complete distribution of discrete particles.*

They will lead to a better understanding of the mechanism of agglomerate breakdown in both laminar and turbulent flow conditions, leading to the development of new dispersion equipment designed to suit the specific rheological properties of the system.

Instrumentation and control

What to instrument and what to control is, like product quality, a subject demanding much consideration and cannot be isolated from a particular product or process line. Most dispersion equipment is provided with meters indicating pressure, temperature and power absorbed from which the performance of the dispersion equipment can be partially monitored by experienced operators.

Automatic monitoring and control of dispersion processes using readily available microprocessor techniques reduces the need for close operator attendance. Both batch and continuously operating equipment can be tied to programmable control computers available at reasonable cost and having high reliability. Such equipment can control and indicate the state of the dispersion operation from the addition of the feed ingredients to actual product discharge. Systems are in use in the plastic compounding, rubber and food industries, and being developed more fully in others.

Figure 7.6 illustrates a typical microprocessor monitoring and control system. The microprocessor system consists of several units which are installed to suit the particular plant situation. For example, the input keyboard and printer may be isolated from the plant and the local access to information limited to the job programmer and display units. Many other arrangements are possible but all systems are likely to contain the following units:

(a) *Input keyboard and printer*
Used to program or reprogram the computer and to connect with the interface unit to obtain an immediate printed status report. It provides a printed record of the monitored and controlled process.

(b) *Job program selector*
Selects the prepared job program specific to each product giving information to the system such as formulation details, operating sequences, time–temperature, time–power characteristics or other process parameters that need to be matched or controlled. New job programs are relatively easy to produce.

(c) *Micro computer and computer interface*
These units store the instructions and carry out the calculation and comparative steps set out in the programs. The input–output interface selects data as required from the process instruments and the equipment and provides signals to the printer, the display panels and back to the equipment to control or check the status of items such as valves, hopper gates, motors, weigh bins, etc.

(d) *Analog data selector*
Input data from many temperature, pressure and motor current sources are in analog form. The analog data selector on instructions from the interface units provides the information from

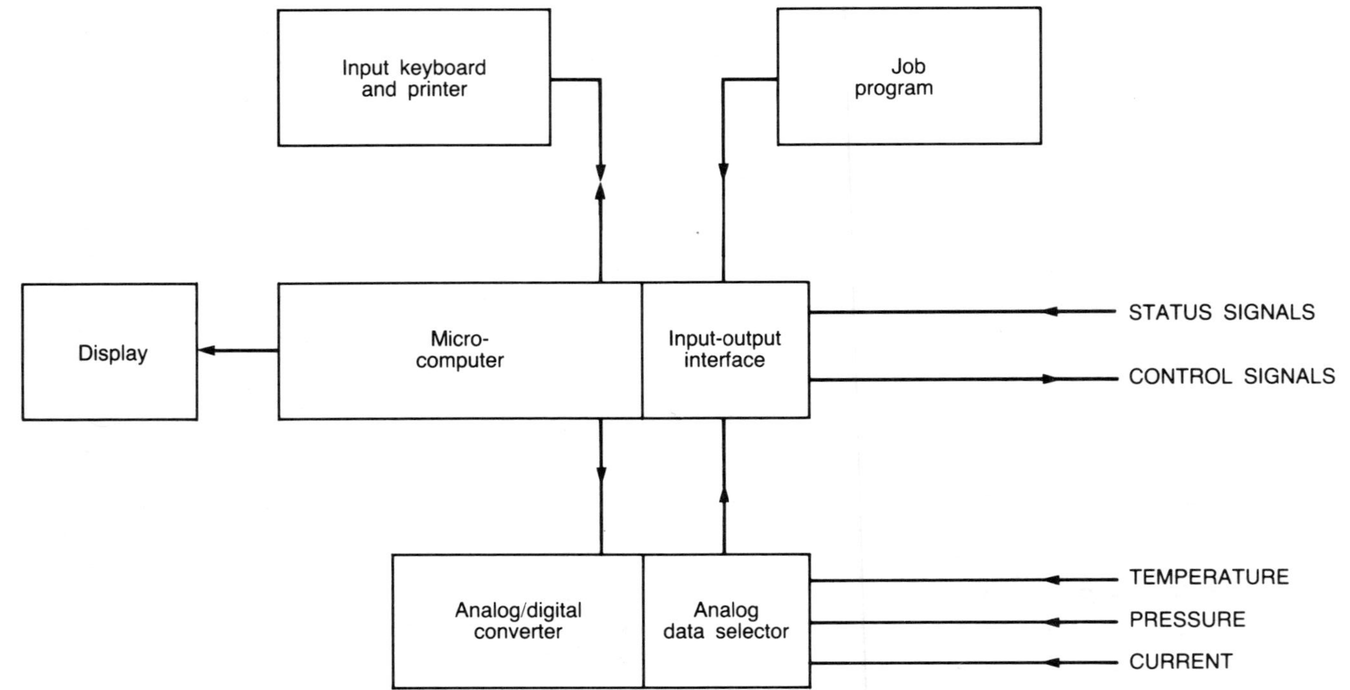

Fig. 7.6. Basic microprocessor monitoring and control system.

the specific input digitally to the computer where any calculations or comparisons with a standard are made and the new data are stored, displayed or printed as pre-programmed.

Such a basic system can have wide applications. In a fully automated plant feed streams for a selected product can be taken in the correct proportions in the selected sequence and fed to the dispersion equipment, opening and closing valves, starting and stopping conveyors and other auxiliary equipment as needed. The system will monitor and compare with a standard, process parameters such as temperature, pressure, energy absorbed and other measurable factors varying with time. This monitoring program is carried out essentially continuously and any deviation outside pre-set limits can be used to warn the operator and/or shut down the process.

The system may be used effectively for quite simple installations. For example, dispersion in a batch mixer having a process operator loading and unloading it could be monitored with success. Automatic loading sequence and possibly load weights might be monitored on a comparative basis by the power absorbed by the mixer, and dispersion end-point monitored by knowing that power absorbed or temperature–time profile characteristics are at all times within limits and that they have been continuously monitored by the computer. The advantages of such systems properly installed and programmed are reduction in operator errors; automated control of the process producing more consistent products; and the ready availability of a printed log, providing process and production data to management leading to improved budgeting, inventory and production control.

EQUIPMENT TYPES

In this section groups of machines have been considered together all having common features showing perhaps empirically how a particular approach to dispersion may have developed. Within each group brief comment is made on typical variations, and application limits with respect to product properties such as viscosity range, heat sensitivity, degree of agglomeration breakdown, etc.

The range of equipment available is immense in terms of design, size range, performance and application. By necessity a somewhat arbitrary approach to selection has been adopted. For those facing the need to specify equipment for new or unfamiliar applications it is recommended they review the equipment and application surveys that have appeared in the press,[8,9] and discuss the problem with equipment users and equipment manufacturers as fully as possible.

Consideration should also be given to the effect of changing legislation concerning the improvement of health and safety standards in manufacturing plants. Discussion documents,[10] leading to possible legislation,[11] or codes of practice indicate that dispersion equipment and processes are not exempt. Both existing and new plant and equipment should be evaluated in the light of these requirements.

Ball and bead mills

In its simplest form the ball mill is a drum with its diameter approximating to its length, and rotating about its horizontal axis. As the mill rotates its charge of materials is subjected to the rotary movements and impacts of the ball, causing agglomerate size reduction and dispersion. This type of equipment (Fig. 7.7) is well known and recent improvements have been restricted mainly to improved drive train arrangements and ball and liner materials of construction.

The grinding and dispersion performance of ball mills is influenced mainly by such factors as

(i) the agglomerate size and hardness,
(ii) weight and volume ratios of pigment and vehicle and the viscosity of the product,
(iii) volume loading in the mills,
(iv) size and specific gravity of the grinding media, and
(v) speed of rotation.

Generally, the higher the density and viscosity of the product the greater should be the size and density of the grinding balls. In addition, larger mills will disperse much more quickly or grind to a finer state than smaller ones, possibly due in part to the greater height of fall increasing grinding pressure between the balls.

Attempts to improve the ball mill rolling and impact action have resulted in the development of planetary ball mills in which the drum rotates about its own axis and around a second axis. Forces of more than ten times those achieved in horizontal mills enable high viscosity products to be ground with larger numbers of small diameter balls, increasing substantially the potentional number of points of contact. For mechanical reasons equipment of this type usually contains several small

Fig. 7.7. A Baker Perkins 3ft (length) × 3ft (diameter), 80 gal(360 litre) ball mill.

pots driven together. Figure 7.8 illustrates a typical four-pot gyratory mill which will grind and disperse an order of magnitude more quickly than conventional horizontal mills. Such mills lend themselves to application in processing highly priced material and for laboratory formulation studies.

Other means of making balls collide than these batch cascading methods have been developed into a range of both batch and continuous equipment. Well established machines in which the product is stirred or vibrated with relatively large balls compete with continuously operating mills using small diameter beads of glass, metallic or ceramic materials. These newer grinding machines have become cheaper and are widely used.

Vibratory mills consist of a vessel given an eccentric or oscillating motion in which the ball medium is forced to oscillate, usually in more than one plane. The contents also tend to rotate slowly as a mass.

Fig. 7.8. A Baker Perkins high speed planetary ball mill.

Though widely employed for dry grinding applications, uses are also being found for it in the viscous products such as inks. Units are available having capacities of up to 750 litres requiring drive motors of about 20 h.p. For these mills a variety of grinding media is available, and for low power demands grinding medium balls typically up to 10 mm or more in diameter are used.

Stirred tank equipment possibly implies non-sophistication but continuing development has proved the value of this approach. A wide variety of designs of tank and agitator combinations, generally called attritors, are available.

Early designs of sand mills had open topped tanks fitted with a screen to retain the beads, and arranged for the product to be fed into the base of the tank and to overflow or be recirculated from the top. Screen plugging and loss of volatiles led to the use of completely sealed units with submerged screens to retain the beads. Rotor and tank designs have been developed to reduce tendencies for the contents to rotate with the rotor. Effective product temperature control during milling is available in some models.

The major factors which affect the efficiency of sand grinding, both from the machine configuration and product type, have been explored.[12] Factors such as disc diameter, tank diameter, pigmentation level and degree of pre-mixing are significant.

The Netzsch–John system illustrated in Fig. 7.9 shows one manufacturer's approach to this type of design. The shaft and tank are arranged for circulating cooling fluids and the agitator pegs rotate between stationary pegs on the sides to prevent the balls and product revolving with the shaft. A narrow annular gap between the rotor and casing ensures separation of grinding beads and product stream as it passes up into the grinding zone to discharge at the top of the machine. It is available in tank sizes of up to 55 litres and depending upon application can have capacities of over 1000 litres/h, demanding motor power ratings of up to 90 kW. Netzsch, like other manufacturers, also offer significantly different rotor and tank configurations in vertical and horizontal arrangements, with different product/bead separation systems and other features designed to improve performance, cleaning, maintenance or applicability to specific products and processes.

The Magmill (Fig. 7.10) represents one of the more highly developed horizontal bead mills and differs from other horizontal bead mills in having a non-cylindrical grinding chamber divided into sections.[13] Each section has different size beads. In the smallest chamber at the feed end large beads are used against the large agglomerates in the feed. As these are reduced the feed flows to the next chamber using smaller beads and so on to the largest chamber, thus grinding the product with a maximum number and weight of fine beads. The rotor is also designed to provide higher energy levels and higher surface speeds in the last chamber, to compensate for the increasing product viscosity and its resistance to the movement of the small beads. Substantially greater volumetric efficiency and lower energy requirements are claimed.

The grinding medium for the impact bead mills is substantially finer than for the vibratory rolls. Vertical sand mills typically use a charge of

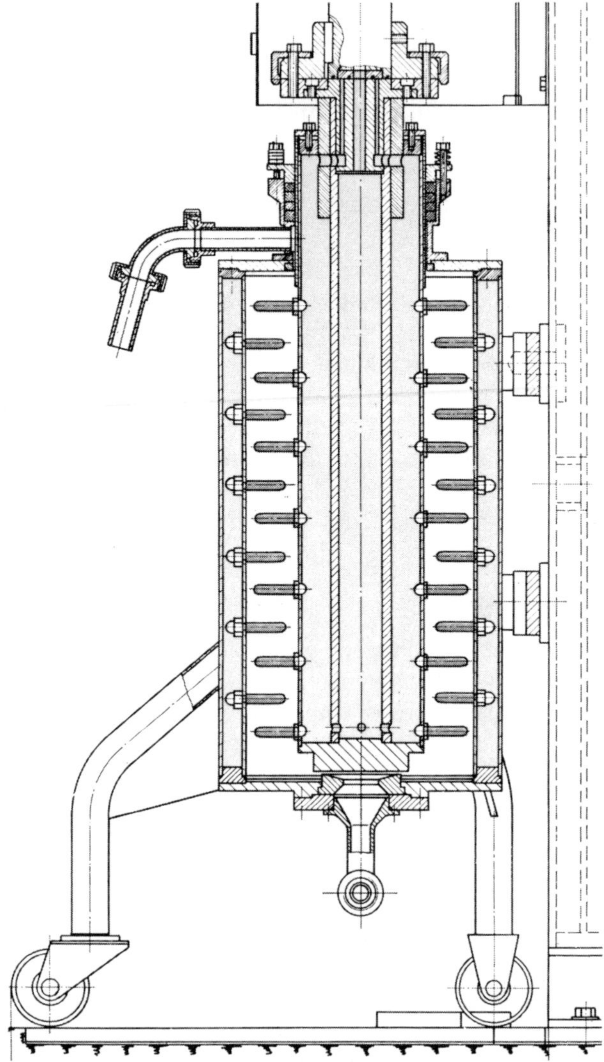

Fig. 7.9. A Netzsch–John sand mill: schematic arrangement.

0·7 mm Ottawa sand, whilst horizontal bead mills use 1 or 2 mm glass beads—some units are in operation using steel or zirconium balls, where their greater mass has advantages.

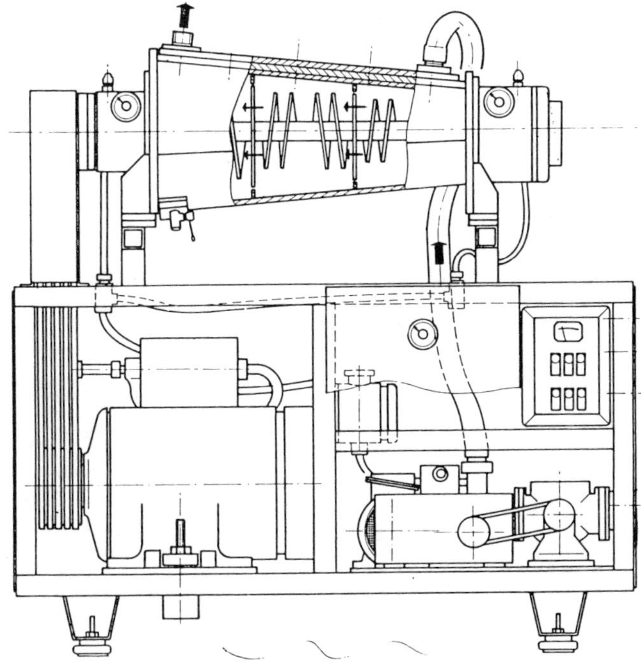

Fig. 7.10. A model P12 Magmill, capacity 40 litres.

The universal mixer

The sigma-bladed Z-shaped twin shaft Universal Mixer is well known and widely used for medium and higher viscosity products. The machine has a rectangular trough curved at the bottom to form two half cylinders and a central saddle. Each blade rotates in its cylindrical section normally towards the saddle on which the material is divided. High shear forces are developed between the blade and trough walls, and the helical nature of the blade form also imparts an axial displacement macro mixing action on the mass of the material being processed. The trough and blades may be jacketed and cored for temperature control. The mix itself can be subjected to vacuum or pressure during the dispersion process and a variety of discharge methods are available.

Where fine dispersions require substantial reduction of agglomerate size in viscous products the dispersion blade is superior to a sigma blade. However, control of batch size viscosity may be necessary to ensure the product is retained in the dispersion zone. A product ram is sometimes required. The double naben blade offers a solution to the difficulties of processing materials that tend to ride over other blade forms, and can also be used to incorporate better the relatively large blocks of feed material into the mix. Wax slabs needed for coating formulations and gum base blocks for chewing gum can be incorporated more quickly in mixers with double naben blades.

Figure 7.11 illustrates the dispersion and double naben blade forms. The method of discharge also varies considerably. Probably the most common is to tip the trough and by reversing the direction of rotation of the blade stiff dough-like products can readily be discharged, while more fluid mixes are poured over a discharge lip with blade rotation. Dis-

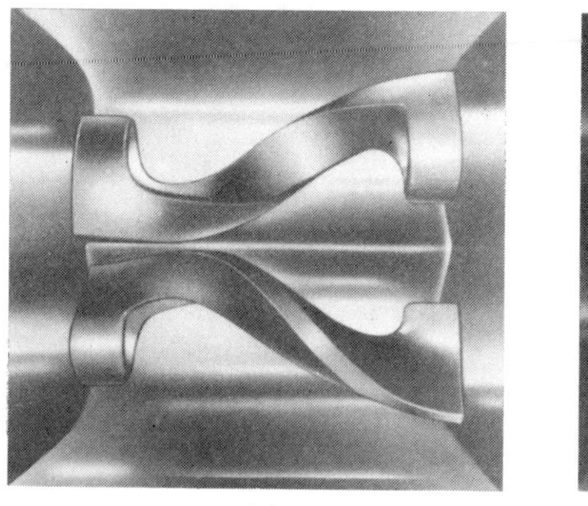

(a)

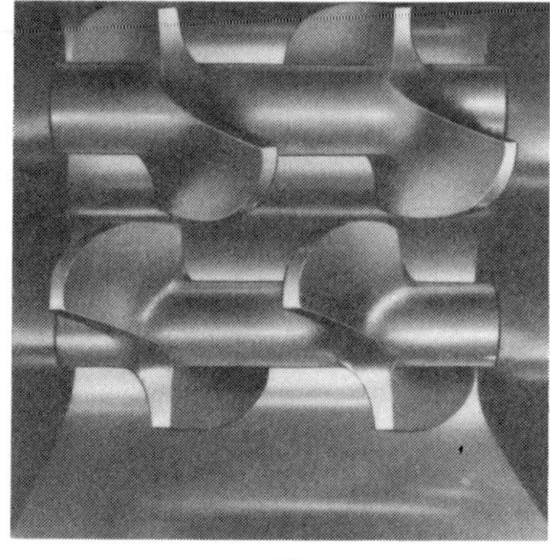

(b)

Fig. 7.11. (a) Single curve dispersion blades; (b) double naben blades.

charge from a fixed trough through valved pipe lines in each trough half cylinder or trough large hinged doors is not unusual. More recently the extruder screw discharge arrangement (Fig. 7.12) has been developed and adapted to a wide size range of trough capacities and input powers.

The flexibility, wide range of possible applications, and specialisation of a particular machine type for a dispersion application is perhaps illustrated by its use as a colour flusher. In this process, more common in the United States, an undried water-wetted pigment as produced by the pigment manufacturer is fed to a version of the universal mixer. To it is added the appropriate varnish, oil or other hydrophobic vehicle required, which displaces the water surrounding the pigment. By tipping the trough the water is decanted off and after several stages the materials are heated under vacuum and the residual water evaporated. At the same

Fig. 7.12. Baker Perkins Guittard M100 Extrumix screw discharge universal mixer.

time as the water is being removed, the pigment is being dispersed in the vehicle to give a readily handled pigmented product for subsequent processing. This product is discharged through valves directly into drums.

Having the performance and mechanical data of a machine it is reasonably straightforward to predict the performance of a larger machine. By measuring the power demand and mix time in a Universal Mixer, the power demand and mix time of a similar machine of different capacity can be estimated. The equipment manufacturer having specific information concerning their range of machines, can usually provide quite detailed information. Figures 7.13 and 7.14 may be used to estimate power requirements and mix times on mixers of the same general type.

Consider the case where a 200 litre working capacity Universal Mixer draws 15 h.p. and requires an 80 min cycle time, excluding loading and discharge times, then the corresponding power demand and mix cycle

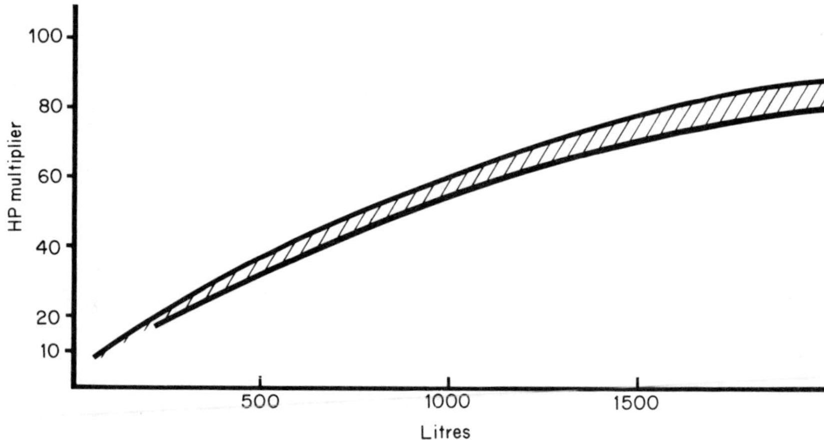

Fig. 7.13. Horsepower multiplying factor.

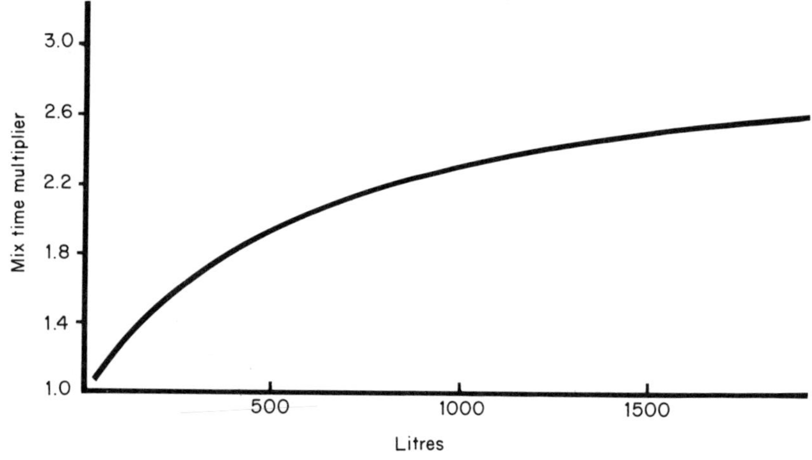

Fig. 7.14. Mix time multiplying factor.

time can be approximated, as follows, for a production machine of 1000 litres working capacity.

Let P_p = power demand for the production machine, P_t = power demand for the test machine, M_p = mix time for the production machine, and M_t = mix time for the test machine.

From Figs. 7.13 and 7.14,

$$H_t = \text{horse power multiplier (test machine)} = 20$$
$$H_p = \text{horse power (production machine)} = 62$$

$$\text{Then } P_p = P_t \frac{H_p}{H_t}$$

$$= 15 \times \frac{62}{20} = 46{\cdot}5 \text{ h.p.}$$

i.e. approximate 50 h.p. will be drawn by the larger machine.

$$T_g = \text{mix time multiplier (test machine)} = 1{\cdot}50$$
$$T_p = \text{mix time multiplier (production machine)} = 2{\cdot}40$$

$$M_p \text{ mix cycle time} = 80 \times \frac{2{\cdot}40}{1{\cdot}50} = 128 \text{ min}$$

i.e. approximately $2\frac{1}{4}$ h will be required to process in the production machine. This time will be extended by the time to add and discharge the product. This method does not include time allowances required for heat transfer, evaporation and other non-mixing parameters.

Roll mills

Roll mills have been with us a long time and though generally their use involves a high labour content having high skills, they are still widespread due to their ability to handle a wide range of product mix, particularly in the intermediate viscosity range such as is required for the dispersion of inks, certain paints and pastes.

Their operation in theory is quite simple; a premixed dispersion base is fed as a thin film between two rollers or between a roller and a bar, and subjected to high hydraulic or mechanically applied pressure between the nip and to high shear forces due to the relative motion of the roll against the bar or second roll. Single, two or three roll versions are common.

In recent years much development work has taken place on these machines both to control and maintain the forces acting on the premix base in the nip, and to control the temperature of the metal surfaces adjacent to the product. This is particularly apparent in machines of the three roll type in which the material from the feed bank rotates, coating the feed roll and central roll and passing on to the apron roll where the dispersed product is removed via an apron. Figure 7.15 typically illustrates this type of equipment.

Manufacturers claim that in the last 10 years or so performance figures of these machines have vastly improved over the earlier models. In particular the later designs offer advantages including the reproducibility

Fig. 7.15. The Torrance 24 in × 12 in streamline triple roll mill with hydraulic pressure control.

of all operating processes, uniformity over the whole operating process and much better control of operating pressures and temperatures, which all provide increased output and a uniformly high quality of end product.

Equipment for dispersion in polymers

The dispersion of fillers, pigments, etc., into polymer systems in which the feed ingredients are solid at room temperature, is of increasing importance. The compounding of rubber and plastic materials prior to final forming is well established and new industries using similar techniques are appearing. The production of powder coatings and new adhesives, and the production of structural foams, use techniques developed from the original compounding lines of the rubber and plastics industries.

As always, the physical characteristics of the materials, particularly particle size and proportions of feed ingredients, are critical. Temperature is, however, the additional factor in such systems in that although in some cases good tumbling or high intensity blending may produce an intimate mixture or pre-mix suitable for direct forming, it is in most cases necessary to raise the temperature of the polymer system to form a viscous liquid in which the solids are dispersed. The dispersion equipment is designed to accept the feed streams, frequently partially pre-mixed, to melt the polymer system, to produce the dispersion, and finally to discharge the dispersed product ready for pre-forming or perhaps shaping into the final end product.

Automatic batch systems, often centred around Banbury type intensive mixers or roll mills, are well established and are not considered here. An expanding range of continuous dispersion equipment is available from a variety of manufacturers.

The simplest form of continuous dispersion equipment is the single screw extruder. Frictional forces, besides transporting the material down the length of the barrel, generally become apparent as heat, resulting in the melting of the polymer system. Dispersion performance of a single screw is generally low but can be enhanced by devices designed to change the shear rate and flow patterns of the melt.

Since friction forces control the conveying action, the heat generation and dispersive flow action, variations in feeding ingredients, particle size and properties significantly effect the throughput and performance of extruders. These limitations must be balanced against the relative cheapness and mechanical simplicity of the equipment.

Multi-screw extruders are available in a wide variety of configurations in which the most flexible in application is probably the twin screw co-rotating intermeshing varieties. Though emphasis is given here to these types of machines, non-intermeshing types such as the Farrell FCM and the multi-screw Eickhoff and Berstorff machines, and many others, have wide applications.

The fundamental difference between single screw extruders and twin screw co-rotating intermeshing machines lies in the positive displacement characteristics of the twin screw machine, which ensure that its conveying capacity is to a large extent only dependent upon being able to accept the feed particles. By suitable screw design, energy inputs and residence time distribution can be controlled, and as static and moving surfaces are arranged to be self-wiped, product hang up within the machine is reduced. The barrels are arranged in various ways to give access to the interior for cleaning, and multiple feed ports may be incorporated and facilities provided to subject the mix to vacuum. The screws themselves are frequently simple shafts on which are fitted kneading blocks, which may be rearranged to suit changing production needs.

These machines also have one other feature not found in most other dispersion equipment, in that provision is made to control the energy input to the system, in many cases actually during the operation. Various techniques are available, the simplest of which is perhaps the variable orifice device (Fig. 7.16) in which tapered blocks on the screw shafts rotate against a tapered section on the barrel. Axial movement of the shafts, relative to the barrel, changes the annular orifice through which the product passes and so controls the degree of fill and residence time of the mix in the machine.

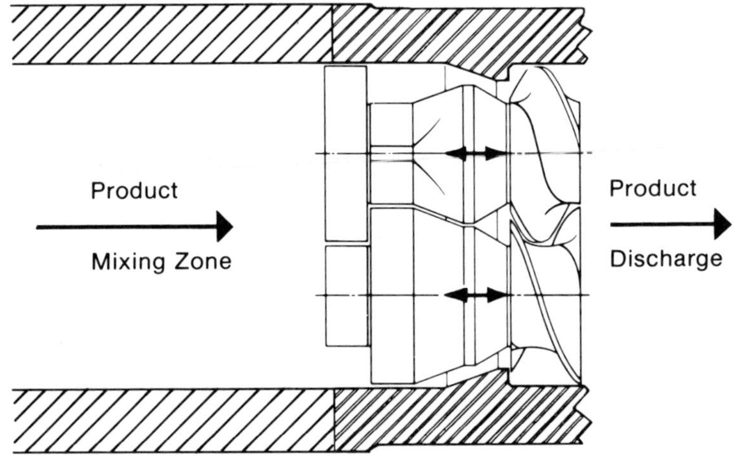

Fig. 7.16. Variable orifice device on twin shaft dispersion machine.

A twin screw machine (Fig. 7.17) operates under production conditions in which a pre-mixed dry feed of polymers, fillers and pigments is simply fed to the continuous mixer, where it is dispersed in the molten state and discharges to a cooling belt for further treatment.

MANUFACTURING PROCESSES

Pharmaceutical and cosmetic product dispersions

The dispersion of various powdery solid ingredients into liquid components to produce smooth creams and pastes is common. Many such cases requiring either batch or continuous manufacturing lines come to mind, and toothpaste manufacture illustrates one of the approaches to such problems. For higher capacity plants producing say over 2000 kg/h, continuous plants are attractive, but for the many

Fig. 7.17. Twin screw compounding extruder ZSK 53 for continuous production of powder coatings.

smaller scale production needs a very wide range of batch equipment is available.

The dozen or so basic ingredients in a typical toothpaste formulation consist of the abrasive materials, binding and stabilising agents, flavours and medical components, carried in a water and glycerol medium. Several of these ingredients total substantially less than 1% of the batch and all must be uniformly dispersed to form the homogeneous dispersion.

The product must be free of air bubbles both to prevent oxidation during storage and to ensure a smooth discharge from the tube. Many formulations are highly thixotropic and final viscosities of 100,000 to 200,000 cP are only achieved after storage for a week or so.

Some feed ingredients in the product are abrasive, which presents wear and seal problems, and the needs to change formulations frequently and to maintain aseptic conditions require that the machine be easily cleaned both of the feed ingredients and the product.

Older plants had open vessels with simple agitators or planetary mixers, which formed fairly good mixes and gave some natural de-aeration but little or no fine homogenisation or dispersion. The poor micro mixing problem was approached both by increasing the thixotropy of the material and by the addition of an in-line colloid mill after the vessel (Fig. 7.18).

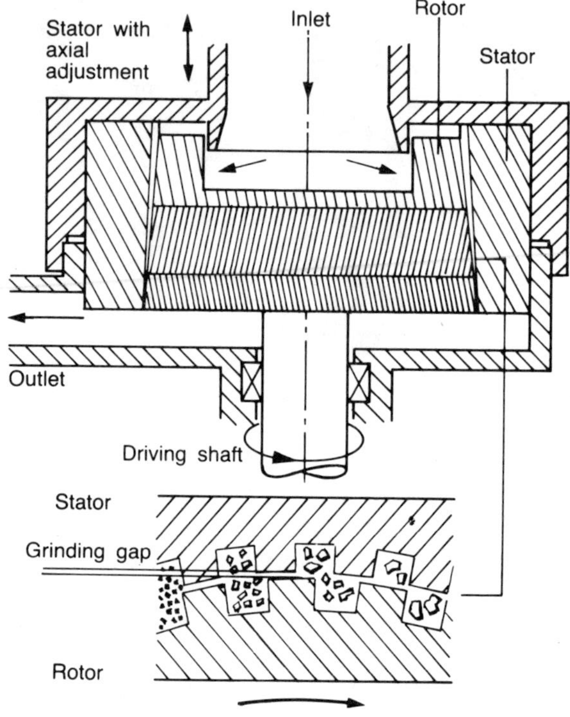

Fig. 7.18. Schematic arrangement of in-line colloid mill.

The Fryma VME (Fig. 7.19) plant consists of a vacuum-tight jacketed working vessel in which is mounted a slow running stirrer system with scrapers, a high speed mixing disc dissolver and an adjustable gap colloid mill as well as a de-aeration system. The various ingredients are pumped or sucked into the vessel by the vacuum system as required, and the mixing and dispersing elements brought into use at the appropriate time during the production cycle time. These plants incorporate product dosing systems and material handling features that reduce cross product contamination and improved cleaning facilities.

Table 7.3 gives some indication of the effect of batch size and product type on processing times in these machines.

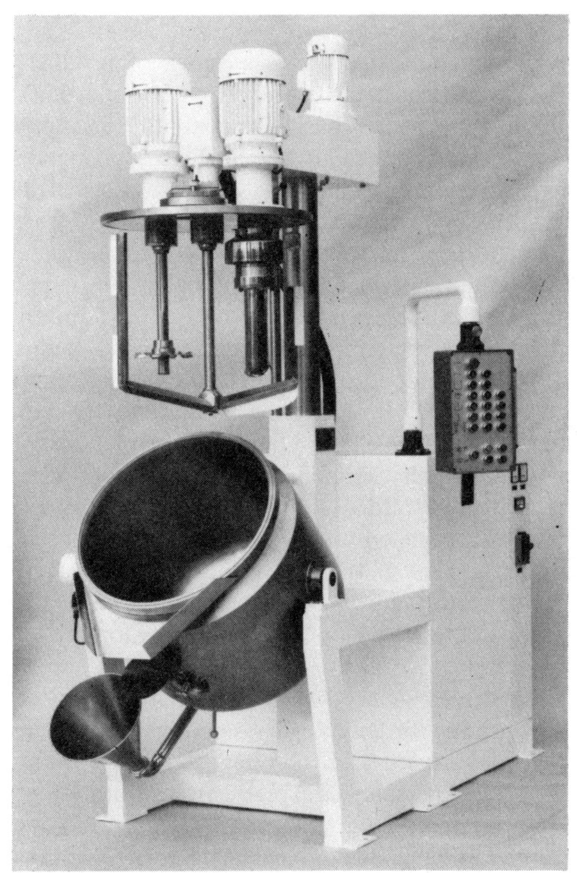

Fig. 7.19. Combination of mixers with tiltable vessel.

TABLE 7.3

STANDARD PROCESSING TIMES (min) IN FRYMA VME MACHINES INCLUDING FEEDING, HEATING, COOLING AND DISCHARGE

Product	*VME 6 to VME 50*	*VME 120 to VME 700*	*VME 1300 to VME 2400*
Toothpaste	40–70	60–90	80–100
Hair shampoo	10–20	10–30	20–40
Salad and spiced sauces (heated to 83°C with steam)	30–60	30–80	40–100
Silicone sealing compounds	40–60	50–70	60– 90

Coatings and adhesives

Most plants endeavour to produce a range of products related by application or raw materials, and many adopt the approach of selecting dispersion equipment on the basis of flexibility in application and production rate. There is much to be gained by having this flexibility; however, the relatively low capital and running costs of high capacity specialised dispersion equipment now available warrant closer examination of the operation of 'one product lines'.

The manufacture of polymer coatings applied to paper, cloth and plastic films to produce such products as wallpaper, floor coverings, and imitation leather is perhaps representative of many plants. Relatively large quantities of dispersions are required in differing resin systems and in a multitude of colours. Typically the pigment is pre-mixed and then fully dispersed in a suitable vehicle. Often the polymer and plasticiser are pre-dispersed and this dispersion is then mixed with the resin. More plasticiser may then be added prior to the final dispersion and de-aeration stages are completed immediately before application. The need for this multi-stage approach, the importance of temperature control and the advantages of a fully automated system have been described using twin shaft pre-mixers followed by three roll mills for the key dispersion equipment.[14]

Figure 7.20 illustrates such a plant. It also illustrates the basic plant needed for the production of powder coatings; in this case the raw materials are dry pre-mixed and then heated to melt the thermosetting resin system in order to disperse the pigments and other ingredients with it, immediately prior to cooling and grinding, to produce a powder later to be electrostatically sprayed and baked to form the finished surface. The temperature at which dispersion can take place is very close to the baking temperature and many resin–hardener systems are highly reactive, severely limiting the time available for the dispersion operation.

Several equipment manufacturers have developed continuous dispersion machines for these materials which, combined with the further development of new polymer hardener systems, have greatly expanded the use of powder coatings. The effect has also been that neither the powder coating product nor the equipment used for its production of a decade or so ago is readily marketable today.

This closer cooperation of raw material suppliers, formulators, producers and equipment manufacturers encourages the more economic manufacture of traditional products and the quicker development of replacement or new products. On the production floor these

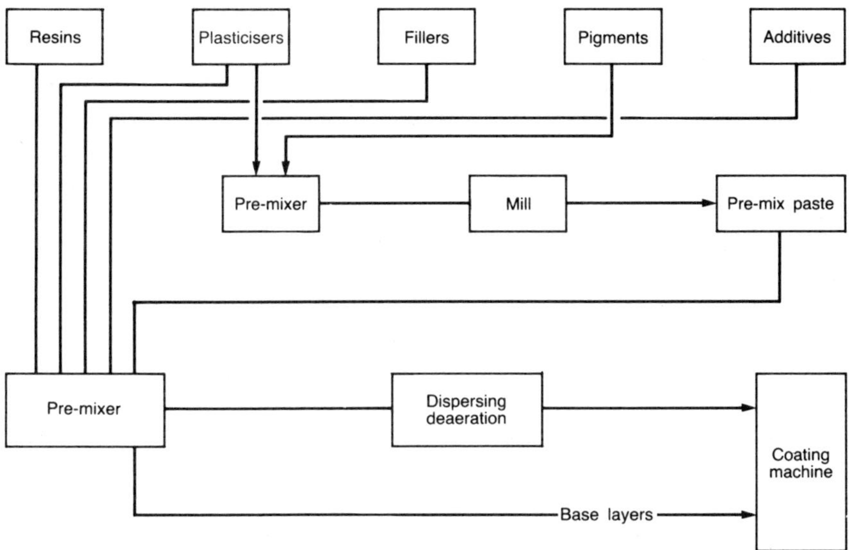

Fig. 7.20. Block diagram of multipurpose coatings and adhesives manufacturing plant.

developments show up as instrumented and fully automated plant and dispersion equipment designed specifically for the application. The paste coating and powder coating lines described are representative of these trends.

The related adhesive industry has rapidly changed from producing rubber cements, starch and casein adhesives, and urea–phenol–formaldehyde based products only, to supplementing them with a multitude of replacement and completely new products. These include two-part epoxy systems, contact adhesives, the pressure sensitive and hot melt types, the acrylics and PVA emulsions, and many others. The dispersions required are generally relatively easy to produce in high speed homogenisers, planetary or sigma-bladed mixers, selected to suit the product viscosity. However, by the nature of the product, its subsequent discharge from the production equipment to coating, filling, chilling or packaging equipment can present problems.

The extruder discharge sigma-bladed mixer (Fig. 7.12) is widely used for the production of many higher viscosity hot melt adhesives. It provides the necessary high power and large surface areas for heat transfer to ensure dispersion of the ingredients in the polymer systems at temperatures low enough to avoid product temperature degradation. On completion the sigma blades carry the material into the extruder screw, which shapes the product as it is forced through a discharge die to the next cooling, prilling, coating or packaging step. The amount of material remaining in the machine is acceptably small and is normally included in the subsequent batch.

Continuous production of such materials is possible in low residence time high shear continuous mixers. These offer the advantages of a lower capital cost than an extruder discharge batch mixer, and require a very small volume of material to be processed at any one time. Offsetting these advantages is the cost of the feed metering and proportioning equipment, and at present the use of continuous production methods is limited. Other industries, notably those producing colour master batches and powder coatings, have solved the problems of handling and formulating a large number of ingredients in different ways, and feeding them to compact continuous dispersion equipment. The same trend can be discerned in the adhesive industry.

Printing inks

These materials, like many coatings and adhesives, are frequently made on the equipment available at the time of the order and so equipment versatility is of prime importance. The typical plant incorporates the raw material or varnish plant. Here the various mineral and vegetable oils, solvents and resins are stored, pre-mixed or dissolved together, filtered and kept as mixed solvents or varnishes. These are pumped to paste mixers, frequently of the change-can type having disc cavitation or stator–rotor mixing heads, where pigments and additives are incorporated to form a paste pre-mix. A semi-finished product is then produced using a ball mill, sand mill or bead mill depending upon the particular ink. Ground product is blended with additional varnish plant products, driers, etc. again using change-can mixers prior to being passed through triple roll mills for final quality corrections, air removal and potting.

The solvent based flexographic and gravure inks are produced in a similar way, except that after dispersing the paste on a roll mill the material on cooling becomes solid, and depending upon the cooling process a chip of varying size is produced. These chips are then redissolved in solvent and additional resin added either by the ink manufacture or printer, to provide the required ink shade and flow characteristics.

For letterpress and offset ink of highest quality produced with frequent colour and product charges ball mills are used to grind the highly concentrated paste economically. Rotary offset and letterpress news inks, having generally lower product quality and a lower viscosity, are frequently ground or dispersed on high capacity bead and shot mills with possible quality and viscosity corrections being made on a roll mill. The solvent based inks require enclosed dispersion machines such as ball mills and bead mills to reduce solvent losses and pollution.

Due to the need to adjust shade, viscosity and other product qualities at the end of most ink production steps, much of the pre-mixing and dispersion equipment can be adapted to manufacture most inks and although the equipment is versatile, production costs are greatly influenced both by the raw material costs and by the working method and equipment selected.

ECONOMICS OF DISPERSION OPERATIONS

This chapter made early reference to the need to establish clear quality standards and assessment methods. Without these the economics of the dispersion operation cannot be assessed. Due to inflation, varying and changing methods of cost accounting and unpredictable material price changes, assessment is further complicated. Similarly the company producing a large variety of related products in relatively small batches must consider the economics of the production processes differently from the one-product one-line producer who is able to consider more deeply the equipment and manning levels, as well as materials and processing costs.

All of these factors cannot be considered here in any depth but comparison of the industry examples already discussed, such as the ink industry with its variety of dispersion equipment adapting formulations to suit sales and production requirements, and the special dispersion lines developed specifically for one product such as those used for powder coating production, may indicate two roads that must be explored in a particular situation.

In considering the actual dispersion operation, material costs are usually the most significant, labour and equipment usage cost being less important. Consideration of material costs should therefore be of next importance after establishing quality standards and assessment methods. In most systems the liquid phase is not changed during the dispersion process, though important exceptions can occur with polymer systems, particularly if they are subjected to high shear rates or prolonged high temperatures. For most liquid systems once acceptable quality standards are met, the economics of the process will depend primarily upon the contract price details and delivery. The solid phase ingredients cannot usually be assessed so simply.

Most dispersions are not complete in that they are commonly sprayed, coated, moulded or shaped onto or into the final commercial product, and it is this commercial product that is subjected to most quality and performance tests. In the case of most coatings and film applications the dispersion must be easy to apply, be of the right colour, provide the required gloss, opacity and texture, and so on. These factors, all defined by the quality standards, are in the main controlled by the solid ingredients in the dispersion and how they are dispersed. A greater degree of dispersion will expose a greater surface area of the dispersed agglomerate to reflected or transmitted light and so change colour and opacity characteristics. If we consider that the degree of dispersion determined by the equipment is optimised for a particular system, then the most significant economic factor again becomes the solid ingredients.

Solid ingredients are usually available from alternative sources, in varying grades and possibly with different surface characteristics, agglomerate size ranges, etc. With so many variables the formulator usually draws upon previous experience to produce the base product, which prior to completion of each batch he will adjust to meet the specification.

In the wet paint industry, computer colour matching systems are being installed. These, in addition to comparing a dry film sample with the standard, can also be programmed to propose what quantity of additional pigments should be added to bring the sample to the colour standard. Typically these programs offer the colour matcher/formulator several alternative groups of additives, and current cost data from which he can best select the group for the application. This final judgement is frequently based upon factors not measured by the computer. A particular pigment system may fade quickly or it may be necessary to control flocculation to prevent premature pigment settlement before use. With this additional knowledge of product quality he can better select the most economically attractive material.

These ever-increasing numbers of raw materials available to the formulator, designed to enhance factors such as viscosity control, colour, gloss, sedimentation and dispersion, make the economic judgement of a particular formulation and dispersion process most difficult. However, with increased instrumentation and microprocessor equipment control combined with computer-aided services such as the colour matcher, and with raw material stock situation and cost data immediately available, the likelihood of selecting the most economic or cost effective formulation and process is enhanced.

ACKNOWLEDGEMENTS

Photographs and equipment data are by courtesy of;
Baker Perkins Chemical Machinery Ltd, Hanley, Stoke-on-Trent, UK
Baker Perkins Guittard SA, Chelles, Paris, France
Fryma Meschinen AG, Rheinfelden, Switzerland
M. and M. Process Equipment Ltd, Hemel Hempstead, Herts, UK
Netzsch-Feinmahltechnik GmbH, D8672 Seld, Germany
Werner and Pfleiderer (UK) Ltd, Stockport, UK
Winkworth Machinery Ltd, Staines, Surrey, UK

REFERENCES

1. I. R. Sheppard, *Dispersion of Powders in Liquids*, 2nd edn., Ed. G. D. Parfitt, Applied Science, London, 1973, Chapter 6.
2. Anon., *Polym. Paint Colour J.*, **5** (1979) 538.
3. T. A. Langstroth, *Colour Eng.*, 7/**8** (1968) 40.
4. D. M. Bigg, *Polym. Eng. Sci.*, **15** (1975) 684.
5. Debal Gupta, *Proc. Soc. Plast. Eng.*, **5** (1979) 311.
6. D. C. H. Cheng, *Proc. 3rd Europ. Conf. on Mixing*, **1** (1979) 73.
7. K. Y. Ng and L. Erwin, *Proc. Soc. Plast. Eng.*, **5** (1979) 241.
8. A. Quillen, *Chem. Eng. Albany*, **6** (1954) 178.
9. J. W. Oldshoe and D. B. Todd, *Chemical Engineers' Handbook*, Ed. R. H. Perry and C. H. Chilton, McGraw-Hill, Tokyo, 1973, Chapter 19.
10. D. J. Tookey, *Proc. 3rd Europ. Conf. on Mixing*, **1** (1979) 269.
11. Anon., Code No. B28.1, Am. Nat. Std. Inst., New York.
12. W. Carr and A. Kelly, *J. Oil Colour Chem. Assoc.*, **62** (1979) 183.
13. K. H. Meller and J. Brenot, UK Patent 1531957, 30 March 1977.
14. U. Wilke, *Polym. Paint Colour J.*, **10** (1978) 921.

CHAPTER 8

ASSESSMENT OF THE STATE OF DISPERSION

S. G. Lawrence
Ciba-Geigy Plastics and Additives Company, Paisley, UK

INTRODUCTION

A dispersion is defined as a two-phase system in which one phase, called the dispersed phase, is distributed as small particles throughout the second phase, called the continuous phase. When the size of the small particles is generally below 1 μm, such systems are referred to as colloidal dispersions and occur on a wide scale in everyday life, as exemplified by paints and inks where the dispersed phase is pigment distributed throughout some suitable resin system, some agricultural and pharmaceutical preparations, and the silver halide sols used in photography.

In the process of preparing, for example, a pigment dispersion, the dry powder is incorporated into the liquid medium in such a way that the individual particles of the powder become separated from each other, or form small clusters reasonably evenly distributed throughout the liquid medium. The final size to which the original clusters in the powder are broken down—referred to as the level or degree of dispersion—depends upon a variety of factors, e.g. the energy put into the system, the forces of adhesion between the particles in the powder, the interaction between pigment and medium, etc. On removal of the dispersion energy, this final level of dispersion need not necessarily be maintained. As a consequence of the attractive force which exists between them, there is a tendency for the dispersed particles to come together to form larger units[1]—flocculates—unless there is some mechanism operating which prevents this. The term dispersion stability defines the resistance of the system to form these flocculates. The various mechanisms which can prevent flocculation from occurring are described in Chapter 1.

The technological properties of dispersions depend markedly both on the size of the dispersed particles and on the stability of the system. Thus, as will be described in more detail later, the colour strength of a pigment dispersion increases as the particle size decreases,[2] and the rheological or flow properties of liquid paints and inks depend critically on both the size of the dispersed phase and on the stability of the system as a whole.[3]

When assessing the state of dispersion we are interested primarily in the size of the dispersed particles. However, in an unstable system these particles are coming together to form flocculates, so the size of the dispersed phase is changing with time. Consequently, it becomes difficult to measure the particle size of such systems—indeed, one has to think carefully about what one is actually interested in measuring. It could be, for example, the size of the dispersed entities immediately after cessation of application of the dispersion energy, or the size of the dispersed phase that exists when no further changes can take place, e.g. in a stoved or dried paint film. Conversely, before undertaking to determine the particle size of a dispersion, some idea of its stability must be obtained. In addition, it should be borne in mind that within a practical dispersion there exists a whole spectrum of particle sizes, from 0·01 μm at the smallest end to perhaps 5 μm at the upper end, i.e. from small individual pigment particles to large undisrupted clusters.

THE IMPORTANCE OF THE LEVEL OF DISPERSION

Many technical properties of a dispersion are dependent upon the degree of dispersion attained. With increasing milling time, or with increased energy input, it is usual for the colour strength of a pigment dispersion to increase. The reason for this is to be found in the decrease in particle size of the dispersed phase. As was first shown some 70 years ago by Mie,[4] the optical properties of a dispersion depend *inter alia* on the size of the dispersed phase. Mie's theory applies to the behaviour of spheres of all sizes and colours and it was applied to pigments initially by Brockes[5] and Chromey.[6] Using this theory, Brockes was able to show that in order to achieve their maximum colour strength, pigments must be dispersed to below a certain minimum size. The actual size depends upon their optical properties (refractive index of the pigment relative to the refractive index of the medium); for strongly absorbing pigments the size must be below 0·05 μm, while for moderately strong pigments it must be below 0·1 μm—both of which predictions agreed well with practical experience. Since this work, it has become generally recognised that the Mie theory may be used to determine the qualitative relationship between colour strength and particle size.[2,7] The general shape of the curves obtained are shown in Fig. 8.1.

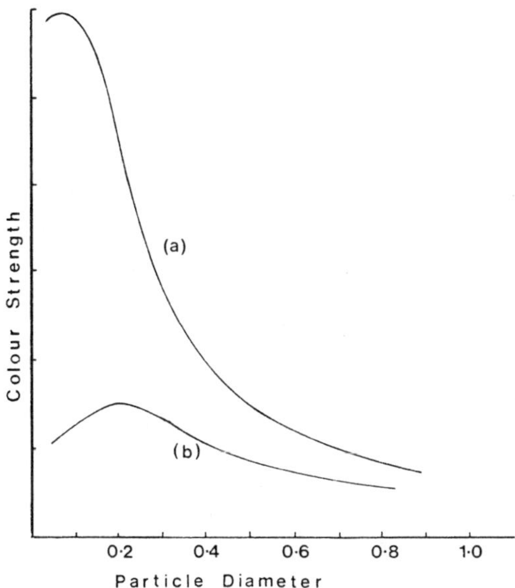

Fig. 8.1. Variation of colour strength as a function of particle size (μm): (a) strongly absorbing pigment; (b) weakly absorbing pigment.

The opacity of a dispersion also undergoes marked changes as the particle size decreases and this trend, too, can be predicted by Mie theory. Initially, as the size is reduced from the micrometre range, there is an increase in opacity which reaches a maximum around 0·2 μm, whilst below this as the particle size is further decreased, an increase in transparency is observed.

From a colouristic point of view, not only the size but also the shape of the dispersed particles is important. Using a model system of optically isotropic particles, Felder[8] has shown that for ellipsoidal particles it is the minimum dimension that is the important one for determining the colour strength. However, organic pigments are rarely optically isotropic[9] and it has been shown that the colour strength of β-copper phthalocyanine pigments dispersed in a paint system increases as the isometric particles are replaced by acicular particles of equivalent size, and in addition the colour becomes redder.[10] This has been attributed to the optical anisotropy of the pigment crystals, the colour becoming more red in shade when viewed along the main axis of the acicular crystals.

It is not only the optical properties of pigmented systems that change as a function of particle size. As the level of subdivision of the dispersed phase increases, the flow properties of the resultant dispersion alter[11] and the gloss of pigmented films generally increases,[12,13] though as Medinger[14] has reported, gloss is also influenced by particle shape and small well dispersed needles can also give rise to distortion in the surface of paint films.

Oversize poorly dispersed particles generally adversely effect the application properties of dispersions. Large particles can disturb the surfaces of paint and ink films leading to poor gloss, they can cause breakage of viscose fibres during spraying and they can cause damage to printing plates.

The above examples have all been taken from the coatings industry with which the author is most familiar. However, within other industries achieving and controlling the required state of dispersion is also of prime importance. In the pharmaceutical industry, uniformity of dispersion and long-term stability are required in order to attain accurate dose rates, and considerable attention has been paid to the degree of dispersion of pharmaceutical materials in relation to the assimilation of drugs.

It is clear, then, that the applicational behaviour of dispersions is a function of the state of dispersion and some means of its assessment is required. On one hand it may be necessary to determine the size as a quality control parameter, or on the other hand it may be required in order to carry out more fundamental research relating size to performance level.

THE NATURE OF POWDERS

Examination of pigment powders using both transmission and scanning electron microscopy shows that they are composed of small, solid particles of a variety of shapes, e.g. spheres, cubes, needles, etc., and that these particles are present either as single, discrete entities or more usually, are gathered together to form some more-or-less coherent assemblies of various sizes. The individual particles generally lie within the size range 0·02–1 μm across, whilst there is—at least from a microscopy view—no limit to the upper size of the clusters. Sieve analysis of the powders indicates that the clusters of individual particles may range from 50 to several hundred micrometres across. The size and the shape of the individual particles may be studied using the transmission electron microscope, and Fig. 8.2 shows micrographs of typical pigments, whilst

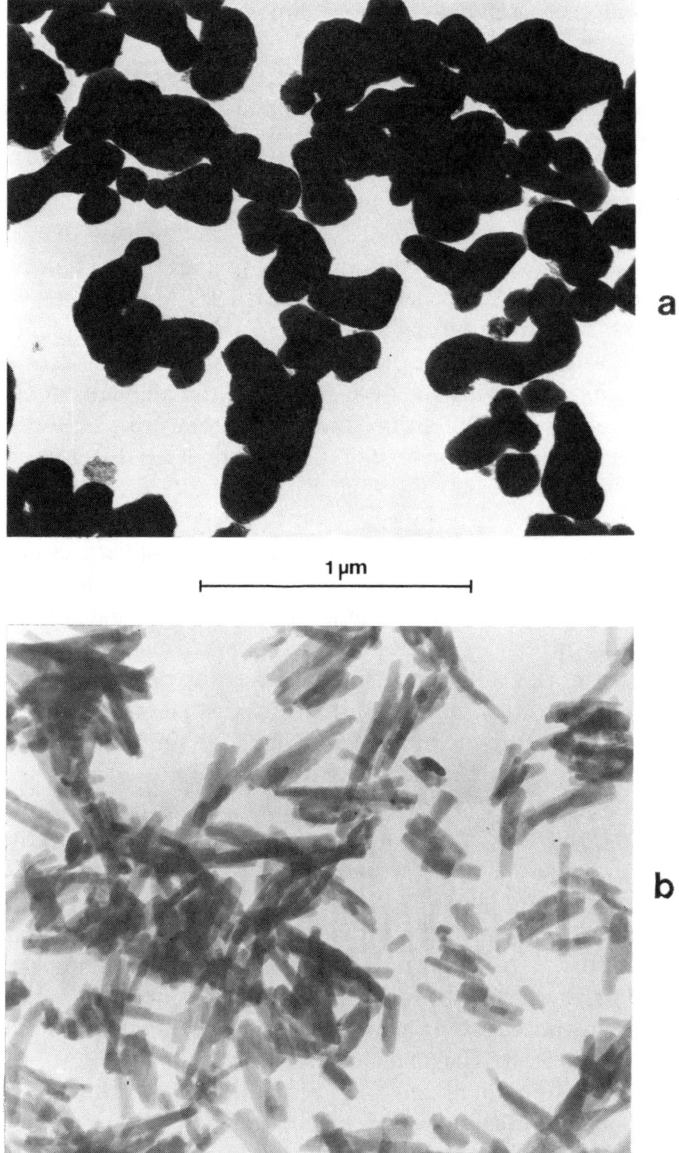

Fig. 8.2. Electron micrographs of pigment powders: (a) CI Pigment White 6; (b) CI Pigment Red 53.

the nature of the assemblies is more conveniently studied using gas adsorption techniques.[15-18]

Clusters of particles are formed during the preparation process. For example, some organic pigments are formed by precipitation from an aqueous medium, the newly formed particles having a high surface area, and when two such particles collide they tend to adhere to each other with a consequent reduction in surface energy. During any subsequent heat treatment, they become fused together as a result of molecular movement and flow. Again, during their preparation, pigments may undergo a dry grinding process when particles with active sites are produced which show a strong tendency to adhere. Hydrophilic aggregation is another cause of cluster formation, in which as the aqueous medium is removed during the drying process, the pigments are brought closer together by surface tension forces until they are so close that the London–van der Waals forces can operate to keep them in contact. Additionally between such faces, unremoved salts formed during the reaction may be present to form a solid bridge.

A powder, then, consists of assemblies of particles of various degrees of complexity which were formed during the preparation and one distinguishes conceptually between primary particles, aggregates and agglomerates.[19] The primary particles themselves are considered to be made up of one or more crystallites which are regions of continuous crystal lattice. For the case of copper phthalocyanine pigments, the areas of continuous lattice structure and the way in which they are fused together to give the primary particles have been studied using high resolution electron microscopy.[20,21] Figure 8.3 shows the lattice planes and the crystallite regions of α-copper phthalocyanine imaged under the high-resolution electron microscope and Fig. 8.4 an idealised model of a collection of primary particles. Groups of these primary particles, in which the constituent particles are bound together by face-to-face contact, are referred to as aggregates (Fig. 8.5(*a*)), while agglomerates are considered to be clusters of aggregates and/or primary particles which touch only at the edges or corners forming a looser or more open structure (Fig. 8.5(*b*)). The main difference between these two sorts of clusters is the strength of bonding between the constituent elements, aggregates being so firmly bound together that they tend to behave as single crystals and pass through the dispersion process undispersed, whereas agglomerates are disrupted during the process. Such a definition is somewhat artificial as it is strongly dependent upon the energy put into the dispersion process.

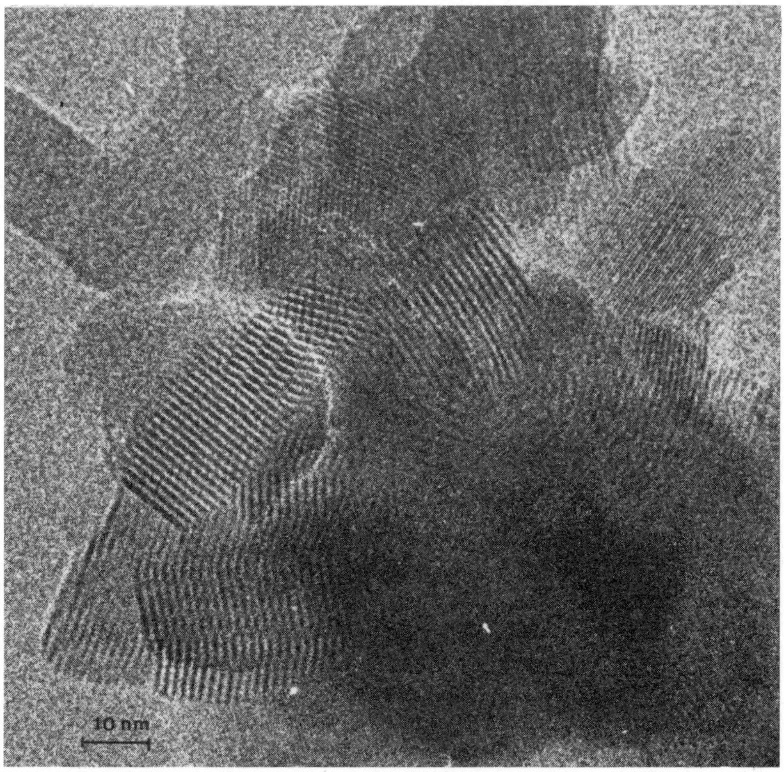

Fig. 8.3. High resolution electron micrograph of α-CuPc showing lattice planes and crystallite regions. (Courtesy of Dr. J. Fryer, Glasgow University)

THE DISPERSION PROCESS

In the dispersion process the dry pigment powder is incorporated into the appropriate application medium, e.g. ink, paint, molten plastic. The original assemblies of particles are broken down and their constituent elements distributed statistically throughout the vehicle. The process of dispersion and the underlying physicochemical principles are described in Chapter 1.

The actual degree to which the clusters are broken down will depend upon a number of factors, e.g. the energy input, the vehicle in question and the strength of adhesion of the aggregates and agglomerates.

In the case of paints and inks, these liquid dispersions are applied to a

Fig. 8.4. Primary particles of a crystalline pigment. (*Note: in this and in subsequent figures, the primary particles are shown in an idealised form.*)

substrate to give, eventually, a dry pigmented film. During the film-forming process a number of forces are at work which can bring about a spatial redistribution of the pigment particles; for example, a pigmented

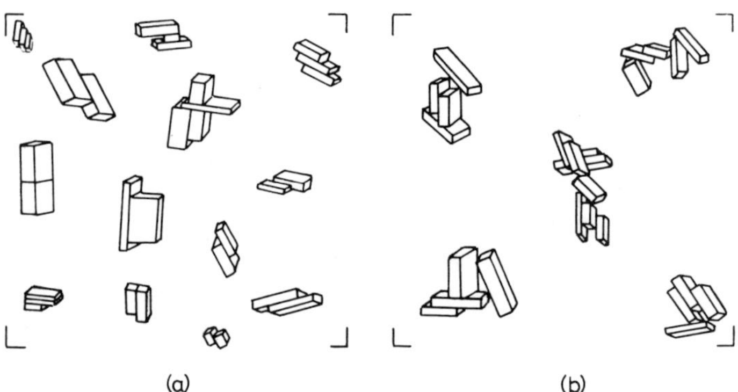

Fig. 8.5. (*a*) *Aggregates, in which the primary particles are joined at crystal faces.* (*b*) *Agglomerates, in which a looser structure is formed, joined at the edges or corners of the primary particles.*

paint system may be stable in the wet state but during drying as solvents evaporate it can become unstable so the particles come together to form loose flocculates or under extreme conditions may form more tightly packed clusters leading to an increase in particle size.

So methods of the assessment of the degree of dispersion are required during both the dispersion process when the system is fluid and also after application when the system is dry (as in the case of paints and inks), or is no longer molten (e.g. PVC), when no further change in the level of subdivision of the particles can take place.

QUALITY CONTROL TECHNIQUES

For quality control the methods for the assessment of the level of dispersion are not required to give absolute values either in terms of a mean particle size or particle size distribution. Instead, generally some characteristic of the test system, which from experience is related to the overall level of dispersion, is compared with some characteristic of a standard dispersion. The methods used must be quick (capable of giving a result in minutes rather hours), robust and capable of being used by non-skilled personnel. The characteristics often used are fineness of grind, colour strength, opacity and flow.

Fineness of grind

A characteristic widely used within the paint and ink industry is the Fineness of Grind (FOG) reading, which is determined on a Hegman or Braive gauge. The gauge consists of a hardened steel block into which a wedge-shaped groove has been cut. The groove, which may be 100 μm deep at one end, tapers uniformly along its length to zero depth at the other end. The depth of the groove is indicated by a linear scale marked alongside. Conventionally, the scale runs from 0 to 8, with 0 corresponding to 100 μm, though gauges are also available in which the depth of the groove is marked directly in micrometres. In practice, a sample of the dispersion is placed at the deep end of the groove and drawn down with a scraper, the long edges of which are rounded to a radius of approximately 0·254 mm, held perpendicularly to the surface of the groove. Immediately afterwards, the film of material is viewed at grazing incidence and that level at which five to ten particles, within an approximately 3 mm band, appear through the surface is recorded.

This method gives no information regarding the actual degree of dispersion; it gives only a useful idea of the number and size of the large, undispersed particles. Nevertheless, as the time of milling increases, the number of clusters decreases and it is assumed that the general degree of dispersion of the finer particles has also been improved.

The use of the Hegman gauge as a means of measuring the oversize particles has been strongly criticised by Schwegmann,[22] and Valentine[23] has pointed out that the size of the aggregates need not correspond exactly to the depth of the gauge. During the drawing down process a certain amount of shear force will be experienced by the particles as they are drawn along by the blade which could cause some extra dispersion.

In spite of all these limitations, the Hegman gauge is widely used in production control and the agreement between trained observers is remarkably good. Part of the problem concerning the use of the Hegman gauge arises from the different 'language' that the observers use. Though the British Standard specifies that the gauge reading is to be taken at the point where the film surface starts to become speckled, certain users take the commencement of the tramlines as the FOG reading.

Though the groove tapers uniformly to zero depth, it is now generally accepted that the limit of sensitivity is around 10 μm, and that differences in the level of dispersion below this value critically influence the application behaviour of pigment dispersions.

Light microscopy

The level of dispersion in the sub-Hegman region can relatively easily be examined using the light microscope; indeed, with the microscope an impression of the size distribution is obtained as both undispersed clusters as well as the fines can be observed. However, it must always be borne in mind that the smallest particle visible under the light microscope is around 0·5 μm. A drop of the dispersion is placed on a glass slide and examined in transmitted light. Rather than carrying out a laborious sizing and counting operation, Gall and Kaluza[24] suggest that the level of dispersion is assessed by comparing the overall appearance of the sample under the microscope with the appearance of standard dispersions.

Colouristic properties

As has been mentioned earlier the colouristic properties of a pigment are markedly dependent upon the degree of dispersion. Generally it is found that on dispersing a pigment into the system, there is initially a fairly rapid change in the colouristic properties, as the weakly adhering agglomerates are broken down, which is followed by a more gentle change as the stronger clusters are disrupted, approaching asymptotically a limiting value for a given system and a given dispersing technique. Even at the limiting value there are still clusters present, but to break these down an uneconomic input of energy would be required.

One means of controlling the degree of dispersion is colour strength, although—depending on the colour coordinate of the pigment in the colour solid—shade and purity are altered as well. Thus, comparison of the colour strength of a sample with a standard, after a definite dispersion process, gives a practically important idea of the level of dispersion. Additionally, by following the strength development curves as a function of time, an idea of the dispersibility of the pigment may be obtained (Fig. 8.6).

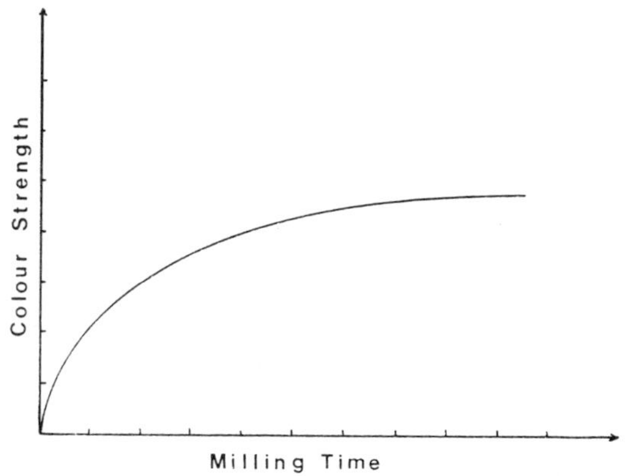

Fig. 8.6. Increase in colour strength as a function of time.

For colour strength assessment to be carried out, the coloured pigment is reduced with a white, usually titanium dioxide, system. In the case of less viscous systems, such as solvent containing inks or paints, it is possible that during the drying period differential separation of the coloured and white pigments can occur, i.e. white or coloured flooding (vertical separation), flotation (horizontal separation) or flocculation. Such effects are more likely to cause problems if pigments with different surface treatments are being compared.

Miscellaneous

In principle, comparison of any property of the system could be used to assess the level of dispersion e.g. gloss, transparency, the number of 'nibs' or oversize particles which disrupt the surface of a paint or spoil the overall visual appearance of a pigmented plastic sheet. Simple viscosity measurements carried out using a flow cup[25] or even a more sophisticated cone and plate viscometer[26] can also give useful results though their interpretation is complicated for unstable systems.

RESEARCH METHODS

For research studies, the level of dispersion is determined by methods which give the distribution of particle sizes or perhaps an average particle size in micrometres. For pigment particles which are in general geometrically anisotropic some consideration has to be given to what is meant by particle size. The size of a sphere or a circle may be defined by its radius or diameter, but for needle shaped particles the situation is more complex. Microscopical methods usually permit the shape of the particle to be seen and so allow some choice in the matter of deciding which measurement to use to define the particle size. Most methods do not allow such a choice, for example, sedimentation methods give a spherical equivalent diameter, that is the diameter of a sphere of the same density which would sediment at the same rate as the particle being studied. The problem of deciding the appropriate way to define the size of irregular particles is fairly complex and is dealt with in the literature.[27,28]

Various methods of particle size analysis may be applied to pigment dispersions, but there is no method which can be used directly for a concentrated dispersion without danger of modifying the system. In many cases the samples have often to be extensively diluted as in the case of optical measurements, or when using the disc centrifuge, with the danger that flocculation of the dispersion could occur leading to erroneous results. It is advisable to dilute in several steps using a resin solution rather than a pure solvent. The samples examined are usually rather small with thc possibility that they may not be totally representative of the whole, and this is especially a problem, as will be described in more detail later, when looking at ultra-thin sections of dried application media under the electron microscope.

These more-or-less basic difficulties associated with particle size analysis should not be over-emphasised; rather the above mentioned factors should be borne in mind when carrying out any size analysis.

Rheological properties

As has been mentioned earlier, with increasing milling time the flow properties of a dispersion change, so conversely it ought to be possible to get some indication of the particle size of a dispersion from the measurement of its flow properties. Unfortunately 'flow' is not solely dependent upon particle size. Frisch and Simha[29] state that the viscosity of suspensions is affected by the following factors:

(i) Shape, size and mass of the suspended particles.
(ii) Size distribution of the particles.
(iii) Volume actually occupied by the particles in suspension.
(iv) Ease of flexibility and ease of deformation of the particles.
(v) Thermodynamic conditions of the system.
(vi) Presence of electric charges in the system.
(vii) Concentration.

Conventionally, as is well known, the flow properties of liquids such as water or oils are characterised by their viscosity, η, which can be thought of as the resistance of a liquid to flow or internal friction. Suppose a liquid is confined between two plates, one fixed and the other [illegible] of area A and separated by a distance x. A constant force [illegible] applied to the moveable plane and after acceleration it a[illegible] constant velocity V. The liquid between the planes is shear[illegible] velocity gradient is set up, intermediate layers moving with int[illegible] velocities. The force F between two adjacent layers of liquid i[illegible]

$$F = \eta A \frac{dv}{dx}$$

where dv/dx is the velocity gradient.

As the shearing force or stress is increased so the rate of flow increases and for liquid media a plot of rate of shear against shearing stress yields a straight line passing through the origin, the slope of the line being the coefficient of viscosity. The rheological behaviour of such systems is described as *Newtonian*. However, for many suspensions there is no such linear relation between the rate of shear and the shearing stress, and the rheological behaviour is said to be non-Newtonian or abnormal. Fig. 8.7 shows the different sorts of basic flow behaviour that can occur.

A Newtonian liquid flows under any applied force, however small. Materials exhibiting *plastic flow*, however, require a certain minimum

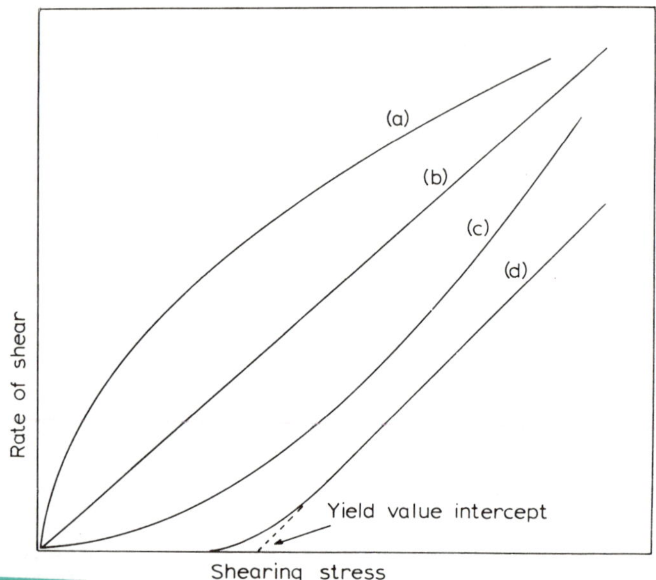

type of rheological behaviour of dispersions: (*a*) *dilatant;* (*b*) *Newtonian;* (*c*) *pseudoplastic;* (*d*) *plastic.*

efore flow commences; this minimum stress is designated , *f*. This type of flow, which is also called *Bingham flow*, is ne dispersions of finely divided solids in liquids such as ng inks, clay pastes used in ceramics, etc. Bingham[30] t in a suspension showing plastic flow the individual h each other in a flocculated array and that in order for flow to commence it is necessary to apply sufficient stress to break some of these interparticle bonds (Fig. 8.8). During shear an equilibrium is set up between the breaking and reforming of these bonds, so that at a constant rate of shear an equilibrium value for the shear stress is attained, and increasing the rate of shear produces a different equilibrium.

Other dispersions exhibit *pseudoplastic flow* (Fig. 8.7*c*) in which there is no initial yield value and where with increasing rate of shear an apparent decrease in viscosity occurs, the shear stress against rate of shear curve being convex to the shear axis. The fourth type of behaviour shown in Fig. 8.7*a* is *dilatant flow*; the apparent viscosity increases with increasing shear rate, the shear stress/shear rate curve being concave to the shear stress axis. This

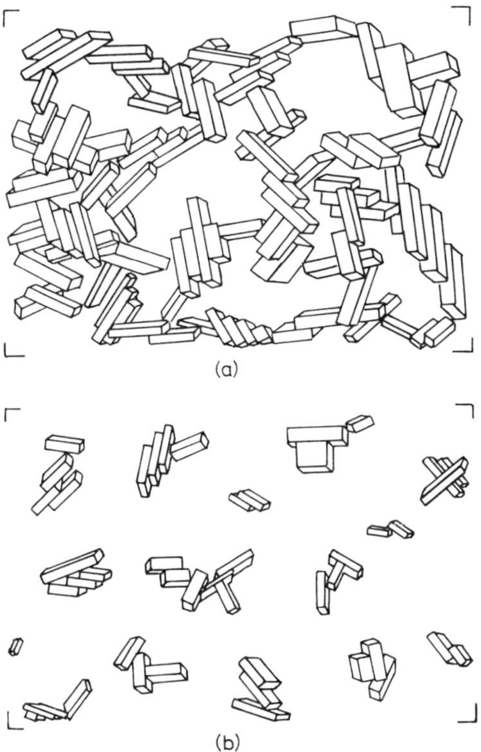

Fig. 8.8. (*a*) *Thixotropic system at rest.* (*b*) *Thixotropic system at shear, showing the breakdown of the structure into smaller flocculates.*

behaviour is shown by deflocculated dispersions of pigments and other powders at high volume concentrations. At these high concentrations of the solid phase, the particles are closely packed and disturbance by shear introduces irregularities into the packing, with bridging effects occurring between the particles. Since the packing becomes looser, the total volume of interparticle space becomes greater and the liquid present is no longer sufficient to fill the space between the particles; consequently, the lubricating effect of the liquid, which enables the particles to slide over one another, is lost.

A further type of rheological behaviour exhibited by certain systems is *thixotropic flow*, which is a time-dependent phenomenon. In this case, the existing structure of the system, for example a flocculated network of particles, is broken down on stirring but does not re-form immediately the stirring action is ceased. The time taken for the structure to reassemble itself may range from seconds to hours. Fig. 8.9 illustrates the type of flow curves which may be obtained using a rotational viscometer with such systems.

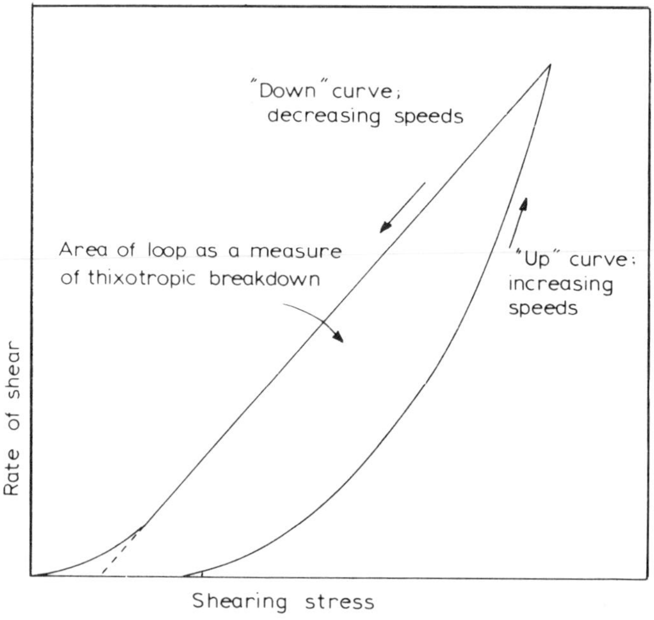

Fig. 8.9. Rheological behaviour of a thixotropic system as measured with a rotational viscometer. The area and nature of the thixotropic loop depends upon the type of instrument and the circumstances of measurement.

Thus, the measurement of flow properties of dispersions is rather complex and at the present, no complete theory exists for the interpretation of flow measurements and their relationship to particle size of the dispersed phase.

For very dilute suspensions, the viscosity may be expressed by the Einstein equation:

$$\eta = \eta_0(1 + 2{\cdot}5\phi) \tag{8.2}$$

where η is the viscosity of the suspension, η_0 the viscosity of the continuous phase and ϕ is the volume fraction of the suspension (i.e. the fraction of the total volume occupied by the suspended particles). The equation is only valid when there is no interaction between the dispersed particles, usually assumed to be when the volume fraction is less than 0·01. Moreover, the Einstein equation indicates that η/η_0 is independent of particle size. The equation has, however, been used by Priel and Silberg[31] to get an idea of the thickness of the adsorbed layer of polymer on titanium dioxide.

As the concentration of the dispersed phase increases, the viscosity of a suspension is increased due to the formation of temporary doublets, triplets and higher multiples which act as single flow units. Much work, both experimental and theoretical, has been carried out attempting to develop equations which describe the viscosity of disperse systems at volume fractions greater than 0·01: these have been reviewed by Frisch and Simha.[29]

Hauser *et al.*[32] found that the rheological behaviour of linseed oil varnishes pigmented with β-copper phthalocyanine, polychloro-copper phthalocyanine or γ-quinacridone of different particle sizes showed a plastic flow behaviour and obeyed the Casson[33] equation. The terminal velocity, i.e. the viscosity extrapolated to a high shear gradient, was found to be independent of particle size, whereas the yield value was strongly dependent on particle size. For systems with a lower viscosity, e.g. paints and liquid inks, McKay[34] found that the viscosity at high shear rates as a function of solid volume fraction could be best expressed by a modified Mooney[35] equation which took account of the high proportion of immobilised liquid associated with the flow units.

Because of the difficulty in interpreting[36] rheological behaviour, flow measurements are often used only in a qualitative manner, observing trends and comparing curves with each other. For example, Jeffries[37] has shown that as milling proceeds, the curves of apparent viscosity against shear rate change their shape giving an indication of the progress of dispersion.

Sedimentation techniques

Both gravitational[38,39] and centrifugal[40] techniques are well established as methods for particle size analysis. The principles and basic suppositions behind both techniques are the same, namely that the settling rates of all particles should be uniquely related to their size, but the range of suitability for the techniques is complementary. Gravitational techniques may be used down to about 2 μm; below that size long settling times are required and it is difficult to maintain the absence of convection currents in the sedimentation fluid.[41] For sizes below 2 μm centrifugal sedimentation is the preferred technique and the lower size determinable here is limited by the Brownian motion of the particles. Brugger[42] suggests that the displacement of the smallest particle by Brownian motion should be ten times shorter than its centrifugal path, which under the centrifuge conditions he was using gave a lower limit of 0·03 μm.

The results of sedimentation analysis are calculated using Stokes' law,[43] which relates the free falling velocity of the particles to the diameter of spheres of the same density with the same rate of fall. For the Stokes equation to be valid the particles must be completely isolated from each other and sediment independently, which means using dilute dispersions. A second criterion is that the condition of viscous flow applies. The upper size limit of particle which can be studied is given by the magnitude of the Reynolds number $(V\,d\rho/\eta)\times 10^{-4}$, where V is the rate of fall (cm/sec), d is the diameter (μm), η is the viscosity of the liquid (poise) and ρ is the density of the pigment (g/cm^3). The Reynolds number is a dimensionless number which should not exceed 0·2, since at higher values turbulent conditions apply and Stokes' law is inapplicable.

For pigment dispersions whose size lies between 0·05 μm and several micrometres, the centrifugal technique is the one of interest. Initially, work was carried out using ordinary laboratory bucket centrifuges, but more recently to overcome some of the limitations of this type of equipment,[44] interest has turned to the disc form of the centrifuge.[45,46] The ICI–Joyce Loebl centrifuge is one such instrument that is commercially available and it has been used to study the size distribution of pigments in both aqueous[47] and non-aqueous[48] systems. The construction and mode of operation of this centrifuge is described in some detail by Beresford.[49] As with any centrifugal method, two modes of size analysis are available:

(i) the homogeneous start technique where the dispersion is introduced into the spinning disc, and
(ii) the line start or two layer technique where a thin layer of the suspension is placed on top of a sedimentation fluid whose density is higher than that of the dispersion medium but lower than that of the pigment particles.

Generally, with the ICI–Joyce Loebl centrifuge the two layer technique is used, for as the particles are all starting to sediment from the same position, it is easy to handle mathematically. But in the case of the homogeneous start all the particles present can commence sedimenting anywhere within the disc, so the raw data require some fairly complicated handling to obtain the desired size information.

However, even with the two layer technique, certain problems occur. Firstly, to ensure that the particles sediment individually the dispersions have to be dilute, usually around 0·5% w/w, and this means that a practical system has to be massively diluted. Care must be taken to ensure that no flocculation or change in the state of dispersion takes place during this stage. Indeed, it must be emphasised that meaningful data can only be obtained on systems that are stable, for if they are unstable, time-dependent effects will be apparent and any handling of the system could bring about deflocculation. Not only must the systems be stable in the dispersion, but also no changes must occur during sedimentation, e.g. for non-aqueous dispersions of organic pigments there must be no tendency for the pigment to bleed into the spin fluid or else the smaller particles will dissolve slowly into it with subsequent redeposition upon the larger particles. The other problem is related to the hydrodynamic instability of the two layer technique which manifests itself as streaming. This is a build-up of particles at the dispersion/spin fluid interface so that the density at this line becomes greater than that of the spin fluid and a density inversion takes place. Turbulence occurs and the particles are dragged down into vortices.

In practice, it is often possible to overcome streaming by using the buffered line start technique,[49] whereby the abrupt interface between the medium and spin fluid is replaced by a region of gradually increasing density and viscosity. The instability of the line start is the reason why some workers prefer the homogeneous technique, though Brugger[50] has argued that hydrodynamically unstable situations can exist also in this case and that a continuous density gradient throughout the whole disc is necessary to obtain stable sedimentation.

The analysis method gives a weight cumulative curve which can be mathematically transformed into an incremental curve. This incremental curve can be determined directly and with a considerable saving of time by monitoring the concentration of material as it passes a fixed radius as a function of time (knowing the radius and time together with the other centrifugation parameters, a Stokes size can be calculated). For pigments this can be done by measuring the attenuation of a beam of light,[51] or (as with rutile) using X-rays.[52] The ICI–Joyce Loebl centrifuge is now fitted with a photosedimentometer attachment, but the difficulty lies in relating the measured attenuation of light at any one time to the weight of pigment present, since for particles below about 1·0 μm the extinction coefficient is a function of particle size. Comparative results, e.g. for following the dispersion of a specific pigment into a system, may be obtained simply by comparing the turbidity versus time (or size) curves. Absolute size distributions can be obtained using calibration curves as suggested by Carr,[53] or by using the Mie theory to determine the extinction coefficient as a function of particle size from a knowledge of the complex refractive index of the pigment and the refractive index of the spin fluid.[54]

Microscopy

Microscopy is perhaps the most direct method for the determination of particle size. For light microscopy the lower limit of applicability is given by the theoretical limit of resolution which is calculated to be about 0·2 μm.[55,56] This is the smallest particle which from theoretical considerations would be capable of giving a sharp, distinct image. Particles smaller than this can be seen in the microscope but their images are diffuse circles whose sizes appear greater than their true sizes. In practice, however, there are additional factors which influence the resolving power of a microscope, e.g. the extent to which the optical system has been corrected for geometric and chromatic aberrations, and also the contrast between particle and background. Thus for practical purposes a lower size limit of 0·5 μm should be assumed for particle size analysis.

These limitations suggest, then, that the optical microscope is not so useful for scanning the state of dispersion of pigments, where the size of the dispersed particles, especially those producing the colour, is often well below 0·5 μm. However, as in general the preparation techniques are relatively simple, the optical microscope is an extremely useful tool for getting an overall impression of a dispersion, especially concerning the undispersed aggregates, and also for use with systems containing fillers whose particle size is sufficiently large. The techniques and precautions necessary for accuracy in the counting and measuring operations using optical microscopy are well documented,[57] but as the size range of particles that may be resolved is generally larger than that of the dispersed particles, they will not be described here.

Where particles are below 1 μm in size, the electron microscope must be used. It has been shown[58] that an accelerated beam of electrons has an effective wavelength which in modern electron microscopes is of the order of 10^{-5} times shorter than that of light, so it is apparent that the resolving power of an electron microscope is potentially far greater than that of the light microscope. Whereas over the years the design and construction of optical microscopes has improved to such an extent that instruments can now be built that approach the limit of resolution, at the present the achievable resolution using an electron microscope is much poorer than the theoretical limitation. The design of the electron microscope and the factors responsible for the limited resolution achievable are clearly described by Agar *et al.*[59] and Fryer,[60] indicating that the resolution obtainable on high performance commercial instruments is now in the range 0·2–0·5 nm.

Because of the vacuum conditions in the electron microscope, in general only dry dispersions can be examined directly, although an indirect method for examining the dispersion in wet paints has been reported.[61] More recently, Larson *et al.*[62] have described a 'wet' cell formed by pressing two coated grids together, which can be used to study suspensions directly in the electron microscope.

In addition, due to the limited penetration of the electron beam, application media cannot be examined directly as they are too thick, and fairly sophisticated techniques have to be used to prepare a sample for examination. The commonest technique for looking at dry films is the film sectioning or ultramicrotomy technique which was first developed for the examination of biological specimens. A comprehensive and detailed description of the technique and some of the general experimental problem is given by Reid,[63] and Kampf[64] describes some of the difficulties associated with its application to pigmented systems. In this method, a thin slither of the pigmented medium is embedded in a thermosetting resin to give a block. After hardening, the block is trimmed to a point and then sections, which are usually between 50 and 100 nm thick, are cut using an ultramicrotome and mounted on grids for examination. This technique is rather time-consuming and calls for some fairly delicate handling, but it can now be considered as a routine research method. Typical cross-sections of pigmented films are shown in Fig. 8.10.

Despite the almost routine method of preparation, the state of dispersion observed on the micrograph might not always give a true indication of the state of dispersion in the bulk. Only a very small volume of the whole of the application medium is being examined—the actual volume of a section is of the order of 10^{-10} cm^3—and it is quite possible to imagine that if the pigment is relatively poorly dispersed with a large number of aggregates distributed throughout the medium, the actual section could be taken from a volume between aggregates giving a false impression of the dispersion. To overcome this, it is always advisable to

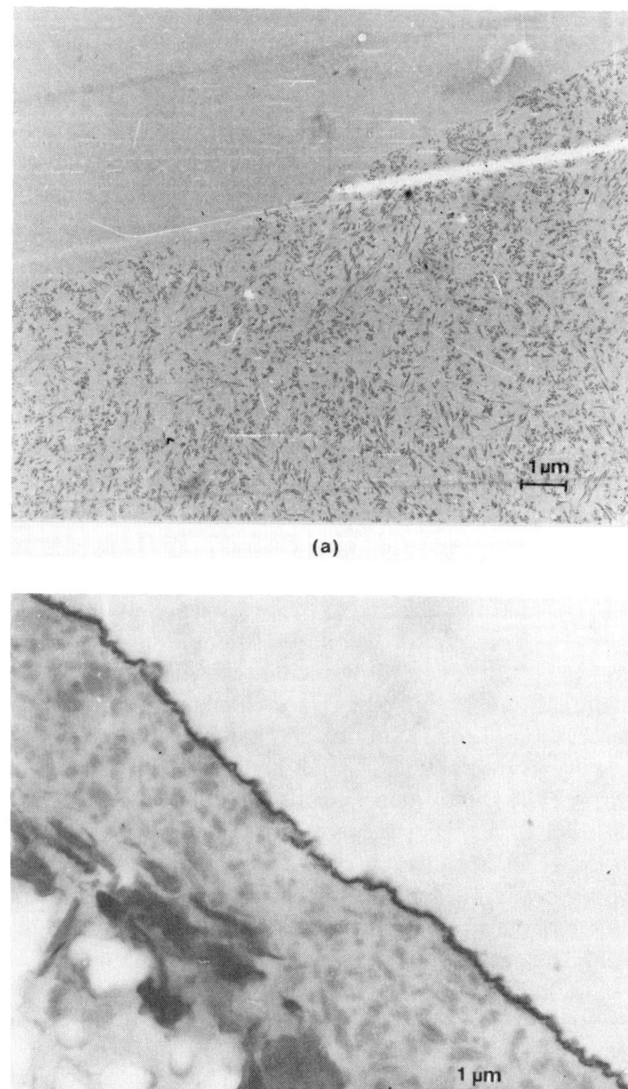

Fig. 8.10. Electron micrographs of ultra-thin sections of pigmented films. (a) Pigmented TSA paint system. The basic particle is needle shaped and the micrograph shows how the orientation of the particles within the section affects the size observed. (b) Pigmented litho ink system. The black line at the ink/embedding resin interface arises from the covering of metallic chromium that was evaporated on to the ink prior to embedding, to prevent the embedding resin from infiltrating and swelling the ink film.

get an overall impression of the state of dispersion using the light microscope, when the large aggregates can be seen.

During the sectioning of films containing organic pigment, not only the application medium but also the individual particles, aggregates and flocculates are being cut, so that in many cases only a part of the dispersed particle will appear within a section, and give rise to the two-dimensional image seen in the final micrograph. The problem of relating the size distribution of images seen on the micrograph to the true size distribution of the dispersed particles—the so-called 'Tomaten Schnitt' problem—requires some rather complex mathematical handling[65] and the section thickness has to be accurately known. For harder inorganic pigments, e.g. titanium dioxide, the ultramicrotome knife tends either to cut around the particles or to tear the particles out of the section giving rather holey films of uneven thickness.[52]

In spite of these drawbacks, electron microscopy is probably the only way of looking at the distribution of particles in a dry film. Furthermore, it has the great advantage that it gives the technologist a visual impression of how the dispersion appears, since the individual particles, aggregates and flocculates can all be seen. Should quantitative estimates of the mean size be required, then automatic counting and sizing techniques[66,67] should be used.

The Coulter counter

The Coulter counter[68,69] sizes particles by monitoring the changes in resistance between two electrodes immersed in an electrolyte as the particles pass through a fine orifice separating the electrodes. During passage through the orifice each particle displaces some liquid, thus altering the resistance across the orifice and causing a momentary electric pulse, the magnitude of which corresponds to the volume of the particle. These pulses are counted and sized by a pulse height analyser, the various settings of which correspond to equivalent spherical diameters.

The dispersions examined by the Coulter counter are extremely dilute, with 10^6 to 10^7 particles per cm^3, and must have a suitable electrical conductivity. The former requirement means that practical dispersions have to be excessively diluted with the risk on one hand that in the absence of sufficient resin or dispersant flocculation may occur, and on the other hand that the original level of dispersion will be altered. With aqueous dispersions the required conductivity can be achieved fairly readily by the addition of some suitable inorganic salt such as sodium chloride, though its influence on the stability of the system must be checked, whereas for non-aqueous systems the choice of electrolyte is more difficult as it must be compatible with the resins present and again not cause flocculation.

The size range of the instrument is generally considered to be from about 0·8 μm to about 200 μm,[70] which is somewhat too high for the majority of pigment dispersions. The lower limit is determined by the ability of the electronic circuits to separate out the signal pulse from the general background thermal and electronic noise. An adaptation of the Coulter technique has been reported[71] making use of extremely small orifices 0·3 μm in diameter which can size particles down to 0·01 μm.

The upper size, for which there is apparently no theoretical limit, is determined practically by the necessity to maintain a homogeneous suspension in the sample beaker. Recently, workers at Coulter Electronics Ltd have developed a large particle system which permits size analyses of about every type of particle with diameters up to 1·5 mm irrespective of density.[72]

The range of sizes over which the Coulter counter operates means that primary pigment particles and small aggregates composed of a few primary particles lie below the limit of detection and only the larger clusters present in a dispersion will be sized. Even so the instrument has been used to size both aqueous[73] and non-aqueous[74] systems. Crowl and Kieran followed the decrease in over-size material, i.e. particles larger than 1 μm, of a number of pigments and extenders as a function of milling conditions in both an aqueous[75] and non-aqueous alkyd medium,[76] where the appropriate conductivity was achieved using a solution of ammonium thiocyanate in methyl ethyl ketone. Interestingly in the latter study, they were able to show a good correlation between the trends obtained from the Coulter counter measurements on diluted systems with the Hegmann gauge readings and development of colour strength measurements, both of which were carried out using undiluted mill base. Pagani *et al.*[77] also used the Coulter counter, to follow the dispersion process of a β-copper phthalocyanine pigment and a red iron oxide pigment into an alkyd medium, by ball milling, pearl milling and high speed stirring. By using small (10 μm) orifices Pagani claims to have been able to measure particles down to 0·2 μm in diameter, though as Beresford *et al.*[73] have pointed out when using small orifices there is always the tendency for them to become blocked by large particles present in the dispersion.

Optical methods

The intensity of a beam of light on passing through a dilute suspension is attenuated as a result of absorption and scattering of light by the particles. The attenuation in intensity is a function of such factors as the refractive index of the particles relative to the medium, the wavelength of light, and the size and size distribution of the particles. The mathematical relationship between these factors was first developed by Mie,[4] and an excellent reference on light scattering and absorption by small particles is the book by Kerker.[78] The variation in absorption and scattering as a function of size is one reason why the colouristic appearance of pigment dispersion is influenced by the level of dispersion. However, in this section it is the converse problem that is of interest, viz. the determination of particle size from light scattering data.

In principle, if the refractive index of the dispersed phase can be determined, then from light scattering measurements it should be possible using the Mie theory to determine the size of the dispersed particles. Though the problem is relatively easily stated, from a practical point of view there are some difficulties to be overcome. Firstly, the refractive indices of pigment powders, which are generally wavelength-dependent and complex, are not easy to measure.[79] Secondly there are some rather involved computations to be carried out which require the use of a computer and as Kerker[78] has pointed out, it is possible for a set of light scattering data to be produced by a number of polydisperse distributions. Practically, too, the dispersion has to be very dilute so that the particles may be considered to be isolated, independent scatterers so that no multiple scattering occurs.

Provided, however, that such difficulties can be overcome, light scattering does offer several advantages over other methods. The measurements are made without disturbing the system, and the theory permits direct calculation of the size without the need for any calibration. The complexity of the calculations, especially when dealing with size distributions, and the general lack of reliable data for the refractive index of the dispersed phase mean that the rigorous use of light scattering and the Mie theory has not found much application to pigment dispersions. In practice simpler optical methods have been used, such as measuring the turbidity of a suspension as a function of wavelength and comparing the experimental results with theoretical curves calculated for various sizes and size distributions. This approach has been used by a variety of workers to size rutile suspensions[80–83] and by Kindervater[84] to size distributions of inorganic pigments. Kindervater made the assumption that the inorganic pigments were non-absorbing, and found that the optically determined results agreed quite satisfactorily with the median value of the surface distribution calculated from the number distribution determined using the electron microscope.

RECENT DEVELOPMENTS

Because of the importance of size analysis to many industries, new methods of analysis are continuously being reported and new instruments developed. It is not the author's intention to give a comprehensive review of the developments which have taken place over the past few years in the field of particle size assessment, this aspect having recently been reviewed by Orr,[85] but rather to highlight one or two of these newer developments which could find application in the analysis of pigment dispersions.

Hydrodynamic chromatography

Small[86] has described a new chromatographic technique, which he has termed hydrodynamic chromatography, for sizing dispersions of colloidal particles in the range from a few tens of nanometres to several micrometres in diameter. If a small quantity of a dilute dispersion of colloidal particles is allowed to pass through a column packed with solid, non-porous particles, size separation takes place with the largest particles eluting first and the smallest last. The exact mechanism whereby separation takes place has not been resolved completely. Based on some work by Dimarzio and Guttman,[87] Small suggests that the bed should be considered as a system of capillaries and the main separation cause as hydrodynamic. Across such capillaries there is a parabolic distribution of flow velocities rising from zero at the wall to a maximum in the centre. On account of their finite size, particles cannot get closer to the walls than their radius and so are not influenced by the slower velocities near the walls. They thus pass through the column faster than the medium, whose molecules are able to experience the whole spectrum of velocities.

Electro-optics

Recent work, notably by Jennings and his coworkers, has shown that electro-optical[88,89] effects may be used to determine the size of colloidally dispersed particles.

Many colloidal particles are optically anisotropic but a dilute dispersion will appear optically isotropic as the individual particles are randomly orientated. When such a dispersion is placed in an electric field its optical properties change, becoming anisotropic as the field interacts with a permanent or induced dipole moment of the constituent particles causing them to align. On removal of the field, under the influence of Brownian motion the system reverts from the orientated state to the natural disordered state and the rate at which the change occurs depends upon the size and shape of the dispersed particles. This change can be followed by monitoring changes in the optical behaviour (e.g. dichroism, birefringence) of the dispersion. It is the analysis of the time-dependent variation of the optical property after cessation of an applied electric field that leads to the size and size distribution of the dispersed particles. Strictly speaking it is the major dimension of the particle that is determined, and some idea of the particle shape is required to enable the calculations to be carried out.

In outline, the apparatus consists of a light source, some sort of optical cell which is equipped with electrodes for applying the electric field, an electric field source and a photocell. Both a.c. and d.c. fields may be applied to the suspension; typically a pulsed d.c. field is used with fields of the order of 10 kV/cm lasting for a duration of 10 ms. For a description of the apparatus and the experimental technique, together with the underlying electro-optic theory, the reader is referred to the original literature.[90–92]

Once the apparatus has been set up, the experimental data can be obtained within a matter of minutes and the technique has been shown to be capable of handling a wide variety of systems, e.g. polymer solutions[92] and suspensions of bacteria[93] as well as aqueous dispersion of solids.[94–96] Generally there appears to be good agreement between the particle size obtained using electro-optics and electron microscopy. Thus, Foweraker and Jennings[95] have found that the average particle length of β-copper phthalocyanine was 130 nm when determined by electro-optics while electron microscopy indicated a most characteristic length of 125 nm. The most recently developed theories yield a distribution curve and using a dispersion of the clay mineral halloysite in water Foweraker *et al.* obtained good agreement between electron microscopy and electro-optic data in the type of distribution, its position and breadth.[94]

Photon correlation spectroscopy

Another recently developed optical technique which shows great potential for sizing particles in the colloidal size range is that of Photon

Correlation Spectroscopy, which is also referred to in the literature as Doppler Shift Spectroscopy, Quasi-electric Light Scattering Spectroscopy or Intensity Fluctuation Spectroscopy.

When a suspension of particles is illuminated by a coherent beam of laser light a complex diffraction pattern results, made up of numerous lighter and darker areas and brought about by the interference of light scattered by each of the individual particles. However, as the individual particles are undergoing Brownian motion their position with respect to each other is continually changing, which results in a continuously changing diffraction pattern. The rate at which the pattern changes reflects the size of the dispersed particles—the smaller particles move more rapidly leading to a more rapidly fluctuating diffraction pattern.

If the scattered light is allowed to fall on a small photomultiplier with a suitably fast response time a fluctuation in the intensity of light around some mean value will be recorded. Analysis of the rate of change of the intensity of scattered light permits the diffusion coefficient of the particles to be obtained and hence by using the Stokes–Einstein[97] equation the particle size may be evaluated. For a more detailed description of this technique the original references should be consulted.[98–100]

Much of the application of these techniques seems to have been for research into biological systems, where the sizes of the particles, e.g. virus[98] or protein,[99] are rather smaller. Its use for larger sized particles has been demonstrated recently by Stenhouse *et al.*[101] and by Lines and Miller,[102] all of whom worked with 'monodisperse' polymer latices and found good agreement between the sizes determined by this technique and those determined by electron microscopy.

Commercial apparatus based on these principles are now coming on the market, e.g. the Nano-Sizer from Coulter Electronics and the 440 Particle Analyser from Malvern Instruments. Results are obtained very quickly, a mean size determination and indication of the width of the distribution taking only a few minutes. As with all optical measurements dilute dispersions (approx. 0·1%) are required, which means that practical dispersions have to be extensively diluted. The method is applicable, in principle, to both aqueous and non-aqueous suspensions and because of the ease of use, highly trained personnel are not required. Together with the speed with which the results are obtained it is suggested that these instruments could be used also in a quality control environment.

CONCLUDING REMARKS

As yet the ideal method for assessing the level of dispersion of pigment systems has not been found. If attention is focused solely on wet systems it can be seen that the specifications that such a method has to meet are extremely stringent, e.g. it should be capable of handling systems at practical levels of pigmentation so that the need for dilution is removed, it should be applicable to both aqueous and non-aqueous systems, the size range required is from at the lowest limit 0·01 μm to an upper limit of several micrometres, it should be rapid, etc. Besides these requirements the more conceptual questions of what we mean by a 'particle' in terms of crystallography, optics or rheology have to be borne in mind.

There are available a fairly large number of techniques for looking at the particle size of dispersions, each one having its advantages and disadvantages. In most cases it is probably better to use a variety of methods to look at any particular problem rather than to rely on a specific technique. For polydisperse systems the different methods will give differing 'absolute' results as often a different 'particle' is being measured by each method, but the trends observed by all methods should be the same.

REFERENCES

1. J. Th. G. Overbeek, *Disc. Faraday Soc.*, **42** (1966) 7.
2. K. Merkle and W. Herbst, *Farbe u. Lack.*, **74** (1968) 1072.
3. R. N. Weltmann, *Rheology*, Ed. F. R. Eirich, Academic Press, New York, 1960, Vol. 3, p. 189.
4. G. Mie, *Ann. Physik.*, **25** (1908) 377.
5. A. Brockes, *Optik*, **21** (1964) 550.
6. F. C. Chromey, *J. Opt. Soc. Am.*, **50** (1960) 730.
7. W. Carr, *J. Paint Technol.*, **42** (1970) 695.
8. B. Felder, *Helv. Chim. Acta.*, **51** (1968) 1224.
9. A. R. Hannam and D. Patterson, *J. Soc. Dyers Colourists*, **79** (1963) 192.
10. P. Hauser, D. Horn and R. Sappok, *XII FATIPEC Congress*, (1974) 191.
11. P. Hauser, M. Herrmann and B. Honigmann, *Farbe u. Lack.*, **76** (1970) 545.
12. S. Wilska, *J. Oil Colour Chem. Assoc.*, **50** (1967) 911.
13. W. Carr, *J. Oil Colour Chem. Assoc.*, **54** (1971) 155.
14. B. Medinger, *XIV FATIPEC Congress*, (1978) 439.
15. J. R. Fryer, R. R. Mather, R. B. McKay and K. S. W. Sing, *J. Chem. Technol. Biotechnol.*, **31** (1981) 371.
16. R. R. Mather, *XIV FATIPEC Congress*, (1978) 433.

17. R. R. Mather and K. S. W. Sing, *J. Colloid Interface Sci.*, **60** (1977) 60.
18. R. Sappok and B. Honigmann, *Characterisation of Powder Surfaces*, Eds. G. D. Parfitt and K. S. W. Sing, Academic Press, London, 1976, p. 231.
19. B. Honigmann, *Ber. Bunsenges. Phys. Chem.*, **71** (1967) 239.
20. J. Stabenow, *Ber. Bunsenges. Phys. Chem.*, **72** (1968) 374.
21. J. R. Fryer, *J. Microsc. (Oxford)*, **120** (1980) 1.
22. B. Schwegmann, *Farbe u. Lack*, **80** (1974) 942.
23. L. Valentine, *J. Oil Colour Chem. Assoc.*, **46** (1963) 674.
24. L. Gall and U. Kaluza, *DEFAZET*, **29** (1975) 102.
25. *Paint Technology Manuals, Part V, The Testing of Paints*, Eds. C. J. A. Taylor and S. Marks, Chapman and Hall, London, 1965.
26. *Printing Ink Manual*, Ed. R. F. Bowles, 2nd edition, W. Heffer and Sons, Cambridge, 1969, p. 194.
27. G. Herdan, *Small Particle Statistics*, 2nd edition, Butterworths and Co., London, 1960, p. 22.
28. C. Orr and J. M. Dallavalle, *Fine Particle Measurement*, The Macmillan Company, New York, 1959.
29. H. L. Frisch and R. Simha, *Rheology, Volume* 1, Ed. F. R. Eirich, Academic Press Inc., New York, 1956, p. 525.
30. E. C. Bingham, *Fluidity and Plasticity*, McGraw Hill, New York, 1922, p. 228.
31. Z. Priel and A. Silberg, *Am. Chem. Soc. Div. Org. Coatings Plast. Chem. Pap.*, **30** (1972) 556.
32. P. Hauser, M. Herrmann and B. Honigmann, *Farbe u. Lack*, **77** (1971) 1097.
33. N. Casson, *Rheology of Disperse Systems*, Pergamon Press, London, 1959, p. 84.
34. R. B. McKay, *XIII FATIPEC Congress* (1976) 428.
35. M. Mooney, *J. Colloid Sci.*, **6** (1951) 162.
36. P. Sherman, *Industrial Rheology*, Academic Press, London, 1970, Chapter 3, p. 97.
37. H. D. Jeffries, *J. Oil Colour Chem. Assoc.*, **45** (1962) 681.
38. British Standard 3406, Methods for the determination of Particle Size of Powders, Part 2, Liquid Sedimentation Methods (1963).
39. Society for Analytical Chemistry, Particle Size Analysis Sub-Committee, *Determination of Particle Size*, Part 1, 'A critical review of sedimentation methods' 1967.
40. C. E. Marshall, *J. Soc. Chem. Ind.*, **50** (1931) 440.
41. A. Muta and S. Watanabe, *Particle Size Analysis 1970*, Eds. M. J. Groves and J. L. Wyatt-Sargent, The Society for Analytical Chemistry, London, 1970, p. 178.
42. K. Brugger, *Powder Technol.*, **14** (1976) 187.
43. G. Herdan, *Small Particle Statistics*, 2nd edition, Butterworths and Co., London, 1960, p. 56.
44. I. Fraser, *Pigment Handbook*. Volume III, Characterisation and Physical Relationships, Ed. C. Patton, John Wiley and Sons, New York, 1973, p. 53.
45. J. G. Steel and R. Bradfield, *Am. Soil Survey Assoc., Rept. 14th Ann. Meeting Bull.*, **15** (1934) 88.
46. W. Bostock and J. K. Donoghue, *Trans. Inst. Chem. Eng.*, **33** (1955) 72.

47. E. Atherton and D. Tough, *J. Soc. Dyers Colourists*, **81** (1965) 624.
48. W. Carr, *Particle Size Analysis 1970*, Eds. M. J. Groves and J. L. Wyatt-Sargent, The Society for Analytical Chemistry, London, 1970, p. 85.
49. J. Beresford, *J. Oil Colour Chem. Assoc.*, **50** (1967) 594.
50. K. Brugger, *Powder Technol.*, **13** (1976) 215.
51. M. J. Groves, B. H. Kaye and B. Scarlett, *Brit. Chem. Eng.*, **9** (1964) 742.
52. M. R. Hornby and R. D. Murley, *Prog. Org. Coat.*, **3** (1975) 261.
53. W. Carr, *Prog. Org. Coat.*, **4** (1976) 161.
54. P. Hauser and B. Honigmann, *Farbe u. Lack*, **83** (1977) 886.
55. E. M. Chamot and C. W. Mason, *Handbook of Chemical Microscopy*, 2nd edition, Volume I, John Wiley and Sons Inc., New York, 1951.
56. M. Born and E. Wolf, *Principles of Optics*, 5th edition, Pergamon Press, Oxford, 1975.
57. British Standard 3406, Methods for the Determination of Particle Size of Powders, Part 4, Optical Microscope Method (1963).
58. L. De Broglie, Dissertation (Paris, Masson), *Phil. Mag.*, **47** (1924) 446.
59. A. W. Agar, R. H. Alderson and D. Chescoe, *Practical Methods in Electron Microscopy*, Volume 2, 'Principles and Practice of Electron Microscope Operation', Ed., A. M. Glauert, North-Holland Publishing Company, Amsterdam, 1974.
60. J. R. Fryer, *The Chemical Applications of Transmission Electronmicroscopy*, Academic Press, London, 1979.
61. K. Merkle and W. Herbst, *IX FATIPEC Congress*, (1968) Sect. 2, 65.
62. R. I. Larson, E. F. Fulham, A. D. Lindsay and E. Matjeivic, *AIChem. J.* **19** (1973) 602.
63. N. Reid, 'Ultramicrotomy', *Practical Methods in Electron Microscopy*, Volume 3, Part II, Ed. A. M. Glauert, North-Holland Publishing Company, Amsterdam, 1974.
64. G. Kampf, *Farbe u. Lack*, **76** (1970) 25.
65. C. Bach, *Quantitative Methods in Morphology*, Eds. E. R. Weibel and H. Elias, Springer-Verlag, Berlin, 1967, p. 23.
66. E. Redman, F. A. Heckman and J. E. Connolly, *Particle Size Analysis*, Ed. M. J. Groves, Heyden and Son, London, 1978, p. 51.
67. L. J. Venuto and W. Hess, *Am. Inkmaker*, **45** (1967) 42.
68. Society for Analytical Chemistry, Particle Size Analysis Sub-Committee, *Determination of Particle Size*, Part 2. 'Electrical sensing zone methods (Coulter Principle)', 1971.
69. H. E. Kubitschek, *Nature*, **182** (1958) 234.
70. S. Kinsman, *Pigment Handbook*, Volume III, Characterisation and Physical Relationships, Ed. T. C. Patton, John Wiley and Sons, New York, 1953, p. 101.
71. R. W. DeBlois and C. P. Bean, *Rev. Sci. Instrum.*, **41** (1976) 909.
72. J. G. Harfield, B. Miller, R. W. Lines and T. Godin, *Particle Size Analysis*, Ed. M. J. Groves, Heyden and Son, London, 1978, p. 378.
73. J. Beresford, W. Carr and G. A. Lombard, *J. Soc. Dyers Colourists*, **81** (1965) 615.
74. V. M. Wood, *Br. Ink Maker*, **8** (1966) 157.
75. V. T. Crowl and L. G. Kieran, *PRS.*, *Res. Memo* 327 (1963).

76. V. T. Crowl, L. G. Kieran, and R. Bolton, *PRS. Res. Memo.* 352 (1966).
77. D. Pagani, P. Moretti and A. Battaglia, *XIII FATIPEC Congress* (1976) 42.
78. M. Kerker, *The Scattering of Light*, Academic Press, New York, 1969.
79. H. Nassenstein, *Ber. Bunsenges. Phys. Chem.*, **71** (1967) 303.
80. C. E. Barnett and G. G. Simon, *J. Colloid Interface Sci.*, **34** (1970) 580.
81. H. K. Dettmar, W. Lode and E. Marre, *Kolloid Z.*, **188** (1963) 28.
82. L. W. Richards, *Pigment Handbook*, Volume III, Characterisation and Physical Relationships, Ed. T. C. Patton John Wiley and Sons, New York, 1973, p. 89.
83. R. A. Slepetys, *J. Paint Technol.*, **44** (1972) 91.
84. F. Kindervater, *XIV FATIPEC Congress*, (1978) 305.
85. C. Orr, *Particle Size Analysis*, Ed. M. J. Groves, Heyden and Son, London, 1978, p. 77.
86. H. Small, *J. Colloid Interface Sci.*, **48** (1974) 147.
87. E. A. Dimarzio and C. M. Guttman, *Macromolecules*, **3** (1970) 131.
88. E. Fredericq and C. Houssier, *Electric Dichroism and Electric Birefringence*, 1st edition, Clarendon Press, Oxford, 1973.
89. B. R. Jennings, *Particle Growth in Suspensions*, Ed. A. L. Smith, Academic Press, London, 1973, p. 95.
90. B. R. Jennings, *Adv. Polym. Sci.*, **22** (1977) 61.
91. V. J. Morris, A. R. Foweraker and B. R. Jennings, *Adv. Mol. Relaxation Processes*, **12** (1978) 65.
92. B. R. Jennings and B. L. Brown, *Eur. Polym. J.*, **7** (1971) 805.
93. V. J. Morris, P. J. Rudd and B. R. Jennings, *J. Colloid Interface Sci.*, **50** (1975) 379.
94. A. R. Foweraker, B. J. Morris and B. R. Jennings, *Particle Size Analysis*. Ed. M. J. Groves, Heyden and Son, London, 1978. p. 147.
95. A. R. Foweraker and B. R. Jennings, *Spectrochim. Acta.*, **31A** (1975) 1075.
96. A. R. Foweraker and B. R. Jennings, *Adv. Mol. Relaxation Processes*, **8** (1976) 555.
97. S. Glasstone, *Textbook of Physical Chemistry*, 2nd edition, MacMillan and Co., London, 1951, p. 261.
98. P. N. Pusey, E. C. Koppel, D. W. Schaefer and R. D. Carmerini-Otero, *Biochem.*, **13** (1974) 952.
99. R. Foord, E. Jakeman, E. J. Oliver, E. R. Pike, R. J. Blagrove, E. Wood and A. R. Peacocke, *Nature*, **227** (1970) 242.
100. D. Munro, A. R. Goodall, M. C. Wilkinson, K. Randle and J. Hearn, *J. Colloid Interface Sci.*, **68** (1979) 1.
101. J. I. T. Stenhouse, B. Scarlett and I. Sinclair, *Particle Size Analysis*, Ed. M. J. Groves, Heyden and Son, 1978, London, p. 177.
102. R. W. Lines and B. V. Miller, *Powder Technol.*, **24** (1979) 91.

CHAPTER 9

DISPERSION OF INORGANIC PIGMENTS

H. D. JEFFERIES
Consultant, Billingham, Cleveland, UK

INTRODUCTION

Classification

Inorganic pigments, including the carbon blacks, are particulate materials that when suspended in a medium in which they are substantially insoluble, can be used to provide a coloured coating on the surface of an object. Alternatively, pigment particles may be dispersed in a polymeric material, such as a rubber or plastic, to colour it and/or to modify its physical properties. The basic hue of a given pigment particle is determined by its chemical composition, but the perceived hue is substantially determined by the particle size and shape. It follows that the degree of dispersion, by determining whether particles are present as entities or groups (aggregates, agglomerates, or flocculates), also determines the perceived hue.

White and black pigments are not coloured in the true sense. A perfect white completely reflects light of all wavelengths, whereas a perfect black absorbs radiation completely. In contrast, a coloured pigment reflects only specific, fairly narrow, spectral bands, the wavelength of the reflected light determining the apparent colour of the pigment. The theory of the colour of inorganic substances has been reviewed by Feitknecht,[1] but the subject falls outside the scope of this chapter. Some appreciation of the variety of chemical compounds that are, or have been, used as pigments can be obtained from the following list:

(i) White (high refractive index): anatase and rutile titanias, zinc sulphide, zinc oxide, lead hydroxycarbonate (white lead), co-precipitated zinc sulphide–barium sulphate (lithopone), basic lead sulphate.

(ii) White (low refractive index): calcium carbonate (whiting, natural or synthetic calcite), silica (natural or synthetic), magnesium–calcium carbonate (dolomite), hydrated aluminosilicates (china clay), barium sulphate (barytes, blanc fixe), magnesium silicate (talc), potassium aluminosilicates (mica).

(iii) Black: natural graphite, carbon blacks, black iron oxide (natural and synthetic).

(iv) Reds, browns: mercuric sulphide (vermilion), lead oxide (Pb_3O_4, red lead), cadmium sulphide, cadmium sulphoselenide, lead molybdochromates, red iron oxide (natural and synthetic), copper oxide, complex ferromanganic, etc. oxyhydrate (umber), hydrated ferric oxide (ferrite browns).

(v) Orange, yellow: lead molybdochromates, basic lead chromate, hydrated ferric oxides (ferrite yellows), alkali–zinc chromates.

(vi) Green: chromium oxide, co-precipitated lead chromate and alkaline ferroferricyanide (struck chrome greens).

(vii) Blues: complex alkaline sulphosilicates (ultramarine green/blue/violet), alkaline ferriferrocyanides (iron blues, Prussian and milori blues, etc.).

(viii) Metallic flakes and powders: lead flake and powder, zinc powder, copper powder, stainless steel powder and flake, aluminium flake, copper bronze flake, copper–aluminium–zinc 'gold' bronzes, nickel flake.

Omitted from the above list is a relatively new type of 'pigmentary particle', namely vesiculated spheroids, also known as microvoids. Each 'particle' consists of an air-bubble of appropriate size encapsulated within a mineral or polymeric shell. They therefore have an effect similar to the 'dry hiding' produced by a paint film that contains insufficient binder to fill the voids between the pigment particles. This phenomenon will be discussed more fully later.

Current coating technology is affected by the proliferation of measures intended to protect the environment. These restrictions on the use of various materials (in the case of pigments, those containing heavy metals such as lead, mercury, and cadmium, as well as arsenic) at present vary from country to country, but those in force in the USA are probably the most comprehensive. One effect has been to promote the use of water as the volatile component of a wide variety of surface coatings.

For fuller information about the chemical composition and physical characteristics of commercial pigments the reader is referred to the brief bibliography at the end of this chapter, and to the technical literature issued by the various suppliers, which by its nature is more up-to-date than the standard texts, though sometimes reticent about certain details that constitute the essential know-how for the manufacture of the product.

Broadly speaking, the basic hue of an inorganic pigment is determined by its chemical composition and crystal structure, but the tone can be varied very considerably by adjustment of the mean particle size and size distribution. Consequently the degree of dispersion of a pigment, by determining its effective particle size distribution, affects the observed colour of a pigmented object, such as a paint film or sheet of plastic.

The achievement of an adequately stable pigment dispersion requires, in addition to some initial mechanical dispersion process, an interaction between the pigment (disperse phase) and the liquid (continuous phase) in which it is dispersed. The interaction most often consists of adsorption on the pigment surface of material from the continuous phase, so that the volume ratios of the disperse and continuous phases are modified. If adsorption is selective, the compositions of the phases will also change. One consequence of the interaction of the phases is that the rheology of the system changes as dispersion proceeds. Further complications ensue because the initial pigment dispersion, if used in paint or ink, is almost always mixed with other materials, some of which may be readily adsorbed by the pigment, possibly displacing material previously absorbed. The nature of these is outlined below.

Pigments in paints and inks

An ordinary liquid paint or ink will usually contain the following components:

(i) *Solid phase.*—A chemically and physically heterogeneous mixture of white, black or coloured materials, organic and/or inorganic.

(ii) *Liquid phase.*—Usually a mixture of a material capable of film formation and a volatile diluent.

(a) Film-former (or binder). A mixture of compounds, either polymeric or capable of polymerisation.

(b) Volatile (solvent). Either water, or one or more organic liquids chosen partly for their ability to dissolve the binder to give the desired solids content and viscosity and partly because their rates of evaporation facilitate application of the coating.

(iii) *Driers.*—One or more soaps, formed from polyvalent metals and organic acids of sufficient molecular size and structure to solubilise the cation in the binder. Traditionally, the metals used were lead, manganese and cobalt, but restrictions on the use of lead and the high price of cobalt have stimulated the search for alternatives. (Manganese has the disadvantage that it gives a brownish tone to the dried paint film and therefore is unsuitable for use in whites or pale shades.) Zirconium driers have been in use for some years but at the present time cerium/zirconium and bismuth seem to be the most effective substitutes for lead driers. Whether calcium soaps are necessary solubilisers with these driers, as they are for lead, is not certain.

(iv) *Anti-skinning agent.*—Usually an oxime or phenolic material, intended to prevent premature oxidative polymerisation (skinning) of the paint or ink during storage. These materials also inhibit oxidative drying after application of the paint or ink so that a balance must be achieved between volatility and effectiveness.

(v) *Dispersing or wetting agent.*—These may be metallic soaps, such as zinc naphthenate, or other surface active materials of natural or synthetic origin. Driers can and do act as pigment dispersants. Many 'dispersing' agents in fact produce a degree of flocculation of the pigment (see (vi), below).

(vi) *Rheology-adjusting agent for the avoidance of hard sediment.*—In the manufacture of a paint or ink, the degree of dispersion attained will have depended on many factors, not all amenable to control, but in the final packed product there is likely to be a proportion of oversize pigment. The oversize may consist of agglomerates or aggregates which were not broken down during manufacture, or flocculates which formed after the milling (pigment dispersion) stage. In certain cases the primary particles of pigment may be too large for a stable dispersion to be made. Whatever the cause, the formulator has to face the fact that some portion of the pigment is likely to settle out if the product is stored long enough; the sediment, when it forms, may be either a *disperse* type or a *flocculant* type.

A disperse type sediment is produced by particles originally in a well-dispersed condition; on settling these pack closely, forming a hard, tough layer, like a stiff clay. The sediment is very difficult to re-disperse.

A flocculate is an entity created in a suspension when previously separate particles come into contact with each other, chiefly at corners and/or edges, forming a skeletal structure, whose gross, or effective, volume may be more than twice that of the total solid volume of its component parts. A paint or ink, in which most of the pigment is flocculated, usually has a high structural viscosity when the system is at rest. When subjected to a shearing stress, flow does not occur until a certain minimum stress (the yield value) is exceeded, after which the stress/strain relationship may, or may not, become linear.

The rate of sedimentation of particles, and in particular flocculated particles, in a liquid such as paint, is not determinable from Stokes' Law, and the viscous resistance to gravitational settling of flocculated pigment may prevent settling, or the flocculates may settle slowly, forming a voluminous sediment beneath an upper layer of substantially unpigmented medium. The voluminous sediment is soft and can be reincorporated in the medium quite easily.

The formulator usually opts for the more easily re-dispersed flocculant sediment, whilst hoping that flocculation-induced structural viscosity will prevent settling altogether. It is for this reason that many paints contain anti-settling agents, which are actually pigment flocculating agents. This type of anti-settling agent must be distinguished from those additives which are expected to achieve their object by inducing a marked degree of structural viscosity in the liquid portion of the paint. These are either materials such as hydrogenated castor oil derivatives, metal soaps (such as aluminium stearate) or specially treated montmorillonite-type clays which create a gelatinous structure in the medium, or a thixotropic alkyd resin formed by reacting a polyamide with an alkyd during resin manufacture to give a material in which hydrogen bonding creates structural viscosity.

The above list of ingredients applies to 'solvent based' paints rather than to 'emulsion based' types. The latter types consist, essentially, of a dispersion of pigments in water (containing appropriate surface active compounds) that is mixed with a dispersion of polymer particles in water, also containing various surface active ingredients to ensure stability. The finished product usually contains other ingredients, added to adjust the rheology of the paint, improve its application properties, and prevent foaming.

Water-thinned paints intended for application by electro-coating vary widely in formulation, the chief determinants being the type of resin and whether the object to be coated serves as anode or cathode. The most accessible source of information about this type of coating material is the literature issued by the raw material suppliers.

The above remarks are intended only to indicate the complexity of a paint or ink considered as a colloidal system, and to emphasise the gulf between the systems studied by most research workers and the more-or-less colloidal systems that are used in practice. In the following pages, the major paint and ink components and their interactions are discussed.

Pigments: manufacturing processes

Inorganic pigments may be synthesised, involving one or more chemical operations, or natural materials that have been physically processed. The chief chemical processes and some of their products are the following:

(i) Oxidation (in the gaseous phase): e.g. fume-process basic lead sulphate, zinc oxide, chloride-process titanium dioxide.
(ii) Reduction (in the gaseous phase): e.g. carbon blacks.
(iii) Precipitation (from solution): e.g. lithopone, chrome yellows, Prussian blue, sulphate-process titanium dioxide.
(iv) Calcination: e.g. ultramarine blue, synthetic iron oxide.

The physical processes chiefly involved are:

(i) Leaching or washing, to remove soluble matter. Often carried out simultaneously with wet grinding of natural pigments.
(ii) Calcination, to obtain particles of pigmentary size. This may involve fusion of crystallites (e.g. lithopones and sulphate-process titanias) or sintering of particles into an agglomerate of pigmentary size (e.g. certain synthetic iron oxides).
(iii) Comminution, either by wet or dry grinding.
(iv) Size classification. There are two chief methods: wet, often used in sequence with water-leaching and grinding of natural pigments; and dry, by entraining particles in a gas stream.

In practice, all inorganic pigments except the carbon blacks usually undergo a final grinding process; if the manufacturing process involves crystal growth, it is not easy, on the industrial scale, to avoid producing a proportion of oversize material. The fundamental problem is that the optimal size of pigmentary particles is smaller than can be easily achieved by any economically viable grinding process, even if the mill incorporates a size classifier. The effect of fluid energy milling, or micronising, is discussed later.

The particles of some kinds of inorganic pigment consist of a core surrounded by a layer of different composition, the thickness of which is dependent on the purpose of the layer and the intended use of the pigment. Certain pigments for use in anticorrosive paints consist of a relatively thick layer of the 'active' component on an inert core, whilst the particles of after- or post-treated titanias have coatings that vary widely in thickness and structure. Thus 'enamel grade' titanias for use in glossy coatings have a comparatively thin, dense coating of oxides, usually of low refractive index, amounting to about 6% by weight of the pigment. In contrast, titania pigments intended for use in emulsion paints (also called latex paints) are given coatings that constitute up to about 20% by weight of the pigment and the coating technique causes the layer to be voluminous and rather 'spiky', thus increasing its effective volume. In the case of lead chromate based pigments, very thin inorganic coatings may be applied to reduce their photosensitivity. In the case of the carbon blacks, the smaller the particle size, the greater is the proportion of the pigment weight that consists of complex organic and metallo-organic compounds. These, however, are deposited on the pigment surface as an integral part of the process of formation of the pigment particle. An as-yet-unresolved problem brought to light by the Toxic Substances Control Act (TOSCA) in the USA is that the low temperature of formation of furnace blacks causes them to contain an unacceptably high concentration of carcinogenic combustion products, whereas the channel blacks do not suffer from this defect but are made by an environmentally unacceptable process.

In order to increase the ease with which the pigment particle is wetted by liquid, a coating, usually of about monomolecular thickness, of an appropriate surfactant compound may be applied. Those most frequently encountered are either amines, amine-soaps or polyhydroxy compounds. Treatment with silicone compounds is also effective, one approach being the exposure of a coated titanium dioxide to methyl polysiloxane vapour, with the object of achieving a hydrophobic pigment surface by covering it with methyl groups linked to the inorganic coating on the pigment through the silicon atom. The effect of these coatings on pigment behaviour is discussed later.

Particle size, opacity and colour of inorganic pigments

The optical properties of a typical paint film cannot be expressed in a rigorously derived mathematical form. Most modern work on this subject is based on the theory developed by Mie in 1908, but the calculations can only be made for grossly over-simplified conditions or for hypothetical light paths. Brockes[2] has considered the relation between staining power and particle size, and Orchard[3] has reviewed the optical properties of opaque pigmented films.

The theory divides the optical properties of a pigment into two components, the colour strength (staining power or tinting strength) and the opacifying (hiding) power. These components are considered as functions of basic pigment properties, namely:

Particle size
Light scattering coefficient S
Light absorption coefficient K
Refractive index n (For a rigorous treatment, the three values of n for the three axes of the crystal must be used.)

The theory of opacity most often used is that of Kubelka and Munk (KM theory).[4] It is sometimes assumed, for simplicity, that for white pigments the value of K can be neglected, but this assumption is only valid if the medium is completely colourless. Conversely, for black pigments the value of S is small. For coloured pigments the value of K is of major importance, but that of S is not necessarily negligible in the case of inorganic pigments, many of which have high refractive indices. Developments in electronics have led to increased use of spectrophotometer/digital computer combinations for colour matching and a comprehensive description of the practical application of such a system has been given by Huey.[5]

For inorganic coloured pigments, the colour strength (staining power, or tinting strength) increases with light absorbing capacity (increasing K) whilst the opacity (hiding power) increases with the light scattering power (increasing S), reaching a maximum at a particle diameter of about 280 nm, since scattering is independent of hue. In this latter respect the coloured and white inorganic pigments are similar and sharply differentiated from organic pigments, whose refractive indices are, in general, much lower.

Opacifying power is only one of several factors that influence the manufacturer's choice of mean particle size for a given pigment. These other factors are related to the pigment colour.

White pigments

A white pigment ideally absorbs no radiation and therefore is colourless in its massive form. The whiteness and opacifying properties result from the scattering of incident radiation by multiple reflection and refraction by particles of pigments and are therefore functions of the ratio of the refractive indices of the pigment and its surrounding medium, and of the particle size. The refractive index is an intrinsic property but the particle size can be deliberately chosen. Whatever the actual size of the primary particles, the state of dispersion in the final pigmented product determines the effective particle size and thus the optical properties.

The dependence of opacifying power on the ratio of the pigment/medium refractive indices causes a number of white pigments to opacify only media of low refractive index; in oily, resinous or plastic media they are translucent or transparent. These low refractive index pigments are usually referred to as 'extenders' or 'fillers', and their chief use is for the adjustment of rheological properties of suspensions and the mechanical and optical properties gloss) of paint and plastic films. They can be used as opacifying and whitening agents in conjunction with low refractive index binders (e.g. glue, casein or egg-white) or very highly pigmented ('under-bound') films containing too little high refractive index binder to reduce substantially the opacity derived from light scattering at pigment/air interfaces.

The apparent opacity of a paint film containing microscopic voids is greater than that of a similar film in which the voids are filled with binder; also, the foam on a substantially colourless liquid appears white. It follows that a film of binder containing bubbles of air will be more or less opaque. The theoretical light scattering power of microvoids has been discussed by Ross,[6] who concluded that the maximum scattering power of bubbles at low concentration in resin is 12% of that of rutile, whilst an ideal system with 50% voids might have half the scattering power of a layer of titania pigmented gloss paint of equal thickness. The practical difficulties of producing a foam coating can be overcome, partially at least, by using vesiculated spheroids of appropriate size, and such products are now commercially available.

White pigment particle size

The greater the number of particles in a unit volume of film, the more effective is the scattering of the radiation entering the film, provided that the particle diameters are not less than half the wavelength of the incident radiation. The optimum particle size for opacification under a monochromatic illuminant is thus readily determined, but this is not the case for illumination by white light. The visible spectrum, which constitutes 'white light', ranges from about 400 to 720 nm, whilst the human eye has a peak response at about 560 nm, in the green portion of the spectrum. There is therefore no single optimum particle size for a white pigment and the apparent optimum size is dependent on the pigment content (opacity) and thickness of the pigmented film.

Light reaching a highly pigmented completely opaque white film is scattered and reflected in the upper layers of the film. If the pigment particles range from about 200 to 1000 nm the film will appear white, unless there is a 'clear layer' of substantially unpigmented medium at the surface of the pigmented layer. (In this latter case, interference effects may affect the apparent hue, even if the medium is completely colourless.)

If the pigmented film, applied to a light absorbing (black) ground is not completely opaque, then the apparent colour and opacity of the film becomes a function of pigment particle size. As an example, we may consider two films, with equal pigment contents and respectively containing uniform particles of 200 nm and of 360 nm diameter, so that there are 5·832 times as many small as large particles per unit volume of film. The refractive index of any pigment increases inversely with the wavelength of the incident radiation, so that blue light is always scattered more efficiently than red.

Some incident light passes through each film and is lost, so the apparent hues of the films are somewhat grey, i.e. less bright, than a fully opaque white film containing either pigment. The greater number of small particles, efficiently scattering the short wavelengths, more than compensate for any loss of brightness due to failure to reflect red light; in addition two or more small particles, close together, may function as a large particle and interfere with the longer wavelengths of incident radiation. The overall effect is that the film containing the small particles appears bluer and brighter than that containing the large, which appears relatively yellower and duller (greyer).

The practical significance of this effect is important in the design of white pigments that are to be used at low concentration in relatively thick films, as for example in plastics or rubber. For this kind of application, a pigment with a mean particle size of about 150 nm is found much superior to one with a mean size of 220 nm or more, such as is found most suitable for use in oleoresinous paints. (The optimum particle size for a general purpose pigment is also influenced by the effect of particle size on the hue of mixtures of white and coloured or black pigment. This is discussed later.)

Postulated models of pigmented film structure

The complex mechanism of opacification has caused various workers to postulate models of film structure in order to determine the hypothetical optimum white pigment concentration. The basis is usually that the uniform spherical particles of a hypothetical pigment are considered as located concentrically with larger spheres arranged in some particular way. The packing fraction or solid volume/gross volume ratio of any given array of uniform spheres is independent of the actual sphere diameter (see Table 9.1).

TABLE 9.1

DIMENSIONS AND SOLID VOLUME FRACTIONS OF UNIFORM SPHERES IN CLOSE-PACKED ARRAYS

	No. of contacting spheres	*Dimensions of unit volume*	*Solid volume fraction*
Rhombohedral	12	$d.(\frac{2}{3})^{1/2}d.(\frac{3}{4})^{1/2}d = d^3/(2)^{1/2}$	0·740
Rhomboidal	8	$d.d.(\frac{3}{4})^{1/2}d = (\frac{3}{4})^{1/2}d^3$	0·605
Cubic	6	$d.d.d = d^3$	0·524

In a given array, it can be assumed that each pigment particle, of diameter $d = \lambda/2$, is located about the centre of a hypothetical sphere of diameter $D = \lambda$ so that each pigment particle is separated from each neighbour by a gap $\lambda/2$ wide. The ratio of the volume of each pigment particle to that of its concentric sphere is given by

$$(d/D)^3 = (\tfrac{1}{2})^3 = \tfrac{1}{8}$$

Thus for various arrays of uniform spheres, the optimum pigment concentration is always an eighth of the solid volume fraction of the chosen array. The appropriate array to choose is arguable; the closest possible packing, with a packing fraction of 0·74, is a highly artificial concept. More appropriate would seem to be either the experimentally determined closest random packing with a packing fraction of 0·64 or, for a given pigment, the packing fraction calculated from its oil absorption (BS method) paste composition. In any case, the initial assumptions are so gross and the hypothetical structure so far divorced from the reality that any apparent agreement between the experimentally determined optimum pigment volume concentration and that calculated in this way can only be coincidental.

Black pigments

An ideal black pigment completely absorbs incident radiation of any wavelength; such a pigment does not yet exist. Actual black pigments vary in the completeness with which they absorb incident light, a fact that has a more marked influence on the undertone, the hue of a mixture of white and black pigments, than on the mass-tone, the hue of the undiluted pigment.

Light absorption is a function of molecular and crystal structure as well as particle size. Very finely divided metal particles are usually relatively black, if the particle arrangement and surface structure do not allow metallic reflection effects. This is true even for metals that are coloured in the massive state, such as gold. Most commercial metallic pigments are however more or less dark grey in hue, or deliberately made in a form that retains metallic reflection properties. The chief truly inorganic black pigment is the black oxide of iron, either natural magnetite, a mixed ferro–ferric oxide usually containing some silica and alumina, or the synthetic product which is similar but may contain additional compounds added to control crystal form and particle growth.

The most important black pigments are those based on graphitic carbon, produced by controlled oxidation of organic matter. The highest quality carbon blacks of maximum jetness were originally made by the channel process, but furnace blacks are now of equal quality, and the nature of the furnace process facilitates production of blacks with specific properties. The present situation in this field is complex: the channel black process, as traditionally operated, produces considerable atmospheric pollution, but in the USA the Toxic Substances Control Act (TOSCA) has led to the discovery that the lower temperature of the furnace process causes the product to contain a higher proportion of carcinogens. (As remarked earlier, the smaller the particle size of carbon blacks, the greater is the proportion of the pigment weight that consists of complex organic and metallo-organic compounds, deposited during particle formation.)

Mass-tone and undertone

The smaller and more uniform the particle size of carbon black, the greater is its jetness and staining power. Conversely, the larger the particles, the washier (greyer) is the mass-tone of the black. Particle size is difficult to define in the case of carbon blacks because of their tendency to form reticulated chains of primary particles; this tendency increases inversely with primary particle size and it is doubtful whether the very small particle blacks are ever completely dispersed in normal use.

The small particle blacks are too small to interfere with the longer wavelengths of the visible spectrum, so that when mixed with white pigment the resultant grey has a marked red–brown tone. Conversely, the larger particle size blacks, with fewer particles per unit weight or volume, do not interfere with the shorter wavelengths and when mixed with white pigment give greys with a 'clean, blue' tone. It should be noted that this effect of particle size is the reverse of that which occurs with white pigments; thus a small particle white with a large particle black gives a clean, blue undertone and a large particle white with a small particle black gives a red–brown, 'dirty', undertone. 'Small' and 'large' in these cases are relative terms; for white pigment a particle diameter of about 200–230 nm would be small and about 300–400 nm large, whereas the effective particle sizes of small and large carbon blacks are about 10–30 nm and 100 nm respectively.

Coloured inorganic pigments

The fundamental theory of colour of inorganic compounds has been developed to a point where in some respects it is in advance of that of the colour of organic pigment dyestuffs.[1] However, in the case of actual inorganic pigments, the situation is complicated by the greater dependence of the colour on crystal structure and particle size. Distortion of the crystal lattice by small amounts of other elements may cause marked variation in hue, as may the degree of hydration of a particular ore. The range of hues of the iron oxide pigments, from yellow through red to black, provides examples of these effects.

The defining characteristics of a coloured pigment are its hue and lightness or saturation. The curve of spectral reflectance from the pigment, illuminated by white light, normally shows a peak in one portion of the spectrum, corresponding to the major hue of the pigment. Some nominally 'pure' pigments have secondary peaks, whilst purples have concave curves with reflectance increasing at both the blue and red ends. The lightness of the colour is defined as the total fraction of the incident light that is reflected, regardless of wavelength, whilst the saturation (chroma or intensity) is defined as the intensity of the colour compared to a neutral grey of equal lightness. Colours with sharp peaks, reflecting substantially completely light of particular narrow bands of wavelengths, the spectral colours, are the most saturated.

The mass-tone of a coloured pigment is defined as the hue of the pigment undiluted with white, and often refers implicitly to the colour of the pigment in an oil or resinous matrix. The undertone is the hue of the pigment when diluted with white or applied to a white substrate in a very thin layer (as may be done with printing inks).

The hue of the mass-tone and the undertone of inorganic pigment may differ considerably, but both are influenced by the mean particle size of the primary particles in a fashion similar to the black pigments. Decreasing particle size causes a shift, especially in undertone, towards the red, increasing size towards the blue. An example is provided by a commercially available range of ten synthetic red iron oxides, said to differ only in primary particle size, ranging from 100 nm, 'strong yellow cast' to 1000 nm, 'strong violet cast'.

The staining power, or tinting strength, of pigments increases inversely with particle size, whereas the scattering power, being independent of colour, reaches a maximum at a particle size that varies according to the pigment concentration and thickness of the pigmented film. The size usually quoted is 280 nm, a half of the wavelength at which the human eye is most sensitive. The scattering effect is more important in the case of inorganic than organic pigments, since many of the former have high refractive indices, a primary cause of the frequently observed higher opacity of inorganic pigments compared with organic pigments of similar hue.

Purity of shade is proportional to the narrowness of the band of reflected light; in the case of inorganic pigments the width of the band may be affected by the particle size distribution. Except for some pigments whose tone is achieved by sintering primary particles to give the necessary pigment particle size, the milling processes used in paint manufacture do not have a significant effect on pigment particle size. The effect on hue of the state of dispersion of the pigment is most often observed in tints in the form of such effects as flooding and floating.

Dispersion media

The liquids in which inorganic pigments are dispersed can be classified as follows:

(i) Thermally liquefied solids.
(ii) Solutions of surface active compounds, either aqueous or non-aqueous.

(iii) Liquid film-forming materials.
(iv) Solutions of film-forming materials (mostly non-aqueous).

The majority of the first class of dispersions are of pigment in thermoplastics, where the dispersion is achieved by mechanical action and maintained by the high viscosity of the medium, that prevents pigment flocculation. Often, the material is rapidly cooled directly after dispersion has occurred, as for example in injection moulding using polystyrene granules previously dry-tumbled with pigment. This type of dispersion will not be discussed in any detail since the major influence factors are mechanical design and rate of cooling (rate of viscosity increase) of the polymer. These are dealt with in detail in a number of books.[7]

The second class of dispersion medium is the subject matter of Chapter 4 and any extended discussion would be repetitious. In passing, it should be noted that most inorganic pigments catalogued as self-dispersing in water contain some water-soluble dispersant, most often a sodium phosphate or similar material. The division between some of the surfactants and the media of the third and fourth types is not sharp; many of the latter are dependent for their dispersing power on the presence of small amounts of surface active material, such as free fatty acids in vegetable oils.

The third class of media includes material such as linseed oil in various forms, low viscosity alkyds and silicone oils and resins. In many cases, media that at first sight appear to belong to this group actually belong to the second class, because their intrinsic pigment dispersing powers are small and some 'wetting agent' is always added.

The fourth class includes the great majority of pigment dispersing media for use in non-aqueous paints. These are mostly alkyd resins, modified alkyd resins or oleo-resinous varnishes in the case of air-drying paints. Many baking enamels are based on a pigment dispersion in alkyd, to which is added a thermosetting amino resin, the principal alternative to these alkyd/amino resin media being the thermosetting acrylic resins.

Of the various media those whose behaviour as pigment dispersants have been most closely studied are linseed oil and alkyds.

Linseed oil

Linseed oil, being a natural product, varies according to the climatic conditions during growth and to soil conditions. The raw oil has limited compatibility with most synthetic resins; it contains a quantity of mucilaginous material, the 'linseed oil foots' that must be removed if the oil is to be thermally processed. Foots removal can be by prolonged storage (possibly the origin of the term 'stand oil') which was the process used by the 16th century Dutch artists. The more usual procedures are treatment with acid or alkali. The first-named, not unreasonably, gives an oil with a higher acid value that is a better pigment dispersant due to its greater content of free fatty acids and it is for this reason that the acid value of oil used for the determination of the 'oil absorption' value of a pigment must be that specified. An increase or decrease of one or two units of acid value can produce a corresponding variation of 10–20% in the oil absorption. A monomolecular film of the oil, assuming a molecular weight of 875 and a density of 0·933 g/cm^3, has a thickness of about 1·56 nm, a surface area of about 1 nm^2 per molecule and weighs about 1·46 mg/m^2.

Linseed stand oil, or linseed lithographic varnish, is made by thermal polymerisation of linseed oil. The viscosity of a stand oil may be from 2 to 1000 poise and therefore should always be clearly specified. The molecular weight distribution is usually fairly broad and both the number average and weight average values should be determined as a guide to the degree of heterogeneity.

Alkyd resins

Alkyd resins are formed by the polycondensation of a polyhydric alcohol (e.g. glycerol or pentaerythritol) with a polycarboxylic acid (usually phthalic) in the presence of a proportion of monocarboxylic fatty acid. The latter may be a vegetable fatty acid (FA) or a mixture of FA and rosin (abietic acid), in which case the alkyd is usually described as 'oil-modified'. Either drying (e.g. linseed), semi-drying (e.g. soya bean) or non-drying (e.g. castor, coconut) oils may be used as the fatty acid source, the drying/non-drying properties of the oil being transferred to the alkyd. If the fatty acid chain is unsaturated then vinyl compounds, most often styrene or α methylstyrene (vinyltoluene), may be introduced.

Oil-modified alkyds are generally classified as short-oil (30–50% FA), medium-oil (50–57% FA) or long-oil (>57% FA) types. Solubility in white spirit (which is usually about an 85/15 w/w mixture of aliphatic and aromatic hydrocarbons) requires at least 55% FA content, solubility in only aliphatic hydrocarbons rather more. Short-oil alkyds are usually dissolved in xylene or a similar type of solvent, whilst some very short (30–35% FA) alkyds require a solvent mixture of xylene with isopropanol or butanol.

Alkyds are complex mixtures, owing to the variety of secondary reactions that can occur. Thus, especially in long-oil alkyds, polymerisation of the fatty acids usually takes place, as well as some formation of esters of glycerol or pentaerythritol. The alcohols may etherify to some degree, especially if the 'monoglyceride' process is used, in which the vegetable oil is heated with glycerol at a molar ratio of 1:2 in the presence of a catalyst. Technical grade pentaerythritol may initially contain some di-, tri- or poly-pentaerythritol. Ring closure during alkyd manufacture is theoretically possible, but if it took place to any significant extent would almost certainly lead to gelation of the resin.

The classification of alkyds in terms of oil-length, based on the weight per cent of fatty acid in the base resin, diverts attention from the molar ratios of the constituents. For example, a typical long-oil alkyd may be described as a pentaerythritol (P/E) esterified alkyd, containing 24% phthalic anhydride (P/A) and 62% soya fatty acids. In terms of molecular ratios the ratio of P/A:FA:P/E is about 3:4:3·15, or in terms of functional groups, 6:4:12·6. In this form the excess of hydroxyl over carboxylic groups is apparent. Similarly, a glycerol esterified short-oil alkyd with 48·5% P/A and 31·0% coconut FA has molecular and functional group ratios of about 9:4:10·5 and 18:4:31·5 respectively. The ratio, 1·43, of hydroxyl groups to carboxylic is even greater in this case.

These values are characteristic, since it is usual to allow an excess of 5–10% by weight of polyol over the stoichiometric amount, because this is found to prevent an excessive rate of increase in viscosity, with its attendant risk of gelation during manufacture. The final alkyd resin always contains some 'free', unreacted, hydroxyl and also carboxyl groups which, through hydrogen bonding, make a considerable contribution to the melt viscosity of the resin. This can be shown by reacting the carboxyl groups with diazomethane, and the hydroxyl groups with acetic anhydride. The effect on the melt viscosity is very great, but on solution viscosity it is much less, and on the intrinsic viscosity virtually negligible.

The structure of alkyd resins is usually investigated by solvent fractionation,[8-11] but the results obtained are dependent on the solvent used. The basic alkyd contains molecules that differ in shape, size and degree of polarity; therefore fractionation using polar solvents will be by polarity, rather than size or shape, and conversely by non-polar solvents. Brett[11] found that a polar solvent mixture produced fractions in which the acid value decreased initially but was thereafter independent of increasing molecular weight. In contrast, with a non-polar solvent, acid value increased with the molecular weight and the weight of resin per carboxylic group correspondingly decreased. The hydroxyl values in general followed the pattern of the carboxyl groups.

Comparison of the number average and weight average molecular weights shows that the fractions are extremely heterogeneous, regardless of the type of fractionating solvents used. This extreme heterogeneity is a basic characteristic of alkyd resins and makes it very difficult to apply the solvent parameter concept to them in a meaningful way. All in all, the evidence from studies of alkyd structure indicate that only the broadest conclusions can be drawn about their behaviour in pigmented systems such as paints or inks. Discrepancies between the findings of various workers are at least as likely to be due to their use of different alkyds as to any other single cause.

Solubility of alkyd resins

A typical modified alkyd resin contains compounds ranging from residues of the initial ingredients up to polymers with a molecular weight of 500 000 or more. Since the monocarboxylic fatty acids are condensation chain-stoppers, the average and maximum molecular weights tend to increase inversely with the fatty acid content. The variety of components causes alkyd solubility to be greatly influenced by the mutual solubilities of similar components and, to a lesser extent, by the mutual insolubility or incompatibility of the most dissimilar compounds. This latter effect may become important if the stability of the resin solution is marginal and certain constituents are effectively removed by adsorption on the surface of pigment.

The fundamental requirement for mutual solubility (spontaneous mixing), according to Hildebrand and Scott,[12] is that the compounds should have similar cohesive energy densities, but it was found that to apply this concept to polymer/solvent combinations required the hydrogen bonding tendency of the solvent be taken into account. A practical method of applying these concepts to some resins and solvents used in the surface coating industry was described by Burrell,[13] since when various workers have refined the original method, generally by subdividing Hildebrand's 'solubility parameter' (the square root of the cohesive energy density) into dispersion, polar and hydrogen bonding components. All the better known methods are however based on the theory of regular solutions proposed by Hildebrand[14] and Scatchard[15] and applied to polymer solutions by Flory[16] and Huggins.[17] This assumes that there is no strong association between molecules, which may be valid for dispersion and polar forces but is probably wrong where hydrogen bonding forces are significant. A method for overcoming this problem has been proposed by Nelson *et al.*[18] which utilises infrared frequency shifts to calculate a 'net hydrogen bond accepting index'.

Determination of this hydrogen bonding index for an alkyd would require the prior separation of the constituent molecules of the resin. This is not feasible, yet alkyd resins are known to have their melt viscosities increased by hydrogen bonding. This effect largely disappears when the resin is in solution, but figures quoted by Burrell[13] for a group of alkyds and soya bean oil suggest that the original solubility parameter concept is not of much use in studying the interaction of alkyd resins with solvents or solvent combinations.

An alternative approach is the study of resin solution viscosities to determine the effect of solute/solvent interaction on the mean equivalent hydrodynamic volume of the polymer molecules.[13] Such data would provide a useful check on information obtained from studies of the effect of the solvent on alkyd adsorption on pigment surfaces, where selective adsorption of certain resin components may occur. The findings of both the solution viscosity and the adsorption experiments are valid only for the resin used; no general rules or equations can be derived from them.

Solvents

The chief functions of the solvent in a pigment/film-former/solvent combination are to adjust the viscosity of the suspension and, at a later stage, to evaporate at a suitable rate. In recent years, the choice of solvent has ceased to be a technical matter; very often, the most important factors are the various regulations enacted to protect health and environment. The regulations vary from country to country, and within countries are liable to change. From a technical viewpoint, their greatest effect has been to promote the use of water as the volatile component of coatings, though in this matter the regulations have been reinforced by economic considerations. Neither the regulatory nor the economic aspects are strictly relevant to the discussion here.

The dissolving of a film-former involves a solute/solvent interaction, one effect of which is to influence the configuration of the film-former molecules. This in turn can affect the configuration and/or amount of film-former molecules adsorbed on the pigment surface and consequently the degree and stability of the pigment dispersion. Also, the solvent itself may be capable of dispersing or flocculating the pigment, though the correlation between pigment behaviour in binary (pigment/solvent) and ternary (pigment/film-former/solvent) systems is not readily predictable.

Solvent selection has been shifted from a wholly empirical to a qualitatively predictable basis chiefly by the proselytising of Burrell[19,20] on behalf of Hildebrand's[12,14] *solubility parameter* concept. The original simple technique suggested by Burrell used three groups of solvents, representing poorly, moderately and strongly hydrogen bonded types, each group containing solvents covering about the same range of solubility parameter. The solubility parameter of a resin was determined from the limiting values of the solvents in which a stable solution was obtained. Various refinements of the method have been proposed, involving division of the solubility parameter into three components, representing (London) dispersion, dipole and hydrogen bonding intermolecular forces. The methods employed by the authors to calculate these forces vary in the soundness of their theoretical foundations; in every case the greater work involved in using the improved method seems largely unrewarded. An excellent review of many of these papers and other aspects of the concept has been provided by Burrell.[20] A critical review of the shortcomings of the basic concept has also been made by Hildebrand.[21]

The solubility parameter permits a rational explanation of a number of apparently anomalous solubility effects, and allows qualitative predictions about solubility in various solvents. The concept is not very helpful in explaining the discrepancy between the postulated influence of the solvent on polymer solution viscosity at low and at high solute concentrations.

It is postulated[22] that in a 'poor' solvent each polymer molecule occupies the minimal volume, its equivalent hydrodynamic volume is therefore minimal and the increase in viscosity due to the presence of a solute is at a minimum. In a 'good' solvent, the polymer molecules are expanded and the converse effects are produced. These relationships have been confirmed for dilute solutions of polymers having fairly narrow molecular weight distributions.[23] From experiments with solutions of a short-oil alkyd, solvent/viscosity relationships that were substantially the reverse of these were reported and alkyd micelle formation in weak solvents was tentatively suggested as a possible cause.[24]

In the particular case of alkyd resins, as mentioned earlier, the heterogeneous nature of the resin must have a significant effect on its behaviour in solution. In the general case, association between solute and solvent molecules would increase the effective volume of the disperse at the expense of the continuous phase, an effect that would be exaggerated at higher solute concentrations. The formation of polymer micelles or aggregates cannot be ruled out, especially near the 'throw-out' point, at which resin molecules coagulate and precipitate from solution; such aggregates could immobilise a large proportion of the continuous (solvent) phase within themselves, with a resultant marked increase in solution viscosity. Useful papers have been published by Shaw and Johnson,[25] Bobalek *et al.*[26] and, in particular, Reynolds and Gebhart,[27] who dealt with solution concentrations of practical interest.

The influence of the solvent on the pigment in a suspension seems largely dependent on the presence or absence of a film-forming material, as well as on the pigment dispersing capacity of the film-former. The interaction of inorganic pigments and solvents was studied by Weisberg,[28] who used dielectric constant as the characterising value for the solvents and also took into consideration the influence of the trace electrolytes inevitably present in a commercial pigment produced by a precipitation process. He was unable to confirm the specific sedimentation volume effect postulated as a characterising constant of a pigment by Dintenfass,[29] but his results showed that in general the sedimentation volume decreased inversely with the dielectric constant of the solvent, with certain exceptions. The latter were provided by the behaviour of the pigments in the solvents with the lowest dielectric constants; the pigments formed dense approximately spherical aggregates. This effect can also be produced very easily by exposing a typical chalk-resistant grade of rutile titanium dioxide to a high humidity atmosphere, then trundling the high moisture content pigment in benzene. The pigment rapidly forms small indented aggregates, reminiscent of miniature golf-balls. (Trundling the moist pigment in ethanol or butanol produces a relatively stable dispersion, as might be expected.)

Attempts[29,30] to correlate pigment dispersion with solvent parameter have not been very successful. Hansen[30] attempted to determine a solvent parameter value for pigments, but was not very successful in obtaining reproducible results. An attempt[31] to correlate pigment dispersion with solvent parameter was not very successful; the influence of intermolecular association of the solvent molecules appears to be greater than that of the solvent parameter, and this aspect of solvent structure seems more readily characterised by dielectric constant.

Sorensen[32] was able to show a quantitative relationship between viscosity of certain printing inks and the hydrogen bonding potential of the vehicle. The relationship varied in form between inorganic pigments (titanium dioxide, chrome yellow) and two groups of organic pigments and also between film-formers (calcium resinate, Versamid 930 polyamide resin and SS grade nitrocellulose). This differentiation between inorganic and organic pigment behaviour was also observed by the Toronto Society for Paint Technology.[31]

It seems likely that the effect of solvents on pigment dispersion behaviour is more greatly influenced by internal bonding forces between the solvent molecules than by the solvent parameter itself. Pending the development of a rigorously founded theory of the structure of liquids, a quantitative theory of solution properties will be difficult to formulate. For ordinary dispersions, made from commercial materials, the influence of trace impurities, batch to batch variations and, in the case of pigments, the variation of adsorbed moisture content with atmospheric humidity, may greatly reduce the practical value of a quantitative treatment when it becomes available. None of the elaborations so far suggested seems to offer a truly significant advance over the original method of using the solvent parameter proposed by Burrell[13] in terms of information obtained for work involved.

CHARACTERISATION OF INORGANIC PIGMENTS: PHYSICAL AND CHEMICAL ASPECTS

It is quite true that the only characteristics of a pigment of importance to the user are those that may be manifested in the final product, whether printed ink, dry paint film or plastic moulding. On the other hand, it is impossible to quantify the factors that determine the success or failure of the pigment in a particular application unless the initial physical and chemical characteristics of the pigment are known. Of these, probably the most important are the particle shape, size and size distribution, since these have a major influence on the rheological characteristics of the pigment suspension. Also, comminution of pigment particles during the dispersion process is unlikely to be achieved by any machine other than the ball mill. Knowledge of the pigment size distribution is therefore necessary, so that before large scale manufacture of a pigment dispersion is undertaken, the probability of obtaining a dispersion of suitable quality, using the proposed equipment, can be assessed.

Assuming that the pigment can be satisfactorily dispersed, the stability of the dispersion will in most cases depend on the adsorption on the pigment surface of components of the liquid phase. This adsorption can occur only on a surface accessible to the adsorbate and will be influenced by the physical structure and chemical composition of that surface.

Particle shape

Individual particles of pigment are visible only if magnified. Inorganic pigment particles are either fragments of natural minerals or the products of a chemical process. In the former case, the variety of shapes and the range of sizes is likely to be larger than in the latter, but materials prepared by precipitation on an industrial scale are rarely perfect crystals. Formerly, the problem was that the magnified image of the particle was a two-dimensional representation of a three-dimensional object, but the development of scanning electron microscopy has been such that it is now possible to obtain stereoscopic pictures of individual particles. One example of the information yielded by such pictures is that particles of chalk whiting are built up from remarkably well-shaped aggregated crystals of calcite.

It is obvious that scanning electron microscopy is a very useful tool for the study of the structure of pigment aggregates and agglomerates, but one problem remains. This is that the number of particles per unit weight of pigment is extremely large, so that a great many particles must be examined before any definitive statement is made about the 'particle shape' characteristic of a given pigment.

Particle size and size distribution

The value described as the particle size of a pigment is almost always an average particle diameter derived in one of the following ways:[33]

(i) Median diameter: d_{med}. The mid-point value of the cumulative size distribution curve, i.e. 50% of the sample is larger and 50% smaller
(ii) Arithmetic-mean diameter: $d_1 = \Sigma nd/\Sigma n$ where n is the number of particles in each class d
(iii) Length-mean diameter: $d_2 = \Sigma nd^2/\Sigma nd$
(iv) Volume-surface mean diameter: the diameter of a particle with the same ratio of $d_3 = \Sigma nd^3/\Sigma nd^2$ surface to volume as the whole of the sample. (This value is usually obtained from surface area measurements.)

The kind of diameter chosen to describe the particle size is often influenced by the method used to determine the size distribution, but in every case the use of an average value involves the almost always incorrect assumption that all the particles are of the same shape. A fact of greater importance is that the numerical values of the various diameters may vary by a factor of two, or more, for the same pigment.

The practical problems of defining the particle shape and the relationship between particle sizes as determined by various methods of sizing analysis were discussed by Heywood.[34] The situation is made more complex by the fact that particulate materials of great industrial interest are, compared to pigments, of very large size indeed (cf. Richards[35]).

Broadly speaking, the size of pigment particles can be measured in terms of the cross-sectional area of a two-dimensional image of the particle, of the specific surface area, of the particle volume, or of the sedimentation velocity. That is to say, by electron microscopy, by the BET gas adsorption method, by a Coulter counter, or a disc centrifuge. There is a considerable literature on these methods, and they will be considered here as known.

Surface characteristics

The preparation of a more or less stable dispersion of very fine particles requires the presence within the system of a stabilising force. The possible stabilising methods are the subject of Chapter 1 and will not be discussed here, except to observe that in many cases a substantial proportion of the surface area of an inorganic pigment may be contributed by pores and fissures that cannot be entered by molecules larger than that of water. The characterisation of pigment surfaces has been discussed very fully by Parfitt and Sing,[36] and of porous solids by Gregg, Sing and Stoeckli.[37] An indication of the effect of micropores and fissures on the specific surface area of a pigment is provided by Urwin.[38] Using a modification of the *t*-plot method originally proposed by Lippens and de Boer,[39] he found that coating increased the surface area of a titania by about $4\,m^2/g$, but only some $1{\cdot}5\,m^2/g$ of that increase would be accessible to molecules larger than that of water. It is this 'accessible' area that should be used when calculating the adsorption of oils or resins per unit area of pigment surface.

It should be noted that, in almost all cases, the first step in measuring pigment particle size is the preparation of a pigment dispersion. The methods are discussed in Chapter 8. The amount of work done on the pigment during the preparation varies considerably, both between methods and in many cases between operators using the same method. The determination of particle size thus may become a method of assessing either the comparative efficiency of different methods of pigment dispersion or the comparative ease of dispersion of pigments by a given method. As such, they have their uses, but the results obtained require careful interpretation

Rugosity, roughness or smoothness factors

There is a need to amplify data such as mean particle size by some indication of the deviation of the particles from sphericity. The deviation may on the one hand be only slight, caused by indentation or roughening of the particle surface, or on the other hand the particles may be far from spherical, rough surfaced and angular. Another problem, for which it is difficult to find a satisfactory solution, is provided by pigments such as carbon black and Prussian blue that consist of agglomerates of fine particles with a considerable internal volume; the structure of the agglomerates may be such that the void spaces are roughly spherical, or at the other extreme they may consist of very narrow pores. Often one sample of pigment may contain both types of structure.

The established methods of characterising such pigment properties require that the pigment size distribution be known to a fair degree of accuracy. Measurement may be by either sieve analysis (not a suitable method for pigments), sedimentation or computation from electron micrographs. From the measured size distribution of the pigment may be calculated either the number of uniform spheres having the same mean diameter, density and total weight as the fraction, or the surface area per unit weight of the pigment.

Robertson and Emödi[40] suggested that the quotient of the specific surface area divided by the surface area of the equivalent spheres should be defined as the coefficient of *rugosity*. In the original work the specific surface was measured by the permeability method of Lea and Nurse,[41] using a modified cell.

Anderson and Emmett[42] proposed that the quotient of the BET nitrogen surface area, divided by the corrected area computed from the electron micrographs, should be termed the *roughness factor*.

The reciprocal of the roughness factor has been termed the *smoothness factor* (or smoothness ratio).

The coefficient of rugosity is reasonably satisfactory only as a means of indicating the surface roughness of individual particles; the problem of determining the specific surface in a more satisfactory manner requires consideration. The coefficient of rugosity is, nevertheless, of practical significance. Discrete particles with a highly indented surface can behave in a disperse system as particles of a size equivalent to the 'shell' which could enclose the particle. The liquid filling the indentations is effectively removed from the continuous phase, so that the effective solid volume fraction in the system is much larger than the calculated value, and the viscosity of the system correspondingly increased. This effect was verified, albeit for particles considerably larger (38–279 μm) than most pigments, by Ward and Whitmore.[43]

The roughness and smoothness factors have each been used to indicate the porosity of carbon black and Prussian blue agglomerates; they cannot be considered as ideally suited to the job. The roughness factor, for example, when applied to material such as carbon black can reach a value of 2–2·5 without requiring any significant degree of true porosity.

The effect of milling (micronising) of pigments

Optimum performance, in any particular application, is given by pigment particles of a particular size, so the problem of producing material having a fairly narrow size distribution has received considerable attention. Dividing the final product into size classes is one way but not the most economic, since it is likely neither that all the fractions will be in equal demand nor that the most copious fractions will be those in most demand. In the case of synthetic products careful control during manufacture may ensure that the size of the primary particles approximates to the optimum, but it is difficult to prevent the formation of some sintered aggregates in the case of materials subjected to calcination, or of cemented aggregates if a precipitation process is used. Products made by the comminution of minerals usually have an undesirably wide spread of particle size, and some form of classification should take place concurrently with the grinding process, so that material ground to the correct size is at once removed from the milling zone.

Pigments present particular problems because of the small size of the particles. For optimum ball milling efficiency the grinding elements should be of similar size to the material being ground, a condition which cannot be conveniently fulfilled if the grinding elements and the milled products are to be separated readily. Contamination of the product by material abraded from the mill also occurs. The logical solution to the problem was to devise a mill in which the feed was ground by impaction on itself, contact with the walls of the mill minimised, the smaller particles separated and removed from the grinding zone, and the larger particles concentrated in the zone in which the milling action was at a maximum.

A number of mills were designed to meet these requirements, of which the best known is the *microniser*, developed at the Dutch State School of Mines. The process is also known as *jet-milling*, or *fluid-energy milling* but, in Europe at least, the term *micronising* is used to describe any process in which milling is carried out by acceleration of the particles of the feed so that, usually in conditions of turbulent flow, they impinge upon each other.

In the case of the true microniser the powder which is to be milled is fed into an annulus, around the outer circumference of which are nozzles through which superheated steam, compressed air or inert gas is injected tangentially to the inner wall. The particles of powder circulate in the high velocity gas stream, centrifugal action ensuring that the largest particles remain near the circumference (in the zone of maximum velocity) whilst ultra-fine material is carried by the escaping gas stream toward the centre and then upward, out of the mill. The maximum velocity and the intersections of the streams of gas from the jets ensure turbulence in the gas stream so that there is very considerable interparticle contact, which is the only substantial source of grinding action. Where the feed to the microniser consists of markedly flocculated or aggregated material then occluded air or moisture may expand rapidly, when heated in the annulus, and cause some disruption of the floccule. This mechanism is not considered to be of great significance and in any case it is essential that the moisture content of the pigment fed to the microniser be kept low, to minimise latent heat losses. If steam, used to drive the microniser, becomes 'wet', then milling efficiency falls; in some cases the pigment may be pelletised, instead of broken down into particles.

The net effect of micronising the pigment is to increase the ease of dispersion by paint milling machinery, and this improvement is particularly notable when the dispersion process is carried out in mills dependent on high speed stirring action to achieve pigment dispersion (e.g. the Kady, Cowles, Torrance cavitation mills, etc.).

Whilst it is accepted that micronised pigments need less milling, there is little physical evidence to indicate the reason for this. The properties of the micronised pigment, relative to the feed, may be summarised as follows.

(i) The milled pigment particles tend to be more rounded, with smoother surfaces.
(ii) The surface area measured by nitrogen adsorption is usually lower, and never greater, than the original surface area.

(iii) The particle size distribution usually shows that oversize particles have been broken down, and the finer particles removed (drawn away in the exhaust gases), so that the size distribution is narrowed and the average size reduced.
(iv) Highly acicular pigments tend to fracture across the main crystal axis.
(v) Pigments such as Prussian blue, which are highly agglomerated when fed to the mill, tend to show the most startling size reductions, but are not normally broken down to primary particles.
(vi) Coated pigments, such as fully processed grades of titanium dioxide, show effects (i) to (iii) above, but do not fracture or lose their coating under normal micronising conditions.

A number of papers have been published on the use of organic additives to improve the efficiency of micronising of minerals. Various grades of titanium dioxide, in particular, are now available which are coated, in part at least, with organic compounds added prior to micronising. Most often used are amines, amine-soaps and polyhydroxy compounds, any of which are readily adsorbed on pigment surfaces. By choosing a compound of suitable structure the organically treated surface may be made hydrophobic or hydrophilic in character and, in addition, the micronised pigment has a slightly narrower particle size distribution than a 'control' without the additive.

Treated pigment is more readily wetted by an appropriate medium so that a smaller proportion of the pigment milling time is required for wetting and disruption of agglomerates and a correspondingly greater proportion is available for the dispersion stage. However, it is not easy to decide whether the greater ease of dispersion of these pigments is due to more rapid wetting because of the organic material already adsorbed on their surfaces, or to the narrower particle size distribution associated with the use of organic additives when micronising.

It is established that the apparent ease of dispersion of most pigments is improved by micronising, but this improvement is very often measured by the reduction in the time required to obtain a particular 'fineness-of-grind' reading. About 40 000 uniform spheres of 0·25 μm diameter in random close packed array could be contained in a sphere of 10 μm diameter, so fineness-of-grind gauge readings do not give any useful guidance to the actual state of dispersion of particles less than 1 μm in diameter. The gloss of alkyd/amino resin based baking enamels provides a more useful guide, and data obtained in this way suggest that the effect of micronising is a physical one. Given adequate milling, the baking enamel containing an uncoated unmicronised rutile will have the same gloss value as that containing a fully treated (coated and micronised) rutile.[44] Data on the comparative hiding powers of very well dispersed uncoated, unmicronised rutile and fully treated (coated and micronised) rutile in an air-drying long-oil alkyd medium have been published.[45] The paint compositions were not adjusted to compensate for the lower content of opacifying pigment in the treated grade (the low refractive index coating oxides do not make a direct contribution to the hiding power) and the plots of hiding power against pigment volume concentration (p.v.c.) showed that after prolonged (7 days) laboratory ball milling the paints containing the untreated grade were slightly superior in hiding power at the higher p.v.c. values. In the lower p.v.c. range the superior particle size distribution of the treated pigment compensated for its lower content of actual opacifying pigment and the opacities of the paints were very similar. These results indicate that both pigments were dispersed to the same extent in the dry paint films.

It seems probable, therefore, that the principal effect of micronising is the breakdown of aggregates and agglomerates, so that only comparatively weakly bound agglomerates, formed by the unavoidable flocculation of the pigment in air, have to be disrupted during the dispersion process. The adsorption characteristics of the pigment surface are probably not much affected by micronising, apart from the changes in apparent characteristics caused by the rounding of the particles. Pending the publication of data correlating the effect of micronising, adsorption characteristics and degree of dispersion, the improvement in ease of dispersion associated with micronising seems attributable to the modification in particle shape and size distribution rather than any other cause.

Structured pigments; carbon blacks

The terms 'low structure' and 'high structure' are of rather obscure origin, but seem to have come into use in the rubber industry to describe the relative effect of a reinforcing carbon black on the consistency of the rubber during compounding. The term is now applied chiefly to carbon blacks, but sometimes also to iron (Prussian) blues and materials such as diatomaceous silica. The common characteristic of these pigments is that their particles consist of agglomerates and/or aggregates of very small primary particles. (Diatomaceous silica, fossil plant skeletons, is a special case.) As a result, when suspended in liquid, each particle has an effective hydrodynamic volume much greater than its nominal size, owing to the quantity of the liquid that is immobilised in the interstices of the pigmentary particle.

The most intense studies of the nature of structured pigments have been made on the carbon blacks, especially the rubber-reinforcing (breaking-strain increasing) grades, where it has been shown[46,47] that the characteristics of the vulcanised rubber can be predicted from the physical characteristics of the black used in it.

The more recent studies of the fine structure of the primary particles of carbon blacks[48,49] have shown these to consist of distorted graphitic carbon layers, without any actual crystals being present. These primary units of carbon black occur in either a fibrous (e.g. channel and furnace blacks) or spheroidal form (thermal blacks). From these primary units are built up three-dimensional aggregates that the more powerful modern electron mciroscopes show to be quite different from the convoluted chain structures formerly proposed.

The highly structured blacks consist of relatively large aggregates that have more 'open' structures and therefore occlude a larger proportion of fluid relative to the volume of black forming the particle. Characterisation of the blacks is carried out either by visual comparison of electron micrographs of the aggregates with a range of standard micrographs, or by an automated scanning method[50] that measures the optical density of the photographs, the darkness of the picture being proportional to the thickness of the aggregate. (This has the disadvantage that a two-dimensional image is being used to characterise a three-dimensional object.) The data are used to characterise the particles in terms of estimated size and number of particles, cross-sectional area and maximum linear dimension. A shape factor is calculated from the relation of the major dimension to the diameter of a circle of equal area. The sphere generated from this circle is described as the *equivalent sphere* and the ratio of the volume of this sphere to the solid volume calculated from the size and number of the particles provides a measure of the 'structure' of the black. Oil and dibutyl phthalate (DBP) absorption values calculated from data obtained in this way show good agreement with the experimentally determined values.[51]

All carbon blacks are of relatively low intrinsic density and have in addition high bulking values (i.e. low bulk densities). For these reasons, densified or pelletised forms of these blacks are popular. The spheroidal thermal process blacks and the low structure furnace and channel blacks will densify quite readily, but the particles of the medium and especially the high structure blacks do not readily pack closely together and therefore resist densification. In all cases, the work done on the pigment in the densification process has to be nullified during the pigment dispersion process. Dispersion methods that allow application of shearing forces powerful enough to disrupt the black pellets are therefore essential. Surfactants may assist pellet disruption, but the secondary effects of such materials on the performance of the product in its intended use should be taken into consideration.

The composition of the pigment surface

Information about the composition and nature of inorganic pigments that is relevant to the behaviour, during the dispersion process and subsequently, of the pigment is, not surprisingly, scarce. A large proportion of the information that is available unfortunately refers to pigments made in the laboratory for the investigation, whilst a still greater proportion of the information is reduced in value because of inadequate characterisation of materials, pigment chemists paying little attention to the media, resin chemists treating pigments in similar cavalier fashion. It is unfortunate, too, that in many cases the behaviour of the pigment is greatly influenced, if not controlled, by trace impurities such as residual electrolytes (from the manufacturing process) that cannot be completely leached out. In addition, in normal industrial conditions the surface of a pigment is covered by about a monolayer of water adsorbed from the atmosphere. The fact that most coloured pigments are likely to be blended to maintain their mass-tone and staining power within specified limits is rarely considered even though in the case of such pigments as lead chromes this may involve quite major variations in chemical composition between one batch and another.

Another factor that is not much considered is the tendency for impurities to concentrate in the surface layers as the crystal grows in its mother liquor, or during a calcination process. The surface layers may therefore differ substantially in chemical composition from the bulk, and the trace impurities may have a disproportionate influence on pigment dispersion behaviour. It is possible that by the deliberate incorporation of 'impurities' during manufacture a pigment with the required surface could be obtained without an additional process such as coating with other oxides.

The composition and structure of the surface layers of a pigment also affect other properties besides dispersibility. Clay[52] showed that the effectiveness of an alumina coating as an improver of lightfastness of lead chromes is reduced as the proportion of coating oxide relative to base pigment is increased. He also stated that a mixture of the hydrated oxides of aluminium and titanium or silicon is superior to either of the additives used singly. An alumina coated chrome was found to be more resistant to darkening when a mercury vapour lamp is used than with a tungsten light as a source; a silica-on-alumina coating gave maximum resistance to mercury light while an alumina-on-silica coating is superior when tungsten lamps are used. Unfortunately Clay did not indicate whether tests were carried out under conditions of equal total energy input to the films.

Coating with other oxides is most firmly established in the titanium dioxide pigment field. These pigments are made either by the sulphate process from a ferrotitanic ore, usually ilmenite, or by the chloride process from natural rutile. The latter method avoids the massive quantities of by-product (ferrous sulphate) and acidic aqueous effluent of the older process; in addition purification by distillation of the intermediate, titanium tetrachloride, is simpler than the wet methods used in the sulphate process and allows production of material of better colour at acceptable cost. The final products are essentially similar, either anatase or rutile types produced in the sulphate process by fusion of crystallites to pigmentary size in a calciner or by controlled crystal growth in the chloride process. The untreated (unrefined) pigments consist of primary particles with no coating oxides, although most rutiles contain a small amount, about 1% w/w, of other oxide, either zinc oxide, alumina, antimony oxide, or alumina and antimony oxide mixture. These latter oxides are added to encourage growth in the rutile form. They usually cause some degradation of the crystal lattice, zinc oxide in particular causing the particle to be somewhat rounded. This is considered desirable, since it prevents the abrupt discontinuity in surface properties that occurs at the crystal plane junctions of a perfect crystal.[53] Both anatase and rutile have the defective lattice structure associated with semiconductor and catalytic properties.

Treated, or coated, titanium pigments have a surface layer of oxides, most often alumina, silica, titanium dioxide and sometimes zinc oxide. These coatings are precipitated on to an aqueous suspension of the primary particles, either simultaneously, one at a time, or in various combinations. The titanium oxide coating is often applied first, since it seems to assist the subsequent adhesion of the alumina and/or silica, though this adhesion-promoting effect seems to function equally well when alumina and titania are co-precipitated.

Apart from the amounts and kinds of oxides used to form the coating, and whether they are precipitated as single oxides, mixed oxides, or part singly, part mixed, the physical form and crystal structure of the oxide coating is markedly dependent on the conditions used, all of the following factors being important:

Temperature
Concentration of reagent solutions
Rate of precipitation (i.e. speed of addition of precipitating reagent)
Time in contact with mother liquor.

The coating technique used for pigments intended for general application is designed to give a relatively smooth uniform thickness of coating oxides, which constitute about 5–7% w/w of the final pigment. More recently introduced types, intended specifically for use in polymer latex based paints ('emulsion paints'), consist of about 85% TiO_2, 15% coating oxides, with coatings applied in such a way that they form dense spiky protrusions from the crystal surface.

The objective in coating application is to ensure that the initial precipitated crystallites adhere to the titania particles and then grow in a controlled fashion. Kämp and Völz[54] have published electron micrographs showing the type of result produced by an unsuitable coating technique. Simple accretion of discrete precipitated crystals does not produce a satisfactory coating, but even a relatively smooth, dense coating of the type applied to 'general purpose' pigments is inevitably fissured by the gaps between adjacent coating crystal growths, and is likely to have also some pores in the coating. These pores and fissures cause a disproportionate increase in the BET surface area, typically from about 8 m^2/g to about 12 m^2/g for a coating that, examined under the electron microscope, is uniform, with a fairly smooth surface, and a thickness of about 5 nm. Of the new surface, some 2·5 m^2/g is contributed by the pores and fissures accessible to a water molecule, but not to larger molecules.

The coating oxides, as applied, carry down with them water, either as hydrates or occluded, as well as small amounts of water-soluble salts, components of the precipitation reaction. The coated pigment, therefore, has to be washed very thoroughly (but some traces of soluble salts are always left) and then dried, before receiving a final milling treatment, usually micronising or an equivalent process.

As noted earlier, some grades of pigment have on their surface residues of an organic compound added prior to micronising. This varies from 0·5 to 1·0% w/w according to the compound and probably forms a monolayer. Another end-treatment involves exposure of the pigment to methyl siloxane vapour, with the object of producing a pigment surface covered in methyl groups, bonded through the silicon atom. This type of treatment is intended to produce a hydrophobic surface with less tendency to adsorb water vapour and with a greater ease of 'wetting' by polymers, such as polyethylene, in the course of a plastic moulding process.

The oxide coating can be heat-treated, causing many of the pores and fissures to close as the coating fuses. Improved 'chalk-resistant' properties are claimed for these pigments, since, in theory at any rate, the semiconductor titania crystals are sealed away from the surrounding film-forming material, and therefore cannot act as oxidation catalysts.

A similar 'annealing' effect also takes place at ordinary temperatures. Aided no doubt by the moisture that in ordinary circumstances is inevitably adsorbed on the particle surface, some crystals join together, so that after about a year there is a small, insignificant drop in specific (BET) surface area. Parallel with this effect is a fall in the so-called 'reactivity' of the pigment, measured as the rate of viscosity increase of a paint made from the pigment and an intrinsically unstable varnish. The viscosity increase seems to be due primarily to an adsorption mechanism, the pigment particles acting either as nuclei or as bridges between polymer particles. The fall in 'reactivity' with time may be due to the reduced surface area, a reduction in the number of crystal plane junctions (high activity zones) or a reduction in surface energy caused by hydration of the pigment surface.

In the absence of organic material, the surface of a coated titanium dioxide will nominally consist chiefly of alumina, silica, or both, although in fact the true surface is most likely to be about a monolayer of water on a surface of hydrated alumina and/or silica. The chemical characterisation of such surfaces has been discussed by Parfitt and Sing.[36]

The surfaces of the pigments just discussed are all essentially inorganic, except for those that have a small amount, possibly a monolayer, of some specific organic compound. The chemical and physical nature of the surfaces of carbon black pigments presents a very different picture. The physical form of these pigments ranges from the spheroidal particles of thermal blacks by way of the fairly simply shaped aggregates of the *low structure* to the very complex shapes of the *high structure* blacks. The very small primary particles consist of graphitic layers, but not graphite crystals, so that the unaligned stacking and the distortion of the surface layers provide a very high energy surface. The distorted or bent layers may be associated with dislocations or even atomic vacancies in the lattice. Since the channel and furnace blacks are produced by burning, these blacks have a surface that is partially oxidised (or incompletely reduced) and consists of various oxyhydrocarbons. The thermal blacks, produced by the high temperature *cracking* of natural gas or acetylene, without combustion, have as a result almost unoxidised surfaces, in marked contrast to the former types.

The physical form of a carbon black is partly dependent on the feedstock, traditionally natural gas for channel blacks and also for some furnace blacks, though a large proportion of modern furnace blacks are made from oil. This difference in the feedstock also affects the chemical composition of the surface. The usual channel black process produces considerable atmospheric contamination and anti-pollution regulations have forced closure of some plants, but some new plants are claimed to be able to meet the strictest requirements. The furnace process is intrinsically cleaner, so that there has been considerable effort to produce blacks by this process that are closer in pigmentary properties to the channel blacks.

One step in this direction was taken when it was found that injection of alkali metal salts into the furnace reduced the flame ionisation potential and resulted in the production of blacks with much less structure, down to the channel black levels. The effect of alkali addition of this type is reflected in the pH value of an aqueous slurry of the black, but this property is also dependent on the volatile content and thus on the manufacturing process.

The volatile content of carbon black pigments is usually expressed as percentage weight loss on heating at about 950°C in the absence of air. It consists of carbon– and hydrocarbon–oxygen complexes and is usually slightly acidic in character. The volatile-free carbon black surface is slightly basic. An aqueous slurry of a thermal black, that has only about 0·5% volatiles, has a pH of about 9·5. Oil furnace blacks, with about 1% volatiles give a pH of 8·5. An original process channel black has about 4–5% volatiles and gives a pH of about 5. Certain channel blacks are heated to red heat in air as a secondary stage in manufacture. This increases the volatile content to 10% or more and the pH drops correspondingly to 3·5–3·0. The treatment also increases the nitrogen adsorption surface area, but chiefly by the formation of internal pores too small to affect the amount of surface available to paint or ink media.

Rivin[55] analysed the volatile constituents of carbon black and postulated a stepwise oxidation process, from phenols and/or hydroquinones (weakly acidic) through quinones (neutral) to either lactones (neutral) or carboxylic groups (strongly acidic). Further oxidation caused carbon dioxide formation, with pore formation as its consequence. Scott[56] has published figures for the contents of the above chemical groupings in the volatile content of various commercial blacks and Dollimore[57] has reviewed the more fundamental aspects of black surface characterisation.

A generalised summary of the relationship between surface composition and ease of dispersion of carbon blacks is not very useful, since the dispersion properties depend in the first instance on the ease with which the black is disrupted into the aggregates that function as pigmentary particles and this in turn depends on the past history of the black, for example whether it has been densified (pelletised) deliberately or accidentally. The rheology of a carbon black suspension is in turn primarily dependent on the physical nature of the aggregates, that is, on the structure, low, medium or high that determines the volume of liquid occluded by a unit volume of black. The chemical nature of the surface is important only insofar as it affects the speed of wetting and the tendency of the particles, subsequent to dispersion, to adsorb components from the suspending fluid that will stabilise the dispersion. The experimental evidence[58] suggests that a relatively high content of acidic volatiles does in fact assist both pigment wetting and dispersion stabilisation.

DISPERSION OF INORGANIC PIGMENTS

In the dispersion of a pigment in a liquid by mechanical action, three processes are involved.

(i) Wetting, in which air is displaced from the surface of and between the particles of the pigment clusters and is replaced by medium, but the form of the clusters may remain unchanged.
(ii) Disruption, by mechanical energy, of the clusters (aggregates and agglomerates) into smaller units dispersed in the medium.
(iii) Flocculation, the process which involves interparticle collisions that reduce the total number of particles.

These processes are not necessarily sequential, all may be occurring concurrently in different parts of the system, but the third process does not become important until the first two are substantially complete. In the absence of some other force, maintenance of the dispersed state will depend on the relative rates of mechanical separation and of flocculation of the particles. These aspects of the dispersion process are discussed in Chapter 1.

If the liquid phase is viscous, pigment mobility and hence rate of flocculation are reduced, a circumstance put to practical use in plastics injection moulding. The pigment is dispersed in the molten plastic by the shearing forces created during the injection process, whereupon cooling in the mould quickly raises the viscosity of the plastic to a level that prevents any significant flocculation of the pigment. For suspensions in more-or-less permanent liquids an alternative method of stabilising the dispersion is required. This is usually the presence in or addition to the liquid of a surfactant that can be adsorbed on the pigment surface and substantially prevent flocculation. (For reasons mentioned earlier in this chapter, an attempt may be made to induce a very loose, voluminous, flocculated pigment phase, but this will not be considered here.)

For various reasons, some of which are discussed later, the pigment dispersion process is often carried out in a liquid with some degree of intrinsic pigment dispersing power. Wetting of the pigment is aided by low viscosity of the medium and in machines using impacting forces, such as the ball mill, the mode of action requires the suspension to have fairly high mobility together with the highest possible pigment content. In other machines such as the triple roll mill, where dispersion is by shearing forces, a high viscosity medium with strong internal cohesive forces is required to transmit the shearing forces produced by the roll speed differential to the pigment. Whilst the method of achieving dispersion varies from one type of machine to another, the efficiency of the process is in every case dependent on the suitability to it of the rheological characteristics of the pigment suspension.

Rheology of pigment suspensions

A liquid subjected to a certain shearing stress will flow in order to relieve that stress; if the stress τ is not too great the flow will be laminar and the rate of shear ϵ will be a simple function of the applied stress, i.e.

$$\tau = \eta\epsilon \tag{8.1}$$

where η is the viscosity of the liquid.

The presence of particles in a liquid to which a stress is applied complicates the response of the system to the stress: the degree of dispersion of the particles will also influence the response. So far as the dispersion of a relatively large volume of pigment in a liquid is concerned, the major stages can be enumerated.

(i) Agglomerated 'dry' pigment with a volume of internal voids equal or even greater than the volume of liquid. Apparent consistency of system very high.

(ii) Pigment wetted and partly dispersed, but with a large number of clumps of particles in which a substantial proportion of the available liquid is immobilised. Consistency lower than (i) but still high.

(iii) Progressive breakdown of the pigment clumps with release of immobilised liquid causing a further reduction in apparent consistency.

(iv) Pigment particles dispersed but tending to flocculate rapidly if shearing stress is reduced or removed. Characterised by reduction in viscosity (increase in rate of shear) with increasing shear stress. The system may have a *yield value*, a certain minimum shear stress being required to induce flow.

(v) Pigment particles substantially dispersed and adsorbing material from the liquid phase, so that the effective volume of the disperse (pigment) phase is increased and that of the continuous phase decreased. The system is more Newtonian than (iv), but its viscosity slightly higher; if the shearing stress is raised sufficiently, adsorbed material may be stripped off.

Of these stages, the first is often critical, since if the medium fails to wet the pigment and so assist in the mechanical disruption of the agglomerates, the machine may be unable to accomplish this task. It is therefore necessary to ensure that the proportions of pigment and liquid are suitable for the accomplishment of this stage, but not such that after wetting-out, the consistency of the mixture is too low for efficient milling. Probably the most useful guide to appropriate proportions is provided by the flow-point test, originally described by Daniel and Goldman.[59]

Pigment/dispersant/liquid ratio

For economic reasons, the greatest possible proportion of pigment should be in a paste or slurry that is to be milled. (In practice, the absolute maximum pigment content suspension may not be the most convenient, and the formulation actually used may be modified to give a greater margin of stability to the dispersion.) The suspension viscosity rises with increasing pigment content, so the initial viscosity of the liquid should be the minimum possible, that is, the amount of dispersant dissolved in it should be minimal. Agglomerated or flocculated pigment particles occlude a large volume of the liquid phase, so the amount of dispersant should not be less than that needed to deflocculate the pigment. The first simple, practical way of determining this relationship for aqueous surfactant solutions was described by Daniel and Goldman[59] and Daniel[60] subsequently described in detail the method of carrying out the determination with alkyd resin solutions, to obtain the optimum pigment/resin/solvent combination for dispersion by ball milling. More recently Daniel[61] has shown how the technique can be applied to the formulation of mill bases suitable for dispersion by high-speed impeller milling ('Cowles milling').

The method of determining the optimum pigment/resin/solvent (or pigment/surfactant/water) combination is to make a paste of pigment and dilute resin solution. In the 1946 and 1966 versions of the method, the resin solution was incorporated in the pigment by pasting with a spatula on a glass plate; the technique is simple, but rapid working is essential because of the loss of solvent. In the 1950 version, the resin solution was incorporated in the pigment by stirring with a glass rod in a small jar or beaker; the end-points of the two methods are not quite the same, the 1950 method usually requiring more resin solution. (The analogy with the work effect in the BS method of oil absorption value determination is very close.) In both methods, two end-points were observed, first the *wet-point* or *ball-point*, at which the pigment just formed a coherent mass which did not (significantly) wet the glass plate or vessel side, and secondly the *flow-point* at which the paste just flowed from the knife (1946) or just flowed from the rod, the paste 'snapping back' to the rod when the stream ruptured (1950). By repeating the determination with a series of resin solutions, curves can be plotted of amount of resin solution per unit weight of pigment required to reach the wet- and flow-points against the resin concentration in solution. Examples are illustrated in Fig. 9.1 for the three systems shown: (a) strong dilatancy (dilatancy will be discussed again below), (b) weak dilatancy, and (c) strong flocculation.[61] In strongly dilatant systems the flow-point is usually only 5–15% above the wet-point, for weakly or non-dilatant systems 15–30% and when flocculation occurs the gap is considerably wider.[61] A good dispersion requires the smallest further addition of resin solution to convert the wet-point to the flow-point paste.

Attempts to classify dispersions only by the pigment volume concentration in the flow-point paste can be misleading. For example, a conventional grade of coated rutile titanium dioxide may give a p.v.c. at the flow-point of about 49% whereas one of the more recently introduced

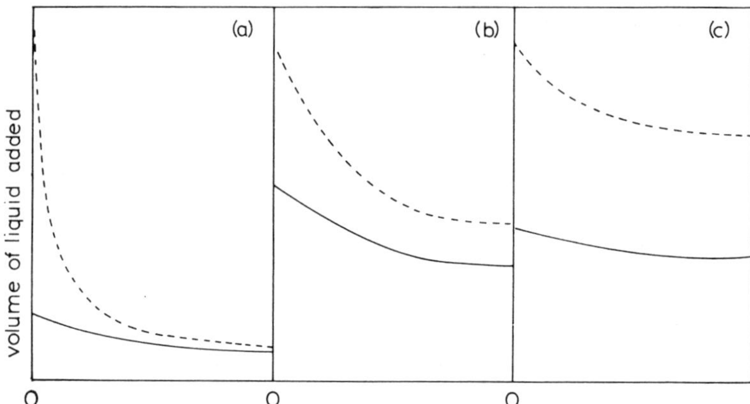

Fig. 9.1. Daniel flow-point method: examples of systems showing (a) *strong dilatancy,* (b) *weak dilatancy and* (c) *strong flocculation (after Daniel*[61]*). Broken line, flow point; continuous line, wet point.*

heavily coated rutiles, consisting of 85% TiO_2 core and 15% w/w coating oxides, might give a flow-point paste with only as little as 33% p.v.c. Both are good dispersions, the extra amount of resin solution required to convert the wet-point to the flow-point paste being identical, about 2·5–3·0 ml per 100 g of pigment. The difference between the pigments is in their behaviour approaching the wet-point stage, where the conventional pigment gives a paste having about 52% and the heavily coated grade gives a paste of about 35% p.v.c. The difference in medium requirements is an indication of the difference in the surface structure, the conventional pigment being relatively smooth whereas the heavily coated grade has a very fissured surface, in which a large proportion of the available medium is immobilised. Once the fissures are filled the particles behave as though their surfaces were smooth, the amount of extra medium needed to convert the wet-point to a fluid being similar to that required by the conventional pigment, since the basic particle size and size distributions are similar, being governed by the size of the titanium dioxide core.

Errors in ranking pigments for ease of dispersion can be avoided if the rheological characteristics of the paste during the flow-point determination are observed. The following classification is that of Daniel and Goldman.[59]

Class I: Good dispersions

(i) At the wet-point
(a) shine without or upon only light tapping,
(b) are dry and hard to knead.

(ii) At the intermediate stage (defined as the mixture containing about a tenth to a fifth more liquid than the amount required to make the wet-point paste, but not having reached the flow-point)
(a) lack a distinct intermediate stage.

(iii) At the flow-point
(a) flow without tapping,
(b) offer considerable resistance to sudden pressure, i.e. behave as dilatant systems.

Class II: Fair dispersions

(i) At the wet-point
(a) shine on sharp tapping.

(ii) At the intermediate stage
(a) flow when tapped or stirred gently,
(b) may offer resistance to sudden pressure.

(iii) At the flow-point
(a) fall, with some elongation at the breaking line,
(b) show no resistance to sudden pressure,
(c) show visible thixotropy.

Class III: Poor dispersions and flocculant systems

(i) At the wet-point
(a) remain dull even when tapped.

(ii) At the intermediate stage
(a) rise on tapping, so that the surface becomes rigid,
(b) show no resistance to sudden pressure,
(c) possess marked plasticity (high yield value).

(iii) At the flow-point
(a) fall without elongation at the breaking point.

(iv) Require a considerable additional volume of medium to convert the wet-point into the flow-point paste. (In the worst cases, addition of extra medium seems to cause scarcely any reduction in the paste consistency.)

Rheological aspects of dispersion

The rheological characteristics of a typical pigment dispersion can be seen to change markedly as milling proceeds and therefore successive rheological measurements should give a measure of the progress of dispersion. Furthermore, there must be an optimum condition for the operation of each type of mill, and this condition should be definable in rheological terms hence leading to a mill base formulated on a sound scientific basis. However, this concept involves many problems.

The production of pigment dispersion in bulk, as for example in paint manufacture, requires the mixture of pigment and medium to have flow properties which are suited to the type of machine used to make the dispersion. The appropriate mill base consistency may be very high for a heavy duty dough mixer or very low for a kinetic dispersion machine. In all cases the formulator is faced with the problem that the consistency, and therefore the power required to drive the mixer, follows a standard pattern of a rapidly developed high initial value, which then decreases as the pigment is dispersed. The consistency may also be reduced as a result of the heat developed in the mill base by the mechanical work done on it. The flow of pigment medium mixture, in response to an applied stress, can be measured and expressed graphically either as the rate of shear induced by a shearing stress, or as the viscosity/rate-of-shear relation.

Rigorous mathematical analysis of practical systems is not usually feasible; the variety of particle shapes and sizes, and the fact that the continuous phase of the dispersion is often a colloidal system in itself provide two almost insuperable problems and a third arises from the nature of the dispersion process. Acceptance of the thesis that dispersion is achieved by the acquisition of an adsorbed layer on the pigment surface requires also acceptance of the concept that material is selectively removed from the continuous phase and added to the discontinuous phase, with a resultant change in the proportions of both phases.

The result of these problems is that study of the rheological characteristics of dispersions usually consists of the measurement of those characteristics at intervals during the dispersion process and consideration of the changes in the observed characteristics. These observations are then interpreted by reference to the observed behaviour of simpler systems or to mathematical models. The precise value of studies of the rheology of dispersions, as a means of studying the dispersion process itself, is therefore arguable. Changes in rheological characteristics may mark the progress of the dispersion process, but do not provide information about the processes involved in achieving that stage. The remarks that follow are therefore only postulates, for the most part.

Pigment in the dry state is usually agglomerated, the particles cohering as a result of van der Waals or London forces, and the cohesive energy increases as the size of the particles decreases. When a liquid, in limited amounts, is incorporated into a mass of dry pigment, there is some reinforcement of the point and line contacts between particles by liquid bridges which are held in place by capillary forces. This has been discussed by Rumpf[62] and other workers, including Pietsch.[63] The effect is not very important in pigment dispersion, since the reinforcement of interparticle adhesion only occurs below a critical ratio of liquid volume to the void volume within the particle mass, and this ratio is quickly exceeded if pigment is added to an agitated liquid, which is the more usual procedure in modern paint and ink manufacture. However, the effect almost certainly helps to increase the initial power requirement above that needed to overcome the cohesive forces between dry pigment particles and interparticle friction.

Once the air has been displaced and the particles have been brought into contact with the liquid phase, the rheological behaviour of the system will usually fall between two extremes, characteristic of highly dilatant and highly flocculant systems.

A *dilatant* system is one in which the particles are so arranged that the solid volume fraction is at or near a maximum value. For such a system to respond to stress, the particles in the system must first move apart from each other. This movement requires time, so the greater the speed with which the force is applied, the greater is the apparent resistance to movement of the system. For this reason a typical dilatant system may flow slowly if its container is tilted but will become rigid and dry in appearance, possibly crumbly, if any attempt is made to stir the paste. Practical mill base formulations are never compounded in such proportions (maximum p.v.c.), but still behave in a manner which is variously described as dilatant or pseudo-dilatant. Much confusion can be avoided by describing their behaviour as shear-increasing viscosity.

A tacit requirement of the dilatant system is that, technically at least, the solid particles are in a dispersed state, because each particle is enveloped by the continuous liquid phase. At the other extreme, a *flocculant* system is one in which each particle is in contact with as many of its surrounding particles as possible. The number of such contacts will depend on the shapes of the particles. In the case of some pigments, and especially mixtures of pigments having widely different shapes (for example, mixtures of natural calcite, which when ground to make a pigment has roughly cubic particles, and mica, which has a lamellar structure and is ground into platelets, or talc, which has a highly acicular, more-or-less fibrous structure), it is often found that the pigment paste, at rest, has a very high consistency, which falls quite rapidly when the paste is agitated. This behaviour is produced by the formation in the paste of a more-or-less rigid skeletal structure of pigment particles having point or line contacts with each other, so that the solid volume fraction of the system may be at or near the minimal value for the system, if a given minimum number of contacts between each particle and its neighbours is specified. In such a system interparticle attraction should exceed the attraction between the particle surface and the medium, but in practice a skeleton of this type may be induced in a system by adding a material, such as water or a suitable surfactant, in sufficient quantity for effective particle–particle contacts to be created by bridges of the additive. Too little additive may have a negligible effect, too much may produce a system in which the particles are dispersed by the action of adsorbed additive, in a continuum of the medium which purports to be the dispersant, but which, in fact, had been unable to perform its task without the further additive.

The formation in a pigment/liquid mixture of a skeletal structure of contacting particles from particles that previously were mostly separate whilst the mixture was agitated, is the preferable definition of the process of flocculation. A high consistency, or structure, induced in a pigment/liquid mixture by the network of particles is best described as *flocculation-induced viscosity*, leaving the term *thixotropic* to be used only for systems which have been tested and found to meet the requirement for such a system, as originally defined by Freundlich.

The term *thixotropy* was coined to describe suspensions in which the solid particles were mutually attractive and in the absence of dispersing forces formed clusters, or flocculates. When a shearing force is applied to the system, the flocculates are dispersed and the system does not display any sign of a residual yield stress. When the shearing force is removed, the structure of the suspension reforms, in a period of time characteristic of the particular system. Plots of rate of shear against shearing stress therefore display a hysteresis loop, the area of which is a measure of the degree of thixotropy of the system.

In the case of pigment suspensions in resin solutions, where the volume concentration of the solid is high, true thixotropic behaviour is rarely exhibited. In most cases the systems are of the kind called *false-*

bodied[64] or *thixotropic plastic.*[65] These are suspensions which do not flow until a certain value of shearing stress is exceeded and which retain a finite yield stress even after agitation. In very few cases, if any, do moderately concentrated suspensions of pigments in resin solutions behave as *Bingham plastic*[66] bodies, having a simple linear relation between rate of shear and shearing stress once the yield stress is exceeded.

It is possible to derive equations that fit a given set of rheological data, but as pointed out by Cheng,[67] we are not yet in a position to predict suspension behaviour from first principles and a knowledge of the composition of the material. However, a qualitative picture has been established.

THE DISPERSION PROCESS IN PRACTICE

The breakdown of aggregates and agglomerates

An aggregate is defined as consisting of particles that are in face-to-face contact, so that the force required for breakdown of an aggregate approximates to that needed for fracture of a crystal. According to Rittinger, the force required must be proportional to the amount of new surface produced, and Rittinger's number is the amount of new surface produced per unit of mechanical energy absorbed by the material. Some values for materials of a pigmentary nature are as follows:

	Specific gravity	*Rittinger's number (cm^2 of new surface per kg-cm of energy applied)*	*Hardness (Moh scale)*
Quartz (SiO_2)	2·66	17·56	7
Sphalerite (ZnS)	4·0	56·2	3·5–6·0
Calcite ($CaCO_3$)	2·71	75·9	3

The Moh scale of hardness is based on the ability of material with a given number to scratch any material with a lesser number. The Moh hardness is *not* a good indicator of the resistance of a material to crushing unless the material is to be crushed very fine, which is the case where pigment comminution is concerned. The Moh hardness *is* a good indicator of abrasivity and so provides a good indication of the probable degree of mill wear.

Assuming that the material is crushed from uniform cubes of 100 μm side length to cubes of 1 μm side length, the new surface created is $5{\cdot}94 \times 10^{-2}$ cm^2 per cube, and the nominal energy needed to crush 1 kg of such cubes of quartz and calcite would be 0·0466 and 0·0106 h.p.h./kg respectively. To split a cube into halves the energy would be 11·18 ergs and 2·59 ergs respectively for quartz and calcite. For a ball mill using high density ($\rho = 3{\cdot}4$) alumina balls of 2 cm diameter in a mill base such as that in Table 9.2, the force available is that provided by a ball descending under the influence of gravity, $f = mg$, where m may be taken as either the true or the effective mass of the ball. In the latter case, $m = V(\rho_1 - \rho_2)$ where V is the volume of the ball and ρ_1 and ρ_2 the densities of the ball and mill base respectively. The two values of m represent the cases of a ball on the face of the mill charge and one just inside it. The value of m would be 14·3 g for the free-falling ball and 4·96 g for the ball immersed in the base, but not influenced by any other part of the charge. The force available would be 14 000 dynes and 4600 dynes respectively, so an impact of 12 ergs would only require the ball to 'fall' approximately 10 μm or 25 μm.

Whilst the breakdown of aggregates in a suspension is within the capability of a ball mill, the task can be carried out more efficiently by the use of a dry grinding process, preferably as a stage in the manufacture of the pigment and not in the making of a dispersion. If the use of an only relatively coarse pigment is being considered, it should be remembered that the fracturing of an aggregate is only a statistical probability and the probability of fracturing a particle of optimum pigmentary size is rather greater, since these will be present in much larger numbers.

In practically all wet milling processes, the liquid component is to some extent a surface active agent, and the rate of breakdown of aggregates should therefore be accelerated by the Rehbinder effect (see Chapter 1). The effect of surfactants on the rate of breakdown is difficult to establish with certainty. There is an optimum type of mill base rheological characteristic at which the highest milling efficiency is attained. The characteristic is transient because the rheology of a suspension must change as the size distribution of the pigment phase alters, and as the degree of interparticle attraction is reduced so that a truly dispersed state is achieved. If the latter state involves selective absorption from the liquid phase there will also be progressive changes in the rheology of the liquid phase itself, an increase in the hydrodynamic volume of the solid phase and a compensating decrease in the effective volume of the continuous phase. A surfactant is usually added to a suspension because it causes a visible change in the rheological characteristics, which also increases the rate of dispersion which changes the rate at which the rheological characteristics of the mill base change. In any typical mill base formulation, deciding whether a surfactant increased the rate of dispersion by causing aggregates to break because of the Rehbinder effect, or because it changed the rheological characteristics of the mill base in a way beneficial to milling efficiency, is a 'chicken and egg' problem.

Agglomerates are held together by weaker forces than those of aggregates, and should be disrupted by shearing forces and not require impaction. In a ball mill breakdown of agglomerates should be swift. In other mills this will only occur if the mill action and the mill base formulation are matched, so that the strongest possible shearing forces are developed. This in its simplest form means that the use of high impeller speeds permits the use of low viscosity suspensions, but low impeller speed necessitates the use of a viscous suspension. Unfortunately suspensions are not Newtonian fluids so a high impeller speed may induce local pseudo-dilatant behaviour, while a low impeller speed may not be great enough for the yield value of a structured suspension to be exceeded. In high speed impeller mills both types of behaviour may occur simultaneously; the rotor initially impels the mill base which then 'sets-up' so that the rotor is soon spinning in a hole that it has formed in the mill base, the walls of the hole being formed by a nearly rigid 'dilatant' suspension. The small amount of mill base relaxing and flowing back to the impeller is thrown back at the wall with sufficient momentum to maintain the dilatant state. Away from the impeller zone the mill base may be quite stiff due to massive flocculation of the pigment, causing a very high viscosity. This kind of situation can be avoided by using the formulating technique described by Daniel,[61] which has been found to give a better indication of the optimum mill base formulation than the method proposed by Guggenheim,[68] which does not allow for the influence on the dispersion process of the intrinsic pigment dispersing power of the milling medium.

Pigment wetting in practice

Typical formulations used for efficient ball milling of pigments in alkyd resin media are given in Tables 9.2 and 9.3 for rutile and carbon black respectively. These pigments have been chosen as representing a typical and a very small particle size pigment, since the wetting of large particles

TABLE 9.2

PROPERTIES OF COATED RUTILE TITANIUM DIOXIDE PIGMENT

Bulk density: Approximately 1 kg/litre
Mean particle size: 220 nm (electron microscope, graticule size-count)
117 nm (from total BET surface area, 12·5 m^2/g)
154 nm (from corrected BET area, 9·5 m^2/g)

	Ball mill base formula			
	Weight (%)	*Density*	*Volume* (*ml*)	(%)
Rutile	80	4·1	19·55	44·6
Alkyd (63% FA)	4	1·06	3·78	8·6
White spirit	16	0·78	20·51	46·8

Gross volume of dry pigment = 80 units
Void volume in dry pigment = 60·45 units
Volume of liquid = 24·29 units
Pigment volume concentration in oil absorption paste = 55·8%

TABLE 9.3

PROPERTIES OF HIGH COLOUR CHANNEL BLACK PIGMENT

Bulk density: Approximately 95 g/litre
Mean agglomerate size: 9 nm (equivalent circle diameter, from electron micrographs)
Mean particle size: ca 3 nm (estimated from electron micrographs)
3·5 nm (from BET total surface area)

	Ball mill base formula			
	Weight (%)	*Density*	*Volume* (*ml*)	(%)
Carbon black	10	1·8	5·55	5·0
Alkyd (63% FA)	30	1·06	28·3	25·5
White spirit	60	0·78	77	69·5

Gross volume of dry pigment = 105 units
Void volume in dry pigment = 99·5 units
Volume of liquid = 105 units
Pigment volume concentration in oil absorption paste = 9·85%

does not present problems (usually) and that of pigments with a wide range of particle size may be considered as the wetting of a series of pigments each of uniform size, covering the range.

In the case of the titanium dioxide the pigment surface accessible to the fluid is about 9·5 m^2/g, significantly less than the total surface area accessible to nitrogen (12 m^2/g). The thermodynamic work done in the wetting process is therefore quite small, relative to that involved in the wetting of the carbon black. The bulk density of the dry pigment is equivalent to a solid volume fraction of about 0·25, indicating an open structure with less than four interparticle contacts per particle. (Four interparticle contacts per particle, as in diamond, give a solid volume fraction of 0·330; a number of possible open packings of spheres with less than four contacts per sphere have been described by Heesch and Laves.[69]) The diameter of the capillary passages between the dry particles can be estimated by considering the pigment as an 'expanded' array of uniform spheres. For random close packing ($F = 0{\cdot}64$) the centre-to-centre spacing of the particles would be 1·37 diameters, or using the value for closest packing derived from the oil absorption paste composition ($F = 0{\cdot}56$) it would be about 1·3 diameters. From these values, the mean pore diameter, calculated from the cross-sectional area of the space between four adjacent 222 nm diameter particles, would be about 260 nm or 235 nm respectively. (Using the smaller particle diameters derived from surface area measurements would significantly reduce these values.) The minimum value for the mean pore diameter (calculated in the same way but assuming that the four particles were in contact) would be 102 nm. These values have no experimental foundation and serve only to provide some reasoned estimate of the magnitude of the pore sizes.

Using Washburn's equation (eqn. 1.15) and assuming a mean capillary radius of 80 nm, a liquid viscosity of 5 cP and that $\gamma_{L/V} \cos \theta = 30$ dyne/cm, the rate at which the fluid would penetrate the capillaries in the pigment mass can be calculated, and is found to be 0·38 cm/min.

In practice, adequate wetting of the pigment particles reduces the interparticle friction that maintains the voluminous structure of the dry pigment. As soon as the interparticle attraction exceeds the friction, a collapse takes place. Gravity alone cannot overcome interparticle friction; if the wetting liquid penetrates the mass of pigment from below then, under static conditions, the wetted pigment may collapse leaving an arch of dry pigment above it and out of reach of the liquid. If penetration is spontaneous, the reduction in pore diameter in the collapsed pigment increases the capillary attraction (Jurin's law) and so partially compensates for the reduced rate of penetration, provided the air can escape freely from the mass of pigment.

The agglomeration of pigment particles in the dry state is a result of van der Waals or London forces, and the cohesive energy is inversely proportional to the particle diameter. If only a limited quantity of liquid is incorporated into a mass of dry pigment, then the point and line contacts between particles may be reinforced by liquid bridges that are held in place by capillary forces. The effect is not very important in pigment dispersion, except in such special cases as the over-rapid charging of pigment to a horizontal trough mixer or pug mill. The very sharp peak in power demand that occurs when pigment is added to medium in a mixer of this type, and that can be demonstrated very readily using a Brabender Plastograph or a similar type of instrument, may be partially contributed by this effect. Fortunately the reinforcement of interparticle adhesion only occurs below a critical ratio of liquid volume to the void volume within the pigment mass, and this ratio is quickly exceeded if pigment is added to an agitated liquid at a sensible rate.

Ideally the pigment should be sifted into the wetting liquid, which should be agitated so that the largest possible proportion of the pigment immediately contacts the liquid. As the pigment falls on to the liquid, penetration of aggregates and agglomerates by liquid can occur easily, the displaced air escaping from the dry upper surface of the pigment.

In less than ideal circumstances, large lumps of pigment may be immersed in the liquid. Particularly in the cases of rutile, chrome yellows or iron oxides it is possible for the outer layers of the lump to be wetted, and to collapse to give a much denser layer of pigment in nearly a close-packed arrangement. The capillary attraction of the liquid into the lump of pigment is then counterbalanced by the pressure required to force air out of the lump, through the dense wetted pigment layer, while the hydrostatic head of the liquid over the lump also opposes the escape of air. Unless the lump is broken by mechanical action little further wetting will occur and the contracted outer layers will exert a considerable compressive force on the dry core.

The formation of partially wetted pellets of pigment is particularly likely to occur during the preparation of pre-mixes for sand mill or bead mill dispersion, if the long-established correct procedure for mixing pigment and medium is ignored. In one case dumping 1000 kg of titanium dioxide, in 25 kg (bag) lots, into about 600 kg of a 16·7% solid content solution of a short-oil alkyd in xylol–butanol, stirred slowly by a simple propeller-type agitator, resulted in the formation of about 30 kg of pellets of diameter 0·5–2·0 cm. The pellets completely blocked the protective grill on the inlet side of the pump feeding the sand mill, and stopped production. Formation of pellets was prevented by reducing the initial charge of liquid and alternating the addition of pigment and liquid so as to keep the pre-mix paste as stiff as the agitator could move, finally thinning the paste (carefully) to the required composition before milling. The extra time required for pre-mixing was more than compensated for by the increased pumping rate at which an acceptable level of dispersion could still be obtained.

A second type of pigment agglomeration is caused by complete incompatibility between the pigment and wetting fluid. Such behaviour can be demonstrated by exposing a quantity of titanium dioxide to an atmosphere saturated with water vapour. The pigment containing about 1% of adsorbed water, when trundled with about 2 volumes of benzene, rapidly forms extremely hard, small (about 2 mm diameter) pellets. As trundling continues the pellets contract slightly until their surfaces become visibly indented, suggesting either that the pellets are formed initially as agglomerates of small pellets which then collect still smaller pellets and/or primary particles, or that water is displaced from within the pellet, causing shrinkage. With toluene or xylene the pellets are usually smaller and of lower density; with lower primary alcohols a stable dispersion is formed.

The converse process in which a titanium dioxide previously coated with, for example, a fatty acid is pelletised by trundling in water has not been observed; perhaps all the coated rutile surface is not available to the acid for adsorption, hence the desired completely hydrophobic surface is not produced. Treatment of the pigment with a methylsilane to replace hydroxyl by methyl groups might produce a surface more analogous to that of the moist pigment.

The wetting of inorganic pigment particles of moderate size (200 nm diameter and above) is a relatively simple process compared to that of material such as channel black (Table 9.3). The black particles look like fused aggregates of spheres and the variety of particle shapes and sizes is much greater than that of titanium dioxide. The electron microscope (EM) particle size is the diameter of the circle of equal area to the projected image of the particle and therefore analogous to the EM size of the titania, as measured using an 'equal areas' graticule. The surface area measured by nitrogen adsorption includes a considerable area, contributed by the walls of pores and fissures, that is probably inaccessible to most if not all of the molecules that constitute the liquid phase. The high

(16%) volatile content and low pH value of the black indicate that the oxy-carbon complexes that form the pigment surface contain a fairly high number of carboxylic groups per unit area. Given the opportunity, the pigment could adsorb a considerable amount of atmospheric moisture. (A monolayer of water on the whole of the nitrogen surface area would amount to about 25 g per 100 g of pigment.)

The bulk density of the dry black is very low, 0·096 g/cm^3, equivalent to a solid volume fraction of 0·053, indicating a very open structure with an average number of interparticle contacts that is not very much above the necessary minimum of two per particle. The structure of the dry black may therefore be visualised as random coils of particle chains. An estimate of the mean pore size in the dry black can be made by considering it in terms of a random close-packed array of uniform spheres, 'expanded' to reduce the bulk density to the observed value. The necessary centre-to-centre spacing of the spheres would be 2·9 diameters, the cross-section of the open space between four adjacent spheres being about 617 nm^2, equivalent to a mean (circular) pore diameter of 28 nm.

The high oil absorption value, 475 g oil/100 g pigment, corresponds to a p.v.c. in the paste of only 9·8%. Calculating, as before, gives a centre-to-centre spacing of 1·87 diameters or if each particle is assumed to have a monolayer of adsorbed oil, of 1·37 diameters. (These figures cannot be used to assess the volume of oil occluded by each particle; the figure obtained can vary from about 1 to 6 particle volumes, depending on the initial assumptions.) For the 1·87 diameters (16·8 nm) separation, the mean pore diameter, calculated as before, would be about 17 nm.

These calculated pore diameters are likely to bear as little relation to reality as do those calculated for titanium dioxide; they merely provide an indication of the order of magnitude.

The rate of penetration of the resin solution into the dry black will be slower than in the case of titanium dioxide, assuming a pore radius of as much as 40 nm (half that assumed for the titania), the additional effect of the higher viscosity, about 15 cP, of the more concentrated resin solution will further reduce the rate of penetration, to 0·6 cm/min. The wetting of the black is likely to take place in two stages, the first consisting of the penetration of the main pores through the mass of pigment and the second of penetration into the interstices of the individual particles. This second stage is likely to be slow, partly because the trapped air will have difficulty in escaping; some form of mechanical milling action is probably essential for this stage to proceed to completion. The expansion of this displaced gas as the ball mill contents are heated by the mechanical energy expended, makes it necessary to vent the mill after the first few hours of running. (This practice is advisable with all pigments, but especially so in the case of carbon blacks, Prussian (iron) blues and similar materials.) The practice of loading a ball mill with a pre-mixed slurry has the advantage that almost all the air has been expelled from the mill base and no milling time is lost in 'wetting-out' the pigment in the ball mill.

Stabilisation of pigment suspensions

A stable disperse state requires the particles in a suspension to be sufficiently widely separated for the repulsive force (electrical or steric; see Chapter 1) to exceed the van der Waals attractive force. Particles in the colloidal size range are subject to Brownian movement, so that a 'buffer' is needed if flocculation is to be avoided. In non-aqueous fluid suspensions it is expected that adsorbed material on the pigment surface will serve this purpose.

Crowl[70] published data for the attraction potential V_A between various sizes of particle, with and without adsorbed layers. By assuming that the particles are uniform spheres, with a closest random packing solid fraction of 0·64, the value determined experimentally by a number of workers, it is possible to calculate the maximum pigment volume concentration in a suspension that will permit adequate separation between the particles for the suspension to be stable. For this purpose it is assumed that at a value of $V_A = -5kT$, flocculation will not occur. An adsorbed layer of resin on the particle surface will reduce interparticle attraction, but to obtain the requisite separation between the outer surfaces of the adsorbed layers, there must be a corresponding increase in the centre-to-centre separation of the particles. This enforces a corresponding reduction in the maximum possible p.v.c. compatible with this minimum particle separation.

In Table 9.4 are given the maximum permissible volume concentrations of spheres, calculated from data given by Crowl for the attraction potential/particle separation relationship calculated using Vold's[71] equation, and assuming that the closest possible packing of the uncoated spheres has a solid volume fraction of 0·64 (64%). The (sphere + adsorbed layer)/sphere volume ratios are also given. These values cannot be converted to weight ratios without assuming a value for the density of the adsorbed layer, a value that in practice is influenced by the resin solution composition, the configuration of the resin in the adsorbed state, and the quantity of solvent associated with the adsorbed resin.

TABLE 9.4

THEORETICAL MAXIMUM VOLUME CONCENTRATION OF UNIFORM SPHERICAL PARTICLES IN DISPERSE SYSTEMS

Particle diameter (nm)	*Adsorbed layer thickness (nm)*	*(Particle + layer)/ particle volume ratio*	*Separation between surfaces (nm)*		*Maximum p.v.c. in suspension (%) (closest packing = 64% p.v.c.*	
			at $-5kT$	*at* $-10kT$	*at* $-5kT$	*at* $-10kT$
40	5	1·953	0·40	(0·15)	32	(32·4)
100	5	1·331	0·82	0·32	47·1	48
200	5	1·158	1·6	0·63	54·1	54·8
400	5	1·077	4·9	1·6	57·3	58·7
1 000	5	1·030	18·8	7·2	58·8	60·8
2 000	5	1·015	41·6	18·4	59·3	61·3
200	0	1·000	4·8	2·6	59·6	61·6
	2·5	1·076	2·6	1·0	57·3	58·6
	5	1·158	1·5	0·82	54·1	54·7
	7·5	1·242	1·2	0·82	50·7	51·0
	10	1·331	1·0	0·60	47·4	47·7
1 000	0	1·000	25·0	12·2	59·5	61·7
	2·5	1·015	21·0	9·2	59·3	61·3
	5·1	1·030	18·5	6·8	58·9	60·9
	7·5	1·045	16·0	5·4	58·4	60·3
	10	1·061	13·9	4·4	57·9	59·5

The effect of an adsorbed layer on the effective volume of particles of less than 200 nm diameter is very marked, and causes a severe limitation on the permissible concentration of such particles if a reasonably low interaction potential is to be achievable. The use of larger particles permits higher volume concentrations, but the effect of gravitational forces on these should be taken into account. Given sufficient time, these particles would settle and form a dense tough sediment, of very dry appearance and very difficult to disperse.

The values for alkyd resin adsorption on titanium dioxide reported by earlier workers were mostly of the order of 2 mg/m^2, but it is now accepted that more realistic values are obtained if the data are evaluated using the method of Rehacek.[72] This requires extrapolation of the straight portion of the adsorption isotherms (mg adsorbed resin/g pigment plotted against polymer concentration). The intercept on the ordinate gives the true weight of polymer adsorbed, and that on the abscissa indicates the concentration of polymer in the adsorbed layer. It appears that for a given alkyd, the resin concentration in the adsorbed layer is substantially independent of the resin solution concentration and usually in the range 30–40% w/w. Adsorbed layer thicknesses up to 15 nm have been reported.[72,73]

In the case of the titanium dioxide mill base (Table 9.2) the adsorbed layer could cause the effective p.v.c. of the suspension to increase from 44·6% to about 55%. In practice, it is likely that the milling action would remove adsorbed material from the pigment almost as soon as it was attached, but the increase in effective p.v.c. would occur once the mill was stopped; this increase in p.v.c. would account for the rapid increase in apparent consistency of the mill base when left undisturbed and provides a more reasonable explanation than 'cooling-out', a very slow process for a mass as great as that of a large pebble mill and its charge. (Pebble mills cannot be water-cooled, unlike steel ball mills.)

The increase in volume of the mill base may also be examined in terms of average particle diameter increase. The enlarged spheres would be 1·07 times the mean diameter of the pigment, equivalent to an adsorbed layer 7·7 nm thick on the mean particle size (220 nm) determined by electron microscopy, and allowing a maximum separation, in terms of uniform spheres, of 12·6 nm between the outer surfaces of the adsorbed layers. From Crowl's data, this would indicate an attraction potential of only just over $-1kT$, so that the dispersion should be quite stable.

In the case of the carbon black dispersion (Table 9.3) deductions about the probable stability of the mill base are much more difficult to make. One reason is that a quantity of the medium, about four times the solid volume of the black, will be occluded by the pigment, with or without any adsorption taking place. Also, the maximum possible separation between the aggregates that are the pigmentary particles will be correspondingly reduced. The volume concentration in the mill base of aggregates plus occluded medium will be about 25%, but the corresponding hypothetical closest packing (derived from the oil absorption paste composition), would have a solid volume content of 49%, giving a maximum separation of 20 nm between pigment aggregates with an average diameter overall of 80 nm. However, the Vold equations indicate that particles of that order of magnitude would have very small attraction potentials, even without adsorbed resin layers, at separations as small as 5 nm.

Work such as that reported by Hess and Garret[58] supports the view that a highly oxidised (16% volatile) black of this type would adsorb a large amount of the medium and probably induce a considerable degree of orientation of the polymer in the vicinity of the pigment particles. The very marked structural viscosity of the system at rest and fairly rapid reduction in apparent viscosity with increasing shear rate that characterise dispersion of this type of black are in accordance with these hypotheses.

The influence of an electrical repulsive force on the stability of the dispersions has not been considered. The zeta potential is to some extent affected by the chemical composition of the pigment surface, that affects the kind of material adsorbed. However, if the adsorbed molecules are large, entropic effects are likely to be more influential than the zeta potential of the particles.

The probable stability of a dispersion can be assessed from its composition, or rather, from the relative proportions of its components and taking into account the adsorption characteristics of the pigment, molecular weight distribution of the polymer and the nature of the solvent. However, the intrinsically stable dispersion may, largely because of its high pigment content, develop a very high structural viscosity if left undisturbed, and be very sensitive to small changes in composition, such as solvent loss by evaporation. For these reasons it is usual to dilute the dispersion with medium, to reduce the pigment content and, often, to increase the intrinsic viscosity of the suspension. (This in itself, by reducing pigment mobility, hinders pigment flocculation and therefore promotes dispersion stability.)

The stabilisation or 'let-down' stage of dispersion manufacture is apparently simple, but contains pit-falls for the unwary. The failure rate is sufficiently high for the term *colloidal shock* (sometimes called pigment shock or resin shock) to have achieved international status in the jargon of paint and ink technologists. Two cases of 'let-down' will be considered.

'Let-down' of pastes prepared by roll milling or heavy duty mixing

Pigment dispersion by heavy duty mixing or roll milling requires a paste of high consistency and internal cohesion, so that the powerful shearing forces act throughout the mixture of pigment and medium in the shearing zones, between the intermeshing blades of the mixer or in the nip of the rolls. A medium that is an effective dispersant for the pigment is an essential for satisfactory results. In the case of roll milling, the paste fed to the rolls must also have good 'tack', to enable the paste to adhere so strongly to the rolls that it is drawn through the roll-nip. A flocculent paste cannot be roll milled satisfactorily; its apparent consistency in the feed-hopper is high but the mill throughput is extremely low. The paste that is collected on the mill apron is of thin consistency, either due to temporary dispersion of the pigment or, more often, to a reduction in the pigment content. The material in the feed-hopper, in the latter case, progressively increases in pigment content and if it is eventually drawn through the rolls, is likely to appear as hard flakes of almost dry pigment on the take-off apron. This type of behaviour is not kin to that involved in 'shock'; the most satisfactory cure of it is reformulation of the product, or at least of the mill base, to provide a suitable medium for pigment dispersion.

In some cases, most often in heavy duty mixer pastes, the pigment/medium ratio may be high and the medium barely able to produce a stably dispersed system. If such a paste is thinned with a liquid lacking in pigment dispersing capacity, the pigment may flocculate at once. If, on the other hand, the paste is diluted with a solvent that has intrinsic pigment dispersing capability (e.g. a high dielectric constant) then pigment flocculation may not become apparent until the film-forming stage, after the paint or ink has been applied and the solvent evaporates. In this case also, reformulation of the product is the only satisfactory solution.

Any high consistency paste should be thinned slowly, to minimise the risk of the paste breaking up into lumps that are subsequently penetrated by solvent at a rate so slow as to be negligible. Ideally the thinning liquid should be added at such a rate that only a thin film spreads on freshly exposed paste surfaces that are then folded in on themselves, so that the thinner is compressed within an envelope of paste. This reduces the time that a marked difference in composition of thinner and paste exists.

Pastes with a high content of pigment (heavy duty mixer formulations) preferably should be diluted with more of the medium used in the paste, to avoid the formation of interfaces between liquids of different composition. The rate of addition of the diluent should be slow; in practice the 'let-down' stage for this type of dispersion may take several times as long as the 'dispersion' stage. 'Shock' in systems of this type is an infrequent occurrence and is almost always due to incompetence.

Under the conditions existing in the roll-nips or the zones of maximum shear between the blades of a heavy duty double sigma-bladed mixer, it is unlikely that any resin is adsorbed on the pigment or if adsorbed, that it is more than momentarily retained. The greater part of the adsorption process must therefore take place in the paste on standing, after the 'dispersion' stage has ended and usually also after the 'let-down' stage. The dispersion process itself is therefore substantially one in which the pigment particles are mechanically separated and reflocculation prevented only by the viscous hindrance to particle movement. In this situation, the reduction of the paste viscosity on 'letting-down' enables the flocculation process to start at the same moment that adsorption of resin by the pigment can take place easily. If the liquid added merely reduces the paste consistency, then the pigment/resin ratio in the original paste may be too high for a stably dispersed system to exist and flocculation is bound to occur. The time for flocculation to become apparent will be longer, the greater the consistency of the diluted paste; a very viscous fluid may delay flocculation but cannot, of itself, prevent it. If the let-down liquid must be one with no pigment dispersing capability, inclusion in it of an appropriate surfactant may be possible, to provide some insurance against flocculation.

'Let-down' of high-pigment, high-solvent, low-resin content dispersions

The 'slurry-grinding' technique is attractive in that it is the best method of milling if pigment aggregates must be broken down, and also, it gives the largest output of milled pigment per unit volume of mill base. For most purposes, it is necessary to reduce the pigment concentration and increase the consistency of the dispersion. It is in carrying out this stage that difficulty is most often experienced.

A typical mill base and 'let-down' solution for the dispersion of rutile titanium dioxide by ball milling is given in Table 9.5.

TABLE 9.5

	Mill base		*'Let-down'*	
	Weight (g)	*Volume (ml)*	*Weight (g)*	*Volume (ml)*
Rutile	183·0	44·6	—	—
Alkyd	9·1	8·6	67·0	63·2
Solvent	36·6	46·8	28·7	36·8
	228·7	100·0	95·7	100·0

At the end of the milling cycle an equal volume of concentrated alkyd solution is added to the mill base in the ball mill and incorporated by milling for about an hour, according to the size of the mill. At equilibrium the composition is as given in Table 9.6.

TABLE 9.6

	Weight		*Volume*	
	(g)	*(%)*	*(ml)*	*(%)*
Rutile	183·0	56·4	44·6	22·3
Alkyd	76·1	23·5	72·3	35·9
Solvent	65·3	20·1	73·4	41·8

The mill could then be discharged and the mill base pumped into storage tanks pending further processing. Generally no problems due to sedimentation should occur even if the stored dispersion is not agitated for three or four weeks.

The trouble that may occur can be easily demonstrated by measuring into a glass jar 100 ml of the unstabilised dispersion of the above composition. On to this is poured 100 ml of the concentrated resin solution. The resin solution will penetrate the layer of dispersion, but is quickly forced upwards and a well-defined interface should be visible. If the jar is closed and allowed to stand, then after about 20 min it will be noticed that the level of the mill base has fallen slightly and a definite 'skin' formed on it, beneath the resin layer. After 24 h the skin has become thicker and stiffer and, with a little care, can be removed intact. If, instead, the jar is shaken thoroughly, the 'skin' is broken up into small hard nibs that settle rapidly on standing; the nibs do not redisperse if left to stand, nor can they be very easily redispersed by milling.

The decrease in volume of the mill base indicates that solvent has passed from the dilute resin solution to the more concentrated. If solvent passed until the resin solution concentrations were equalised, then for the examples quoted the equilibrium compositions would be as given in Table 9.7.

TABLE 9.7

	Mill base			*'Let-down'*
	Weight	*Volume*		*Weight*
	(g)	*(ml)*	*(%)*	*(g)*
Rutile	183·0	44·6	70·6	—
Alkyd	9·1	8·6	13·6	67·0
Solvent	7·8	10·0	15·8	57·5

These values indicate that the postulated model is over-simplified; the pigment volume concentration markedly exceeds that of the closest packing value that it is reasonable to assign to the titanium dioxide, but the general indications are in agreement with the practical results.

The skin formation is therefore a function of the time that an interface exists between mill base and concentrated resin solution, and of the difference in concentration of the two solutions. The use of a dilute stabilising resin solution is desirable, but most important is the minimising of the time during which an interface can exist. For this reason if a large quantity, say 500 litres, of concentrated resin is to be added to a ball mill, it is best to do it in stages, since the time required to get in such a quantity of liquid, even by pumping, means that skinning is unavoidable. The time taken by the stepwise stabilisation is not as great as is usually necessary to completely redisperse the highly pigmented skin.

Osmotic transfer of solvent can also be reduced by using a solvent with low affinity (little solvent power) for the resin in the mill base and one with high affinity for the stabilising resin solution. But in most cases other restrictions, such as required evaporation rate, prevent the use of solvents sufficiently different in character to have a significant effect, if there is a considerable difference in the resin concentration of the two solutions.

Pigment mixtures

It is possible to correlate theory and practice in the case of simple dispersions, such as those of titania and carbon black just discussed, but such a correlation is hardly possible in the case of pigment mixtures of the type used in flat or semi-gloss paints. A 'simple' pigmentation for a white tint base will probably contain rutile titania blended with two or more extenders. The following comments are therefore very general.

Pigment packing

The oil absorption value has been discussed at some length in earlier editions of this book. Here, a simple assertion will be made, that the oil absorption paste should contain the pigment with a monolayer of oil on each particle and sufficient additional oil to just fill the voids between the closely packed coated particles. In certain cases, such as the larger and the smallest particle size carbon blacks, the theoretical oil absorption, calculated using the particle diameter derived from the BET surface area and assuming a packing fraction of 0·64, is in good agreement with the postulated model.

Generally, the oil absorption of a single pigment is of less interest than that of a mixture of two or more pigments, in particular, of a prime pigment and an extender. If a plot of the oil absorption values of binary mixtures against mixture composition deviates from a straight line, it is a sign that the two kinds of particle pack together more closely than they can do individually. Pigments that have a wide particle size range are usually capable of packing themselves, and a mixture of two pigments of widely different particle size is likely to show packing. Very high oil absorption values are characteristic of certain types of particle, such as diatomaceous silica, because of their structure, whereas other particle shapes, such as acicular particles, tend to pack closely and therefore have a low oil absorption value.

The oil absorption value, though quite easily determined, has two fundamental defects, namely that the result obtained is to some extent dependent on the operator and, secondly, that the value is, strictly, only valid for linseed oil. An absorption value determined using a plasticiser such as dibutyl phthalate, for example, does not usually give the same value for the pigment volume concentration. (As the oil absorption values determined using linseed oils of different acid value, or pigments with different moisture contents, can vary considerably, this is not surprising.)

Critical pigment volume concentration (c.p.v.c.)

The critical pigment volume concentration was originally defined by Asbeck and van Loo[74] as a fundamental physical transition point in a pigment/binder system at which the appearance and behaviour of paint films changed considerably. Asbeck *et al.*[75] subsequently described a slightly modified version of the original method of c.p.v.c. determination, and compared the method with the oil absorption value determination. They considered that, by mechanising the pigment dispersion stage, the reproducibility of the test results was greatly improved. Also the modified test could be used to determine the pigment packing factor, but in calculating this factor no allowance was made for the effect of an adsorbed layer of oil on the pigment particles. The influence of the adsorbed layer becomes increasingly significant as the specific surface area of the pigment increases.

In the 30 years since the c.p.v.c. concept was first introduced, it has been widely accepted and much use made of it because the determination can be made using the relevant paint medium. Another change during the same period has been the trend toward micronised and size-classified inorganic pigments of all types, so that, overall, the average particle size of commercial pigments has almost certainly decreased and the specific surface area increased (notwithstanding the comments on the effects of micronising made earlier). The c.p.v.c. determination therefore indicates the effective packing fraction, rather than the absolute closest packing of the pigment(s), a value that can only be determined by correcting for the adsorbed layer thickness.

The adsorbed layer thickness can be determined from the adsorption isotherm, but attempts, such as that of Saarnak,[76] to use the less time-consuming method of calculating the adsorbed layer thickness from rheological data, are subject to the criticism levelled by Cheng,[67] that the hundreds of equations relating the viscosity of a suspension to the viscosity of its continuous phase are basically empirical.

The trend toward water-thinnable paints has created an additional problem. Whereas a film-forming binder in solution constitutes a continuous phase, a PVA or similar type of 'emulsion' is a suspension of more-or-less rigid particles that are intended to coalesce during the paint film-forming stage. This creates a need to determine the 'pigment binding power' of the emulsion.

Berardi[77] found that there was a constant relationship between the oil absorption value of a pigment and the volume of polymer in the c.p.v.c. film. He called this relationship the 'binding power index'. Ramig[78] later used the binding power index and the oil absorption value (expressed as g(oil)/g(pigment)) to develop a mathematical model for the calculation of the c.p.v.c. of an emulsion paint. He found evidence that the c.p.v.c. is a property of both the pigmentation and the binder, a result that might be expected, since the emulsion particles cannot be considered as true fluids.

The importance of the c.p.v.c. of highly pigmented paints is that it marks the transition from a 'fully bound' film, in which all the voids between pigment particles are filled with film-forming material, to an 'underbound' film, in which there are pigment/air interfaces. At these interfaces, the low refractive index particles can also refract and thus scatter light in the same way as a high refractive index particle. The total opacity of the film is thereby increased, but only at the expense of film porosity, with resultant poor 'enamel hold-out'. It must be emphasised that any comparison of different emulsion paint formulations should be made at equal p.v.c./c.p.v.c. ratios, the c.p.v.c. having been determined by a suitable method, such as that described by Ramig.[78]

In paints made with oleo-resinous binders, the 'binding power index' is not a relevant factor, but it is worth noting that in such media, the opacity provided by titania is due to the core, not the silica and alumina in the coating. Consequently, it may be possible to formulate a flat or semi-gloss paint with a moderately coated grade of titania together with a fine particle extender at a lower raw material cost than when using a heavily coated grade (about 83% TiO_2) in combination with a smaller amount of extender. If the extender in the second paint is relatively coarse, the likelihood of success is greater, but in any case it must be pointed out that heavily coated titanias are not intended to be used in oleo-resinous paints.

PIGMENT DISPERSION IN THE FINAL PRODUCT

Paint film structure

The structure of a paint film can be examined by either of two methods. The longer established method involves sectioning the paint film to determine the degree of dispersion of the pigment within it, usually by means of electron micrographs. The newer method consists of controlled erosion of the paint film surface, combined with scanning electron micrography. Obviously, both methods can be combined, to provide a three-dimensional picture of the film. In addition, electron beam microprobe analysis can be used to determine the distribution of the components of a mixture of inorganic pigment particles, parallel to the surface of the film Unfortunately, the necessary equipment for the above types of investigation is not widely available and no substantial body of work has as yet been published.

Murley and Smith[79] reported that in sectioned films there was a layer of medium 100–150 nm thick, almost without pigment, at the surface of a paint film containing rutile titania at about 10% p.v.c. Kawabata[80] reported that a film pigmented at 30% p.v.c. had a 60 nm clear layer, measured by an interference method. Wilska[81] determined the clear layer thicknesses of rutile pigmented (14% p.v.c.) alkyd and acrylic media by measuring the weight lost by etched films and reported clear layer thicknesses of 540–1000 nm, but assumed that the removed material was completely free of pigment.

Murley and Smith suggested that pigment near the paint/air interface was repelled from the paint surface as a result of the electrostatic 'image force' and derived an equation for the distance travelled in a given time. Their equation gave a value about six times too great, probably because they assumed that the viscosity of the film remained constant. Recalculation, using higher values of film viscosity and correcting for the effect on suspension viscosity of the increasing solid volume concentration, gives values that are nearer to those found and suggests that the postulated mechanism of clear layer formation is correct.

Discussion of paint film structure in general terms is inhibited because the available data are valid for the system investigated, but not necessarily applicable to other systems. The complexity, both physical and chemical, of the constituents of a typical paint film, referred to earlier in this chapter, causes any observed phenomenon to be interpretable in a variety of ways. Gordon[82] has discussed the use of measurements of the permeability of plastic films as an aid to elucidation of polymer structure, and the effect of pigment on the diffusion of ions through protective films has been studied using a radiochemical technique[63] and the result interpreted in terms of the lattice structure of polymeric films. Funke[84] has continued his work on permeability of paint films and reviewed the special and unsolved problems in this field so far as anti-corrosive paints are concerned.

Seiner[85] has described the evaluation of coatings containing microvoids, and titanium dioxide, singly and in combination. The microvoids in this case were the interstices between polymer particles, instead of vesiculated microspheroids, and it was found that, by reducing the flocculation or agglomeration of the titanium dioxide, the opacity of the film was greatly enhanced.

Balfour and Hird[86] devised a method for quantitative assessment of flocculation of titania-pigmented films, based on the scattering of infrared radiation. In a subsequent paper Balfour[87] showed that, because whiting did not scatter infrared radiation, it was possible to measure the flocculation of titanium dioxide in $TiO_2/CaCO_3$ pigmented emulsion paints. However, no attempt was made to elucidate the flocculation phenomenon.

It seems that detailed studies of paint film structure are more likely to be made in the course of attempts to produce improved paints for special purposes, such as the painting of nuclear power stations[88] or of off-shore oil and gas platforms,[89] than as 'pure research' projects. It may then be necessary to carry out more fundamental studies in order to interpret the results obtained.

Paint film gloss

Subjective rating of the relative gloss of a series of paint films presents difficulties that arise from the fact that observers assess gloss by personal techniques that are often quite dissimilar, so that one observer may rate best a film rated worst by others, and that the eye tires, limiting the number of panels that can be reliably rated in a session. In addition, the human observer, often subconsciously, integrates a variety of factors in making his assessment, of which not all are necessarily to do with the gloss, or lack of it, of the film under examination. To be set against these defects is the fact that on occasion the subjective assessment and the instrumental rating of gloss are grossly at variance.

Paint film 'gloss' may be defined as the extent to which the surface of a paint film functions as a specular reflector, or conversely as the extent to which it does not act as a diffuser of incident radiation. The latter definition is a more helpful aid to an understanding of the problem. A perfect diffuse reflector would scatter the incident light ray equally in all directions, so that the locus of equal intensity of reflected radiation would form a hemisphere around the point of incidence. In practice, this hemisphere is slightly distorted at its lower edge and markedly distorted by a protrusion from its surface at the angle of reflectance of the incident ray. The cross-section of the locus can be conveniently plotted two-dimensionally using a gonio-reflectometer, for any given angle of incidence. For a perfect mirror, the emergent ray should have the same dimensions and intensity as the incident ray, with no dispersion of light by diffuse reflection. This is scarcely achieved by the finest metallic mirrors. For a surface such as a paint film, that is quite likely to have a thin layer of clear medium over the pigmented bulk, true 'mirror image' gloss is unattainable. Even if the paint/air interface were optically flat, the pigmented layer below the clear layer is bound to be uneven, to a degree dependent on the intrinsic particle shape and the degree of dispersion (aggregate and/or agglomerate shape), and also non-uniform in refractive index, depending on the concentration of pigment particles in the medium. Pigments that have refractive indices greater than that of the medium, especially the white 'hiding' pigments, will scatter the incident light and thus inevitably cause some diffusion. In addition, the relationship of the clear layer thickness to its refractive index and the wavelength of the incident light may be such that an interference effect is set up; such an effect, especially in the case of white pigments, can cause a variation in the apparent hue of the film.[90]

If the paint film surface is not smooth, its effect on the gloss of the film will depend on the size of the irregularities and the wavelength of the incident light. This has been described by Colling *et al.*[91] who found that large irregularities caused angular divergence of the specular beam of light, whereas small irregularities, of the same order of size as the incident light wavelength, caused scattering (diffusion) of the light. The relationship between the state of dispersion of the pigment and the gloss of the paint film is a fairly simple one, but not always that which might be expected; it is affected by the type of medium and the pigment concentration in the film.

In the case of alkyd/amino resin based baking enamels, shrinkage during film formation arises not only from the loss of solvent but also from the further condensation of the butylated amino–formaldehyde resin. This latter effect can be considerable but is a function of the resin composition, and therefore specific to the resin used. The effects of whether the pigment was dispersed in the alkyd or the amino resin, the kind of alkyd and amino resin used, the alkyd/amino resin ratio and the grade of titanium dioxide have been shown to be very marked.[44] McCausland[92] showed that there was a marked correlation between the size of pigment agglomerates in the paint and the major irregularities of the surface of the baked film of a rutile pigmented alkyd/amino paint. The effect of improved dispersion (increased ball milling time) on the gloss, and the gloss/p.v.c. relationship of alkyd/amino resin based films is quite different from those for the same pigments in air-dried long-oil alkyd based paints.[44] This difference might be attributed to the extra shrinkage of the baked film due to condensation of the amino resin, but if the state of pigment dispersion is very good, there is a marked correlation between gloss loss on overbaking and the presence or absence of coating oxides on titania pigments, both anatase and rutile. This may be due to a greater amount of adsorbed resin, giving an increased 'clear layer' thickness and therefore less surface roughening on added shrinkage due to overbaking of the film.

In all cases, there is a steady fall in gloss values as the pigment concentration in the film increases. Improved dispersion reduces the gradient of the gloss/p.v.c. curve and the gloss value at any given p.v.c. increases. The improvement in gloss is more marked when measured using 20°/20° incident/reflected light beams than when 45°/45° beams are used, indicating that the chief cause of the improvement is a reduction in diffusion of the incident light, or, in other words, a reduction in the haziness of the film.

In the case of long-oil alkyd based paints, the influence of pigment content on gloss is not marked until a pigment content of 15% v/v is exceeded. Even then, the effect is more marked in the case of a more readily dispersed grade of rutile than in that of one less easily dispersed. The gloss of the films containing the less easily dispersed pigment falls slightly with increasing p.v.c.; longer milling to give improved dispersion results in slightly lower gloss and also causes the reduction in gloss with increasing p.v.c. to take place a little more rapidly. The easily dispersed pigment, in marked contrast, gives films with much reduced gloss, whether measured at 20°/20° or 45°/45°, when the films have more than about 15% p.v.c. At 20% p.v.c. and above, the pigment content/gloss relationship is to some extent dependent on the medium; in a long linseed alkyd the 20° gloss levelled out at about 78% at 20 and 25% p.v.c. (compared with a 15% p.v.c. gloss of 90%). Increased milling to improve the dispersion only resulted in very slightly reduced gloss of the linseed alkyd paint films containing less than 15% p.v.c. and a more marked reduction, to about 73%, of the gloss values at 20 and 25% p.v.c. In the long soya alkyd based paints the more easily dispersed pigment gave films with markedly superior 20° and very slightly superior 45° gloss up to 15% p.v.c., above which the gloss of the films dropped markedly, the 20° gloss being equal to that of the paint containing the less well dispersed pigment at 20% p.v.c. The gloss of the films containing the less well dispersed pigment was substantially unaffected by the p.v.c., up to 20% p.v.c., but sharply declined at higher p.v.c. values, reaching a minimum (20° gloss = 16%) at about 50% p.v.c., whereas the better dispersed pigment had a 20° gloss of 40% at that level. At 60% p.v.c., the 20° gloss value was independent of the pigment, both films giving readings of 80%.

In all cases, the fineness-of-grind was better than could be measured with a Hegman gauge, so that the effect was primarily due to the difference in the size distributions of the pigments below a maximum agglomerate size of about 1000 nm. From electron micrographs of paint film sections, one apparent difference was that the less easily and the easily dispersed pigments contained respectively 6·6% and 1·5% by weight of aggregates each formed from 10–12 primary particles, the larger of these being about 600 nm in diameter.

The gloss/p.v.c./dispersion relationship may be explained on the basis that in the lower p.v.c. films the clear layer thickness is similar for both pigments. At higher pigment contents, the larger total number of particles and the larger number of small particles, that will be less strongly repelled from the paint/air interface, combine to turn the statistical probability that there will be some particles in the clear layer into a certainty, so that the gloss is reduced. At more than 30% p.v.c. the effective volume of the particles, assuming a closest possible packing solid fraction of about 0·54, exceeds 55% and the pigment particles start to form a significant proportion of the paint film surface. The clear layer is likely to be reduced to the adsorbed resin layer on the pigment and ceases to make a significant positive contribution to the specular reflection from the surface. The combination of unpigmented medium and pigment particles forming the surface will increase the proportion of the incident light that is scattered. As the pigment particles form an increasing proportion of the paint surface at still higher p.v.c. values diffusion will increase, but the more dispersed pigment, with fewer large aggregates, will cause less distortion of the residual specularly reflected light beams, so that its measured gloss will be higher. At 54% p.v.c. the void volume between the particles will exceed the total volume of the medium and the paint surface will consist almost entirely of pigment particles. The number average particle diameter, about 180 nm (less than the weight average of about 220 nm) is less than the half-wavelength value for the radiation forming the visible spectrum (400–720 nm), so that only those surface irregularities due to aggregates will cause light scattering, though some of the incident light will be scattered after it has entered the pigment layer. (The number of aggregates of 10–12 particles is about $17 \times 10^6/\text{cm}^2$, assuming that distribution of the various sizes of aggregate is uniform throughout the film (20 μm thick). The total number of primary particles is about $24 \times 10^8/\text{cm}^2$.) The proportion of incident light that is scattered is relatively small, however, so the measured specular reflectance becomes quite high (60% at 20°/20°) for both pigments.

Flooding and floating

If a paint film contains two or more pigments, it is possible that, during film formation, these pigments may be substantially separated from each other. Pigment segregation into strata parallel to the paint film surface is described as *flooding* (the film colour being uniform, but the hues of the upper and lower surfaces of a stripped film differing considerably). Pigment segregation into 'columns' perpendicular to the surface is described as *floating* and is characterised by the appearance on the paint film surface of more or less hexagonal Bénard cells. These are the more visible the more marked the contrast in hue of the two pigments.

Assuming the paint to be thoroughly mixed before application, segregation can only occur in the period during which the film is fluid enough to allow pigment particle movement; likelihood of the defect therefore increases with film thickness and with reduction in the rate of solvent loss from the film.

A vast number of papers have been published describing these defects and suggesting methods for their prevention. All of the latter can be summed up as devices for reducing the tendency to separate and/or achieving conditions in the film such that pigment separation will be hindered. The latter condition is most readily achieved by inducing an acceptable degree of flocculation; the immobilisation of a large volume of the liquid phase within the flocs causes a marked increase in the viscosity of the suspension and in addition the relatively large size of the flocs makes them much less mobile than the individual pigment particles. A corollary to this is that if one pigment is flocculated and the other dispersed, then flooding or floating will be accentuated. Also, if both pigments are fully dispersed but of considerably different mean particle size, flooding or floating is very likely.

One of the chief mechanisms causing separation is thermal (convection) currents, set up as a result of the cooling action of solvent evaporation on the paint film surface. Convection in the film may be increased by application of the paint to a warm substrate, or reduced by application to a cold one, especially if the substrate is a good conductor of heat. In this latter case, the reduced rate of solvent loss and extended period of paint film fluidity are more likely to assist flooding or floating than the increase in the viscosity, due to the lower film temperature, is to hinder them.

A second mechanism is electrical in nature. The dispersed particles are likely to have some amount of electrical charge,[70,93] and if both pigments have charges of the same sign, their mutual repulsion will cause separation; this in itself could not cause flooding or floating, but an electrical charge is usually built up at the paint/air interface, through solvent evaporation, and this could cause horizontal stratification, that is, flooding. Unlike charges would cause association, pigment flocculation, the most reliable preventive of flooding or floating.

The Bénard cells that are associated with floating are almost certainly present in all paint films, but without pigments of highly contrasted hue to mark their boundaries, they are not visible. Exposure of the paint film to ultraviolet radiation and a humid atmosphere will sometimes make the Bénard cell structure apparent in the very early stages of paint film erosion. A problem of interpretation can arise here, because a very similar Bénard cell pattern may be produced by a stress-relieving action in a paint film. This is especially important in the case of alkyd/amino resin blends, where the shrinkage of the amino resin (as it is further condensed during baking) may set up quite considerable stresses in the film.

It is perhaps worth remarking that the popular association of flooding and/or floating with particular pigment combinations is not entirely justified. Floating, in particular, is the more visible, the greater the contrast in hue of the pigments. White and black or blue offer maximum contrast and therefore are more noticeable. Careful examination of films, under a good light, and especially using a low-power wide-field microscope, will often reveal the characteristic cell pattern of floating, even though the contrast in hue of the two pigments is too small to be apparent to an ordinary observer.

Pigment dispersion and hiding power

Early in this chapter it was pointed out that there is no one optimum pigment particle size for either white, black or coloured pigments. For white pigments, for example, optimum size varies with the pigment concentration, and the thickness of the film to be opacified. The optimum particle size *for a given application* must be determined by experiment, since there is no theory of opacification able to do more than indicate the probable size range as being large, medium or small. However, before making a positive statement concerning the best particle size, it is best to know the particle size distribution of the pigment in the test-piece. This may or may not correlate with the particle size distribution determined using a pigment dispersion prepared for that specific purpose.

There is no absolutely satisfactory method of determining the pigment particle size distribution in its intended end-use. Paint-film sectioning, though often used, has the disadvantage discussed earlier, that the film-forming process can give a flocculated appearance to the particles. The higher the pigment volume concentration, the worse is this effect and the more difficult the sizing and counting of the particles. Methods such as those described by Hess and Garret[58,94] have the disadvantage, in the one case of being very time consuming and in the other that relatively low pigment concentrations (3% p.v.c. of carbon black) must be used, or else films pressed-out very thin. The latter procedure could easily disperse agglomerates and so give an incorrect result.

It is difficulties such as these that make especial care necessary to ensure that any report dealing with opacity is not of the *ipse dixit* or *ipso facto* type. 'A set of dispersions were prepared and property "X" was measured; the ranking order of the dispersions in this test was paralleled by the ranking order in respect of opacity of films made from the dispersions; therefore it is concluded that an increased value of property X is productive of higher opacity.' The conclusion may be valid, but some causality should be demonstrated.

Franklin[93] showed that hiding power of titanium dioxide dispersions in alkyd increased with the zeta potential (calculated from electrophoretic mobility) but that the effect of zeta potential became less important as the molecular weight of the resin increased. The problem of the adsorbed layer thickness/adsorbed resin weight per unit area was discussed by Doorgeest.[95] Thus, whilst it is reasonable to consider that a thicker adsorbed layer on the surface of pigment particles will reduce the likelihood of pigment flocculation, and that a high molecular weight resin should give a thicker adsorbed layer, it is necessary to take into account the influence of the pigment surface on resin configuration in the adsorbed layer. Goldsbrough and Peacock[73] have suggested that the configuration of alkyd molecules adsorbed on alumina and on silica coatings on rutile is markedly different and that a negligible amount of solvent is associated with the resin adsorbed on silica. As the two surfaces are respectively basic and acidic in character, it seems possible that on alumina adsorption is through carboxylic groups that are by definition end groups for alkyd molecules, whilst adsorption on silica is through the residual hydroxyl groups that are distributed along the polymer chain. The alkyd molecules could thus be envisaged as being 'end-on' to the alumina-coated and lying flat on the silica-coated pigment surface. The alumina-coated pigment would be expected to have superior opacifying power, because the thicker layer should help to prevent excessive crowding and consequent reduced effectiveness of the particles as light scatterers. Evidence in support of this hypothesis has been presented by Franklin *et al.*[96]

The simplistic view of the pigment dispersion and paint film opacity relationship outlined above does not deal adequately with the problems of pigment/medium interactions in commercial paint formulations. Some of the effects on paint properties of various additives have been described by Blakey.[97] Quite often two or more such additives are used in one paint; a study of their interaction and effects on the properties of the suspension might provide a useful insight into some frequently encountered problems.

BIBLIOGRAPHY

Inorganic pigments

T. C. PATTON (Ed.), *Pigment Handbook*, Vols. 1–3, John Wiley, New York (1973).
H. KITTEL (Ed.), *Die Pigmente*, Wissenschaftliche Verlag., Stuttgart (1960).
C. J. A. TAYLOR (Ed.), *OCCA. Paint Technology Manuals*, Vol. 6, *Pigments, Dyestuffs, and Lakes*, Chapman and Hall, London (1966).
O. LUKERT, *Pigment und Füllstoff Tabellen*, 2nd Edition, M. and O. Lukert, Laatzen, BRD (1980).

Pigment characterisation

T. ALLEN, *Particle Size Measurement*, 3rd Edition, Chapman and Hall, London (1980).
S. LOWELL, *Introduction to Powder Surface Area*, John Wiley, New York (1979).
S. J. GREGG and K. S. W. SING, *Adsorption, Surface Area and Porosity*, Academic Press, London (1967).
G. D. PARFITT and K. S. W. SING (Eds.), *Characterization of Powder Surfaces*, Academic Press, London (1976).

REFERENCES

1. W. Feitknecht, in *Pigments, an introduction to their physical chemistry*. Ed. D. Patterson, Elsevier, London, 1967.
2. A. Brockes, *Optik*, **21** (1964) 550.
3. S. R. Orchard, *J. Oil Colour Chem. Assoc.*, **51** (1968) 44.
4. P. Kubelka and F. Munk, *Z. Tech. Phys.*, **12** (1931) 593; P. Kubelka, *J. Opt. Soc. Am.*, **38** (1948) 448.
5. S. Huey, in *Industrial Color Technology*, Adv. Chem. Series No. 107, ACS, Washington, 1972.
6. W. D. Ross, *Ind. Eng. Chem. Prod. Res. Develop.*, **13** (1974) 45.
7. For example G. Schenkel, *Plastics Extrusion Technology and Theory*, Iliffe, London, 1966. The Plastics Institute publications (Iliffe, London): P. D. Ritchie (Ed.), *Physics of Plastics*, 1965; E. G. Fischer, *Extrusion of Plastics* (2nd edn.), 1965; W. R. Groves and M. G. Munns, *Plastics Moulding Plant*, Vols. 1 and 2, 1963 and 1964.
8. R. Wilson and A. H. Robson, *Off. Digest*, **27** (1955) 111.
9. A. R. H. Tawn, *J. Oil Colour Chem. Assoc.*, **39** (1956) 223.
10. A. J. Seavell, *J. Oil Colour Chem. Assoc.*, **42** (1959) 319.
11. R. Brett, *J. Oil Colour Chem. Assoc.*, **41** (1958) 428.
12. J. H. Hildebrand and R. Scott, *The Solubility of Nonelectrolytes*, 3rd edn., Reinhold, New York, 1949.
13. H. Burrell, *Off. Digest*, **27** (1955) 726 and 1069.
14. J. H. Hildebrand, *J. Am. Chem. Soc.*, **38** (1916) 1452.
15. G. Scatchard, *Chem. Rev.*, **8** (1931) 321.
16. P. Flory, *J. Chem. Phys.*, **10** (1942) 51.

17. M. L. Huggins, *J. Phys. Chem.*, **46** (1942) 151; *Ann. NY Acad. Sci.*, **43** (1942) 1; *J. Am. Chem. Soc.*, **64** (1942) 1712.
See also: P. J. Flory, *Principles of Polymer Chemistry*, Cornell Univ. Press, New York, 1953, Chapter 12. H. Tompa, *Polymer Solutions*, Butterworths, London, 1956, Chapter 2.
18. R. C. Nelson, R. H. Hemwall and G. D. Edwards, *J. Paint Technol.* **42** (1970) 636.
19. H. Burrell, *VI FATIPEC Congress* (1962) 21.
20. H. Burrell, *J. Paint Technol.*, **40** (1968) 197.
21. J. H. Hildebrand, *Chem. Rev.*, **44** (1949) 37.
22. H. Tompa, *Polymer Solutions*, Butterworths, London, 1956.
23. T. G. Fox and P. J. Flory, *J. Phys. Colloid Chem.*, **53** (1949) 197.
24. Los Angeles Society Technical Committee 10, *Off. Digest*, **33** (1961) 1329.
25. C. M. Shaw and J. F. Johnson, *Off. Digest*, **25** (1953) 339.
26. E. G. Bobalek, R. L. Savage and M. C. Schroeder, *Ind. Eng. Chem.*, **48** (1956) 1956.
27. W. W. Reynolds and H. B. Gebhart, Jr., *Off. Digest*, **29** (1957) 1174.
28. H. E. Weisberg, *Off. Digest*, **34** (1962) 1154.
29. L. Dintenfass, *J. Oil Colour Chem. Assoc.*, **41** (1958) 125.
30. C. Hansen, *J. Paint Technol.*, **39** (1967) 505.
31. Toronto Society for Paint Technology, *Off. Digest*, **35** (1963) 1211.
32. P. Sorensen, *J. Oil Colour Chem. Assoc.*, **50** (1967) 226.
33. E. K. Fischer, *Colloidal Dispersions*, John Wiley, New York, 1950. R. D. Cadle, *Particle Size Determination*, Interscience, New York, 1955.
34. H. Heywood, *Symposium on Particle Size Analysis*, Supplement to *Trans. Inst. Chem. Engineers*, **25** (1947) 14.
35. J. C. Richards (Ed.), *The Storage and Recovery of Particulate Solids*, Inst. Chem. Engineers Working Party Report, London, 1966.
36. G. D. Parfitt and K. S. W. Sing (Eds.), *Characterization of Powder Surfaces*, Academic Press, London, 1976.
37. S. J. Gregg, K. S. W. Sing and H. F. Stoeckli (Eds.), *Characterisation of Porous Solids*, Soc. Chem. Ind., London, 1979.
38. D. Urwin, *J. Oil Colour Chem. Assoc.*, **52** (1969) 697.
39. B. C. Lippens and J. H. de Boer, *J. Catal.*, **4** (1965) 319.
40. R. H. S. Robertson and B. S. Emödi, *Nature (London)*, **152** (1943) 539.
41. F. M. Lea and R. W. Nurse, *J. Soc. Chem. Ind.*, **58** (1939) 277.
42. R. B. Anderson and P. H. Emmett, *J. Appl. Phys.*, **19** (1948) 367.
43. S. G. Ward and R. L. Whitmore, *Brit. J. Appl. Phys.*, **1** (1950) 325.
44. H. D. Jefferies, *J. Oil Colour Chem. Assoc.*, **45** (1962) 681.
45. H. D. Jefferies, *Verfkroniek*, **37** (1964) 436.
46. B. B. Boonstra and A. I. Medalia, *Rubber Age*, **92** (1963) 802 and **93** (1963) 82.
47. E. Micek, F. Lyon and W. M. Hess, *Rubber Chem. Technol.*, **41** (1968) 1271.
48. L. L. Ban and W. M. Hess, Paper given at Ninth Biennial Conf. on Carbon, Boston, 1969.
49. D. F. Harling and F. A. Heckman, *Materie Plastiche ed Elastomeri*, **35** (1969) 80; Cabot Corp. Research Paper 37–72.
50. C. Fisher and M. Cole, *The Microscope*, **16** (1968) 81.

51. A. I. Medalia, *J. Colloid Interface Sci.*, **32** (1970) 115.
52. H. F. Clay, *J. Oil Colour Chem. Assoc.*, **40** (1957) 935.
53. A. W. Evans and R. D. Murley, *VI FATIPEC Congress* (1962) 125.
54. G. Kämp and H. G. Völz, *Farbe u. Lack*, **74** (1968) 37.
55. D. Rivin, *Proc. 4th Rubber Techn. Conf. 1962*, Inst. of Rubber Industry, 1964, p. 261.
56. N. Scott, *J. Oil Colour Chem. Assoc.*, **49** (1966) 559.
57. D. Dollimore, *J. Oil Colour Chem. Assoc.*, **54** (1971) 616.
58. W. M. Hess and M. D. Garret, *J. Oil Colour Chem. Assoc.*, **54** (1971) 24.
59. F. K. Daniel and P. Goldman, *Ind. Eng. Chem., Anal. Edn.*, **18** (1946) 26.
60. F. K. Daniel, NPVLA *Sci. Sect. Circ.* No. 744, 1950.
61. F. K. Daniel, *J. Paint Technol.*, **38** (1966) 534.
62. H. Rumpf, *Agglomeration*, Ed. W. A. Knepper, Interscience, New York, 1962, p. 379.
63. W. P. Pietsch, *Nature (London)*, **217** (1968) 736.
64. J. Pryce-Jones, *J. Oil Colour Chem. Assoc.*, **19** (1936) 295.
65. J. M. Burgess and G. W. Scott-Blair, *Report on Nomenclature*, North Holland, Amsterdam, 1949.
66. E. C. Bingham, *Fluidity and Plasticity*, McGraw-Hill, New York, 1922.
67. D. C.-H. Cheng, *Chem. Ind. (London)*, 1980, No. 10, 403.
68. S. Guggenheim, *Off. Digest*, **30** (1958) 729.
69. H. Heesch and F. Laves, *Z. Krist.*, **85** (1933) 443.
70. V. T. Crowl, *J. Oil Colour Chem. Assoc.*, **50** (1967) 1042.
71. M. J. Vold, *J. Colloid Sci.*, **6** (1951) 492.
72. K. Rehacek, *Farbe u. Lack*, **76** (1970), 656; *Ind. Eng. Chem. Prod. Res. Dev.*, **15** (1976), 1 and 75.
73. K. Goldsbrough and J. Peacock, *J. Oil Colour Chem. Assoc.*, **54** (1971) 506.
74. W. K. Asbeck and M. van Loo, *Ind. Eng. Chem.*, **41** (1949) 1470.
75. W. K. Asbeck, D. D. Laiderman and M. M. van Loo, *Off. Digest*, **30** (1952) 326.
76. A. Saarnak, *J. Oil Colour Chem. Assoc.*, **62** (1979) 455.
77. P. Berardi, *Paint Technol.*, **27** (7) (1963) 24.
78. A. Ramig, *J. Paint Technol.*, **47** (602) (1975) 60.
79. R. D. Murley and H. Smith, *J. Oil Colour Chem. Assoc.*, **53** (1970) 292.
80. A. Kawabata, *Shikizai Kyokaishi*, **42** (1969) 98.
81. S. Wilska, *J. Paint Technol.*, **43** (1971) 65.
82. M. Gordon, *The Structure and Physical Properties of High Polymers*, Plastics Monograph No. C10, The Plastics Institute, London, 1957.
83. A. L. Glass and J. Smith, *J. Paint Technol.*, **39** (511) (1967) 490.
84. W. Funke, *J. Oil Colour Chem. Assoc.*, **62** (1979) 63.
85. J. A. Seiner, *J. Oil Colour Chem. Assoc.*, **60** (1977) 335.
86. J. G. Balfour and M. J. Hird, *J. Oil Colour Chem. Assoc.*, **58** (1975) 331.
87. J. G. Balfour, *J. Oil Colour Chem. Assoc.*, **60** (1977) 365.
88. P. Walker and A. Haighton, *J. Oil Colour Chem. Assoc.*, **62** (1979) 337.
89. D. Atherton, *J. Oil Colour Chem. Assoc.*, **62** (1979) 351.
90. W. G. Armstrong and W. D. Ross, *J. Paint Technol.*, **38** (1966) 462.
91. J. H. Colling, W. E. Craker and J. Dunderdale, *J. Oil Colour Chem. Assoc.*, **51** (1968) 526.

92. R. J. McCausland, *Double-Liaison*, **91** (1963) 53.
93. M. J. B. Franklin, *J. Oil Colour Chem. Assoc.*, **51** (1968) 499.
94. M. D. Garret and W. M. Hess, *J. Paint Technol.*, **40** (1968) 367.
95. T. Doorgeest, *J. Oil Colour Chem. Assoc.*, **50** (1967) 1079 and discussion, pp. 1106–14.
96. M. J. B. Franklin, K. Goldsbrough, G. D. Parfitt and J. Peacock, *J. Paint Technol.*, **42** (551) (1970) 740.
97. R. R. Blakey, *J. Paint Technol.*, **43** (559) (1971) 65.

CHAPTER 10

DISPERSION OF ORGANIC PIGMENTS

R. B. McKay
Ciba–Geigy Plastics and Additives Company, Paisley, UK

and

F. M. Smith
Ciba–Geigy Plastics and Additives Company, Manchester, UK

INTRODUCTION

Organic pigments are finely divided crystalline solids. They are essentially insoluble in the media in which they are applied, principally inks, paints, plastics and artificial fibres, and usually have to be dispersed by mechanical means. Their primary function is to impart colour, but in some instances they can have additional beneficial effects, for example on the flow properties and hence distribution and transfer of printing inks and on the opacity of surface coatings. Their colour has obvious decorative appeal, visual impact for communication and usefulness for identification purposes, for example in electrical cable coatings. When compared to coloured inorganic pigments they tend in general to be more finely divided, with greater tinctorial strength, brightness and resistance to chemical attack, but lower fastness to heat and light and lower resistance to organic liquids.

The selection of organic pigments for particular applications is made in the first instance on the bases of fastness properties and cost, both of which depend upon chemical constitution. This leads to two broad categories of pigments, namely high grade and classical. The former of more recent origin are generally the more durable and expensive and can be regarded as speciality products for the most demanding applications.

Classical pigments are long established and have been the subject of more intensive development work designed to optimise their colouristic performance. Particularly noteworthy among the classical types are copper phthalocyanine pigments which combine high durability with comparatively low cost.

The colouristic performance of a given type of organic pigment in the final application system is highly dependent upon the state of dispersion achieved therein. This and the ease with which pigments can be dispersed are in most cases the main criteria in choosing between pigments of a particular chemical type for a given purpose. The ease of dispersion and state of dispersion achieved are dependent upon the preparative history, composition and method of dispersion of the pigments in the particular application medium. Thus, performance is controlled partly by the pigment manufacturer and partly by the pigment user. The manufacturer designs pigments to optimise performance in specific types of application system that use specific dispersion methods. By development and control of manufacturing processes and additive treatments he can produce a multitude of pigments with widely varied application behaviour from a given chemical compound. The main controllable variables are crystal lattice type, crystal size, shape and state of assembly, additive content and disposition, wettability, dispersibility and ability to adsorb components of application media that affect flocculation resistance and hence rheological properties of dispersions. The problem is to isolate the finely divided material from the carrier liquid, usually water, in a conveniently handleable form, in such a way as to avoid irreversible damage and to enable the pigment user to reconstitute substantially the high degree of subdivision in his particular systems.

Organic pigments are sold mainly in the forms of powder or granules for dispersion by user industries. They are, however, also sold predispersed in suitable carriers. Concentrated aqueous pastes are long established for use, for example, in emulsion paints, but more recent introductions are non-aqueous dispersions for use in inks and paints, and plasticiser dispersions for use in flexible plastics. Concentrated dispersions in polymeric carriers or nitrocellulose are sold as solid chips for use in plastics and polar liquid inks, respectively.

Before considering dispersion of organic pigments in application media it is necessary to have an appreciation of the physical nature of organic pigments, of what the most important characteristics are, and of the means by which these characteristics can be modified and controlled—in other words an appreciation of the skills of organic pigment technology. Numerous general articles have been published dealing with various aspects of the technology, for example refs. 1–3.

If the present chapter appears preoccupied with copper phthalocyanine pigments, it merely reflects the importance of these pigments and the fact that they have been the subject of much more reported investigative work than any other class of pigment. They have been particularly suitable for use in establishing principles.

CHEMICAL CLASSIFICATION OF ORGANIC PIGMENTS

The chemical constitution of organic pigments has been dealt with elsewhere.[4,5] It is appropriate here to outline the most salient features. Organic pigments may be classified into the following groups.

Phthalocyanines

Commercially available phthalocyanine pigments are predominantly based on the copper chelate of tetrabenzoporphorazine, known as copper phthalocyanine blue (CI Pigment Blue 15) (Fig. 10.1) and the polychlorinated derivative (14–16 Cl atoms per molecule) known as copper phthalocyanine green (CI Pigment Green 7). Metal-Free phthalocyanine (CI Pigment Blue 16) is of some, but minor, significance. They have outstanding colour strength, brightness and all-round fastness properties and are by far the most important pigments in the blue–green colour area.

Polycyclic pigments

This group consists of a wide variety of chemical types featuring aromatic and heterocyclic condensed ring systems. They are expensive high quality pigments, colouristically strong and bright, and with excellent fastness properties. The most important types are quinacridone red (CI Pigment Violet 9), carbazole dioxazine violet (CI Pigment Violet 23) and the isoindolinones (CI Pigment Red 180 and Yellow 109).

Metal-free azo pigments

This group consists of benzenoid, naphthalenoid or heterocyclic aromatic ring systems linked by azo (or azomethine) groups. The simplest types, monoazos (e.g. yellow 10G, CI Pigment Yellow 3) and bisazos (e.g. metaxylidide yellow, CI Pigment Yellow 13), are very important commercially but are in general not the most resistant to solvents, heat and light. More expensive, higher-molecular-weight types have improved

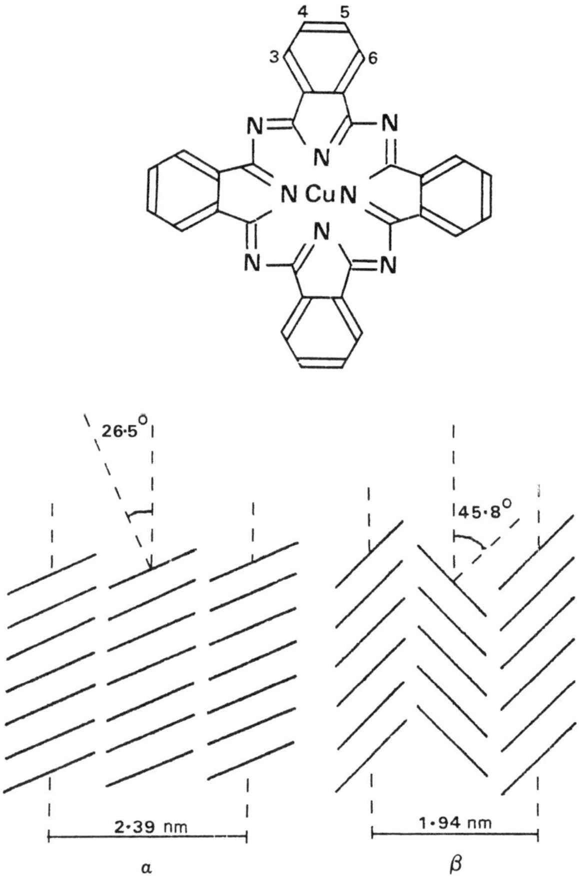

Fig. 10.1 Chemical constitution and crystal lattice structures of copper phthalocyanine.

fastness properties, for example, the azo condensation series as represented by CI Pigment Yellow 95 and CI Pigment Red 144.

Metal salts

These are calcium, barium, strontium or manganese salts of sparingly soluble anionic azo dyes. Usually these dyes contain one sulphonic acid group per molecule and sometimes an additional carboxylic acid group. Important examples are calcium red 4B (CI Pigment Red 57), barium lake red C (CI Pigment Red 53) and calcium, barium and manganese red

2B toners (CI Pigment Red 48). In general, these salts have high colour strength and brightness, good resistance to heat and solvents, but poor resistance to chemical attack, especially by alkali, and have moderate lightfastness.

Metal chelates

Also known as metal complex pigments, these consist of metal atoms covalently linked to organic ligands. The metal atom is an integral part of the chromophoric system and, unlike that in metal salts, has a profound effect on the colour of the molecule. Metal chelate pigments have good solvent resistance and better chemical resistance than metal salts. Examples are nickel azo yellow (CI Pigment Green 10) and more recent azomethine chelates (CI Pigment Yellow 117). Copper phthalocyanine pigments are really members of this class.

Vat pigments

Vat pigments are derived from insoluble vat dyes. They have good fastness properties, but tend to be colouristically dull and have been virtually superseded by polycyclic pigments. Still important, however, are flavanthrone (CI Vat Yellow 1) and indanthrone (CI Vat Blue 4).

Cationic dye complexes

Cationic (basic) dye complexes, also known as fanal pigments, consist of well-known basic dyes, such as rhodamine B, methyl violet, insolubilised by interaction with complex inorganic acids such as phosphomolybdotungstic acid or with ferrocyanide ions. They are colouristically very strong and bright, but have poor fastness properties.

Lakes

Lakes are metal salts of readily soluble anionic (acid) dyes of azo, triphenylmethane or anthraquinone types. They are usually precipitated by a calcium, barium or aluminium salt on to a substrate of alumina or blanc fixe or a mixture of both. They are cheap, bright, have high colour strength, good solvent fastness, but poor resistance to light and chemical attack.

The characteristic colour and fastness properties of organic pigments are determined primarily by the chemical constitution of the pigment molecules. The colour is dependent upon an extensively conjugated chromophoric system, which must have moderate to high symmetry, especially for violets and blues. Consequently the molecules tend to be large, planar and reasonably symmetrical, the perfect example being copper phthalocyanine. This shape has an important influence on the stacking of molecules in the crystal lattice and thus has secondary effects on colour, semiconductor properties, lightfastness, lattice energy, deformability and solubility, and leads to anisotropic characteristics, such as acicular crystal shape and dichroism.

CRYSTALLOGRAPHY OF ORGANIC PIGMENTS

Almost all commercially available organic pigments are crystalline in that they show characteristic X-ray diffraction patterns. Those for which electron micrographs have been published appear to consist of crystals varying in shape from plate-like, isometric or brick-shaped (aspect ratio $\lesssim 2$) to rod-shaped (maximum aspect ratio >2, typically 4 to 10) and varying in size from about 0·02 to 0·5 μm in minimum and maximum dimensions respectively. Examples are given in Fig. 10.2. However, crystal size, and indeed shape, can be assessed only approximately by normal methods. One reason is the fact that organic pigment crystals are susceptible to damage in the electron beam,[6] some more than others, typical effects being chemical degradation, ablation (rounding of edges and corners) and loss of crystallinity. Orientation and superposition of crystals and their tendency to exist in various states of assembly cause further difficulty. High-resolution electron microscopy using special techniques has been able to resolve the lattice planes in suitably orientated crystals of copper phthalocyanine pigments,[6-9] thus enabling detection of lattice defects. The higher the resolution, however, the less statistically representative is the area of the sample examined and the greater the risk of beam damage. As a result high resolution is used only for special purposes.

Most attention has been focused on copper phthalocyanine pigments partly because of their particular importance as pigments, but also because they provide excellent examples of polymorphic character in molecular crystals. Copper phthalocyanine occurs in two distinct crystal types, namely the α-(red-shade blue) and β-(green-shade blue) forms, and several modified α-forms[10] (sometimes referred to as γ, δ and ϵ forms[11]). Only the α- and β-forms are of commercial significance and are sufficiently different in colour to be regarded as distinct pigment types. The β-form is thermodynamically the more stable, its lattice energy being lower by 10.3 kJ/mole.[12] Conversion of α

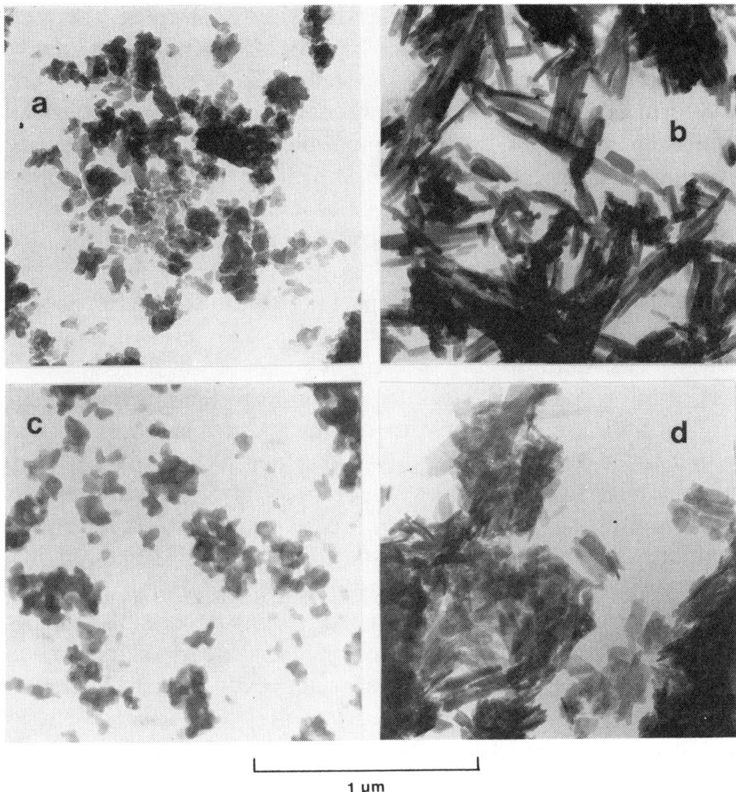

Fig. 10.2. Transmission electron micrographs of a selection of commercially available organic pigments. (a) A β-copper phthalocyanine pigment prepared by a dry attrition process; (b) a solvent-treated β-copper phthalocyanine pigment containing resin (8% w/w) (resin not visible); (c) a metal-free azo pigment (Pigment Yellow 13); (d) a calcium 4B toner (metal salt) containing resin (20% w/w) (resin not visible).

to β can be brought about either by heating or by recrystallisation in suitable liquid media.

The lattice structure of β-copper phthalocyanine has been determined from single crystal X-ray diffraction data.[13] The lattice structure of the α-form has been determined indirectly from single crystal studies of 4-monochloro-copper phthalocyanine with which it is isomorphous.[14] The two lattice structures have recently been compared and the essential differences highlighted.[15]

Both forms consist of columnar stacks of molecules, the axes of adjacent stacks being in parallel (Fig. 10.1). Within each stack the molecules are tilted at an angle to the stack axis, this angle being greater in the β-form as a result of octahedral coordination of the copper atoms with nitrogen atoms in adjacent molecules. Octahedral coordination does not occur in the α-form. Whereas in the α-form the molecules in a given stack are tilted parallel to those in adjacent stacks, in the β-form the molecules are tilted in the opposite direction to those in adjacent stacks, thus giving a 'herring-bone' array. This array and the greater angle of tilt lead to more compact packing of the stacks in the β-form, thereby resulting in a greater number of contacts between molecules in adjacent stacks than in the α-form, together with interlocking that restricts rotation round the stack axis. As a result the β-form is the more stable form and less susceptible to mechanical deformation. It is significant that the β-form has a slightly higher density than the α-form.[16]

Intermolecular forces within a stack are stronger than those between stacks more especially in the β-form. As a result crystals tend to grow preferentially in the direction of the stack axis and thus to be rod-shaped. Attempts to grow α-form crystals by treatment with solvents result in conversion to the β-form, so that large α-form crystals cannot be obtained. It has been shown, however, that in the early stages of solvent treatment, growth of the α-form does in fact occur prior to nucleation of the β-form,[17,18] and that these grown α-crystals are rod-shaped.[18] Elongated α-form crystals have also been observed following heat treatment of copper phthalocyanine deposited on cleaved muscovite by vacuum sublimation.[19]

Copper phthalocyanine as synthesised by conventional processes in a carrier fluid consists of large rod-shaped β-form crystals with length up to about 100 μm,[20] size depending upon the particular process and carrier liquid used. This crude non-pigmentary material can be reduced to pigmentary size (0·02–0·5 μm) either by attrition or precipitation processes. Dry attrition in the presence or absence of inorganic salt promotes β- to α-form conversion, up to about 99% α-form being attainable in normal practice. Precipitation from solution in concentrated sulphuric acid (acid pasting) produces 100% α-form. Pigmentary β-form is produced by attrition processes in the presence of inorganic salt containing a small amount of a crystallising liquid,[20] but still essentially dry, or by grinding in the presence of bulk crystallising liquid.[21] It is also produced by grinding crude material in various liquid–salt mixtures containing a Lewis acid, such as ferric chloride,[22] or in aqueous surfactant solutions.[23] Alternatively, pigmentary β-form is produced from α, β-form mixtures (from dry attrition processes) by treatment with crystallising liquids,[21,24] such as aliphatic alcohols, by treatment with acids with dissociation constants of 5×10^{-2} or less,[25] or by treatment with aqueous media containing either emulsified high boiling solvents[26] or aqueous surfactant solutions without added solvent.[27]

α and β-form pigments from dry attrition processes tend to consist of small brick-shaped crystals typically 0·03 μm in maximum dimension. As their size is close to the limits of resolution of X-ray diffraction, their diffraction patterns tend to be diffuse. β-form crystals produced by solvent, emulsion or surfactant treatment tend to be rod-shaped. Processes optimised to give satisfactory tinctorial properties yield crystals of varying sizes up to about 0·2–0·3 μm in length, but usually consistently around 0·03 μm in width. This is evident from the many electron micrographs in the literature. Such crystals tend to give sharp X-ray diffraction patterns indicating a high degree of crystal perfection. Acid-pasted α-form crystals may be brick-shaped, short rods or plates, depending upon the particular conditions used. They are noticeably more resistant to conversion to β-form in liquids than are α-form samples produced by dry attrition, apparently due to absence of β-nuclei, but conceivably also due to traces of sulphonated copper phthalocyanine that poison growth surfaces. It is evident that crystal shape is influenced not only by molecular interactions within the lattice, but also by external factors during the size reduction stage in pigment manufacture.

The reddish shade blue of α-copper phthalocyanine is desirable in paints and plastics. Unfortunately, the conditions encountered in these systems tend to cause conversion to the greener β-form. The α-form must therefore be stabilised. This is achieved by partial chlorination of the copper phthalocyanine molecule in the 4-position so that there is at least the equivalent of just over one half chlorine atom per molecule.[20] The resulting mixed species has colour similar to that of α-copper phthalocyanine and does not revert to a β-form even under severe conditions. Detailed studies have shown that 4-monochloro-copper phthalocyanine itself does not exist in a β-form although it can exist in various modified α-forms.[10] Stabilised α-copper phthalocyanine may be produced in pigmentary crystal size by the attrition and acid-pasting processes described above. It tends to be brick-shaped like α-copper phthalocyanine itself, but is different in that it can grow in crystallising liquids, without phase conversion, to give long rods or filaments. Conversion of α- to β-form copper phthalocyanine pigments can also be retarded by adsorbing soluble substituted copper phthalocyanine derivatives to the pigment surface, some types of derivative being more effective than others.[20,28,29] These additives also inhibit crystal growth. Evidently they poison growth surfaces, but their detailed mechanism of action has not been established.

Further examples of polymorphism in organic pigments are provided by indanthrone (CI Pigment Blue 60),[10] linear *trans*-quinacridone (CI Pigment Violet 19)[4,30] and certain azo pigments,[30] but these have not been investigated in the same degree of detail as copper phthalocyanine.

Recently, detailed studies have been reported on the crystal structures of six monoazo pigments involving β-naphthol (CI Pigment Reds 1, 2, 3 and 6 and chlorinated derivatives of CI Pigment Red 9 and CI Pigment Brown 1).[31] It has been shown that although intramolecular hydrogen bonding occurs, intermolecular hydrogen bonding does not and the crystal lattice is held together by van der Waals forces. The molecules are arranged in columnar stacks, but tilted at an angle to the stack axis. Whereas in some pigments the molecules in adjacent stacks are in parallel (cf. α-copper phthalocyanine), in others they are tilted in the opposite direction, giving a 'herring-bone' array (cf. β-copper phthalocyanine). Detailed crystallographic data have also been reported for CI Pigment Yellow 1 and its chlorinated analogue,[32] CI Pigment Yellows 1, 3 and 4, and brominated derivatives of CI Pigment Yellow 3.[33]

WHY EFFECTIVE DISPERSION IS ESSENTIAL

Whereas the characteristic colour of an organic pigment is determined primarily by its chemical constitution and secondly in some cases by its crystal lattice type, its precise hue, intensity and brightness exhibited in an ink or paint film or a plastic are dependent upon the degree of subdivision achieved.

The colour strengths of dispersions of a given organic pigment in a given medium increase with decrease in mean particle size in the submicron region. This has been shown theoretically by applying the Mie theory to values of the complex refractive index of the pigment as determined from ellipsometric measurements made on discs of compressed pigment powder.[34-36] The application of the theory has been simplified by assuming isotropic particles, an assumption which is certainly not strictly valid for organic pigments, and it is doubtful if the surface of a pressed disc truly represents the state of pigment that exists in a dispersion. Nevertheless the relationship is firmly established qualitatively, if not quantitatively. A similar trend has been demonstrated experimentally in paint systems by relating reflectance data from coated and dried paint films (containing pigment in admixture with a large excess of titanium oxide) to mean particle size measurements made by centrifugal sedimentation of diluted samples of the paint concentrates (without titanium oxide).[36–39] Despite the indirectness of comparison of the variables, the trend is clear qualitatively. The dependence of colour strength on particle size is marked at sizes around 0·1 μm, particularly with pigments that are inherently strong.

It has also been shown theoretically, at least with some major pigment types, namely copper phthalocyanine blues, quinacridones and certain classical azos, that colour strength does not continue to increase as particle size is further reduced until the pigment is molecularly dispersed. Instead, colour strength passes through a maximum within the particle size range from 0·1 to 0·01 μm.[34,35] This means that there is an optimum potential colour strength of dispersions of a given pigment and it is important to know how close existing pigment types approach this optimum in application systems.

Brightness of shade is also sensitive to particle size variation at sizes around 0·1 μm, and diminishes with increase in size.[2,36] This has been attributed to broadening of particle size distribution and hence absorption spectrum with increase in size.

Particle shape is another important consideration. It has been shown using isotropic model particles that when particles are non-spherical it is the minimum dimension that is relevant to colour strength.[40] Independently it has been shown that the colour strength of β-copper phthalocyanine pigments in a paint system increases as particles consisting of isometric crystals are replaced by rod-shaped particles of comparable mean size.[41] The change in shape, however, had the further effect of causing a change in shade to a more reddish blue. This has been attributed[41] to optical anisotropy (dichroism in this case) of the pigment crystals, the colour being more reddish when viewed along the main axis of the rod-shaped crystals, than when viewed at right angles to this axis. The shade was dominated by this major axis even when the crystals were randomly orientated. The magnitude of the shade difference between the isometric and acicular crystals although significant was much less than that between the β- and α-crystal forms.[42] It follows that mixtures of isometric and rod-shaped crystals of β-copper phthalocyanine will exhibit less cleanliness of shade than will homogeneous fractions.

Organic pigment dispersions in ink or paint media with particle size of the order of 0·1 μm or less give coated films that are highly transparent. Although this is desirable in many circumstances, especially for example in multicolour process printing where successive layers of ink are superimposed, in other circumstances it is not. There is a need for red and yellow pigments that give paint films with good hiding power or opacity. Optimum inherent opacity requires pigment particles of larger size than usual, [43–45] indeed with a large proportion of isometric particles of size in the range from 0·2 to 0·6 μm,[44] thus to produce optimum light scattering. This aim can be achieved during pigment manufacture by treatments that promote crystal growth in a controlled manner, such as heating aqueous slurries of precipitated pigment in the presence of immiscible or partially miscible organic liquids at over 100°C under pressure.[46] Inherent opacity is obviously achieved at the expense of colour strength.

The surface appearance of ink and paint films can be adversely affected by inadequate dispersion of organic pigments. For example, it is common experience that gloss improves with increasing time or energy of dispersion. Indeed there is evidence to suggest that improvement continues until the mean particle size is of the order of 0·1 μm (oil ink system[36]) to 0·4 μm (decorative paint system[38]). Whether or not there is any real significance in the specific size values quoted is not clear in view of limitations of the methods used to determine them. It has also been suggested that improvement of gloss below a mean particle size of about 0·4 μm in the decorative paint system is due to reduction in haze associated with light scattering from particles within the film, rather than surface roughness.[47] Particle shape is important in this context also and it has been shown that acicular particles can give rise to surface imperfections considerably larger than themselves.[48]

The above considerations have been concerned with the level of dispersion that it is desirable to achieve. Also of importance, however, is the small proportion of residual oversize particles (greater than about 5 μm) that have not been effectively dispersed. These can appear as visible specks in plastics, can cause surface defects in ink and paint films, and can even cause damage to printing plates. Organic pigment dispersions are specifically examined for oversize particles, for example, by means of the Hegman gauge or similar instrument, or increasingly the optical microscope, to indicate their maximum size in inks and paints, and by means of the optical microscope for plastics. Failure to meet established criteria can lead to rejection of an otherwise satisfactory dispersion.

AGGREGATION

Aggregate structure

It has long been recognised that there are various types of assembly of organic pigment crystals. This follows from observations of the consistency of pigment powders and granules, and the ease or difficulty with which they can be dispersed in application media. Following investigations of copper phthalocyanine pigments by X-ray diffraction, electron microscopy and specific surface area determination by nitrogen adsorption, Honigmann and Stabenow[49,50] proposed an hypothetical system of classification of assembly on the basis of geometrical structure. Three types of building units were identified, namely individually disperse crystals, impenetrable aggregates of crystals joined face-to-face, and porous aggregates of ill-defined crystals. Agglomerates were defined as assemblies of those units loosely joined at edges and corners. Four types of agglomerate were proposed on the basis of their constituent building units, namely assemblies of (a) individual crystals, (b) impenetrable aggregates, (c) individual crystals and impenetrable aggregates and (d) porous aggregates. It was implied that the spaces between the building units in the agglomerates are freely penetrable by nitrogen. The impenetrable character of aggregates was indicated by BET specific surface areas much lower than those determined from crystal sizes estimated from electron micrographs. Despite their weakly coherent nature, agglomerates (considered[50] to be synonymous with air flocculates) must of course be substantial enough to produce a powder or granular consistency.

Although conceptually useful, the above classification over-simplifies the situation. Such precise distinction between aggregates and agglomerates is seldom possible in practice. Moreover, many if not most commercial organic pigments contain substantial amounts of additives ranging typically from 5 to 50% by weight, and these additives by design influence the mode of packing of crystals and subsequent ease of dispersion. Indeed, the additives may form an integral part of the assembly structure.[51] It is therefore perhaps more realistic to describe assemblies of crystals by the term 'aggregate' in the general sense adopted by the IUPAC council,[52] that is as a group of particles held together in any way. This terminology has therefore been adopted here.

The disposition of additives cannot be distinguished satisfactorily by electron microscopy, at least using established techniques. Indeed, even with pigments free from additives, electron microscopy is of limited usefulness for investigating the state of assembly of crystals in powders. The problem lies in artefacts arising from the deposition of a sample of the powder on the microscope grid. Consequently, other methods of approach are necessary.

Gas adsorption using nitrogen at $-196°C$ is well established as a means of determining the specific surface area of organic pigment powders. Recently, however, more detailed examination of adsorption/desorption isotherms has provided a further insight into aggregate structure.[15] Both specially prepared[53,54] and commercial[55] copper phthalocyanine pigments free from additives have been examined after outgassing (at room temperature to avoid thermal modification).

Certain β-form pigments with rod-shaped crystals gave reversible adsorption, in the sense that adsorption and desorption isotherms coincided (e.g. Fig. 10.3A); there was no hysteresis. They therefore behaved as non-porous solids. The BET specific surface area of the pigment discussed in ref. 55 is comparable to that calculated from the approximate crystal size data quoted, thus indicating that a large proportion of the crystal surfaces are accessible to nitrogen. This particular pigment appears to correspond to agglomerate structure (a) as designated earlier in the Honigmann and Stabenow classification.

In contrast, the other pigments examined gave marked hysteresis (e.g. Fig. 10.3B and C) indicative of mesoporous character (pores of size 2–20 nm). The hysteresis characteristics were consistent with slit-shaped pores of a non-rigid type between adjacent crystals in an aggregate structure. All of those pigments consisted of small brick-shaped crystals of size around 0·03–0·05 μm in maximum dimension (N.B. sizes quoted in ref. 53 are low by a factor of 3) irrespective of their chemical constitution (unchlorinated, monochloro or fully chlorinated) or crystal lattice type (α or β). It is evident that a large proportion of the crystal surface is not accessible to nitrogen, which penetrates only the outer regions of the aggregate structure. The Honigmann and Stabenow classification does not adequately describe such cases.

Clearly the nature of the aggregate structure formed by these copper phthalocyanine pigments is determined primarily by their crystal morphology (size and shape), rather than crystal lattice or chemical type.

It has been stated by Honigmann that porous character is not found in high quality organic pigments,[50] although no gas adsorption/desorption data have been produced in support. In our experience many commercially important azo and copper phthalocyanine pigments exhibit

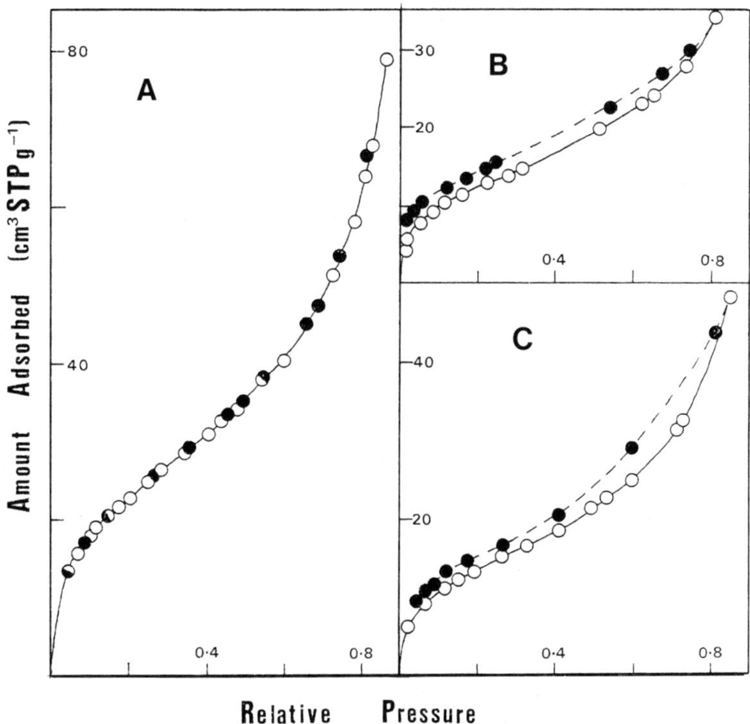

Fig. 10.3. Adsorption/desorption isotherms for nitrogen at −196°C (courtesy of Dr R. R. Mather). (A) A β-copper phthalocyanine pigment with well developed rod-shaped crystals (approx. 0·2 μm × 0·03 μm); (B) a β-copper phthalocyanine pigment with brick-shaped crystals (0·03–0·05 μm long); (C) an α-copper phthalocyanine pigment with crystals similar in size and shape to those of (B). ○, Adsorption; ● desorption.

hysteresis indicative of porous character, indeed to such an extent that non-porous character may be the exception rather than the rule.

Control of aggregation

Aggregation of crystals is inevitable during the manufacture of organic pigments, at stages where finely divided crystals are being produced, for example during attrition or precipitation processes. Since friction, crystal fracture and compaction mechanisms operate in the latter, all tending to give extensive distortion of the crystal lattice, the aggregation occurring is liable to be of a particularly strongly coherent nature. For example, crude β-copper phthalocyanine when dry milled with steel balls produces finely divided α,β-form mixtures in a severely compacted state. In a specific example the crystals produced were brick-shaped of size around 0·03 μm, yet had an extremely low BET specific surface area of 5 m^2/g, more than an order of magnitude less than that expected from the crystal dimensions, and exhibited marked porosity.[15] Aggregation is also liable to occur during the drying of aqueous filtercakes, particularly in the later stages where removal of water from the capillaries between adjacent crystals can lead to the crystals being pulled into close contact by capillary attraction.[51] This effect is commonly referred to as hydrophilic aggregation. In practice aggregation operates against attainment of high colour strength by increasing the difficulty of achieving small particle size in dispersion in application media.

Aggregation can be controlled in three ways, namely recrystallisation, use of additives that affect the properties of the pigment–water interface, and methods that impose a mechanical barrier between crystals in situations where strong interaction is prone to occur. All of these approaches are important in practice. Although there have been few reported investigations of their mechanisms, nevertheless they are extensively exemplified in the patent literature.

Recrystallisation gives crystal growth and in appropriate cases induces a change in crystal lattice type to the most stable form. It is extensively used in the manufacture of easily dispersible β-copper phthalocyanine pigments. It features in the treatment with organic liquids of dry-ground α,β-form mixtures[21,24,56,57] and probably also in analogous treatments involving emulsified organic liquids[26] or aqueous surfactant solutions.[27] For example, a badly compacted α,β-form mixture with small brick-shaped crystals (0·03 μm in size) when refluxed with an aliphatic alcohol was converted into rod-shaped β-form crystals (0·1–0·2 μm in length).[15] The BET specific surface area increased from 5 to 55 m^2/g with virtual removal of hysteresis in the adsorption/desorption isotherm. The product dispersed much more easily in an oil ink system to give much higher colour strength than did the initial material. The marked decrease in aggregation induced by the solvent treatment had more than offset the potentially detrimental effect of increase in crystal size on specific surface area and eventual colour strength. Similar beneficial effects can be achieved by treatment of other types of organic pigments with organic liquids, for example partially chlorinated α-copper phthalocyanine and quinacridone.[57]

Deaggregation by recrystallisation can also be achieved by heating pigment in water or aqueous media at temperatures up to 300°C under pressure. This type of procedure has been used to produce β-copper phthalocyanine pigments with controlled crystal shape[36] and inherently opaque metal-free azo pigments.[2,46]

A wide variety of additives have been claimed in the patent literature as means of inhibiting aggregation of organic pigments and thereby improving their dispersibility and/or colour strength. The most important of these are abietyl resins,[58] aliphatic amines,[59] amides[60] and substituted derivatives of pigments themselves.[61,62] The effect and effectiveness of additives depend upon the stage of pigment manufacture at which they are added. Their presence during precipitation and attrition or ripening processes can affect crystal size and even lattice type, as well as the state of aggregation; an example is substituted pigment derivatives present during precipitation of metal-free azo pigments.[62] Additives added at later stages when the crystals have already been formed tend to affect only the interaction between crystals such as that which tends to occur during drying. This is common in the resination of, for example, many types of metal-free azo pigments. The aqueous pigment slurry is mixed with an alkaline solution of wood rosin or a related compound, which is then precipitated in intimate admixture with the pigment.

Some investigative work has been reported on the use of additives in preventing hydrophilic aggregation on drying of aqueous pigment filter-cakes and thus easing subsequent dispersion in application media. Treatment of certain types of azo pigments, for example, CI Pigment Yellow 12, with long-chain aliphatic amines promotes chemical interaction of the amine groups with the pigment molecules and renders the crystal surfaces hydrophobic.[51] It is believed that an oleophilic dyestuff is formed *in situ* and that this forms an adsorbed layer at the pigment surface.[51] Capillary attraction and associated aggregation on drying are thus virtually eliminated and the resulting pigment powder is dispersible simply on shaking in media rich in aromatic hydrocarbon, such as certain gravure inks.[51]

Treatment with resins, on the other hand, is believed to form a mechanical barrier of precipitated resin between adjacent crystals,[51] thus preventing their close approach during drying, but subsequently removable in the application system. This is the rationale behind various patents on so-called easily dispersible pigments. It has been suggested that special treatments, such as controlled heating and cooling or treatment with solvents, applied after precipitation of the resin and designed to soften and redistribute it, lead to an open-textured powder in which the pigment particles are cemented together by resin into a network structure.[51] The presence of this structure could be identified by resistance to compression under low pressures (138 kN/m^2) as indicated by bulk volume measurements, and appeared to be responsible for improved dispersibility in oil-based media, such as paints and inks. The structure appeared neither to improve wettability nor to lead to enhanced dispersion stability and its effect therefore appeared to be in easing separation of the pigment particles.[51]

Other purely mechanical means claimed to reduce aggregation on drying are freeze drying,[63] which avoids capillary attraction by freezing the water and removing the ice by sublimation, freeze drying from organic solvents in which capillary attraction is small anyway,[64] and drying methods which deposit pigment in admixture with solid compounds, such as ammonium carbonate, that keep crystals apart, but that are subsequently removed by decomposition into gases when the temperature is increased.[65]

DISPERSION

During the process of dispersion the composition of the pigment inevitably undergoes change. Under the influence of the liquid medium and mechanical energy input the aggregate structure is substantially broken and the disposition of additives altered. The additives may have dissolved away either partially or completely and they may have been re-adsorbed to other regions of the crystal surfaces. Likewise components of the medium may have been adsorbed. Ideally there should be no changes in the size and shape of the pigment crystals and in the crystal lattice type. With commercially acceptable pigments this is substantially so, although fracture of large crystals, irreversible aggregation, recrystallisation and crystal phase change are liable to occur to appreciable extents with less satisfactory pigments.

Wetting

The theory of wetting has been described in Chapter 1. Wetting of organic pigments is in practice a complex series of processes as yet incompletely understood. It involves not only displacement of air from the surfaces of the pigment crystals, but also penetration of liquids into the interior of the aggregate structure. Furthermore, in the case of pigments containing additives, likely complicating factors are solvation and dissolution of the additives and alteration of their disposition. It is evident that a thorough characterisation of aggregate structure and additive disposition are prerequisites for investigation of wetting properties. Published investigations have been concerned with pigments free from additives.

Vapour adsorption studies using outgassed pigment samples at room temperature have provided a useful link between aggregate structure, as determined by gas adsorption, and events that happen when the pigment is brought into contact with liquid. Benzene vapour has been shown to penetrate into the aggregate structure of various mesoporous β-copper phthalocyanine pigments, thereby prising the crystals apart and exposing an internal surface previously inaccessible to nitrogen.[66] In fact, after removal of the benzene vapour the BET specific surface area as determined by gas adsorption had increased. This indicated that the opening of the structure was at least partly irreversible. Similar studies of benzene on an α-copper phthalocyanine pigment produced a marked loss in BET specific surface area with evidence of α to β conversion.[66]

Studies have also been reported using toluene and *n*-propanol vapours on a non-porous β-copper phthalocyanine pigment with well-developed rod-shaped crystals, and a mesoporous α-form pigment with small brick-shaped crystals.[67] The pigments had comparable specific surface areas, but the α-form gave much the greater uptake of either vapour due to penetration and opening up of the inner aggregate structure. Whereas successive adsorption/desorption cycles produced relatively little change in the β-form pigment, they produced a marked drop in vapour uptake and in BET specific surface area of the α-form pigment. In the latter case, however, there was no evidence of α to β conversion and the mechanism was considered to involve fusion of crystals through condensed vapour within the porous structure.

A further study of two similar types of pigment led to the same conclusions.[55] This study, however, also included a mesoporous β-copper phthalocyanine pigment with small brick-shaped crystals. Successive adsorption/desorption cycles produced a progressive reduction in vapour uptake and in BET specific surface area. It thus behaved in a manner analogous to the α-form pigment.

These vapour adsorption studies collectively affirm that the BET specific surface area is not a fundamental property of the pigment. Whereas nitrogen can gently probe the external regions of the aggregate structure, penetrating and prising the crystals apart only to a limited extent, the organic vapours are more vigorous in their action, penetrating further and opening up the internal regions. Furthermore it is evident that vapour adsorption behaviour is influenced primarily by aggregate structure, rather than crystal lattice type.

Immersion calorimetry of outgassed pigment samples has enabled direct investigation of wetting by bulk liquids and has thrown further light on the relationship between aggregate structure and wetting. Heats of wetting of substantially unaggregated β-copper phthalocyanine pigments with well-developed rod-shaped crystals and free from additives have been reported for a variety of organic liquids, namely toluene, aliphatic hydrocarbons and alcohols.[15,55] They are all within the narrow range from 76 to 104 mJ/m^2, results being expressed per unit BET specific surface area determined by nitrogen adsorption. Independent investigations using a similar pigment and the same types of liquid produced comparable results.[68,69] Results with other liquids also fell within this range or slightly higher, except that with *n*-butylamine which was significantly higher at 145 mJ/m^2.[68] With this last exception the interactions involved appeared to be fairly weak.

Markedly different heats were obtained using α-, β- and monochloro-copper phthalocyanine pigments, all with small brick-shaped crystals and porous aggregate structure as determined by gas adsorption/desorption.[15,55,70] They were of much greater and widely varying magnitude. This has been attributed to the liquid molecules penetrating and opening up the interior of the aggregate structure to a much greater extent than nitrogen and thereby gaining access to crystal surfaces that were not accessible to nitrogen. Thus, although the BET nitrogen specific surface area is not a fundamental property of the pigment it is a valuable reference parameter. Ageing effects occurred during the immersion experiments, namely crystal growth and, in the particular case of one α-form pigment, some α to β conversion.[15,55] It has been shown that in addition to recrystallisation, fusion of crystals can occur when pigments are contacted with liquids.[15]

Other investigations of a β-copper phthalocyanine pigment with 'isometric' crystals have yielded heats of immersion only marginally larger than those obtained with rod-shaped crystals. It would appear that this particular pigment is substantially unaggregated and non-porous, although no gas adsorption/desorption data have been reported. These isometric crystals are substantially larger than the brick-shaped crystals of the above studies and are believed to have been prepared by recrystallisation of ground material in water above 100°C under pressure (cf. ref. 36). Whereas one school of thought[15,54,55] has been concerned with the influence of aggregate structure on interaction of pigments with liquids, a topic of importance both in manufacture and application of pigments, another school[68] has been concerned exclusively with characterising the surfaces of pigments thought to be essentially unaggregated. In the latter case attempts are being made to resolve the measured heats into interactions with the different crystal faces[68] and into the various component interaction parameters of the pigment surface itself.[71]

There has been some activity on determination of surface parameters of organic pigments by other methods, notably water vapour sorption,[72] contact angle measurements using aqueous and non-aqueous sessile drops on compacted pigment discs[73,74] or sublimed crystalline layers,[75] and contact angle measurements using the Wilhelmy plate principle applied to large single crystals of β-copper phthalocyanine.[76] Where comparison is possible there is remarkable agreement of the data, despite the differences in methods and condition of surfaces.

Separation

In ink and paint systems separation of the constituent particles of the aggregate structure starts on wetting, but is completed by the application of mechanical energy. Entrapped air is released and the system is mixed to give a homogeneous dispersion. Energy input involves shearing and impaction, the relative importance of those two mechanisms depending upon the type of equipment used.

Equipment design is influenced by the viscosity and the volatility of the system. Ball mills, attritor mills, sand mills, Perl mills, and cavitation mixers are commonly used for the most fluid systems, namely aqueous systems, alcohol-rich or ester-rich flexographic inks, gravure inks and moderately viscous systems, such as non-aqueous decorative and industrial paints. In all of these systems the pigment concentration during milling is normally below 15% w/w. Viscous systems such as oil inks are usually dispersed at pigment concentrations in excess of 20% w/w in a two-stage process, featuring high-speed stirring followed by three-roll milling. Activated bead mills, however, such as the Netzsch John Mill, are of growing importance for oil inks and use pigment concentrations of less than 20% w/w.

Whereas the primary aim is separation of the pigment powder into constituent particles, the smallest of which are individual crystals, machines such as the sand mill and Perl mill and even ball milling can cause crystal fracture. Indeed they can, for example, be used to reduce crude copper phthalocyanine to pigmentary form. Care is therefore required in the selection of milling conditions. Generation of elevated temperatures can lead to crystal growth, with loss of colour strength and increase in opacity, that is particularly undesirable in inks for multicolour process printing where successive layers of ink are superimposed. For example, temperatures in excess of 100°C can be generated readily in high-speed stirring or bead milling of oil inks. Certain diarylamide yellow azo pigments when dispersed in oil ink media by bead milling give inks with gloss inferior to that of their three-roll milled counterparts.[77] This effect has been attributed to crystal growth,[2] but the mechanism is likely to be more complex. Although additives are used to improve dispersibility some additives actually inhibit dispersion.[78]

In highly viscous systems such as molten plastics, adhesion of the polymer molecules to the outer surfaces of the aggregate structure replaces wetting in the normal sense. The structure is in effect torn apart by shearing in the narrow gap between the rollers of two-roll mills, between the screw and barrel of extruders or between the blade and wall of Z-arm mixers. Heat is applied in all cases to soften the plastic and reduce the viscosity to a workable level. Temperatures up to about 350°C are used depending upon the polymer and this puts demands on the chemical stability of the pigment. Usually the organic pigment is less than 1% w/w of the pigmented plastic.

A common practice is to pre-mix pigment with polymer granules at room temperature, either by tumbling or by a mechanical stirring action. The aim is to cause preliminary breakdown of the aggregate structure and to promote adhesion of the pigment to the surface of the polymer granules. Unfortunately, however, impaction of the polymer granules is liable to cause compaction of the pigment and thus to inhibit its subsequent dispersion in the molten polymer.[79] It has been suggested that crystal lattice type could be important in determining the compactibility of a pigment, in that α- and stabilised α-copper phthalocyanine pigments appear to be more susceptible to compaction than β-form pigments.[15] The latter type have a less deformable crystal structure, plastic deformation of crystals being an important factor influencing compactive aggregation of powders in general. Compactibility of pigments has been the subject of some investigation.[79]

As described earlier one of the means of producing easily dispersible pigments is the use of additives such as resins which form an integral part of the aggregate structure and keep the pigment particles apart. It has been suggested that although an open textured structure as described elsewhere[51] is desirable for ease of dispersion in ink and paint systems, it is not desirable for plastics as it is susceptible to damage by compaction.[80] There is no published evidence to indicate whether or not this is borne out in practice.

Whereas resins are used to facilitate separation in non-aqueous systems, combinations of surfactants and hydrophilic polymers feature in pigments for aqueous systems.[81,82] For example, one approach features the use of reversible complexes formed by certain hydrophilic polyacids and polyoxyalkalene compounds as an essential part of the separation stage.[82] The complexes break down when the pigment is gently stirred into slightly alkaline emulsion paint media, thus giving virtually spontaneous separation.

Dispersion stability

The smallest individual particles produced in dispersions of organic pigments in ink and paint media are predominantly small crystal aggregates, with a few individually dispersed crystals and a small proportion of larger aggregates that have survived the dispersion process. This is evident from electron micrographs of ultra-thin sections of coated films (see e.g. Fig. 8.10). In exceptional cases the particles are predominantly large individual crystals.

The particles are subject to influences that strive to reduce their high degree of subdivision, namely van der Waals attraction (and possibly in certain cases electrostatic attraction[83]) between neighbouring particles, and the tendency of the particles to recrystallise. Attractive interaction is aided by factors that cause close approach of particles, namely high pigment concentration, agitation, Brownian movement in systems where viscosity permits, and sedimentation under gravity. Recrystallisation is aided by solvent action and elevated temperature. Effective measures have to be adopted to counteract these influences so that dispersions remain stable over prolonged periods of time.

Recrystallisation leads to crystal growth with undesirable effects already mentioned and in specific instances, for example α-copper phthalocyanine, to crystal phase change with loss of colour strength and change in shade. It is particularly prone to occur on storage in systems with a high aromatic hydrocarbon content, such as industrial paints and certain types of gravure ink. Careful selection of the correct chemical type of pigment is necessary in the first instance. As described earlier recrystallisation can be controlled by incorporating additives, usually substituted derivatives of the pigment itself, into the pigment during its manufacture.

Interaction between particles leads to the formation of loose assemblages known as flocculates, which prevail in the absence of applied shear. This can be readily seen by optical microscopy of dilute dispersions, and in more concentrated dispersions it is manifest as shear thinning and thixotropic flow behaviour, even in media that are themselves Newtonian. Flocculation also leads to sediment volume fractions very much larger than the pigment volume fraction, sometimes by more than an order of magnitude.[84] It is usually reversible in a mechanical sense. For example, optical microscopy of dilute dispersions of β-copper phthalocyanine pigments in gravure ink media enables flocculates to be seen breaking and reforming on application and removal of fingertip pressure on the coverslip. Likewise in more concentrated systems flocculated structure can be broken and reformed reversibly by alternating from high to low shear rates in a rotational viscometer.[78] This mechanically reversible flocculation appears to be of the weak or secondary minimum type.

Very large proportions of liquid are entrapped and effectively immobilised within the flocculates (in the voids between constituent particles[84]), as distinct from the liquid on or within the constituent particles themselves.[85] Consequently there is a profound effect on the rheological properties of the dispersion and the formation of extensive gel-like structures in systems of sufficiently high pigment concentration.

Some degree of flocculation is desirable in printing inks because its effect on the rheological properties of the inks is beneficial in the printing process. Excessive flocculation, however, is highly undesirable because the strength and speed of structure formation causes problems in handling ink and paint concentrates and leads to surface defects in coated or printed films. It is also detrimental to colour strength, either directly by in effect increasing particle size or indirectly by accentuating white flooding in paint films containing titanium oxide.

Flocculation of organic pigment dispersions in non-aqueous ink and paint systems can be controlled by the use of additives that are derivatives of the pigment itself. This aspect of technology is well developed with copper phthalocyanine pigments. The additives most commonly used have been derived from chloromethylated and chlorosulphonated copper phthalocyanine intermediates, i.e. CuPc $\text{+}(CH_2Cl)_n$ and CuPc $\text{+}(SO_2Cl)_n$, respectively, where $2 \leqq n \leqq 4$. The substituent groups are in the outer benzenoid rings of the copper phthalocyanine residue. Particularly important examples have been aminomethyl derivatives,[86] i.e. CuPc $\text{+}(CH_2NRR')_n$, including 'polymeric' types,[87] the 4-monosulphonic acid derivative,[88] amine salts of sulphonic acids,[89] i.e. CuPc $\leftarrow SO_3^- \overset{+}{N}HRR'R'')_n$, and sulphonamide derivatives,[90] i.e. CuPc $\leftarrow SO_2NRR')_n$, where R may be H, R may be equal to R′ or R″, and R′ and R″ range from simple aliphatic hydrocarbon to more complex species. These additives are effective when used at 5 to 10% w/w of pigment. Many other types have been claimed in the patent literature.

The additives are best incorporated into the pigment during manufacture at stages where access to the pigment crystal surfaces is favoured, for example during attrition, precipitation or liquid treatment, where there is also the possibility of beneficially affecting the crystallographic properties. Some additives can be added to the ink or paint system along with the pigment or even after dispersion of the pigment, but generally their effectiveness is less than when incorporated in the pigment during its manufacture. A given additive may be effective in some types of ink or paint system, but not in others. Likewise in a given system some additives of a given type are effective, whereas others of the same type, but differing in the detailed constitution of the substituent groups, are not. For example, certain copper phthalocyanine derivatives containing sulphonic acid or sulphonic acid/amine salt substituents are particularly effective in controlling flocculation of β-copper phthalocyanine pigments in alcohol-rich flexographic inks.[91,92] On the other hand, simple aminomethyl and certain sulphonamido derivatives are particularly effective with β-copper phthalocyanine pigments in toluene-based gravure inks and with partially chlorinated α-copper phthalocyanine pigments in hydrocarbon-based industrial and decorative paints.[86,87]

Little has been published on the mechanism of action of those types of additives. Simple aminomethyl copper phthalocyanine derivatives have been shown to adsorb to the surface of copper phthalocyanine pigments from hydrocarbon solution.[93] In this way, functional groups are anchored at the pigment–medium interface. Alkyd resin of paint type media interacts with these functional groups in such a way as to give an adsorbed layer capable of conferring flocculation resistance.[93] The key role of surface modification and interaction with resin in achieving flocculation resistance has been affirmed by hindered settling studies with copper phthalocyanine pigments in model phenolic/toluene gravure ink systems.[84] Likewise electrophoretic mobility measurements (electrodeposition method) have shown that a simple aminomethyl derivative promotes adsorption of phenolic resin in a similar ink system.[83] The substituted aminomethyl derivatives are effective not only with β- and partially chlorinated α-copper phthalocyanine pigments, but also with other types of pigment. Their ability to adsorb and anchor functional groups at surfaces in general has been attributed to the extensive planar character of the copper phthalocyanine residue.[93]

Although adsorption of the bodying resin of the ink or paint system is generally necessary for flocculation resistance it is not in itself a sufficient criterion. This is evident, for example, from consideration of adsorption and flocculation resistance data given in ref. 78. There all of the pigments examined adsorbed resin, but not all were flocculation resistant. Consideration has to be given to the composition of the bodying resins which are heterogeneous both in chemical constitution and molecular size. Application of gel permeation chromatography to adsorption studies with titanium oxide in paint systems has shown that certain resin size fractions are preferentially adsorbed to the pigment and that this can be a factor of importance to flocculation resistance of these systems.[94] It is probable that a similar situation exists with dispersions of organic pigments.

A number of investigations have examined the relative importance of electrical charge and steric stabilisation mechanisms in improved flocculation resistance of copper phthalocyanine pigment dispersions in hydrocarbon-based systems. These systems ranged from idealised model systems consisting of pigment dispersed in solutions of simple surface active agents[95–97] to dispersions in actual ink or paint media containing resins of relatively high molecular weight.[98–100] Surface electrical charge was assessed from electrophoretic mobility measurements usually by equating these measurements with zeta potentials, although in the more recent investigations the mobility measurements themselves have been considered a more reliable indication of surface charge. Although in the model systems flocculation resistance could be correlated with surface charge, this was not the case in the ink and paint systems. In these latter cases it was concluded that steric stabilisation involving the resin predominated over electrical charge stabilisation. This conclusion is consistent with the critical importance of adsorbed resin for resistance to flocculation in hydrocarbon systems discussed earlier.[83,84,93]

Discussion so far has been concerned with stability of organic pigment dispersions in ink and paint media—the users' systems. Recently, however, concentrated non-aqueous dispersions of organic pigments have been marketed by pigment manufacturers for use in inks and paints.[101] They contain up to 40% w/w of pigment dispersed in hydrocarbon-rich media (cf less than 20%, usually much less, in an ink or paint system) and a different approach is necessary to achieve free-flowing character with a satisfactory shelf life. The essential high degree of flocculation resistance is achieved by the use of low-molecular-weight 'copolymeric' additives of amphipathic constitution. Such additives can be either block copolymers with discrete units of insoluble and soluble segments[102] or copolymers with insoluble polar residues spaced at intervals along the polymeric chain and with pendant aliphatic chains or polymer chains as solubilising groups, for example, polyureas,[103] polyurethanes,[104] esterified polyacids[105] and polymers containing ammonium or acid salt groups.[106] These additives are designed to adsorb to the pigment surface, the insoluble residues being envisaged as effectively anchoring the additive molecules to the surface and the soluble segments or aliphatic chains extending into the liquid phase. This is believed to give a well solvated, yet firmly anchored, adsorbed layer with the essential properties for steric stabilisation, namely sufficient thickness, solvation density and entropy. No firm experimental evidence to support this mechanism has been published for the specific systems involving organic pigments, but there is no reason to doubt it, and of course an analogous mechanism is well established in latex dispersion technology.

The critical importance of the balance between effective solvation and effective anchoring of the adsorbed additive layer is illustrated in the following example. Figure 10.4 shows apparent (composite) adsorption isotherms obtained using solutions of an additive of the polyurethane type in various liquids, a well crystallised β-copper phthalocyanine pigment, and the method detailed in ref. 78. The true adsorption in each case is the intercept on the c_A axis obtained by back extrapolation of the linear part of the isotherm (after Rehacek[107]). It decreases with increasing solubility of the additive which is least in the *n*-heptane/toluene mixture (slightly turbid solutions) and greatest in toluene. The greatest flocculation resistance, however, (as indicated by fluid character of dispersions containing 30% w/w pigment) is given in the *n*-heptane/toluene/*n*-propanol mixture (80:20:20). The other three systems gave pseudo-solid gels at this pigment concentration. Clearly the amount of adsorption is not a sufficient criterion for effectiveness of a given additive. Instead, the solubility/insolubility balance is critical. If the additive is too insoluble, then although adsorption is high, the state of solvation of the adsorbed layer is inadequate for effective steric stabilisation of the dispersion.

More detailed studies of adsorbed polymer or resin layers at the solid–liquid interface in organic pigment dispersions are inhibited by the difficulty of defining the specific surface area available to large solute

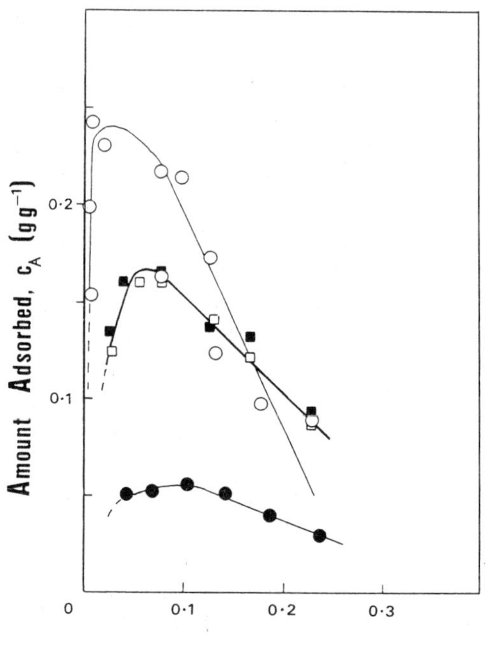

Fig. 10.4. Apparent adsorption isotherms obtained with an amphipathic polyurethane-type additive in the liquids shown and a β-copper phthalocyanine pigment with well developed rod-shaped crystals. ○, *n-Heptane + toluene* (80:20, *w/w*); ■, *n-heptane + n-propanol + toluene* (80:20:20, *w/w*); □, *n-heptane + n-propanol + toluene* (80:40:20, *w/w*); ●, *toluene.*

molecules. Although it is possible to prepare dispersions of special pigments that can be dispersed substantially as individual crystals, for example additive-free β-copper phthalocyanine with well developed rod-shaped crystals, in many cases the types of pigment of interest exist as small porous aggregates,[85] the size of the aggregates depending on the efficiency of dispersion. The practice of using the BET specific surface area determined by nitrogen adsorption is not satisfactory. This is evident from the vapour sorption and immersion calorimetry work described earlier where it was shown clearly that solvent molecules can penetrate and open up an internal pore structure inaccessible to nitrogen. Likewise, for the same reason, specific surface areas determined by adsorption of phenol[108] or *p*-nitrophenol[109] from dilute solution have in many cases given specific surface areas several times greater than the

BET nitrogen values, especially with pigments prone to aggregation. Although the areas determined by solution adsorption may be a better approximation to the area of the solid–liquid interface, they are still not satisfactory for systems containing large solute molecules, which are liable to be excluded from pores readily penetrable by small molecules, such as phenol.

Flocculation and rheological behaviour of organic pigment dispersions

The rheological behaviour of organic pigment dispersions in ink and paint media is important for three main reasons. It changes as the process of dispersion itself proceeds, thus affecting the relative contributions of impaction and shearing mechanisms, and the efficiency of the process. This is why two-stage dispersion processes are sometimes used, for example, high-speed stirring followed by three-roll milling of oil inks. The rheological behaviour affects the handling and application properties of inks and paints. It also provides a means of investigating the state of dispersion, especially in concentrated dispersions.

Organic pigment dispersions in ink and paint media are often prepared at a high pigment concentration and then diluted with other ingredients to give the final ink or paint. The concentrates can contain pigment at up to about 30% w/w, for example, in viscous oil ink systems, but usually considerably less in more fluid ink and paint systems. Their rheological behaviour tends to be dominated by the effects of flocculation of the pigment particles and resulting structure formation, the main effects being shear thinning character and thixotropy.

Under low shear conditions such as those encountered in pouring and handling dispersions, strength of structure is the most important property. It tends to set the upper limit to the pigment concentration that can be used, and hence throughput in the dispersion process in ink and paint manufacture. Where significant structures are formed, the rate of structure formation can also be important, for example in the printing of inks, because of its effect on dot size in half-tone printing and on surface roughness of the ink film, and hence on print definition and gloss, respectively.

Under high shear conditions flocculation is substantially broken down and the most important flow property is the viscosity, or more strictly the limiting viscosity (also known as plastic, terminal or equilibrium viscosity). For example, in a high-speed printing press operating at a given roller speed, viscosity controls ink tack and transfer properties.[110]

Organic pigment dispersions in highly and moderately fluid systems, such as hydrocarbon-based gravure inks and paints, can be characterised, using a rotational viscometer, for example the Weissenberg rheogoniometer fitted with a Mooney cell. Measurement of shear stress τ as a function of rate of shear D produces rheograms of the types shown in Fig. 10.5. Non-flocculating systems give Newtonian behaviour. Most systems, however, flocculate in a mechanically reversible manner and are

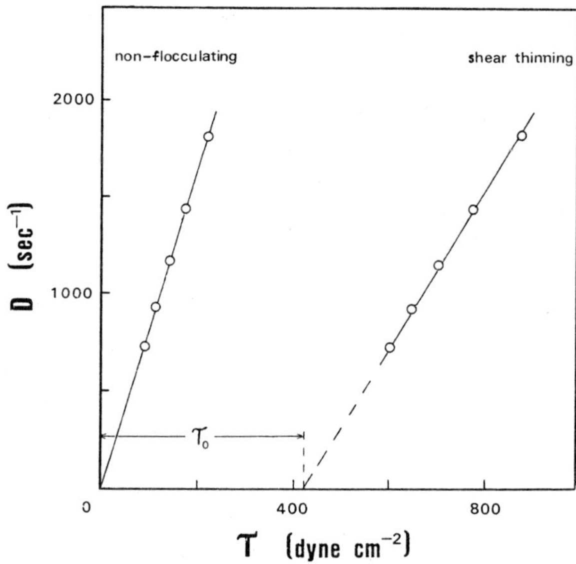

Fig. 10.5. Typical rheological data obtained using dispersions of copper phthalocyanine pigments in a phenolic/toluene ink system (method of ref. 78).

shear thinning. Increasing D progressively disintegrates flocculated structure, eventually giving the linear or approximately linear 'upper Newtonian' region at high rates of shear. In this region the system is essentially deflocculated. The upper limit of measurement is determined by the onset of marked frictional heating, for example at about 1800 s^{-1} for very fluid gravure ink systems; the lower limit is determined by marked deviation from linearity. Calculation of the line of best fit enables evaluation of the slope and intercept on the τ axis. The reciprocal of the slope is the limiting viscosity η; the intercept τ_0 is the extrapolated shear stress[111] (sometimes called the yield value). Some authors[83] have preferred to use the Casson equation[112] to evaluate η and τ_0 (i.e. plotting $D^{1/2}$ versus $\tau^{1/2}$ and evaluating the intercept on the τ axis), but where the upper Newtonian region is well defined and linear there is no advantage in this procedure, which is based on an arbitrarily chosen theoretical model. It gives numerically lower values of τ_0.

The limiting viscosity η may be regarded as the viscosity of the dispersion when flocculated structure has been broken down. It increases with increase in the effective pigment volume fraction, which is the pigment volume fraction plus the volume fraction of immobilised liquid associated with the pigment particles in adsorbed or boundary layers, surface irregularities and pores. The proportion of immobile liquid in the resulting 'flow units' is high. Rough estimates made from independent rheological and solution adsorption data are well in excess of 50% in some cases, including an inherently non-flocculating, Newtonian dispersion.[78]

The yield value τ_0 is an indication of the strength of flocculated structure present in the system at rest. It can also be regarded as the force per unit area of the shear plane required to keep structure from reforming once it has been broken down. Since it refers by extrapolation to zero rate of shear, then it represents to a first approximation nett residual attraction between particles rather than hydrodynamic attraction.

Intuitively τ_0 is some direct function of n and f, where n is the number of individual particles per unit volume of the deflocculated system and f is the mean net force of attraction of each particle for its neighbours. This simple relationship is borne out in practice with dispersions of a β-copper phthalocyanine pigment in a phenolic/toluene ink system:[78] τ_0 increases with increase in pigment concentration, i.e. with increase in n; τ_0 is zero with non-flocculating systems, i.e. when f is zero; τ_0 can be reduced by the use of additives that either improve flocculation resistance, i.e. reduce f, or inhibit dispersion, i.e. reduce n, these two cases being distinguishable in practice by centrifugal hindered settling.[78]

A more recent investigation of a series of β-copper phthalocyanine pigments dispersed in a similar type of gravure ink medium has focused attention on the effect of particle shape on τ_0 and electrophoretic mobility.[83] It has been concluded that τ_0 increases with increase in axial ratio of the particles due to a corresponding increase in bimodal charge distribution enhancing electrostatic interaction between rod-shaped particles. Consideration, however, has been given only to the strength of interaction between particles. It has been tacitly assumed that all the pigments disperse completely into individual crystals in the gravure ink system, so that on the one hand n is the same for all dispersions, while on the other crystal shape can be equated with particle shape. Whereas these assumptions are unlikely to be entirely justified, they may be sufficiently valid to justify the conclusion in the broadest terms. The effect of particle shape on n has not been considered.

Highly viscous oil ink systems with pigment content usually in the range 15–30% w/w exhibit marked shear thinning and thixotropic behaviour. Usually they cannot be examined at sufficiently high rates of shear to attain a well defined upper Newtonian region. For example, inks with high pigment content tend to undergo cohesive failure and split at relatively low rates of shear on a rotational viscometer.[113] Consequently η and τ_0 cannot be determined as confidently as with very fluid systems and in some cases they cannot be determined at all. In an investigation of suitable systems featuring β-copper phthalocyanine,[36] copper phthalocyanine green[37] and quinacridone pigments,[37] η and τ_0 have been evaluated by means of the Casson equation.[112] The results have been related to pigment volume concentration and mean particle size d as determined by disc centrifugation after dilution. Whereas η increased markedly with increase in pigment volume concentration, i.e. with increase in n, at a fixed concentration there appeared to be little dependence on d. In this latter case, however, data were too few to be conclusive. The value of τ_0 increased markedly with pigment volume concentration, i.e. with increase in n, but also at a given concentration with decrease in d in the range 0·3–0·1 μm. In fact τ_0 appeared to be proportional to d^{-3}, which is proportional to n. Clearly τ_0 is some direct function of n in this oil ink system.

REFERENCES

1. D. Easton, *Can. Paint Finish.*, **49** (1975) 20.
2. W. Herbst and K. Hunger, *Prog. Org. Coat.*, **6** (1978) 105.
3. R. B. McKay, *Rev. Prog. Color. Relat. Top.*, **10** (1979) 25.
4. E. R. Inman, Roy. Inst. Chem. Lecture Series, 1967, No. 1.
5. J. Lenoir, in *The Chemistry of Synthetic Dyes*, Ed. K. Venkataraman, Academic Press, New York and London, 1971, Vol. 5, p. 313.
6. J. R. Fryer, *The Chemical Applications of Transmission Electron Microscopy*, Academic Press, London, New York and San Francisco, 1979, Chapter 1, Section G4, and Chapter 6, Section D.
7. J. W. Menter, *Proc. Roy. Soc.* (*London*) **A 236** (1956) 119.

8. J. Stabenow, *Ber. Bunsenges. Phys. Chem.*, **72** (1968) 374.
9. J. R. Fryer, *Inst. Phys. Conf. Ser.*, No. 36 (1977) 423.
10. B. Honigmann, *J. Paint Technol.*, **38** (1966) 77.
11. D. Horn and B. Honigmann, *XII FATIPEC Congress*, (1974) 181.
12. J. H. Beynon and A. R. Humphries, *Trans. Faraday Soc.*, **51** (1955) 1065.
13. C. J. Brown, *J. Chem. Soc. (A)*, (1968) 2488.
14. B. Honigmann, H-U. Lenné and R. Schrödel, *Z. Kristallogr.*, **122** (1965) 185.
15. J. R. Fryer, R. R. Mather, R. B. McKay and K. S. W. Sing, *J. Chem. Technol. Biotechnol.*, **31** (1981) 371.
16. F. H. Moser and A. L. Thomas, *Phthalocyanine Compounds, ACS Monograph 157*, 1963, Reinhold, New York and London, Chapter 2, pp. 15 and 22.
17. B. Honigmann and D. Horn, in *Particle Growth in Suspensions*, Ed. A. L. Smith, Academic Press, London and New York, 1973, Chapter 18, p. 283.
18. E. Suito and N. Uyeda, *Kolloid Z. Z. Polym.*, **193** (1963) 97.
19. M. Ashida, N. Uyeda and E. Suito, *J. Cryst. Growth*, **8** (1971) 45.
20. F. M. Smith and D. Easton, *J. Oil Colour Chem. Assoc.*, **49** (1966) 614.
21. Du Pont, USP 2,556,726.
22. Du Pont, USP 3,954,795.
23. Ciba–Geigy, USP 3,775,149.
24. Du Pont, USP 3,944,564; Ciba–Geigy, UKP 1,140,836.
25. Du Pont, USP 3,051,718.
26. Ciba–Geigy, UKP 1,541,699; Du Pont, USP 3,017,414 and USP 4,024,154.
27. Ciba–Geigy, Ger. Patent 2,745,893.
28. L. Cabut, J.-C. Hardouin, H. Ciceron and A. Chapelle, *Double-Liaison*, **21** (1974) 159/59.
29. Bayer, UKP 1,422,834; BASF, USP 4,069,064; Francolor, USP 3,985,569 and USP 3,985,570.
30. W. Herbst and K. Merkle, *Dtsche. Farben-Z.*, **24** (1970) 365.
31. A. Whitaker, *J. Soc. Dyers Colour.*, **94** (1978) 431.
32. H. C. Mez, *Ber. Bunsenges. Phys. Chem.*, **72** (1968) 389.
33. S. J. Chapman and A. Whitaker, *J. Soc. Dyers Colour.*, **87** (1971) 120.
34. A. R. Hanke, National Printing Ink Research Institute Technical Conference, June 1968, Lehigh University, Bethlehem, Pennsylvania, USA; Report of a lecture, *J. Opt. Soc. Amer.*, **56** (1966) 713.
35. M. A. Maikowski, *Ber. Bunsenges. Phys. Chem.*, **71** (1967) 313.
36. P. Hauser, M. Hermann and B. Honigmann, *Farbe u. Lack*, **76** (1970) 545.
37. P. Hauser, M. Hermann and B. Honigmann, *Farbe u. Lack*, **77** (1971) 1097.
38. W. Carr, *J. Oil Colour Chem. Assoc.*, **54** (1971) 1093.
39. M. A. Maikowski, *Prog. Colloid Polym. Sci.*, **59** (1976) 70.
40. B. Felder, *Helv. Chim. Acta*, **51** (1968) 1224; B. Felder, *J. Color Appearance*, **1** (1971) 9.
41. P. Hauser, D. Horn and R. Sappok, *XII FATIPEC Congress*, (1974) 191.
42. R. Sappok, *J. Oil Colour Chem. Assoc.*, **61** (1978) 299.
43. K. Merkle, W. Herbst and G. W. Eulitz, *Farbe u. Lack*, **82** (1976) 801.
44. O. Hafner, *J. Paint Technol.*, **47** (609) (1975) 64.
45. W. Gerstner, *J. Oil Colour Chem. Assoc.*, **49** (1966) 954.

46. Hoechst, UKP 1,406,797, UKP 1,455,653 and UKP 1,456,331.
47. J. Beresford and F. M. Smith, in *Dispersion of Powders in Liquids*, 2nd edition, Ed. G. D. Parfitt, Applied Science, London, 1973, Chapter 9, p. 383.
48. B. Medinger, *XIV FATIPEC Congress*, (1978) 439.
49. B. Honigmann and J. Stabenow, *VI FATIPEC Congress*, (1962) S89; B. Honigmann, Disk. Bunsen-ges. Frankfurt a.M.–Höchst, (1966) 1.
50. B. Honigmann, *Farbe u. Lack*, **82** (1976) 815.
51. J. Moilliet and D. A. Plant, *J. Oil Colour Chem. Assoc.*, **52** (1969) 289.
52. IUPAC, Manual of symbols and terminology for physicochemical quantities and units, Appendix II, Part 1, *Pure Appl. Chem.*, **31** (1972) 579.
53. R. R. Mather and K. S. W. Sing, *J. Colloid Interface Sci.*, **60** (1977) 60.
54. C. R. S. Dean, R. R. Mather, D. L. Segal and K. S. W. Sing, in *Characterisation of Porous Solids*, Eds. S. J. Gregg, K. S. W. Sing and H. F. Stoeckli, Society of Chemical Industry, London, 1979, p. 359.
55. R. R. Mather, *XIV FATIPEC Congress*, (1978) 433.
56. BASF, UKP 1,402,011.
57. Du Pont, USP 2,857,400.
58. ICI, UKP 978,242; KVK, UKP 1,245,268; Chemetron, USP 3,615,812.
59. Du Pont, UKP 499,334; KVK, UKP 1,080,115; ICI, UKP 1,096,362; Bayer, UKP 1,162,036.
60. KVK, UKP 1,325,371.
61. Ciba–Geigy, UKP 1,486,117; BASF, Ger. Patent 2,554,252.
62. Ciba–Geigy, USP 3,776,749.
63. American Cyanamid, USP 3,159,498.
64. BASF, Ger. Patent 2,013,672; Eastman Kodak, USP 3,731,391.
65. Chemetron, Can. Patent 835,264.
66. V. Ya. Davidov, A. V. Kiselev and T. V. Silina, *Kolloidn Zh.*, **36** (1974) 945. English translation available from British Library.
67. C. R. S. Dean, R. R. Mather and K. S. W. Sing, *Thermochim. Acta*, **24** (1978) 399.
68. R. Sappok and B. Honigmann, in *Characterisation of Powder Surfaces*, Eds. G. D. Parfitt and K. S. W. Sing, Academic Press, London, New York, San Francisco, Chapter 6, p. 231.
69. O. J. Schmitz, P-J. Sell and K. Hamann, *Farbe u. Lack*, **79** (1973) 1049.
70. D. L. Segal and K. S. W. Sing, Discussion on Session 5, in *Characterisation of Porous Solids*, Eds. S. J. Gregg, K. S. W. Sing and H. F. Stoeckli, Society of Chemical Industry, London, 1979, p. 387.
71. J. Schröder, *J. Colloid Interface Sci.*, **72** (1979) 279.
72. K. Apel, W. Gückel and B. Honigmann, *Dtsche. Farben-Z.*, **21** (1967) 626.
73. S. Wu and K. J. Brzozowski, *J. Colloid Interface Sci.*, **37** (1971) 686.
74. P. R. Buechler, G. L. Brown, V. P. Parikh and H. J. Salmon, *J. Paint Technol.*, **45** (577) (1973) 60.
75. G. E. H. Hellwig and A. W. Neumann, *Farbe u. Lack*, **73** (1967) 823.
76. P.-J. Sell and D. Renzow, *Farbe u. Lack*, **83** (1977) 265.
77. C. G. Crawforth, *Paint Manuf.*, (Oct. 1978) 20.
78. R. B. McKay, *XIII FATIPEC Congress*, (1976) 428; *Br. Ink Maker*, **19** (1977) 59.
79. M. J. Smith, *J. Oil Colour Chem. Assoc.*, **57** (1974) 36.

80. R. Deverell-Smith, *Polym. Age*, **2** (1971) 267.
81. Hercules, UKP 1,489,693 and USP 4,056,402.
82. Ciba–Geigy, UKP 1,499,660.
83. W. Ditter and D. Horn, *Proc. 4th Int. Conf. Org. Coatings Sci. and Technol.*, (1978) 251.
84. R. B. McKay, *J. Appl. Chem. Biotechnol.*, **26** (1976) 55.
85. R. B. McKay, in *Particle Size Analysis*, Ed. M. J. Groves, Heyden, London, Philadelphia, Rheine, 1978, p. 421.
86. BASF, UKP 949,739; ICI, UKP 972,805.
87. ICI, UKP 1,082,967 and UKP 1,149,778.
88. Interchemical, USP 2,526,345; Du Pont, USP 2,799,594.
89. Ciba–Geigy, Ger. Patent 2,720,464.
90. Du Pont, USP 2,861,005; Siegle, Ger. Patent 1,260,433; BASF, Belg. Patent 615,287; ICI, UKP 1,465,972; Ciba–Geigy, UKP 1,535,434.
91. American Cyanamid, UKP 1,230,525.
92. Ciba–Geigy, UKP 1,263,684.
93. W. Black, F. T. Hesselink and A. Topham, *Kolloid-Z. Z. Polym.*, **213** (1966) 150.
94. V. T. Crowl, *J. Oil Colour Chem. Assoc.*, **55** (1972) 388.
95. D. N. L. McGown, G. D. Parfitt and E. Willis, *J. Colloid Sci.*, **20** (1965) 650.
96. W. D. Cooper and P. Wright, *J. Chem. Soc., Faraday Trans. I*, **5** (1974) 858.
97. W. D. Cooper and P. Wright, *J. Colloid Interface Sci.*, **54** (1976) 28.
98. V. T. Crowl, *J. Oil Colour Chem. Assoc.*, **50** (1967) 1023.
99. V. T. Crowl and M. A. Malati, *Disc. Faraday Soc.*, **42** (1966) 301.
100. W. Carr and J. A. Long, *J. Oil Colour Chem. Assoc.*, **57** (1974) 316.
101. A. Topham, *Prog. Org. Coat.*, **5** (1977) 237.
102. ICI, UKP 1,429,934.
103. ICI, UKP 1,313,745.
104. Ciba–Geigy, UKP 1,445,104.
105. Ciba–Geigy, UKP 1,489,493; Du Pont, USP 3,704,255.
106. ICI Australia, UKP 1,350,284 and UKP 1,355,412.
107. K. Rehacek, *Farbe u. Lack*, **76** (1970) 656.
108. B. Honigmann and W. Gückel, *IX FATIPEC Congress*, (1968) Section 1, p. 22; M. Herrmann and B. Honigmann, *Farbe u. Lack* **75** (1969) 337.
109. C. H. Giles and S. N. Nakhwa, *J. Appl. Chem.*, **12** (1962) 266.
110. H. Green, *Industrial Rheology and Rheological Structures*, Wilcy, New York and Chapman & Hall, London, 1949, Chapter 9, p. 112; A. C. Zettlemoyer and R. R. Myers, in *Rheology*, Ed. F. R. Eirich, Academic Press, New York and London, 1960, Vol. 3, Chapter 5, p. 145.
111. British Standard 5168:1975, Glossary of Rheological Terms.
112. N. Casson, in *Rheology of Disperse Systems*, Conf. of Br. Soc. Rheol., Swansea, 1957, Pergamon Press, London, New York, Paris and Los Angeles, 1959, p. 84.
113. J. Ferguson, G. F. Bradley and J. Beresford, *Rheol. Acta*, **10** (1971) 479.

INDEX